Pro/ENGINEER® WILDFIRE™ 3.0

Louis Gary Lamit

with **James Gee**

De Anza College

THOMSON

Australia   Canada   Mexico   Singapore   Spain   United Kingdom   United States

Pro/ENGINEER® WILDFIRE™ 3.0

by Louis Gary Lamit

Associate Vice-President and Editorial Director
Evelyn Veitch

Publisher:
Chris Carson

Developmental Editor:
Hilda Gowans

Manufacturing Coordinator:
Joanne McNeil

Creative Director:
Angela Cluer

Cover Design:
Johanna Liburd

Cover Image:
J. Helgasou/Shutter Stock

Printer:
Thomson West

COPYRIGHT © 2007 by Nelson, a division of Thomson Canada Limited.

Printed and bound in the United States
1 2 3 4   09 08 07 06

For more information contact Nelson, 1120 Birchmount Road, Toronto, Ontario, Canada M1K 5G4. Or you can visit our Internet site at
http://www.nelson.com

Library of Congress Control Number:
20069306344

ISBN: 0-495-24468-6

ALL RIGHTS RESERVED. No part of this work covered by the copyright herein may be reproduced, transcribed, or used in any form or by any means—graphic, electronic, or mechanical, including photocopying, recording, taping, Web distribution, or information storage and retrieval systems—without the written permission of the publisher.

For permission to use material from this text or product, submit a request online at
www.thomsonrights.com

Every effort has been made to trace ownership of all copyrighted material and to secure permission from copyright holders. In the event of any question arising as to the use of any material, we will be pleased to make the necessary corrections in future printings.

North America
Nelson
1120 Birchmount Road
Toronto, Ontario M1K 5G4
Canada

Asia
Thomson Learning
5 Shenton Way #01-01
UIC Building
Singapore 068808

Australia/New Zealand
Thomson Learning
102 Dodds Street
Southbank, Victoria
Australia 3006

Europe/Middle East/Africa
Thomson Learning
High Holborn House
50/51 Bedford Row
London WC1R 4LR
United Kingdom

Latin America
Thomson Learning
Seneca, 53
Colonia Polanco
11560 Mexico D.F.
Mexico

Spain
Paraninfo
Calle/Magallanes, 25
28015 Madrid, Spain

About the Authors

Louis Gary Lamit is currently a full time instructor at De Anza College in Cupertino, Ca, where he teaches Pro/ENGINEER, Pro/SURFACE, Pro/SHEETMETAL, Pro/NC, Expert Machinist, and Unigraphics NX. He is the founder of Scholarships for Veterans at www.scholarshipsforveterans.org.

Mr. Lamit has worked as a drafter, designer, numerical control (NC) programmer, technical illustrator, and engineer in the automotive, aircraft, and piping industries. A majority of his work experience is in the area of mechanical and piping design. He started as a drafter in Detroit (as a job shopper) in the automobile industry, doing tooling, dies, jigs and fixture layout, and detailing at Koltanbar Engineering, Tool Engineering, Time Engineering, and Premier Engineering for Chrysler, Ford, AMC, and Fisher Body. Mr. Lamit has worked at Remington Arms and Pratt & Whitney Aircraft as a designer, and at Boeing Aircraft and Kollmorgan Optics as an NC programmer and aircraft engineer. He also owns and operates his own consulting firm (CAD-Resources.com- Lamit and Associates), and has been involved with advertising, and patent illustration. He is the author of over 30 textbooks, workbooks, tutorials, and handbooks.

Mr. Lamit received a BS degree from Western Michigan University in 1970 and did Masters' work at Wayne State University and Michigan State University. He has also done graduate work at the University of California at Berkeley and holds an NC programming certificate from Boeing Aircraft.

Since leaving industry, Mr. Lamit has taught at all levels (Melby Junior High School, Warren, Mi.; Carroll County Vocational Technical School, Carrollton, Ga.; Heald Engineering College, San Francisco, Ca.; Cogswell Polytechnical College, San Francisco and Cupertino, Ca.; Mission College, Santa Clara, Ca.; Santa Rosa Junior College, Santa Rosa, Ca.; Northern Kentucky University, Highland Heights, Ky.; and De Anza College, Cupertino, Ca.). His textbooks include:

- ***Industrial Model Building,*** with Engineering Model Associates, Inc. (1981),
- ***Piping Drafting and Design*** (1981),
- ***Piping Drafting and Design Workbook*** (1981),
- ***Descriptive Geometry*** (1983),
- ***Descriptive Geometry Workbook*** (1983), and
- ***Pipe Fitting and Piping Handbook*** (1984), published by Prentice-Hall.
- ***Drafting for Electronics*** (3rd edition, 1998),
- ***Drafting for Electronics Workbook*** (2nd edition 1992), and
- ***CADD*** (1987), published by Charles Merrill (Macmillan-Prentice-Hall Publishing).
- ***Technical Drawing and Design*** (1994),
- ***Technical Drawing and Design Worksheets and Problem Sheets*** (1994),
- ***Principles of Engineering Drawing*** (1994),
- ***Fundamentals of Engineering Graphics and Design*** (1997),
- ***Engineering Graphics and Design with Graphical Analysis*** (1997), and
- ***Engineering Graphics and Design Worksheets and Problem Sheets*** (1997), published by West Publishing (ITP/Delmar).
- ***Basic Pro/ENGINEER in 20 Lessons*** (1998) (Revision 18) and
- ***Basic Pro/ENGINEER (with references to PT/Modeler)*** (1999) (Revision 19 and PT/Modeler), published by PWS Publishing (ITP).
- ***Pro/ENGINEER 2000i*** (1999) (Revision 2000i), and
- ***Pro/E $2000i^2$ (includes Pro/NC and Pro/SHEETMETAL)*** (2000) (Revision $2000i^2$), published by Brooks/Cole Publishing (ITP).
- ***Pro/ENGINEER Wildfire*** (2003) (Revision Wildfire) published by Brooks/Cole Publishing (ITP).
- ***Introduction to Pro/ENGINEER Wildfire 2.0*** (2004) (Revision Wildfire 2.0) published by SDC.

James Gee is currently a part time instructor at De Anza College, where he teaches Pro/ENGINEER, Pro/MECHANICA, Pro/CABLE, and Pro/MOLD. Mr. Gee graduated from the University of Nevada- Reno with a BSME. He has worked in the Aerospace industry for Lockheed Missiles and Space Company, Sunnyvale, Ca.; Space Systems/Loral, Palo Alto, Ca.; and currently for BAE Systems in San Jose, Ca. Mr. Gee has assisted in checking and writing the Pro/ENGINEER series of textbooks.

Contents

Introduction   1

- Parametric Design   1
- Fundamentals   5
- Part Design   5
- Establishing Features   6
- Datum Features   7
- Parent-Child Relationships   8
- Capturing Design Intent   10
- Assemblies   13
- Drawings   15
- Using the Text   17
- Text Organization   18

Lesson 1   Pro/ENGINEER Wildfire 3.0 Overview   19

- Creating the Pin Part   20
- Creating the Plate Part   28
- Creating the Assembly   41
- Creating Drawings   53

Lesson 2   Pro/ENGINEER Wildfire 3.0   65

- Pro/ENGINEER's Main Window   65
- Catalog Parts   67
- ProductView Lite   73
- File Functions   78
- Help   79
- View and Display Functions   81
- Using Mouse Buttons to Manipulate the Model   83
- System Display Settings   84
- Information Tools   85
- The Model Tree   88
- Working on the Model   89
- About the Dashboard   92

Lesson 3   Direct Modeling   93

- Modeling   93
  - Extrude Tool   95
  - Round Tool   98
  - Shell Tool   99
  - Draft Tool   101
  - Chamfer Tool   102
  - Hole Tool   104
  - Extrude Tool (Cut)   106
  - Mirror Sketch   108
  - Revolve Tool   110
  - Cross Sections   112
  - Revolve Tool (Cut)   115

Lesson 4 Extrusions   119
- The Design Process   120
- Material Files   121
- Sketch Tool   124
- Environment   126
- Constraints   127
- Sketching   128
- Dimensioning   129
- Modifying Dimensions   131

Lesson 5 Datums, Layers, and Sections   141
- Navigator   143
- Colors   147
- Appearance Editor   148
- Datum Plane Tool   156
- Layers   158
- Geometric Tolerances   161
- Hole Tool (Sketched)   167
- Suppressing and Resuming Features   170
- Cross Sections   171
- View Manager   172
- Relations   175
- Info   177
- Feature List   178

Lesson 6 Revolved Features   179
- Revolve Tool   179
- Chamfers   180
- Threads   180
- Standard Holes   180
- Navigation Window   181
- Manipulating Folders   182
- Folder Browser   182
- Model Tree   199
- Holes   204
- Dimension Properties   206
- Materials   210
- Cosmetic Threads   216
- Using the Model Player   220
- Printing and Plotting   224

Lesson 7 Feature Operations   225
- Ribs   226
- Relations   226
- Parameter Symbols   227
- Operators and Functions   227

v

- Arithmetic Operators — 227
- Assignment Operators — 227
- Comparison Operators — 227
- Mathematical Functions — 228
- Failed Features — 229
- Family Tables — 229
- Copy — 232
- Paste Special — 232
- Rib Tool — 238
- Flexing the Model — 247
- Measuring Geometry — 248
- Standard Holes — 250
- Family Tables — 251
- Pro/MANUFACTURING — 264

Lesson 8   Assemblies — 265
- Assembly Constraints — 265
- Placing Components — 266
- Pro/Library — 268
- Catalog Parts — 268
- Layer Tree — 273
- Add a Component to the Assembly — 274
- Regenerating Models — 282
- Copy and Paste Components — 300
- Bill of Materials — 302
- Assembly Sections — 303
- Creating Components in the Assembly Mode — 306
- Top-Down Design — 306
- Global Reference Viewer — 314
- Pattern — 320
- Analysis — 335
- Interference — 335
- Rotating Components — 336
- Component Operations — 337
- Bill of Materials — 339
- Edit Definition — 341
- Edit — 344

Lesson 9   Exploded Assemblies and View Manager — 347
- Creating Exploded Views — 348
- Component Display — 349
- Types of Representations — 350
- View Manager — 350
- URLs and Model Notes — 352
- Views: Perspective, Saved, and Exploded — 358
- Saved Views — 359
- Default Exploded Views — 360

- View Manager 361
- Explode View 361
- View Style 368
- Model Tree 372

Lesson 10 Introduction to Drawings 375
- Formats, Title Blocks, and Views 376
- Specifying the Format Size 377
- Standard Formats 379
- Drawing Templates 380
- Template View 380
- Views 382
- Page Setup 386
- System Formats 387
- Show/Erase 390
- Cleanup Dimensions 391
- Move Views 392
- Delete Views 393
- Dimension Options 394
- Drawing Scale 399
- Section Views 401
- Drawing View Properties 402

Lesson 11 Part Drawings 411
- Page Setup 413
- Drawing Options 414
- Drawing Options File 415
- Drawing Views 416
- Auxiliary Views 420
- Section Views 422
- Show/Erase Drawing Items 425
- Detail Views 430
- Dimension Properties 439
- Text Style 440
- Geometric Tolerances 442
- Pictorial Drawing Views 444
- Title Block Notes 446
- Drawing View Information 448

Lesson 12 Assembly Drawings 449
- Format Options 452
- Format Notes 453
- Tables 454
- Text Style 457
- Report Symbols 458
- Adding Parts List Data 460
- Parameters 461

- Relations 464
- Assembly Drawings 475
- Bill of Materials 476
- Assembly Drawing Views 477
- Assembly Section 481
- BOM Balloons 487
- Exploded Assembly Drawings 497

Lesson 13   Patterns   508
- Fill Pattern 509
- Axial Pattern 512
- Copy and Paste Special 514
- Group 515
- Insert Mode 516
- Scale Model 517
- Sheetmetal 518
  - Flat Walls 520
  - Flange Walls 521
  - Directional Pattern 525
  - Flat Pattern 527

Lesson 14   Blends   529
- Blend Options 531
- Parallel Blends 532
- Rotational and General Blends 533
- Blend Tool 538
- Polar Grid 539
- Axial Pattern 544
- Analysis Measure 546
- Section 546

Lesson 15   Sweeps   549
- Sweep Forms 550
- Sweep Options 551
- Sweep Tool 557
- Trajectory 558

Lesson 16   Helical Sweeps and 3D Model Notes   569
- Helical Sweeps 570
- 3D Model Notes 571
- Helical Compression Spring 572
- Helical Sweep Tool 574
- Model Notes 583
- URL Links 586
- Springs 587

Lesson 17   Shell, Reorder, and Insert Mode   589
- Creating Shells   590
- Reordering Features   591
- Inserting Features   591
- Draft Tool   600
- Shell Tool   601
- Pattern Table   610
- Reorder   616
- Insert Mode   616

Lesson 18   Drafts, Suppress, and Text Extrusions   621
- Drafts   622
- Suppressing and Resuming Features   623
- Text Extrusions   624
- Draft Tool   629
- Shell Tool   631
- Group   643
- Mirror   644
- Surface Round   645
- Section   648
- View Manager   649
- Solidify Tool   650
- Suppress   653
- Extrude Tool   655
- Resume   658
- ModelCHECK Geometry Check   662

Appendix   663
- Customizing the User Interface (UI)   663
- Customize Screen   666
- Navigation   667
- Browser   668
- Mapkeys   669
- Record Mapkeys   670
- Customize Mapkeys   675
- Quick Reference Cards   677

Index   683

Extras from www.cad-resources.com

- o Pro/SHEETMETAL   Click CDI73
- o Pro/NC   Click CDI72
- o Expert Machinist   Click CDI72
- o Pro/SURFACE   Click CDI74

Preface

Pro/ENGINEER® is one of the most widely used CAD/CAM software programs in the world today. Any aspiring engineer will benefit from the knowledge contained herein, while in school or upon graduation as a newly employed engineer.

The text involves creating a new part, an assembly, or a drawing, using a set of Pro/E commands that walk you through the process systematically.

Projects are not included in the text to keep the length and cost to the user down. For instructors and students wanting more material, compressive supplemental lessons (Pro/SHEETMETAL, Pro/NC, Expert Machinist, and Pro/SURFACE) can be downloaded at www.cad-resources.com.

If you wish to contact the author concerning orders, questions, changes, additions, suggestions, comments, or to get on our email list, please send an email to one of the following:

- **CAD-Resources.com** (Louis Gary Lamit and Associates)
- Web Site: **www.cad-resources.com**
- Email: **cad@cad-resources.com**

Pro/E files are available from the author for <u>**instructors**</u> (not individuals) who adopt this text. The Pro/E files will open only on Academic and Commercial versions *(not Student Edition or Tryout Edition)* of Pro/ENGINEER software. *Please contact us from your academic email account.*

CADTRAIN's ***COAch for Pro/ENGINEER***®, www.cadtrain.com, is a computer-based training (CBT) product designed to provide a comprehensive training program for Pro/E users in their actual CAD environment. COAch for Pro/ENGINEER, has been referenced in the book's figures with the authorized use of illustrations. COAch is one of the best ways available for expanding your knowledge of Pro/E software.

> # Dedication
>
> This book is dedicated to the men and women in the armed services who give their effort and sometimes their lives for freedom around the world.
>
> *Om Mani Padme Hum*

Acknowledgments

I want to thank the following people and organizations for the support and materials granted the author:

Leslie Minasian	Parametric Technology Corporation
Gar Smyth	Parametric Technology Corporation
Mike Campbell	Parametric Technology Corporation
Larry Fire	Parametric Technology Corporation
Thuy Dao Lamit	Lamit and Associates
Dennis Stajic	CADTRAIN
Chris Carson	Thomson Engineering

Max Gilliland- is the instructional associate for CAD at De Anza College has been essential to the CAD program. Besides assisting in the classroom, he maintains the software and hardware for the program and in my home office.

Letha Jeanpierre- the dean of the Business and Computer Systems Division at De Anza College, has been the most supportive and professional administrator I have had the pleasure of working with out of the nine schools and colleges at which I have taught.

> **Donations and Scholarships for Veterans (SFV)**
>
> A portion of this text's profits go to my scholarship fund at Foothill-De Anza Community College District (FHDA Foundation). We fund one or more scholarships per year. Your contributions provide extra scholarships as funds are available.
>
> Scholarships for Veterans hope to motivate individuals and businesses to assist in providing scholarships for returning veterans. Money donated goes directly into the designated college's scholarship fund and is collected and awarded by the college foundation. SFV provides funding for a 2-year AS, or AA degree, which covers tuition and fees (or applied to expenses) for 90-quarter or 60-semester units up to $2000.00 (US).
>
> Scholarships are available to any qualified veteran of the Army, Navy, Air Force, Marines, or Coast Guard. Scholarships are administered by the local college foundations. **No administration fees are taken by *Scholarships for Veterans*.** Committee members of Scholarships for Veterans make final selections. **All costs associated with Scholarships for Veterans are borne by Lamit and Associates and CAD-Resources.com.**
>
> For more information, see Scholarships for Veterans at www.scholarshipsforveterans.org
>
> We encourage you to establish a scholarship at your local community college or university, or contribute to an existing scholarship for veterans.

Downloads

Extra material can be downloaded from *www.cad-resources.com* ⇒ *Downloads*

- Part, Assembly, and Drawing Projects
- Lessons
- Tutorials
- Pro/E Files
- Pro/E related articles and information

Projects from CAD-Resources.com

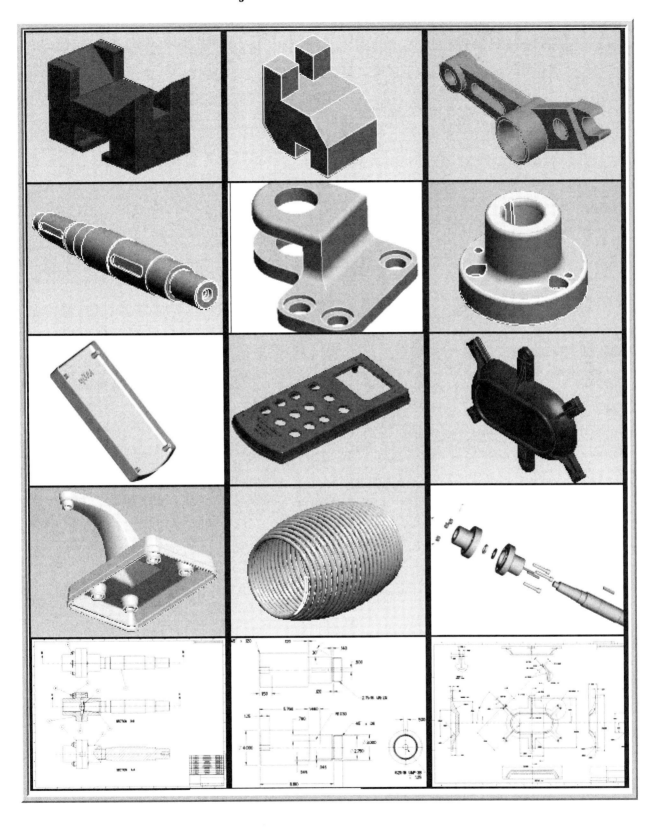

Pro/ENGINEER® Wildfire™ 3.0

Introduction

This text guides you through parametric design using Pro/ENGINEER® Wildfire 3.0™. While using this text, you will create individual parts, assemblies, and drawings.

Parametric can be defined as *any set of physical properties whose values determine the characteristics or behavior of an object.* **Parametric design** enables you to generate a variety of information about your design: its mass properties, a drawing, or a base model. To get this information, you must first model your part design.

Parametric modeling philosophies used in Pro/E include the following:

Feature-Based Modeling Parametric design represents solid models as combinations of engineering features (Fig. 1).

Creation of Assemblies Just as features are combined into parts, parts may be combined into assemblies (Fig. 1).

Capturing Design Intent The ability to incorporate engineering knowledge successfully into the solid model is an essential aspect of parametric modeling.

Figure 1 Parts and Assembly Design

Parametric Design

Parametric design models are not drawn so much as they are *sculpted* from solid volumes of materials. To begin the design process, analyze your design. Before any work is started, take the time to ***tap*** into your own knowledge bank and others that are available. Think, Analyze, and Plan. These three steps are essential to any well-formulated engineering design process.

Break down your overall design into its basic components, building blocks, or primary features. Identify the most fundamental feature of the object to sketch as the first, or base, feature. Varieties of **base features** can be modeled using extrude, revolve, sweep, and blend tools.

Sketched features (*extrusions, sweeps, etc.*) and pick-and-place features called **referenced features** (*holes, rounds, chamfers, etc.*) are normally required to complete the design. With the SKETCHER, you use familiar 2D entities (points, lines, rectangles, circles, arcs, splines, and conics) (Fig. 2). There is no need to be concerned with the accuracy of the sketch. Lines can be at differing angles, arcs and circles can have unequal radii, and features can be sketched with no regard for the actual objects' dimensions. In fact, exaggerating the difference between entities that are similar but not exactly the same is actually a far better practice when using the SKETCHER.

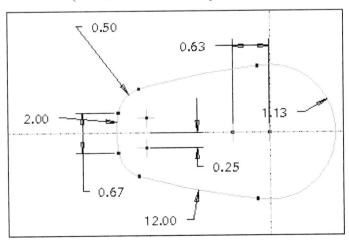

Figure 2 Sketching

Geometry assumptions and constraints will close ends of connected lines, align parallel lines, and snap sketched lines to horizontal and vertical (orthogonal) orientations. Additional constraints are added by means of **parametric dimensions** to control the size and shape of the sketch.

Features are the basic building blocks you use to create an object (Fig. 3). Features "understand" their fit and function as though "smarts" were built into the features themselves. For example, a hole or cut feature "knows" its shape and location and the fact that it has a negative volume. As you modify a feature, the entire object automatically updates after regeneration. The idea behind feature-based modeling is that the designer constructs an object so that it is composed of individual features that describe the way the geometry is supposed to behave if its dimensions change. This happens quite often in industry, as in the case of a design change. Feature-based modeling is diagramed in Figure 4.

Figure 3 Feature Design (Courtesy CADTRAIN)

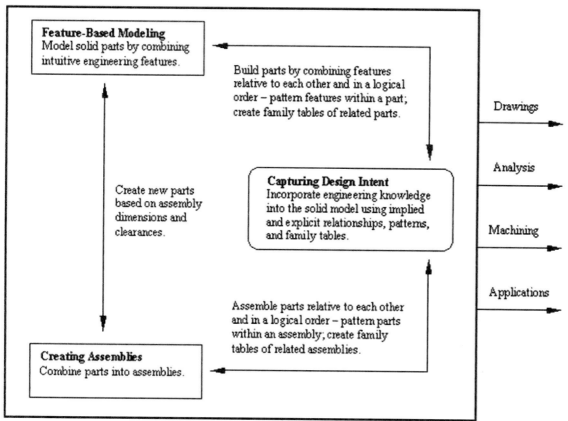

Figure 4 Feature-based Modeling

Parametric modeling is the term used to describe the capturing of design operations as they take place, as well as future modifications and editing of the design. The order of the design operations is significant. Suppose a designer specifies that two surfaces be parallel, such that surface two is parallel to surface one. Therefore, if surface one moves, surface two moves along with surface one to maintain the specified design relationship. The surface two is a **child** of surface one in this example. Parametric modeling software allows the designer to **reorder** the steps in the object's creation.

Various types of features are used as building blocks in the progressive creation of solid objects. Figures 5(a-c) illustrates base features, datum features, sketched features, and referenced features. The "chunks" of solid material from which parametric design models are constructed are called **features**.

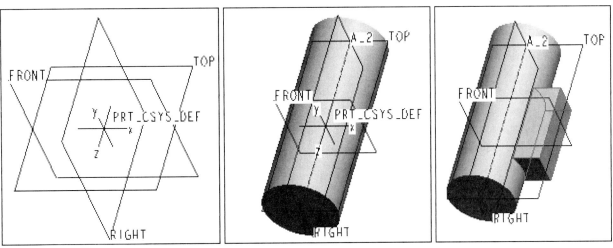

Figures 5(a-c) Features

Features generally fall into one of the following categories:

Base Feature The base feature is normally a set of datum planes referencing the default coordinate system. The base feature is important because all future model geometry will reference this feature directly or indirectly; it becomes the root feature. Changes to the base feature will affect the geometry of the entire model [Figs. 6(a-c)].

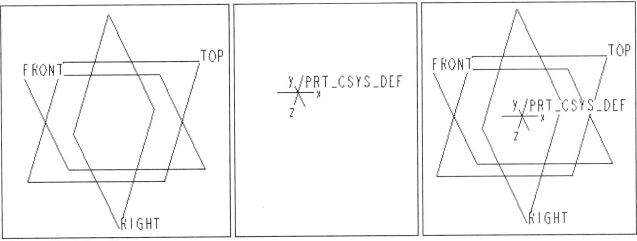

Figures 6(a-c) Base Features

Datum Features Datum features (lines, axes, curves, and points) are generally used to provide sketching planes and contour references for sketched and referenced features. Datum features do not have volume or mass and may be visually hidden without affecting solid geometry (Fig. 7).

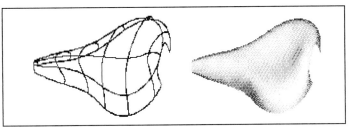

Figure 7 Datum Features

Sketched Features Sketched features are created by extruding, revolving, blending, or sweeping a sketched cross section. Material may be added or removed by protruding or cutting the feature from the existing model (Fig. 8).

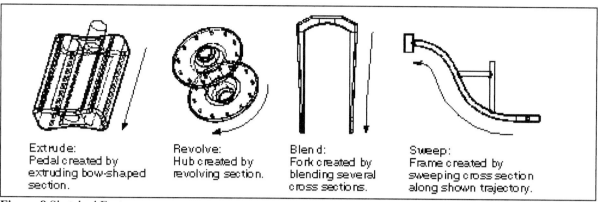

Figure 8 Sketched Features

Referenced Features Referenced features (rounds, holes, shells, and so on) utilize existing geometry for positioning and employ an inherent form; they do not need to be sketched [Figs. 9(a-b)].

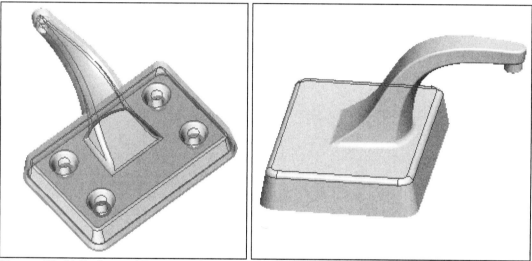

Figures 9(a-b) Referenced Features- Shell and Round (Spout is a Swept Blend Feature)

    A wide variety of features are available. These tools enable the designer to make far fewer changes by capturing the engineer's design intent early in the development stage (Fig.10).

Figure 10 Parametric Designed Part

Fundamentals

The design of parts and assemblies, and the creation of related drawings, forms the foundation of engineering graphics. When designing with Pro/ENGINEER, many of the previous steps in the design process have been eliminated, streamlined, altered, refined, or expanded. The model you create as a part forms the basis for all engineering and design functions.

The part model contains the geometric data describing the part's features, but it also includes non-graphical information embedded in the design itself. The part, its associated assembly, and the graphical documentation (drawings) are parametric. The physical properties described in the part drive (determine) the characteristics and behavior of the assembly and drawing. Any data established in the assembly mode, in turn, determines that aspect of the part and, subsequently, the drawings of the part and the assembly. In other words, all the information contained in the part, the assembly, and the drawing is interrelated, interconnected, and parametric (Fig. 11).

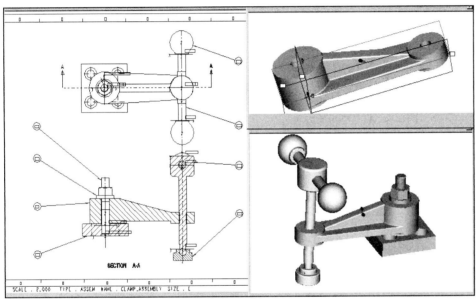

Figure 11 Assembly Drawing, Part, and Assembly

Part Design

In many cases, the part will be the first component of this interconnected process. The *part* function in Pro/E is used to design components.

During part design (Fig. 12), you can accomplish the following:

- Define the base feature
- Define and redefine construction features to the base feature
- Modify the dimensional values of part features (Fig. 13)
- Embed design intent into the model using tolerance specifications and dimensioning schemes
- Create pictorial and shaded views of the component
- Create part families (family tables)
- Perform mass properties analysis and clearance checks
- List part, feature, layer, and other model information
- Measure and calculate model features
- Create detail drawings of the part

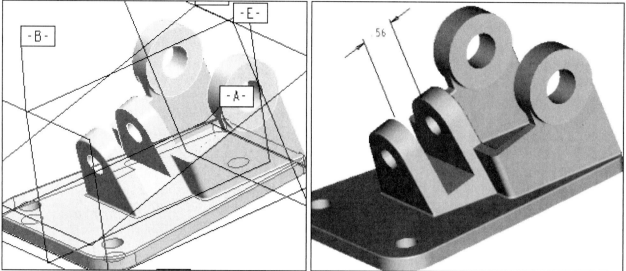

Figure 12 Part Design

Figure 13 Pick on the **.56** dimension and type in a new value

Establishing Features

The design of any part requires that the part be *confined*, *restricted*, *constrained*, and *referenced*. In parametric design, the easiest method to establish and control the geometry of your part design is to use three datum planes. Pro/E automatically creates the three **primary datum planes**. The default datum planes (**RIGHT**, **TOP**, and **FRONT**) constrain your design in all three directions.

Datum planes are infinite planes located in 3D model mode and are associated with the object that was active at the time of their creation. To select a datum plane, you can pick on its name or anywhere on the perimeter edge. Datum planes are *parametric*--geometrically associated with the part. Parametric datum planes are associated with and dependent on the edges, surfaces, vertices, and axes of a part.

Datum planes are used to create a reference on a part that does not already exist. For example, you can sketch or place features on a datum plane when there is no appropriate planar surface. You can also dimension to a datum plane as though it were an edge. In Figure 14, three **default datum planes** and a **default coordinate system** were created when a new part was started using the default template. Note that in the **Model Tree** window they are the first four features of the part, which means that they will be the *parents* of the features that follow. The three *default datum planes* and the *default coordinate system* appear in the Model Tree as the first four features of a new part (**PRT0003.PRT**).

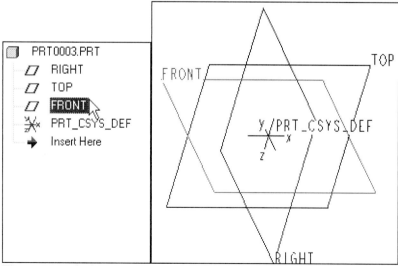

Figure 14 Default Datum Planes and Coordinate System

Datum Features

Datum features are planes, axes, and points you use to place geometric features on the active part. Datums other than defaults can be created at any time during the design process.

As we have discussed, there are three (primary) types of datum features (Fig. 14): **datum planes**, **datum axes**, and **datum points** (there are also *datum curves* and *datum coordinate systems*). You can display all types of datum features, but they do not define the surfaces or edges of the part or add to its mass properties. In Figure 15, a variety of datum planes are used in the creation of the cell phone.

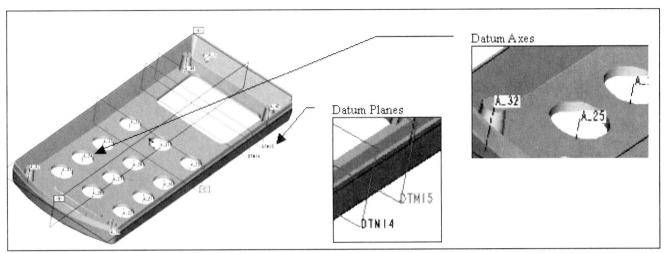

Figure 15 Datums in Part Design

Specifying constraints that locate it with respect to existing geometry creates a datum. For example, a datum plane might be made to pass through the axis of a hole and parallel to a planar surface. Chosen constraints must locate the datum plane relative to the model without ambiguity. You can also use and create datums in assembly mode.

Besides datum planes, datum axes and datum points can be created to assist in the design process. You can also automatically create datum axes through cylindrical features such as holes and solid round features by setting this as a default in your Pro/E configuration file. The part in Figures 16(a-b) shows **A_1** through the hole. **A_1** is the default axis of the circular cut.

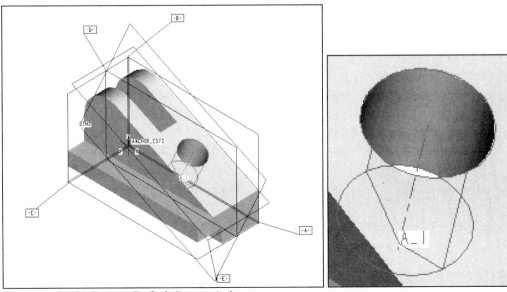

Figures 16(a-b) Feature Default Datum Axis

Parent-Child Relationships

Because solid modeling is a cumulative process, certain features must, by necessity precede others. Those that follow must rely on previously defined features for dimensional and geometric references. The relationships between features and those that reference them are termed *parent-child relationships*. Because children reference parents, parent features can exist without children, but children cannot exist without their parents. This type of CAD modeler is called a history-based system. Using Pro/E's information command will list the models' information as shown in Figure 17.

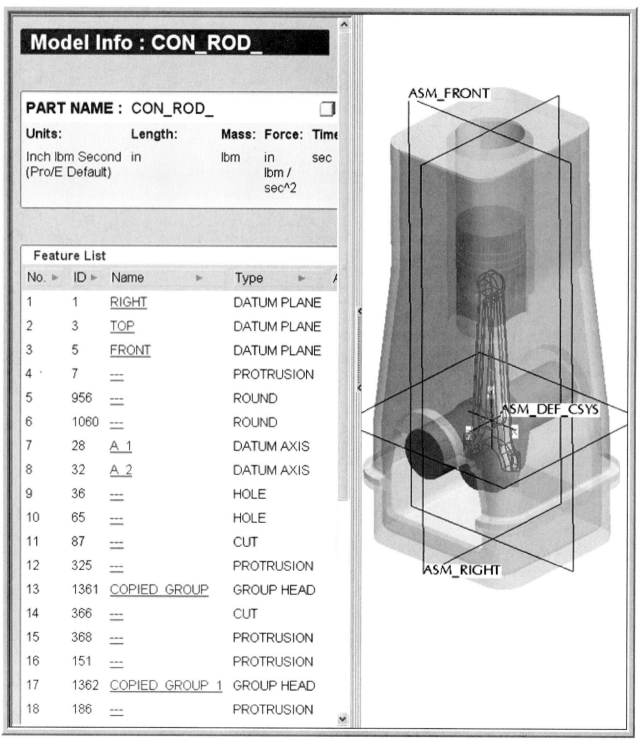

Figure 17 Model Information

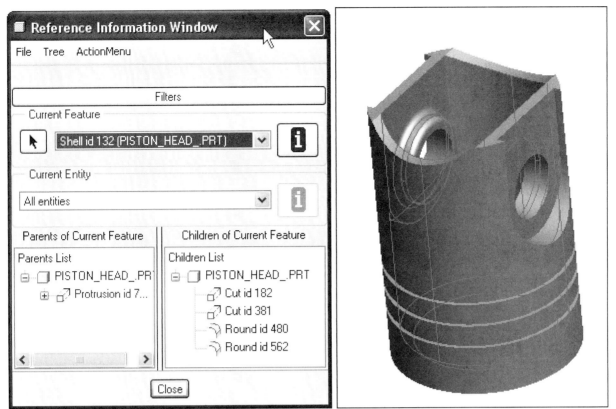

Figure 18 Parent-Child Information

The parent-child relationship (Fig. 18) is one of the most powerful aspects of parametric design. When a parent feature is modified, its children are automatically recreated to reflect the changes in the parent feature's geometry. It is essential to reference feature dimensions so that design modifications are correctly propagated through the model/part. Any modification to the part is automatically propagated throughout the model (Fig. 19) and will affect all children of the modified feature.

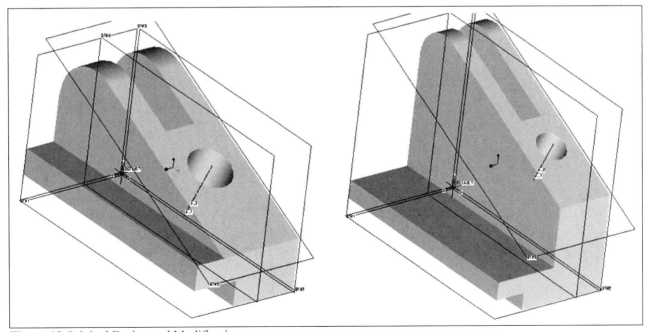

Figure 19 Original Design and Modification

Capturing Design Intent

A valuable characteristic of any design tool is its ability to *render* the design and at the same time capture its *intent* (Fig. 20). Parametric methods depend on the sequence of operations used to construct the design. The software maintains a *history of changes* the designer makes to specific parameters. The point of capturing this history is to keep track of operations that depend on each other. Whenever Pro/E is told to change a specific dimension, it can update all operations that are referenced to that dimension.

For example, a circle representing a bolt hole circle may be constructed so that it is always concentric to a circular slot. If the slot moves, so does the bolt circle. Parameters are usually displayed in terms of dimensions or labels and serve as the mechanism by which geometry is changed. The designer can change parameters manually by changing a dimension or can reference them to a variable in an equation (**relation**) that is solved either by the modeling program itself or by external programs such as spreadsheets.

Features can also store non-graphical information. This information can be used in activities such as drafting, numerical control (NC), finite-element analysis (FEA), and kinematics analysis.

Capturing design intent is based on incorporating engineering knowledge into a model by establishing and preserving certain geometric relationships. The wall thickness of a pressure vessel, for example, should be proportional to its surface area and should remain so, even as its size changes.

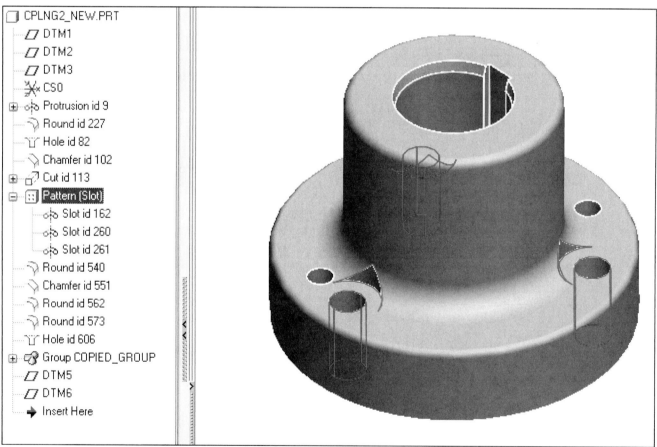

Figure 20 Capturing Design Intent

Parametric designs capture relationships in several ways:

Implicit Relationships Implicit relationships occur when new model geometry is sketched and dimensioned relative to existing features and parts. An implicit relationship is established, for instance, when the section sketch of a tire (Fig. 21) uses rim edges as a reference.

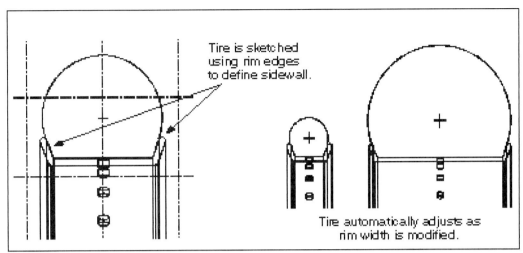

Figure 21 Tire and Rim

Patterns Design features often follow a geometrically predictable pattern. Features and parts are patterned in parametric design by referencing either construction dimensions or existing patterns. One example of patterning is a wheel hub with spokes (Fig. 22). First, the spoke holes are radially patterned. The spokes can then be strung by referencing this pattern.

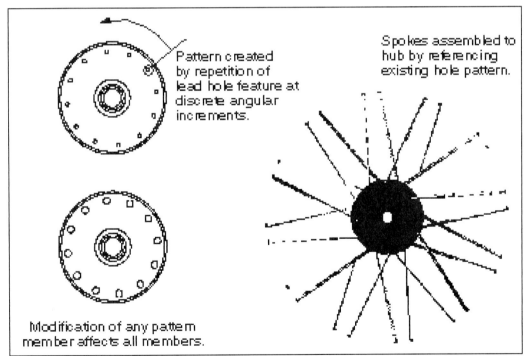

Figure 22 Patterns

Modification to a pattern member affects all members of that pattern. This helps capture design intent by preserving the duplicate geometry of pattern members.

The modeling task is to incorporate the features and parts of a complex design while properly capturing design intent to provide flexibility in modification. Parametric design modeling is a synthesis of physical and intellectual design (Fig. 23).

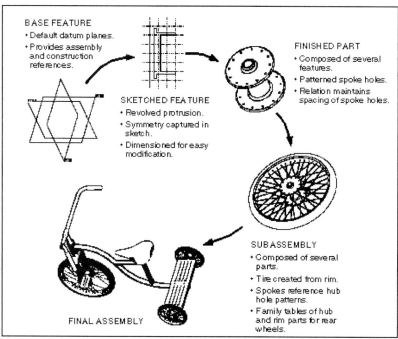

Figure 23 Relations

Explicit Relations Whereas implicit relationships are implied by the feature creation method, the user mathematically enters an explicit relation. This equation is used to relate feature and part dimensions in the desired manner. An explicit relation (Fig. 24) might be used, for example, to control sizes on a model.

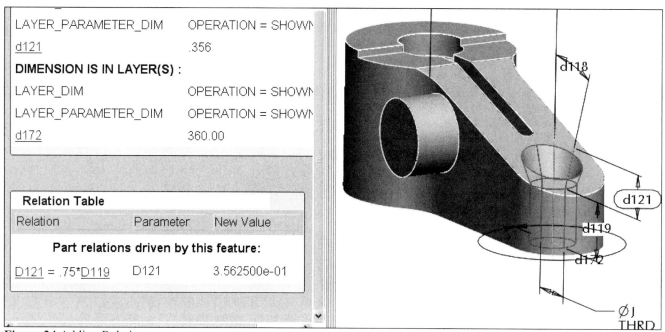

Figure 24 Adding Relations

Family Tables Family tables are used to create part families [Figs. 25(a-c)] from generic models by tabulating dimensions or the presence of certain features or parts. A family table might be used, for example, to catalog a series of couplings with varying width and diameter as shown in Figure 26.

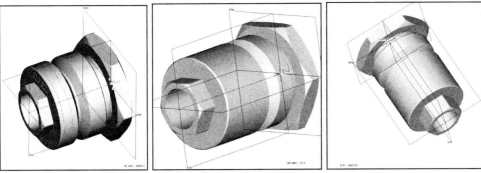

Figures 25(a-c) Family of Parts- Coupling-Fitting

Type	Instance Name	Common Name	d30	d46	F1068 [CUT]	F829 [SLOT]	F872 [SLOT]
	COUPLING-FITTI...		2.06	0.3130	Y	Y	Y
	CPLA		3.00	0.5000	N	N	N
	CPLB		3.25	0.6250	N	Y	N
	CPLC		3.50	0.7500	Y	N	Y

Figure 26 Family Table for Coupling-Fitting

Assemblies

Just as parts are created from related features, **assemblies** are created from related parts. The progressive combination of subassemblies, parts, and features into an assembly creates parent-child relationships based on the references used to assemble each component (Fig. 27).

The *Assembly* functionality is used to assemble existing parts and subassemblies.

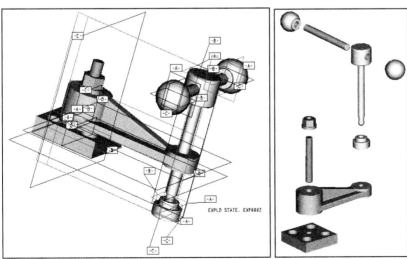

Figure 27 Clamp Assembly and Exploded Clamp Assembly

During assembly creation, you can:

- Simplify a view of a large assembly by creating a simplified representation
- Perform automatic or manual placement of component parts
- Create an exploded view of the component parts
- Perform analysis, such as mass properties and clearance checks
- Modify the dimensional values of component parts
- Define assembly relations between component parts
- Create assembly features
- Perform automatic interchange of component parts
- Create parts in Assembly mode
- Create documentation drawings of the assembly

Just as features can reference part geometry, parametric design also permits the creation of parts referencing assembly geometry. **Assembly mode** allows the designer both to fit parts together and to design parts based on how they should fit together.

In Figure 28, an assembly *Bill of Materials* report is generated.

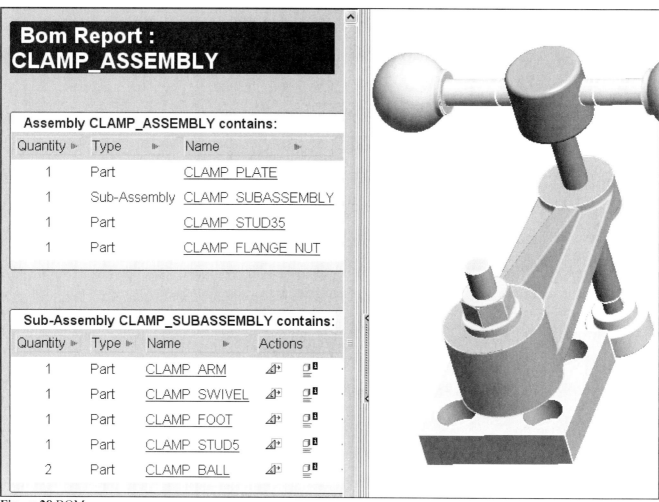

Figure 28 BOM

Drawings

You can create drawings of all parametric design models [Figs. 29(a-b)]. All model views in the drawing are *associative:* if you change a dimensional value in one view, other drawing views update accordingly. Moreover, drawings are associated with their parent models. Any dimensional changes made to a drawing are automatically reflected in the model. Any changes made to the model (e.g., addition of features, deletion of features, dimensional changes, and so on) in Part, Sheet Metal, Assembly, or Manufacturing modes are also automatically reflected in their corresponding drawings.

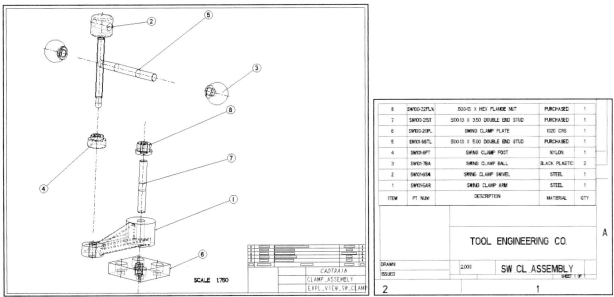

Figures 29(a-b) Ballooned Exploded View Assembly Drawing with Bill of Materials (BOM)

The **Drawing** functionality is used to create annotated drawings of parts and assemblies. During drawing creation, you can:

- Add views of the part or assembly
- Show existing dimensions
- Incorporate additional driven or reference dimensions
- Create notes on the drawing
- Display views of additional parts or assemblies
- Add sheets to the drawing
- Create draft entities on the drawing
- Balloon components on an assembly drawing (Fig. 30)
- Create an associative BOM

You can annotate the drawing with notes, manipulate the dimensions, and use layers to manage the display of different items on the drawing. The module **Pro/DETAIL** can be used to extend the drawing capability or as a stand-alone module allowing you to create, view, and annotate models and drawings.

Pro/DETAIL supports additional view types and multi-sheets and offers commands for manipulating items in the drawing and for adding and modifying different kinds of textural and symbolic information. In addition, the abilities to customize engineering drawings with sketched geometry, create custom drawing formats, and make numerous cosmetic changes to the drawing are available.

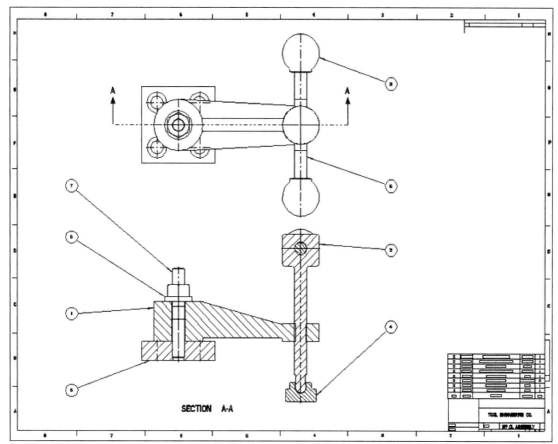

Figure 30 Ballooned Assembly Drawing

      Drawing mode in parametric design provides you with the basic ability to document solid models in drawings that share a two-way associativity.

      Changes that are made to the model in Part mode or Assembly mode will cause the drawing to update automatically and reflect the changes. Any changes made to the model in Drawing mode will be immediately visible on the model in Part and Assembly modes. The model shown in Figure 31 has been detailed in Figure 32.

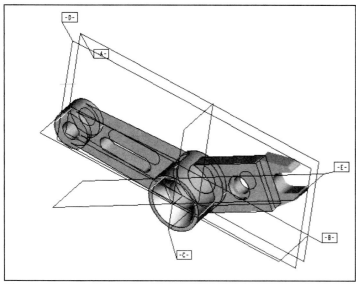

Figure 31 Angle Frame Model

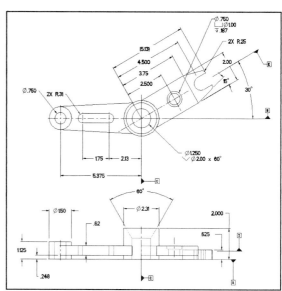

Figure 32 Angle Frame Drawing

Using the Text

The text utilizes a variety of command boxes and descriptions to lead you through construction sequences. Also, see the Appendix for Pro/ENGINEER WILDFIRE 3.0 Quick Reference Cards. The following icons, symbols, shortcut keys, and conventions will be used *(command sequences are always in a box)*:

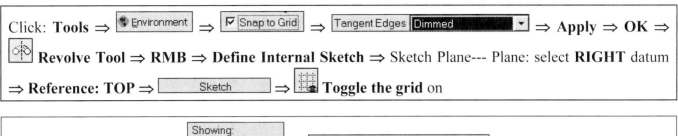

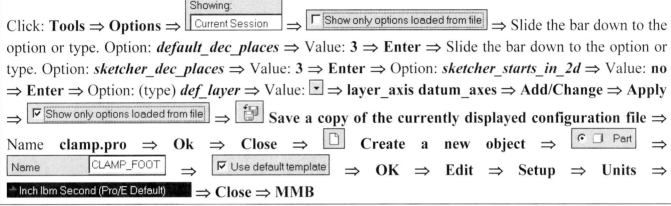

Commands:

- ⇒   Continue with command sequence or screen picks using **LMB**
- ◩ **Create lines**   icon (with description) indicates command to pick using **LMB**

Mouse or keyboard terms used in this text:

- **LMB**   **L**eft **M**ouse **B**utton
  - or "**Pick**"   term used to direct an action (i.e., "Pick the surface")
  - or "**Click**"   term used to direct an action (i.e., "Click on the icon")
  - or "**Select**"   term used to direct an action (i.e., "Select the feature")
- **MMB**   **M**iddle **M**ouse **B**utton (accept the current selection or value)
  - or **Enter**   press **Enter** key to accept entry
  - or ✓   Click on this icon to accept entry
- **RMB**   **R**ight **M**ouse **B**utton
  (toggles to next selection, or provides a list of available commands)

Shortcut Keys:

Ctrl+A (Activate)	**Ctrl+C** (Copy)	**Ctrl+D** (Standard Orientation)
Ctrl+F (Find)	**Ctrl+G** (Regenerate)	**Ctrl+N** (New)
Ctrl+O (Open)	**Ctrl+P** (Print)	**Ctrl+R** (Repaint)
Ctrl+S (Save)	**Ctrl+V** (Paste)	**Ctrl+Y** (Redo)
Ctrl+Z (Undo)		

(Refer to the Pro/ENGINEER Wildfire 3.0 Quick Reference Cards in the Appendix)

Text Organization

Text Lessons (Parts, Assemblies, and Drawings)

- **Lesson 1** has you complete two simple parts, an assembly, and a drawing using default settings.
- **Lesson 2** introduces Pro/ENGINEER's Wildfire 3.0 interface and embedded Browser.
- **Lesson 3** provides uncomplicated instructions to model a variety of simple-shaped parts.
- **Lessons 4** and **5** involve part modeling, using a variety of commands and tools.
- **Lessons 6** through **12** involve modeling the parts, creating the assembly [Fig. 33(a)], and documenting a design with detail and assembly drawings [Fig. 33(b)].
- **Lessons 13-18** provide instructions to model various parts using other features.
- **Appendix** includes Screen Customizing, Mapkeys, and *Pro/E Quick Reference Cards*.

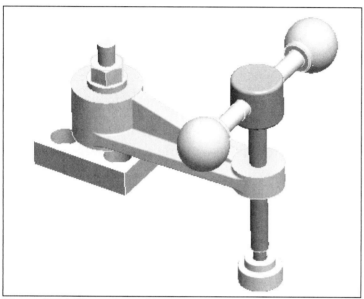

Figure 33(a) Clamp Assembly

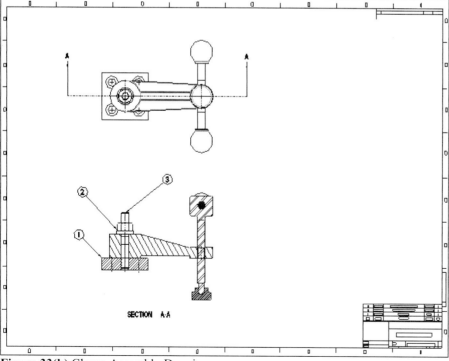

Figure 33(b) Clamp Assembly Drawing

Lesson 1 Pro/ENGINEER Wildfire 3.0 Overview

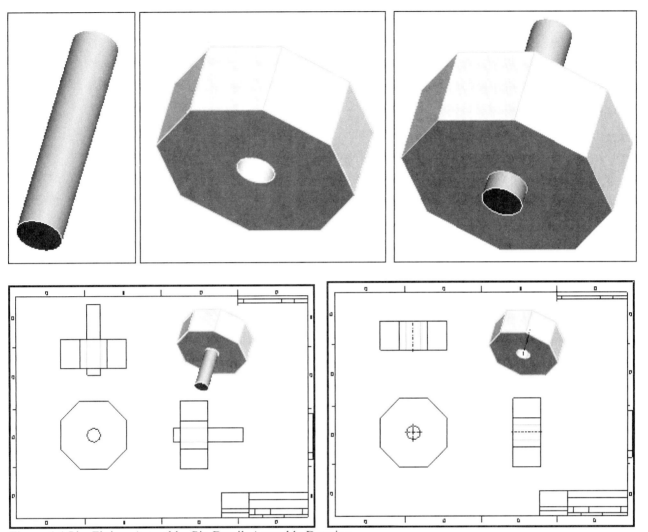

Figure 1.1 Pin, Fitting, Assembly, Pin Detail, Assembly Drawing

OBJECTIVES

- Create two parts
- Assemble parts to create an assembly
- Create a part drawing
- Create an assembly drawing

Pro/ENGINEER Wildfire 3.0 Overview

This lesson will allow you to experience the part, assembly, and drawing modes (Fig. 1.1) of Pro/ENGINEER Wildfire 3.0 by creating two simple parts (Figs. 1.2 and 1.5), assembling them, and creating drawings.

    A bare minimum of explanation is provided here. This lesson will quickly get you up and running on Pro/ENGINEER Wildfire 3.0.

Lesson 1 STEPS

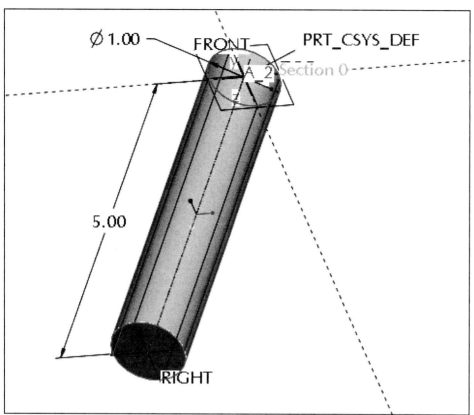

Figure 1.2 Pin

Creating the Pin Part

In this lesson most commands and picks will be accompanied by the window or dialog box that will open as the command is initiated. After this lesson, the commands will not show every window and dialog box, as this would make the text extremely long. Appropriate illustrations will be provided.

Throughout the text, a box surrounds all commands and menu selections.

Open **Pro/ENGINEER Wildfire 3.0** using a shortcut icon on your Desktop *(or with WINDOWS, click: Start ⇒ Programs ⇒ proewildfire3.0)* Pro/E will open on your computer.

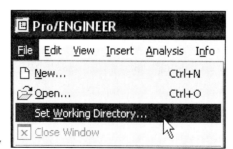

File ⇒ **Set Working Directory** ⇒ select the desired working directory or accept the default directory [Fig. 1.3(a)] ⇒ **OK** ⇒ **Create a new object** from Top Toolchest ⇒ [○ □ Part] ⇒ Name **PIN** ⇒ [☑ Use default template] [Fig. 1.3(b)] ⇒ **OK** [Fig. 1.3(c)]

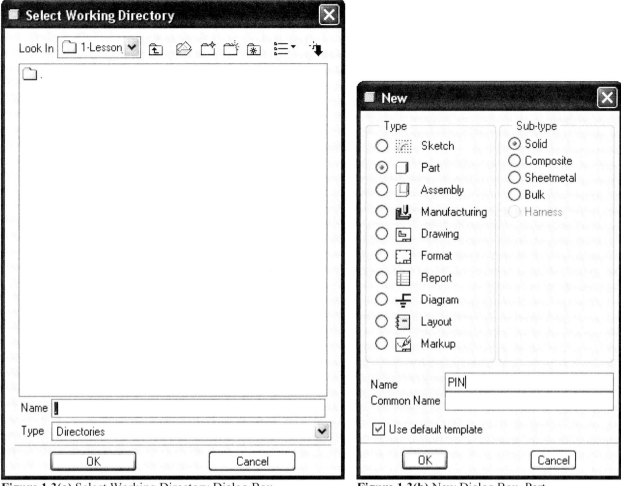

Figure 1.3(a) Select Working Directory Dialog Box

Figure 1.3(b) New Dialog Box, Part

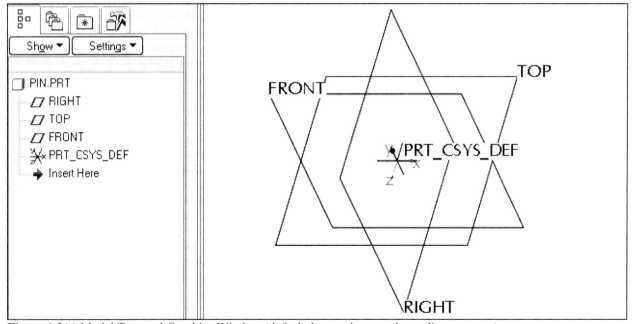

Figure 1.3(c) Model Tree and Graphics Window (default datum planes and coordinate system)

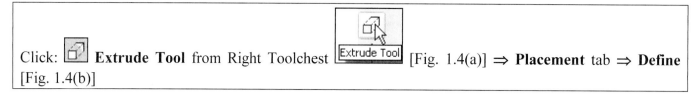

Click: [icon] **Extrude Tool** from Right Toolchest [Extrude Tool icon] [Fig. 1.4(a)] ⇒ **Placement** tab ⇒ **Define** [Fig. 1.4(b)]

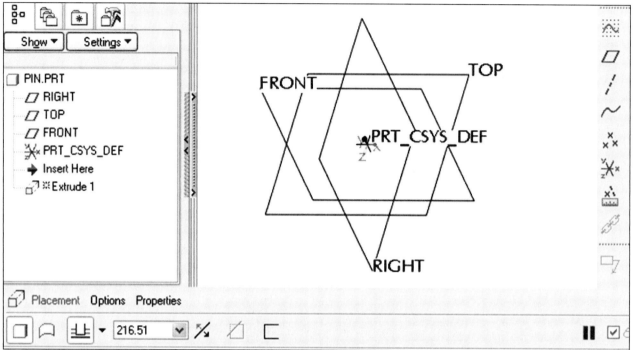

Figure 1.4(a) Extrude Dashboard Opens

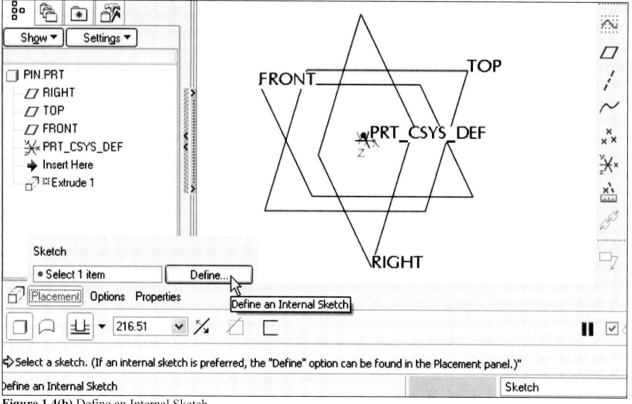

Figure 1.4(b) Define an Internal Sketch

Select the **FRONT** datum plane from the graphics window (or Model Tree) [Fig. 1.4(c)]. The FRONT datum plane is the Sketch Plane and the RIGHT datum plane is automatically selected as the Sketch Orientation Reference [Fig. 1.4(d)].

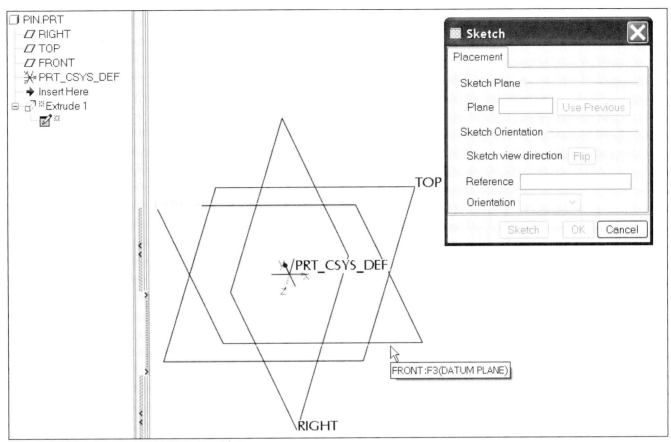

Figure 1.4(c) Select the FRONT Datum Plane

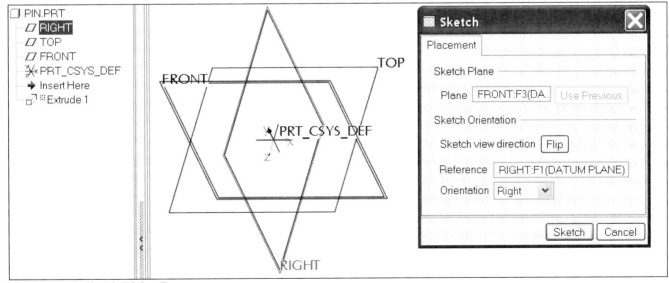

Figure 1.4(d) Sketch Dialog Box

Click: **Sketch** [Fig. 1.4(e)] ⇒ ⭕ **Create circle by picking the center and a point on the circle** from Right Toolchest [Fig. 1.4(f)]

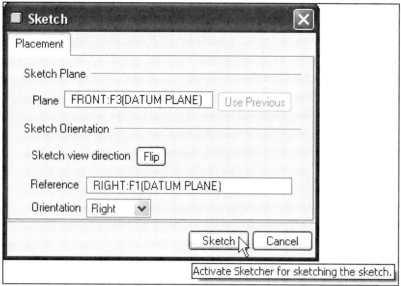

Figure 1.4(e) Activate the Sketcher

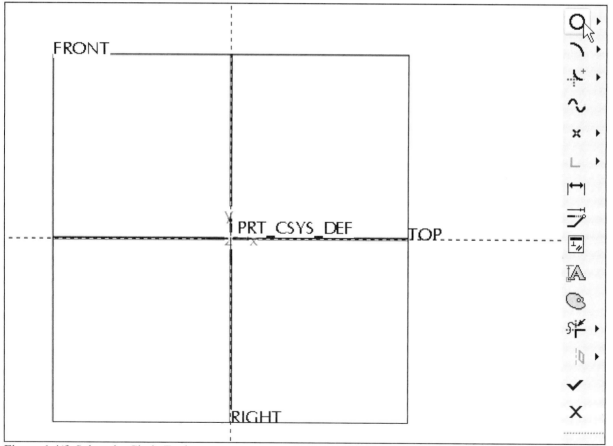

Figure 1.4(f) Select the Circle Tool

Pick the origin for the circle's center [Fig. 1.4(g)] ⇒ pick a point on the circle's edge [Fig. 1.4(h)] ⇒ **MMB** (**M**iddle **M**ouse **B**utton) [Figs. 1.4(i-l)] ⇒ double-click on the dimension ⇒ type **1.00** ⇒ **Enter**

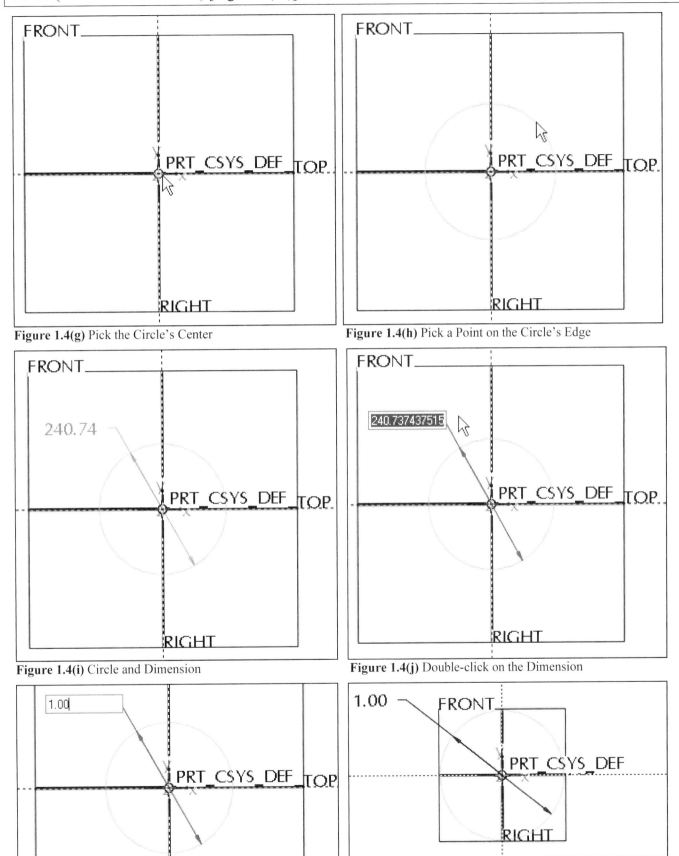

Figure 1.4(g) Pick the Circle's Center

Figure 1.4(h) Pick a Point on the Circle's Edge

Figure 1.4(i) Circle and Dimension

Figure 1.4(j) Double-click on the Dimension

Figure 1.4(k) Type 1.00

Figure 1.4(l) Completed Sketch

Click: ☑ **Continue with the current section** from Right Toolchest [Fig. 1.4(m)] ⇒ click in value field [Fig. 1.4(n)] ⇒ type **5.00** [Fig. 1.4(o)] ⇒ **Enter** ⇒ ⇒ **Standard Orientation** [Fig. 1.4(p)]

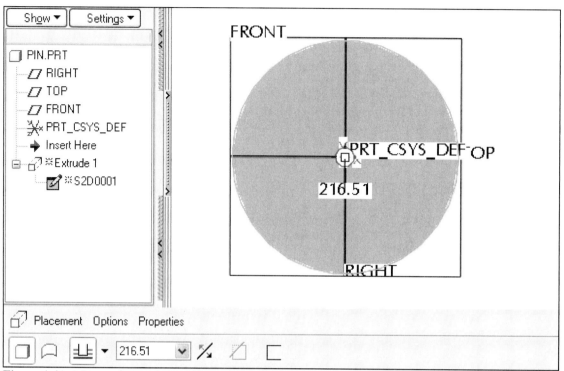

Figure 1.4(m) Dynamic Preview of Geometry

Figure 1.4(n) Extrude Dashboard Depth Value Field

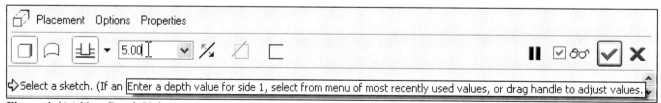

Figure 1.4(o) New Depth Value

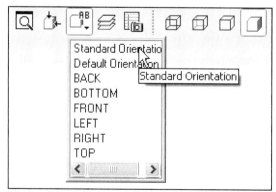

Figure 1.4(p) Standard Orientation

Click: 🔍 **Refit object to fully display it on the screen** from Top Toolchest ⇒ hold down the **Ctrl** key and your **MMB**, move your mouse downward to zoom in [Fig. 1.4(q)] *[if you have three-button mouse with a thumb wheel, simply rotate (not click) the thumb wheel to zoom]* ⇒ click **MMB** [Fig. 1.4(r)] ⇒ 🔍 ⇒ **File** ⇒ **Save** [Fig. 1.4(s)] ⇒ **MMB** Model Tree displays the Extrude feature [Fig. 1.4(t)]

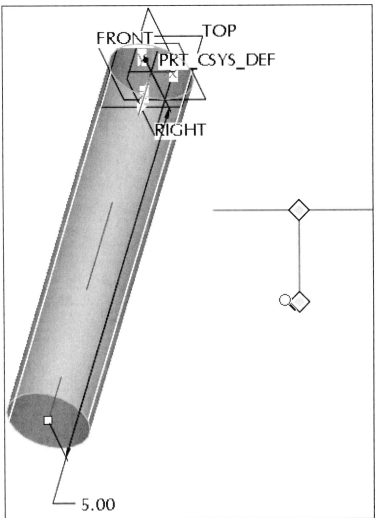

Figure 1.4(q) Ctrl+MMB to Zoom In   **Figure 1.4(r)** Completed Pin

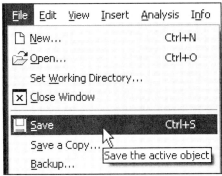

Figure 1.4(s) Save the active object

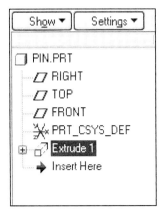

Figure 1.4(t) Model Tree

You have completed your first Pro/ENGINEER component.

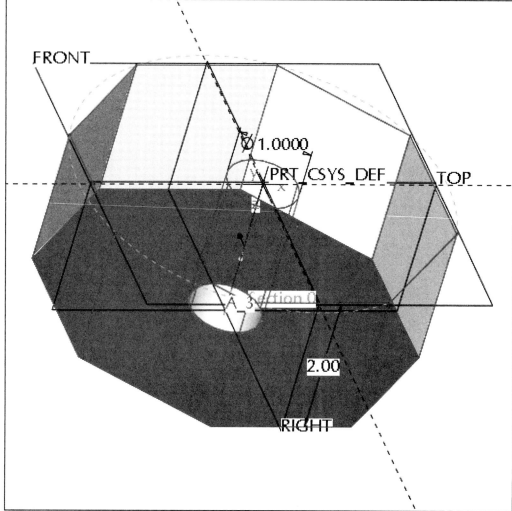

Figure 1.5 Plate

Creating the Plate Part

In this lesson most commands and picks will be accompanied by the window or dialog box that will open as the command is initiated. After this lesson, the commands will not show every window and dialog box, as this would make the text extremely long. Appropriate illustrations will be provided.

> *If you are continuing from the last section, then skip this set of commands*:
>
> Open **Pro/ENGINEER Wildfire 3.0** using a shortcut icon on your Desktop ⇒ **File** ⇒ **Set Working Directory** ⇒ select the working directory or accept the default directory (make sure this working directory is the same directory where the PIN was saved) ⇒ **OK**

Click: ⬜ **Create a new object** ⇒ ⦿ ▢ Part ⇒ Name **PLATE** ⇒ ☑ Use default template [Fig. 1.6(a)] ⇒ **OK** ⇒ **Window** from menu bar [Fig. 1.6(b)] (note: you now have two parts *"in session"*) ⇒ **Activate** ⇒ 💾 **Save the active object** *(Hint: Pro/E does NOT save for you- ever! Nor does it prompt you to save- ever! So save often)* ⇒ **OK** (or **MMB**) ⇒ 🔍 [Fig. 1.6(c)]

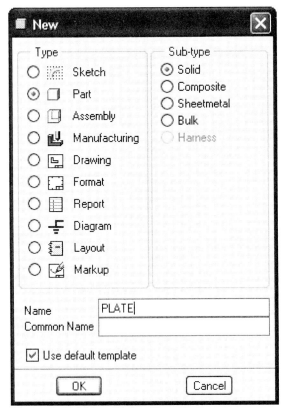

Figure 1.6(a) New Dialog Box, Part

Figure 1.6(b) Window from the Menu Bar

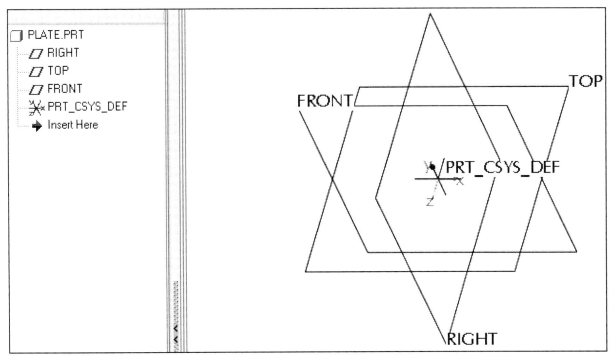

Figure 1.6(c) Plate

Click: **Tools** from menu bar [Fig. 1.6(d)] ⇒ **Environment** [Fig. 1.6(e)] ⇒ ☑ Snap to Grid *(your system may have this already set as the default)* ⇒ **Apply** ⇒ **OK**

Figure 1.6(d) Tools ⇒ Environment

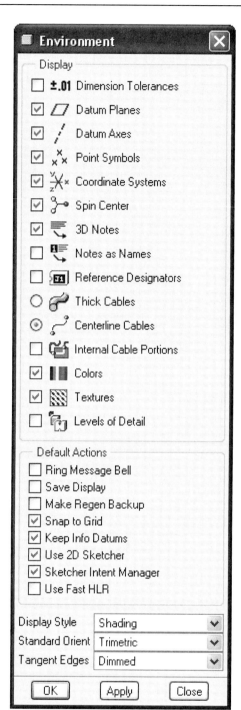

Figure 1.6(e) Environment Dialog Box

Pick/select the **FRONT** datum plane [Fig. 1.7(a)] ⇒ 🗔 **Extrude Tool** ⇒ **RMB** (**R**ight **M**ouse **B**utton) ⇒ **Define Internal Sketch** [Fig. 1.7(b)] ⇒ **Sketch** [Fig. 1.7(c)]

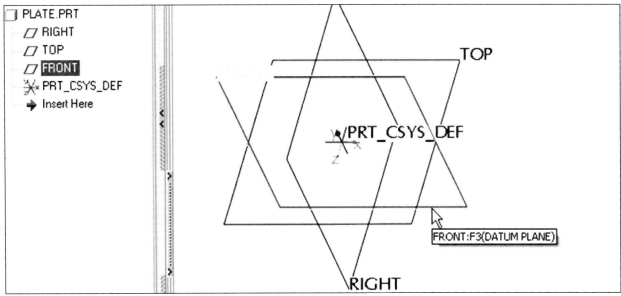

Figure 1.7(a) Select the Front Datum Plane

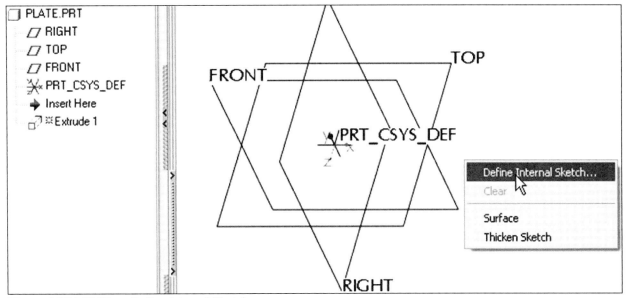

Figure 1.7(b) RMB ⇒ Define Internal Sketch

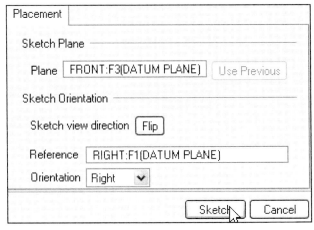

Figure 1.7(c) Sketch Dialog Box

Click: **Insert foreign data from Palette into active object** [Fig. 1.7(d)] ⇒ scroll down to see the octagon ⇒ double-click **Octagon** [Figs. 1.7(e-f)]

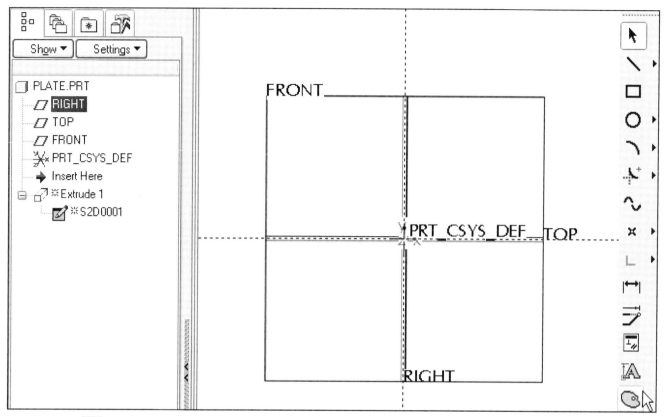

Figure 1.7(d) Insert Foreign Data from Palette into Active Object

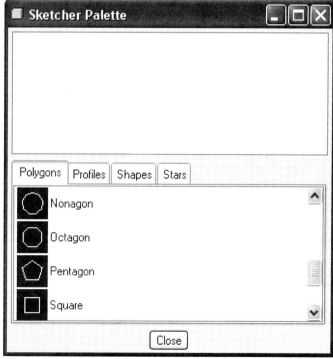

Figure 1.7(e) Sketcher Palette

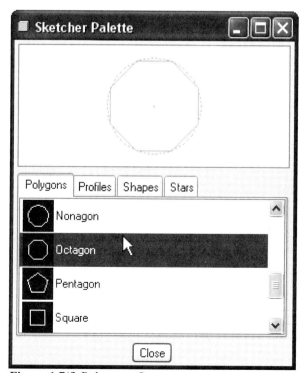

Figure 1.7(f) Polygons: Octagon

Place the octagon on the sketch by picking a position [Fig. 1.7(g)] ⇒ with the **LMB,** drag and drop the center of the octagon at the origin [Fig. 1.7(h)] ⇒ modify Scale value to **2** 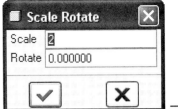 ⇒ Enter ⇒ 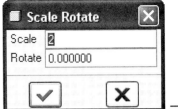 from the Scale Rotate dialog box

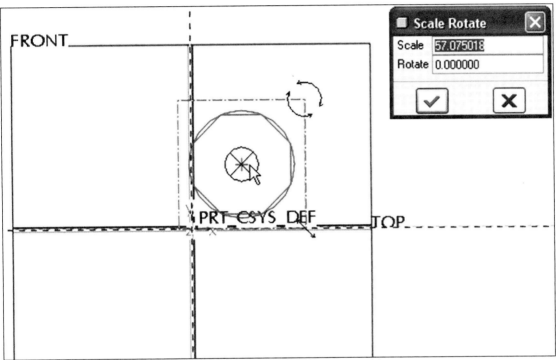
Figure 1.7(g) Place Octagon anywhere on the Sketch by picking a Position

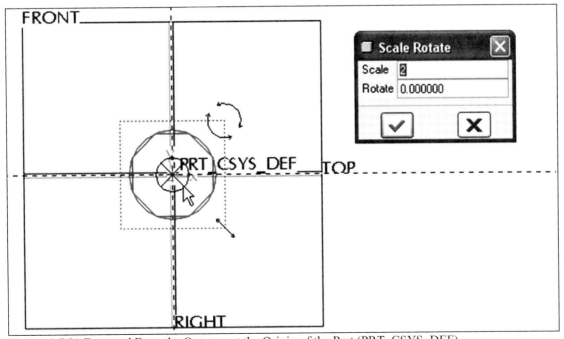
Figure 1.7(h) Drag and Drop the Octagon at the Origin of the Part (PRT_CSYS_DEF)

Click: **Close** the Sketcher Palette ⇒ [icon] [Fig. 1.7(i)] ⇒ [✓] from Right Toolchest [Fig. 1.7(j)] ⇒ modify the depth in the dashboard to **2.00** [2.00] [Fig. 1.7(k)] ⇒ **Enter** ⇒ [icon] ⇒ **Standard Orientation** from Top Toolchest [Fig. 1.7(l)] ⇒ [✓] from dashboard [Fig. 1.7(m)] ⇒ **LMB** to deselect

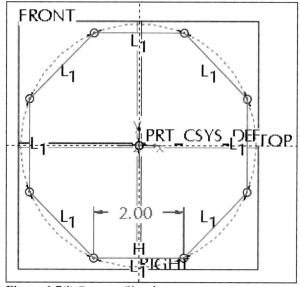

Figure 1.7(i) Octagon Sketch

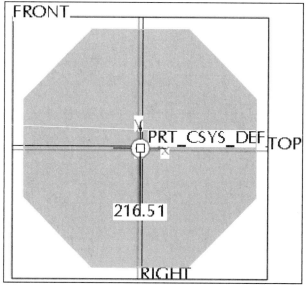

Figure 1.7(j) Octagon Dynamic Preview

Figure 1.7(k) Modify Depth to **2.00**

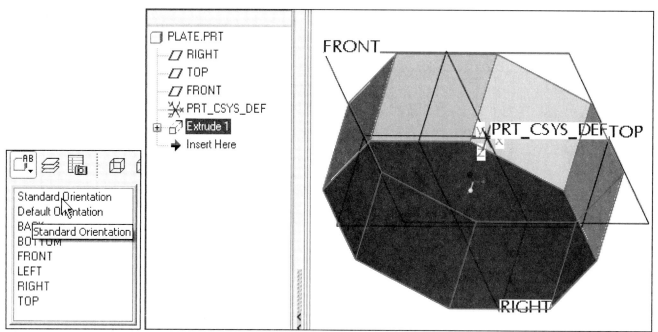

Figure 1.7(l) Orientation   **Figure 1.7(m)** Extruded Octagon

Click: **Hole Tool** from Right Toolchest ⇒ pick on the top surface [Fig. 1.8(a)] ⇒ **Hidden Line** from Top Toolchest [Fig. 1.8(b)]

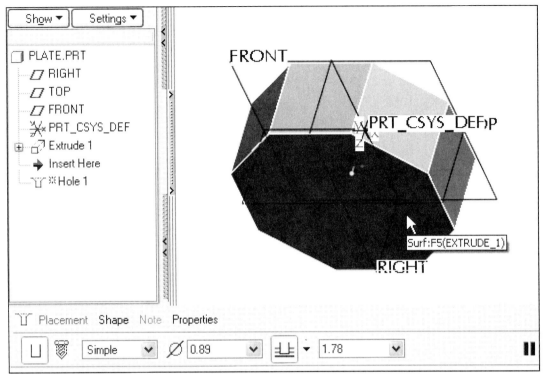

Figure 1.8(a) Pick Top Surface

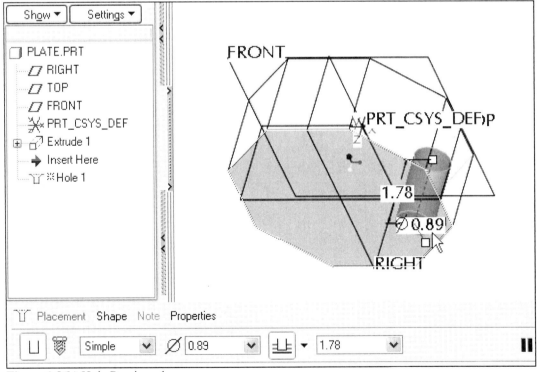

Figure 1.8(b) Hole Previewed

Click: **RMB** [Fig. 1.8(c)] ⇒ **Secondary References Collector** ⇒ press and hold the **Ctrl** key ⇒ select the **TOP** datum plane from the Model Tree ⇒ with the **Ctrl** key still pressed, select the **RIGHT** datum plane from the Model Tree [Fig. 1.8(d)] ⇒ **Placement** tab from the dashboard [Fig. 1.8(e)]

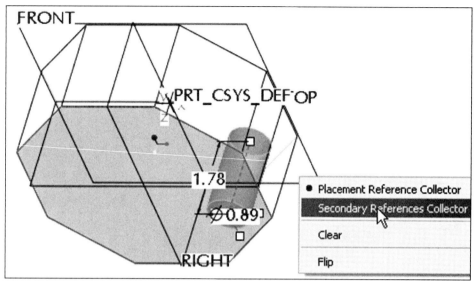

Figure 1.8(c) RMB ⇒ Secondary References Collector

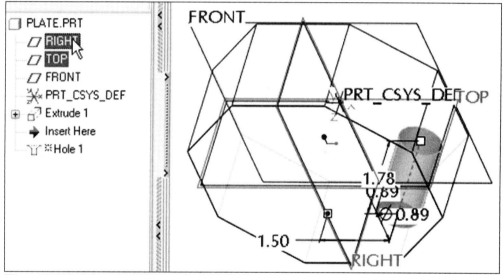

Figure 1.8(d) Select the TOP and the RIGHT Datum Planes as Secondary References

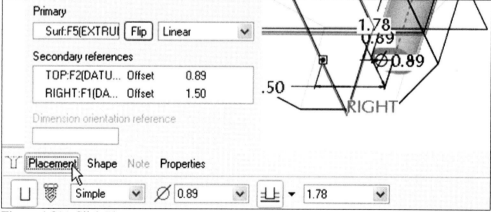

Figure 1.8(e) Click Placement

In the Secondary references collector, click on **TOP** [Fig. 1.8(f)] ⇒ click on Offset ⇒ **Align** [Fig. 1.8(g)] ⇒ click on **RIGHT** [Fig. 1.8(h)] ⇒ click on Offset ⇒ **Align** [Figs. 1.8(i-j)]

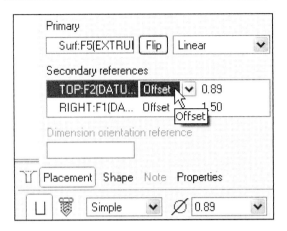

Figure 1.8(f) Click on TOP ⇒ Offset

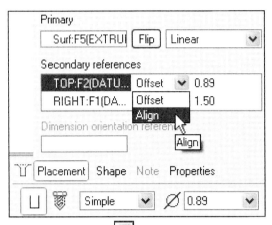

Figure 1.8(g) Click ⇒ Align

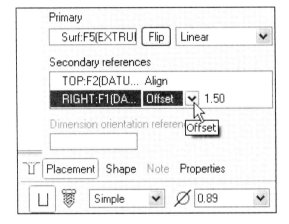

Figure 1.8(h) Click on RIGHT ⇒ Offset

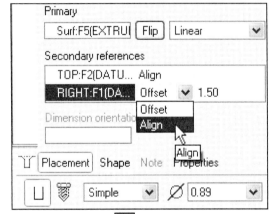

Figure 1.8(i) Click ⇒ Align

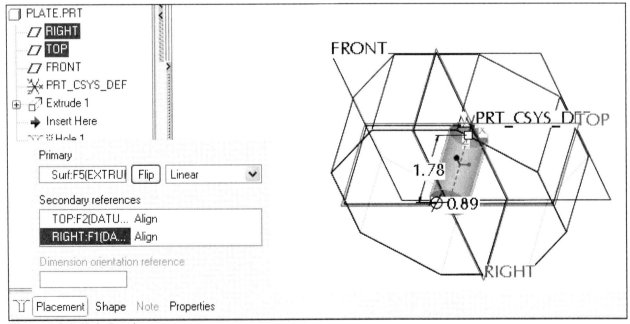

Figure 1.8(j) Hole Preview

Click: **Placement** tab to close the Placement panel ⇒ [icon] ⇒ **Drill to intersect with all surfaces** [Fig. 1.8(k)] ⇒ modify the hole's diameter, type **1.00** ⇒ **Enter** [Fig. 1.8(l)] ⇒ [✓] ⇒ [Fig. 1.8(m)] ⇒ **Ctrl+S** to save the part ⇒ **OK**

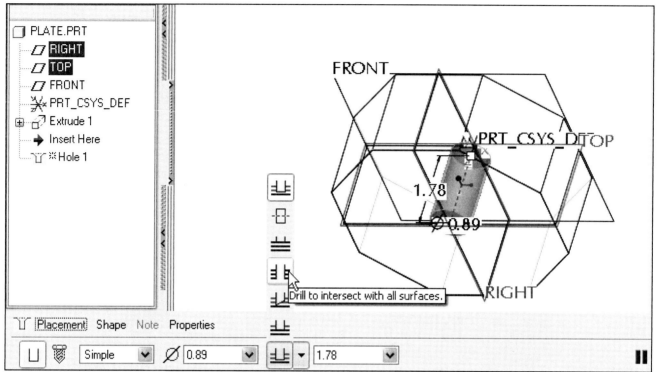

Figure 1.8(k) Drill to intersect with all surfaces

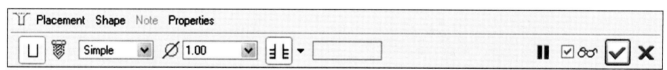

Figure 1.8(l) Diameter **1.00**

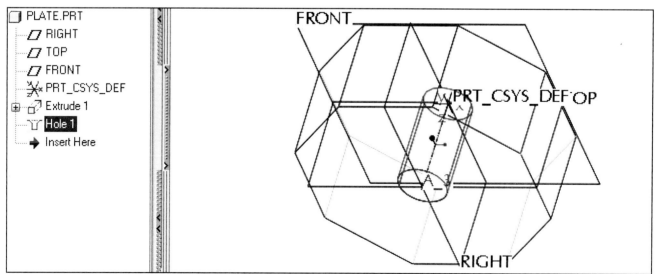

Figure 1.8(m) Completed Hole

Click: ⬜ **Shading** ⇒ **View** from menu bar [Fig. 1.9(a)] ⇒ **Color and Appearance** ⇒ ptc-metallic-gold [Fig. 1.9(b)] ⇒ **Apply** ⇒ **Close** the dialog box

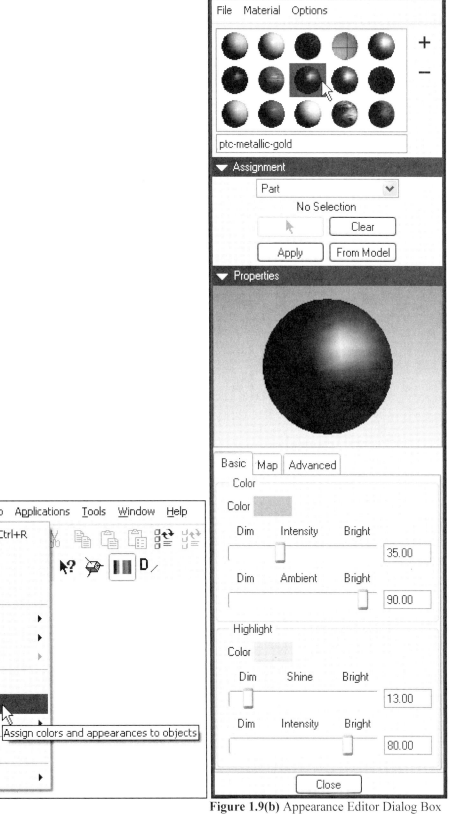

Figure 1.9(a) Color and Appearance

Figure 1.9(b) Appearance Editor Dialog Box

Click: 🔍 ⇒ 💾 ⇒ **OK** (Fig. 1.10)

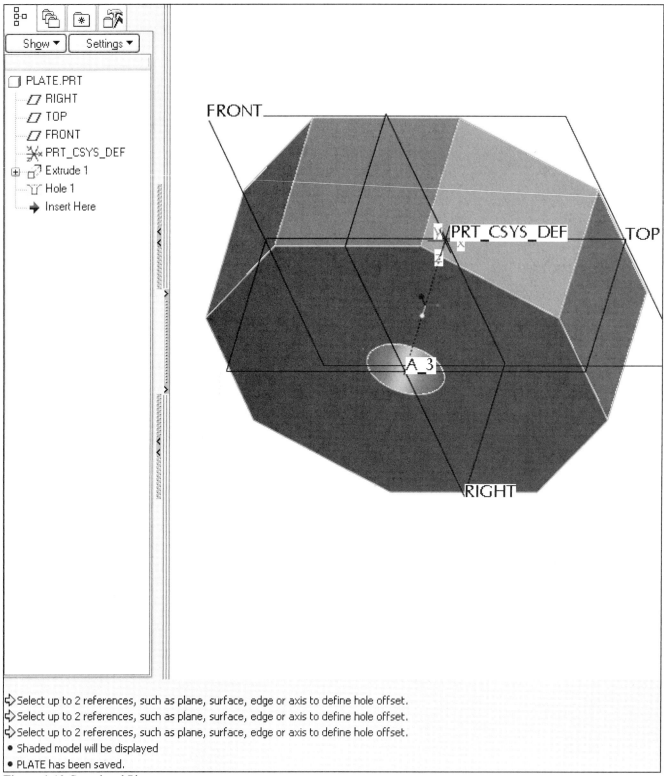

Figure 1.10 Completed Plate

You have completed your second Pro/ENGINEER component. In the next section, you will put the parts together to create an assembly.

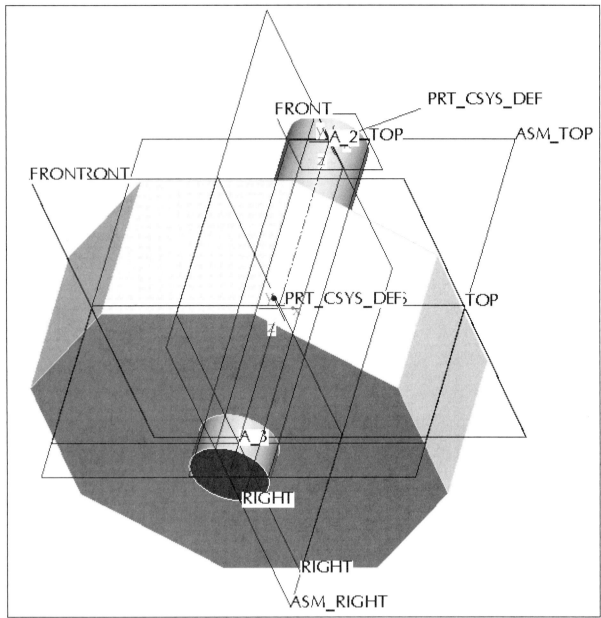

Figure 1.11 Assembly

Creating the Assembly

Using the two simple parts just created you will now create an assembly (Fig. 1.11). Just as you can combine features into parts, you can also combine parts into assemblies. The Assembly mode in Pro/E enables you to place component parts and subassemblies together to form assemblies.

As with a part, an assembly starts with default datum planes and a default coordinate system. To create a subassembly or an assembly, you must first create datum features. You then create or assemble additional components to the existing component(s) and datum features.

If you are continuing from the last section, then skip this set of commands:

Open **Pro/ENGINEER Wildfire 3.0** using a shortcut icon on your Desktop ⇒ **File** ⇒ **Set Working Directory** ⇒ select the working directory or accept the default directory (make sure you set the same working directory as where your parts were saved) ⇒ **OK**

Click: **Create a new object** ⇒ Assembly ⇒ Design ⇒ Name **CONNECTOR** ⇒ Use default template [Fig. 1.12(a)] ⇒ **OK** ⇒ **Settings** from Navigator [Fig. 1.12(b)] ⇒ **Tree Filters** ⇒ toggle all on [Fig. 1.12(c)] ⇒ **Apply** ⇒ **OK** [Fig. 1.12(d)] ⇒ **File** ⇒ **Save** ⇒ **OK**

Figure 1.12(a) New Dialog Box, Assembly

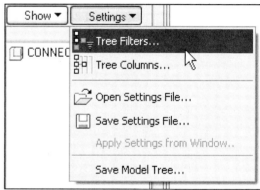

Figure 1.12(b) Tree Filters

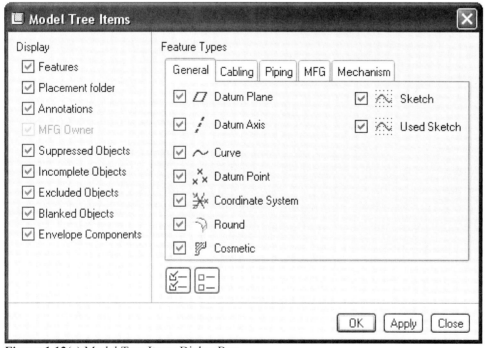

Figure 1.12(c) Model Tree Items Dialog Box

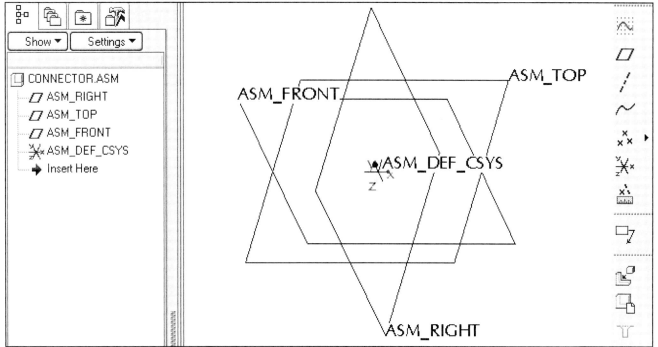

Figure 1.12(d) Assembly Datums and Coordinate System

Click: 🖼 **Add component to the assembly** from Right Toolchest ⇒ **plate.prt** [Fig. 1.13(a)] ⇒ **Open**

Figure 1.13(a) Open Dialog Box

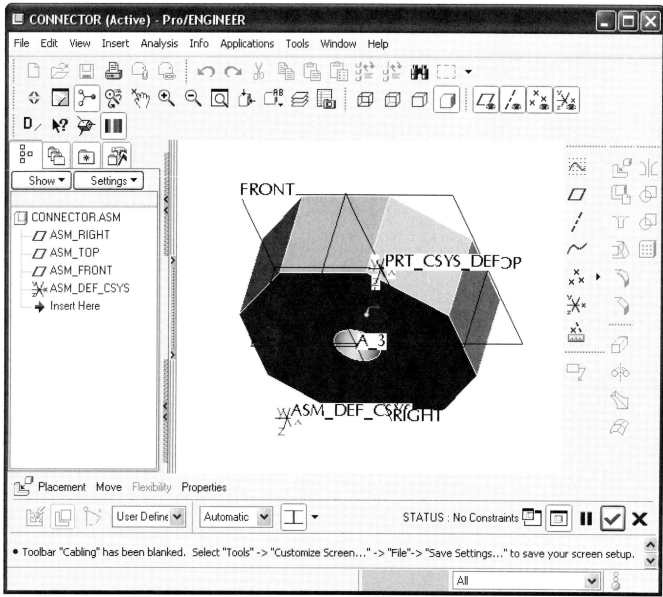

Figure 1.13(b) Assembly Dashboard

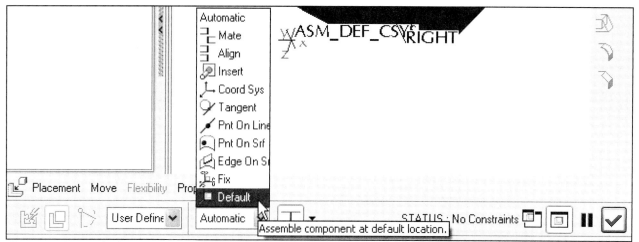

Figure 1.13(c) Default Constraint

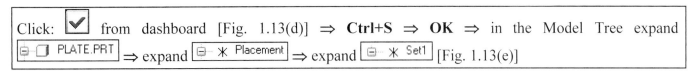

Click: ☑ from dashboard [Fig. 1.13(d)] ⇒ **Ctrl+S** ⇒ **OK** ⇒ in the Model Tree expand PLATE.PRT ⇒ expand ✻ Placement ⇒ expand ✻ Set1 [Fig. 1.13(e)]

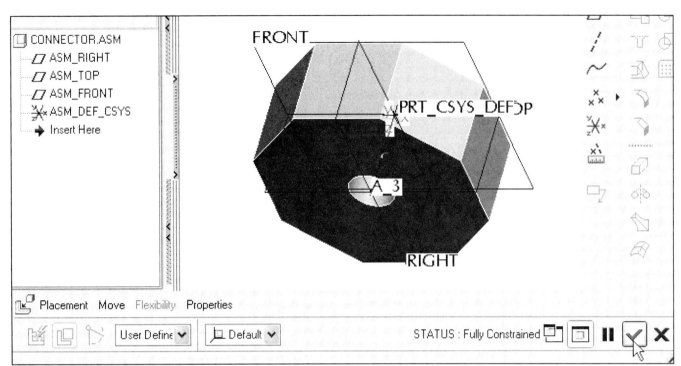

Figure 1.13(d) Fully Constrained Component

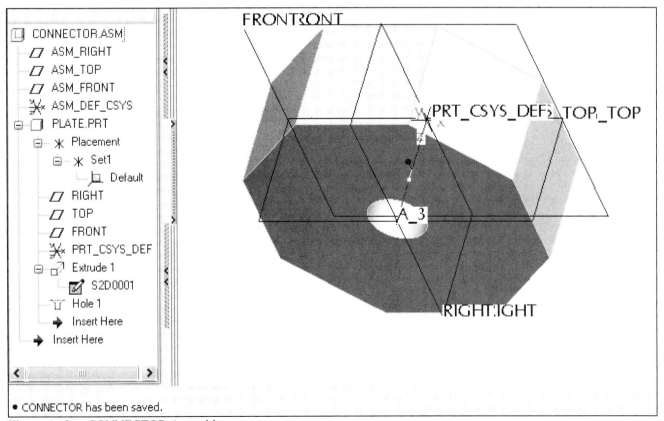

Figure 1.13(e) CONNECTOR Assembly

Click: 🖳 **Add component to the assembly** from Right Toolchest ⇒ **pin.prt** ⇒ **Preview** >>> [Fig. 1.14(a)] ⇒ **Open** [Fig. 1.14(b)]

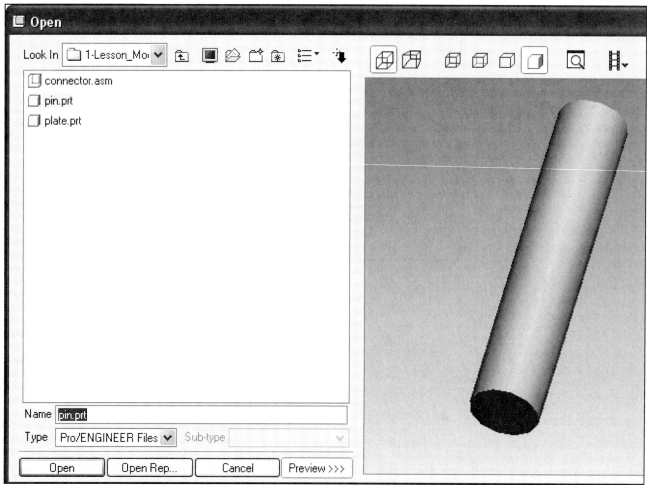

Figure 1.14(a) Previewed Pin

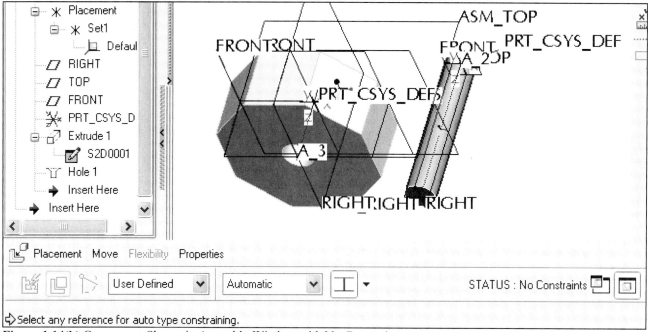

Figure 1.14(b) Component Shown in Assembly Window with No Constraints

Pick on the Pin's cylindrical surface [Fig. 1.14(c)] ⇒ pick on the Plate's cylindrical hole surface [Fig. 1.14(d)] ⇒ **Placement** tab from dashboard ⇒ **New Constraint** [Fig. 1.14(e)]

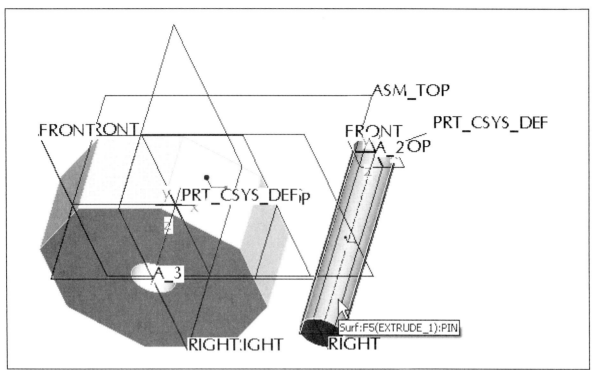

Figure 1.14(c) Pick on the Pin's Cylindrical Surface

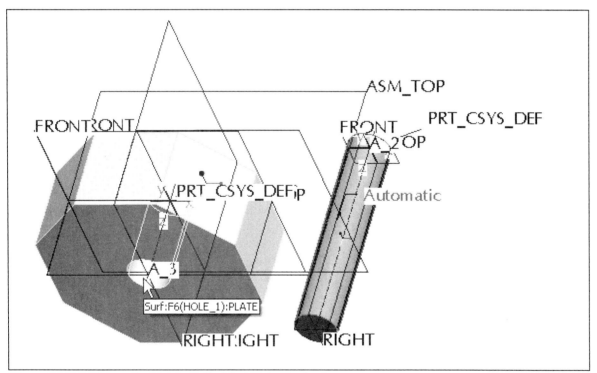

Figure 1.14(d) Pick on the Plate's Cylindrical Hole (Surface)

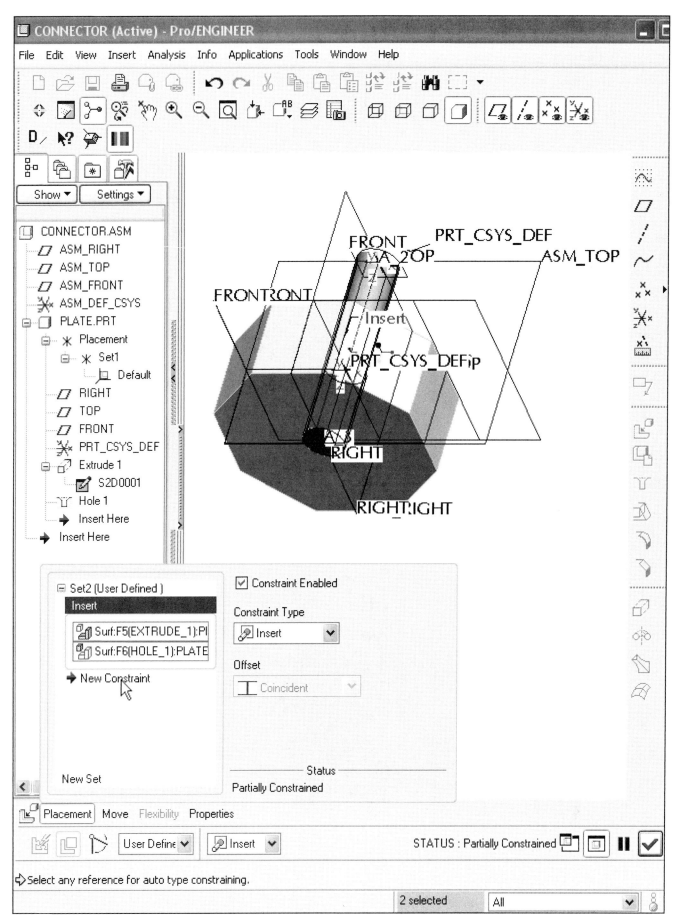

Figure 1.14(e) New Constraint

Pick on the Plate's front surface [Fig. 1.14(f)]

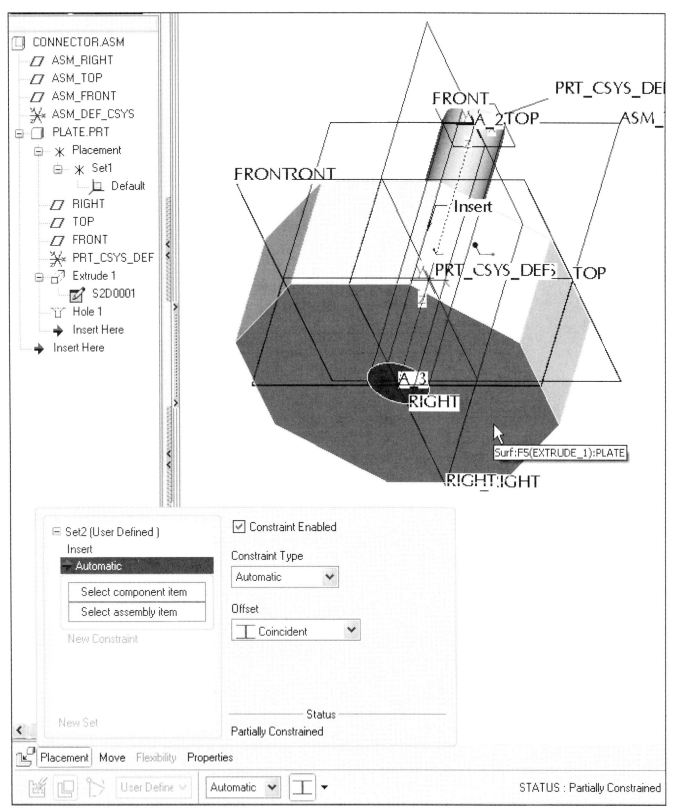

Figure 1.14(f) Pick on the Plate's Front Surface

Pick on the Pin's end face surface [Fig. 1.14(g)]

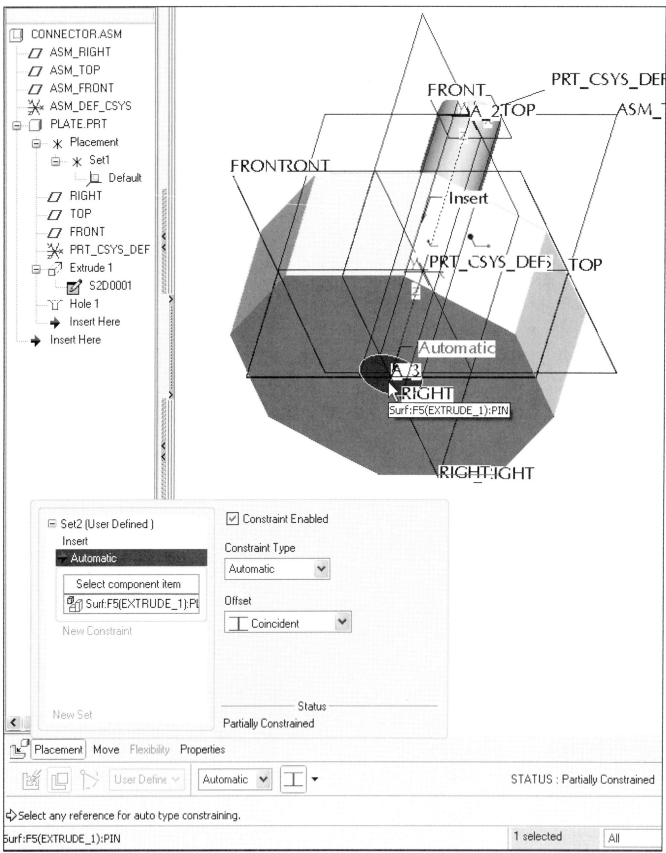

Figure 1.14(g) Pick on the Pin's End Face Surface

Click: [Coincident] [Fig. 1.14(h)] ⇒ [Offset] [Fig. 1.14(i)] ⇒ click in value field next to Offset, type **.50** [Offset] [0.50] [Fig. 1.14(j)] ⇒ **Enter**

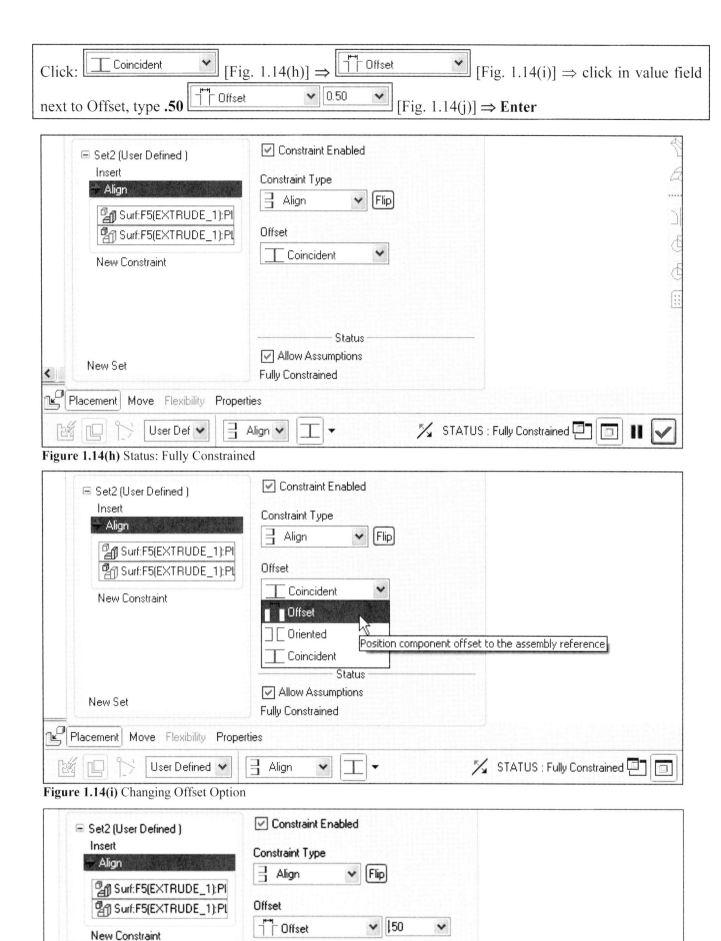

Figure 1.14(h) Status: Fully Constrained

Figure 1.14(i) Changing Offset Option

Figure 1.14(j) Offset **.50**

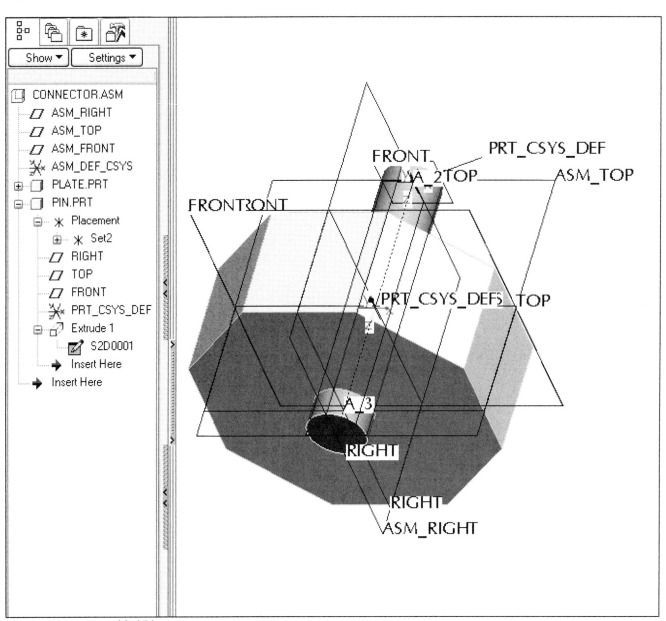

Figure 1.14(k) Assembly Dashboard: Fully Constrained

Figure 1.14(l) Assembled Pin

You have completed your first Pro/ENGINEER assembly. In the next section you will create a set of drawings for the Plate component and the Connector assembly.

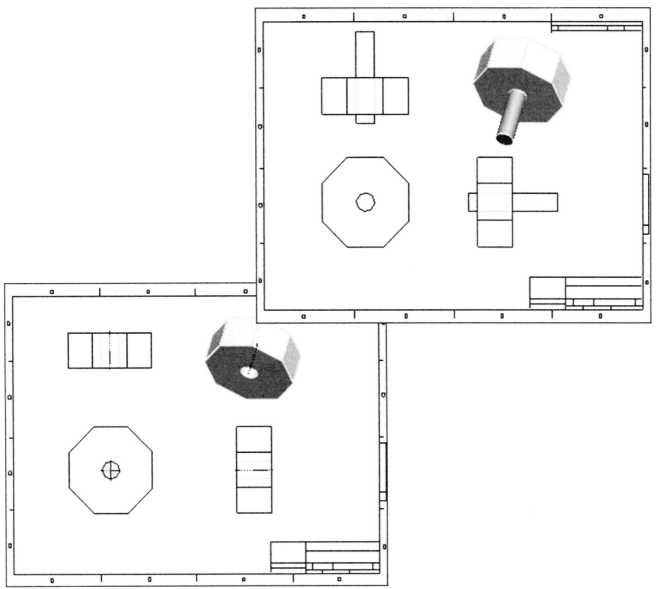

Figure 1.15 Drawings

Creating Drawings

Using the parts and the assembly just created, you will now create a detail drawing of the Plate and an assembly drawing of the Connector (Fig. 1.15).

The Drawing mode in Pro/E enables you to create and manipulate engineering drawings that use the 3D model (part or assembly) as a geometry source. You can pass dimensions, notes, and other elements of design between the 3D model and its views on a drawing.

If you are continuing from the last section, then skip this set of commands:

Open **Pro/ENGINEER Wildfire 3.0** using a shortcut icon on your Desktop ⇒ **File** ⇒ **Set Working Directory** ⇒ select the working directory or accept the default directory (make sure you set the same working directory as where your parts and assembly models were saved) ⇒ **OK** ⇒ **File** ⇒ **Open** ⇒ **Pin** ⇒ **File** ⇒ **Open** ⇒ **Plate** ⇒ **File** ⇒ **Open** ⇒ **Connector** *[you now have all three objects (the two parts and the assembly) "in session"]*

Click: ☐ **Create a new object** ⇒ ⊙ 🗏 Drawing ⇒ Name CONNECTOR_ASM ⇒ ☑ Use default template [Fig. 1.16(a)] ⇒ **OK** [Fig. 1.16(b)] *[Note: if the Default Model is not the CONNECTOR.ASM, then click: Browse ⇒ Preview >>> ⇒ connector.asm [Fig. 1.16(c)] ⇒ Open]*

Figure 1.16(a) New Dialog Box, Drawing

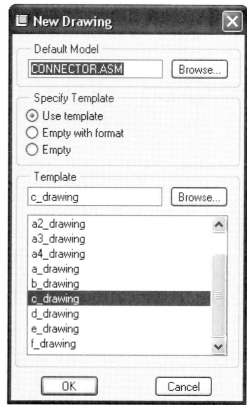

Figure 1.16(b) New Drawing Dialog Box

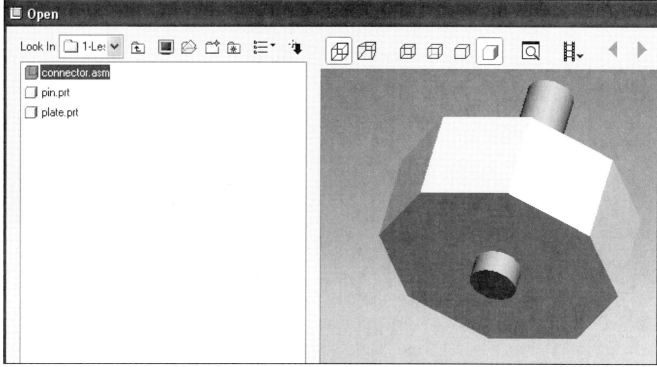

Figure 1.16(c) Connector Assembly Preview

Click: **OK** with the default model set as [connector.asm] the drawing will display [Fig. 1.16(d)] ⇒ **RMB** ⇒ **Page Setup** [Fig. 1.16(e)]

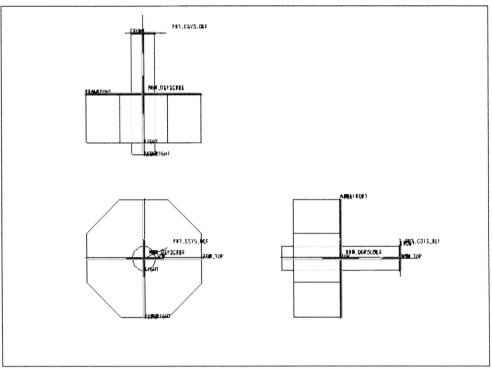

Figure 1.16(d) Drawing

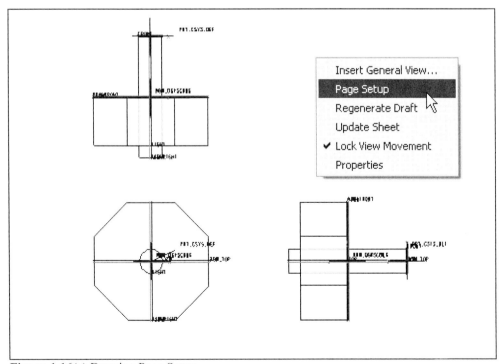

Figure 1.16(e) Drawing Page Setup

Page Setup dialog box displays [Fig. 1.16(f)] ⇒ click in the **C Size** field [Fig. 1.16(g)] ⇒ to see options [Fig. 1.16(h)] ⇒ **Browse** [Fig. 1.16(i)] ⇒ **c.frm** [Fig. 1.16(j)] ⇒ **Open** [Fig. 1.16(k)] ⇒ **OK**

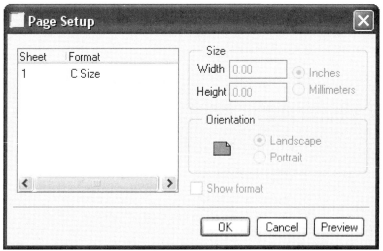
Figure 1.16(f) Page Setup Dialog Box

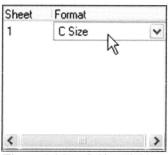
Figure 1.16(g) C Size Field

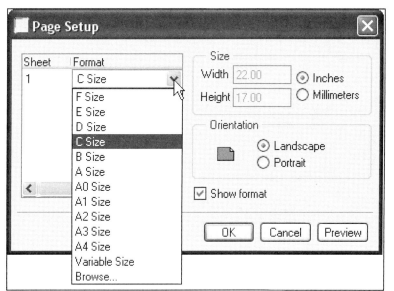
Figure 1.16(h) Format Options

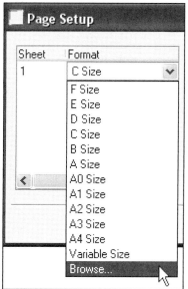
Figure 1.16(i) Browse

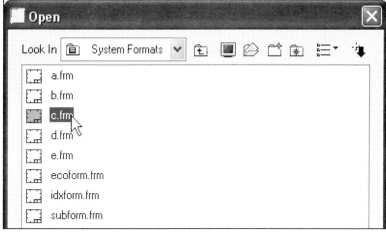
Figure 1.16(j) c.frm from the System Formats Library

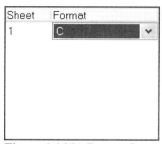
Figure 1.16(k) Format C

⌞🗗 /⚏ ˣxₓ ᵞxₓ⌟ off ⇒ 🗗 ⇒ **RMB** ⇒ **Insert General View** [Fig. 1.16(l)] ⇒ **DEFAULT ALL** [Fig. 1.16(m)] ⇒ **OK** ⇒ pick a position for the new view [Fig. 1.16(n)] ⇒ **OK** from the Drawing View dialog box [Fig. 1.16(o)] ⇒ **Ctrl+S** ⇒ **OK** the drawing is now complete [Fig. 1.16(p)] ⇒ 🗗 ⇒ **LMB**

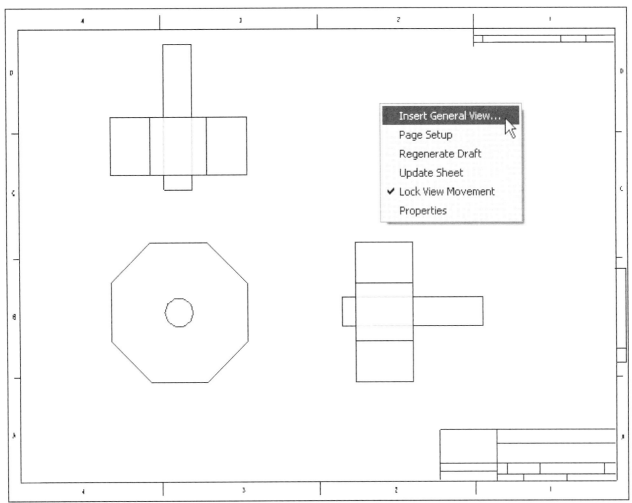

Figure 1.16(l) Insert General View

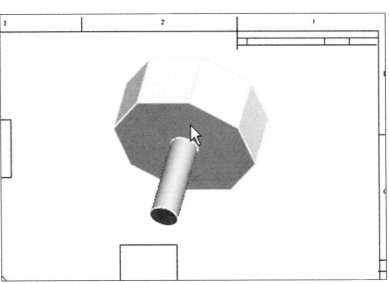

Figure 1.16(m) Select Combined State   **Figure 1.16(n)** Pick a Position for the New View

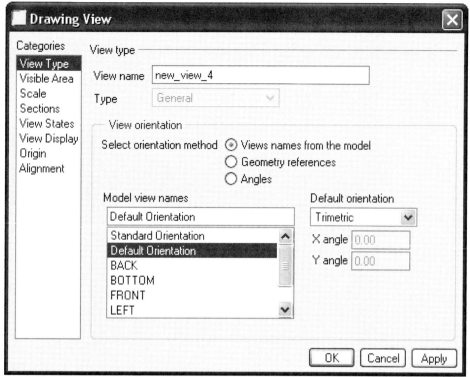

Figure 1.16(o) Drawing View Dialog Box

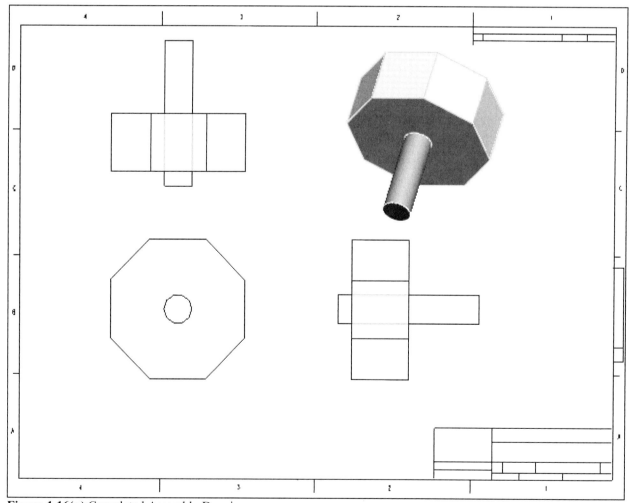

Figure 1.16(p) Completed Assembly Drawing

Click: 🗎 **Create a new object** ⇒ ⦿ 🗎 Drawing ⇒ Name PLATE ⇒ ☑ Use default template [Fig. 1.17(a)] ⇒ **OK** [Fig. 1.17(b)] ⇒ Default Model **Browse** ⇒ **plate.prt** ⇒ **Preview** >>> [Fig. 1.17(c)] ⇒ **Open**

Figure 1.17(a) New Dialog Box, Drawing

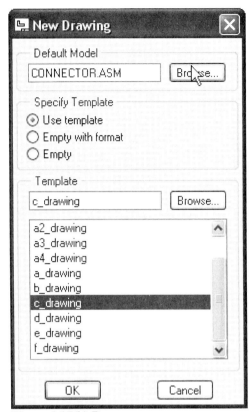

Figure 1.17(b) Browse

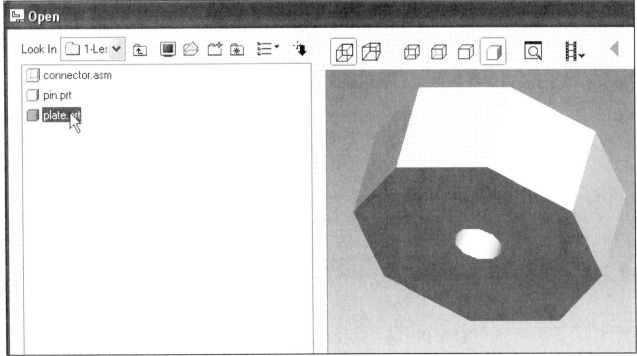

Figure 1.17(c) Plate Preview

Click: **OK** [Fig. 1.17(d)] drawing opens ⇒ **RMB** ⇒ **Page Setup** [Figs. 1.17(e-f)] ⇒ click in the **C Size** field [Fig. 1.17(g)] ⇒ to see options [Fig. 1.17(h)] ⇒ **Browse** [Fig. 1.17(i)] ⇒ **c.frm** [Fig. 1.17(j)] ⇒ **Open** ⇒ **OK** formatted drawing is displayed [Fig. 1.17(k)]

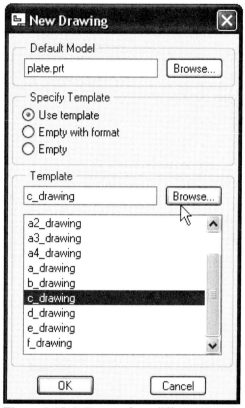
Figure 1.17(d) Browse for a different Template

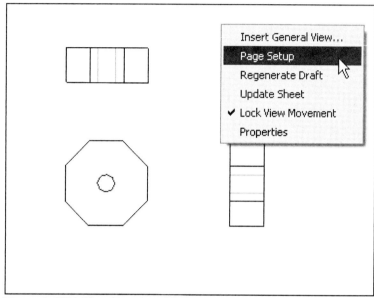
Figure 1.17(e) Plate Drawing, RMB ⇒ Page Setup

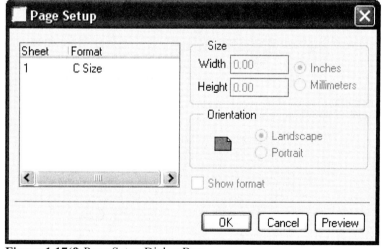
Figure 1.17(f) Page Setup Dialog Box

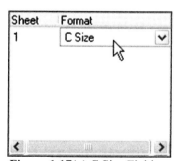
Figure 1.17(g) C Size Field

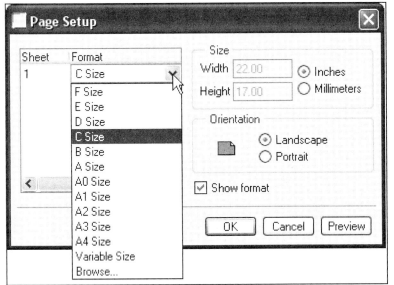

Figure 1.17(h) Page Setup Dialog Box, Format Options

Figure 1.17(i) Browse

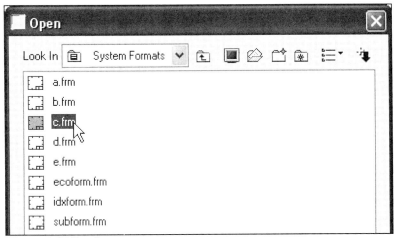

Figure 1.17(j) c.frm

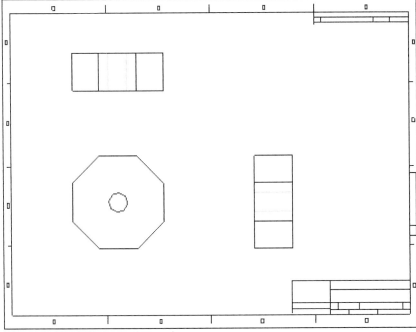

Figure 1.17(k) Formatted Drawing

Click: **RMB** ⇒ **Insert General View** [Fig. 1.17(l)] ⇒ pick a position for the new view ⇒ **OK** from the Drawing View dialog box [Fig. 1.17(m)]

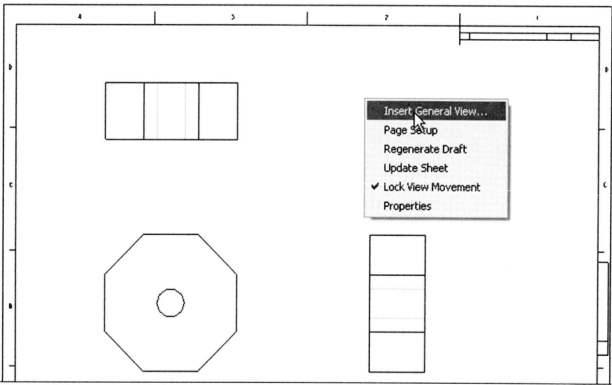

Figure 1.17(l) Insert General View

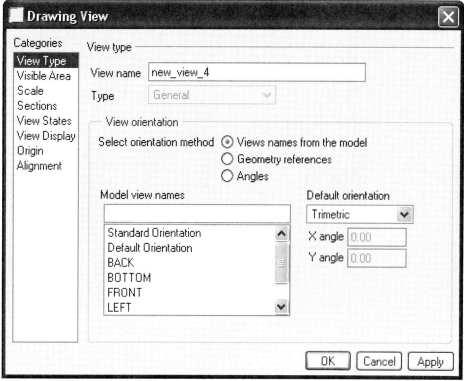

Figure 1.17(m) Drawing View Dialog Box

Click: **LMB** ⇒ [icon] ⇒ [icon] **Open the Show/Erase dialog box** [Fig. 1.18(a)] ⇒ **Show** ⇒ [icon] **Axis** [Fig. 1.18(b)] ⇒ [Show All] **Show all items** [Fig. 1.18(c)] ⇒ **Yes** [Fig. 1.18(d)] ⇒ **Accept All** [Fig. 1.18(e)] ⇒ **Close** ⇒ **LMB** ⇒ [icon] [Fig. 1.18(f)] ⇒ [icon] ⇒ **MMB**

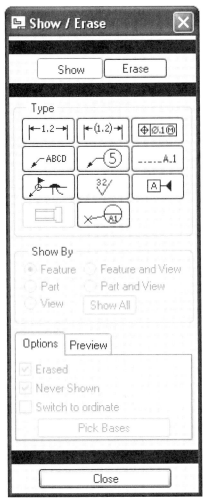

Figure 1.18(a) Show/Erase   **Figure 1.18(b)** Type: Axis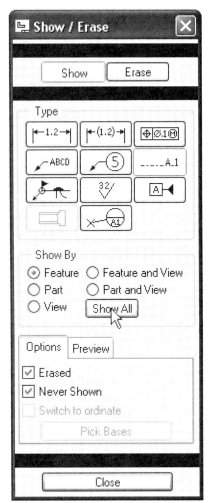

Figure 1.18(c) Show All

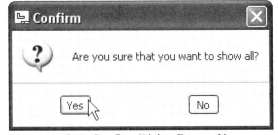

Figure 1.18(d) Confirm Dialog Box ⇒ Yes

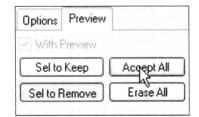

Figure 1.18(e) Accept All

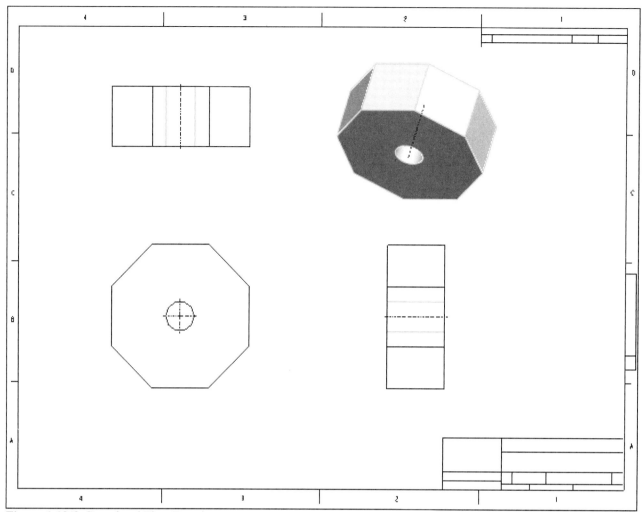

Figure 1.18(f) Completed Plate Drawing

This concludes the overview of Pro/ENGINEER Wildfire 3.0.

Spend some time exploring the website *www.cad-resources.com* for downloads, materials, and other CAD related information. To get on the CAD-Resources email list for job opportunities, software updates, tutorials, and new lessons, send an email to *cad@cad-resources.com*.

Lesson 2 Pro/ENGINEER Wildfire 3.0

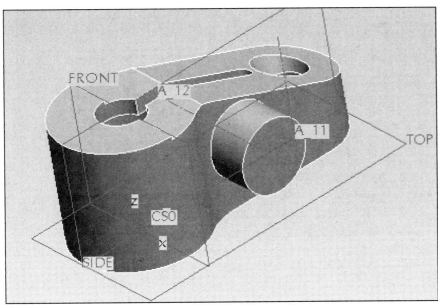

Figure 2.1 Swing/Pull Clamp Arm (SPX Fluid Power Part)

OBJECTIVES

- Understand the **User Interface (UI)**
- Download **Catalog Parts** using the **Browser**
- Master the **File Functions**
- Learn how to **Email** active Pro/ENGINEER objects
- Become familiar with the **Help** center
- Be introduced to the **Display** and **View** capabilities
- Use **Mouse Buttons** to **Pan**, **Zoom**, and **Rotate** the object
- Change **Display Settings**
- Investigate an object with **Information Tools**
- Experience the **Model Tree** functionality

Pro/ENGINEER Wildfire 3.0

This lesson will introduce you to Pro/ENGINEER's working environment. An existing part model (Fig. 2.1) will be downloaded from the Catalog and used to demonstrate the UI (user interface) and the general interaction required to master Pro/E.

You will be using a part available thru the **Catalog** using the **Browser**. If you are not connected to the Internet, your instructor will provide you with a simple start part. In addition, if you have Pro/Library installed, you may use any library part that you wish.

Pro/ENGINEER's Main Window

The Pro/ENGINEER user interface consists of a navigation window, an embedded Web browser, the menu bar, toolbars, information areas, and the graphics window (Fig. 2.2).

- **Navigation Window** Located on the left side of the Pro/E main window, it includes tabs for the Model Tree and Layer Tree, Folder Browser, Favorites, History and Connections.
- **Browser Window** An embedded Web browser is located to the right of the navigation window. This browser provides you access to internal or external Web sites.

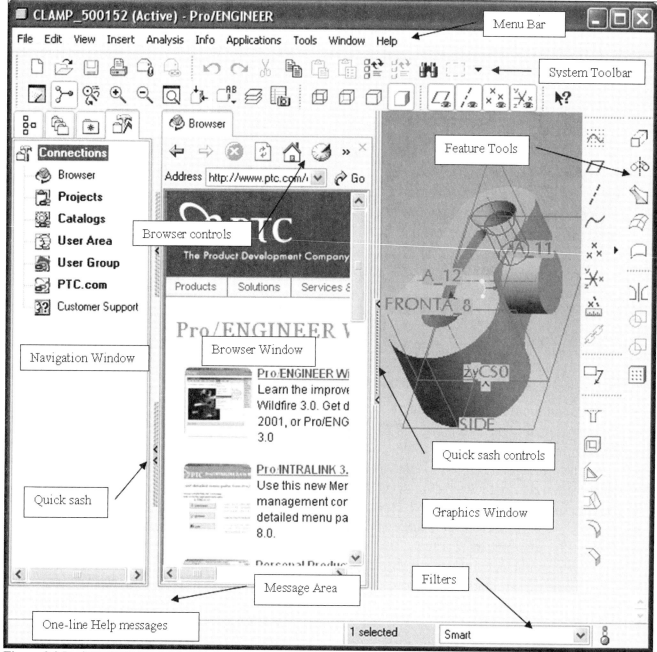

Figure 2.2 Pro/ENGINEER's Main Window

- **Menu Bar** The menu bar contains menus with options for creating, saving, and modifying models, and also contains menus with options for setting environment and configuration options.
- **Information Areas** Each Pro/E window has a message area near the bottom of the window for displaying one-line Help messages.
  - **Message Area** Messages related to work performed are displayed here.
  - **Status Bar** At the bottom of the window is a one-line Help area that dynamically displays one-line context-sensitive Help messages. If you move your mouse pointer over a menu command or dialog box option, a one-line description appears in this area.
  - **Screen Tips** The status bar messages also appear in small yellow boxes near the menu option or dialog box item or toolbar button that the mouse pointer is passing over.
- **Toolbars** The toolbars contain icons to speed up access to commonly used menu commands. By default, the toolbars consist of a row of buttons located directly under the main menu bar. Toolbar buttons can be positioned on the top, left, and right of the window. Toolbar buttons can be added or removed from the Toolchest by customizing the layout.
- **Graphics Window** The graphics window is the main working space (main window) for modeling and is to the right of the embedded Browser. Normally the Browser is collapsed when modeling.

Lesson 2 STEPS

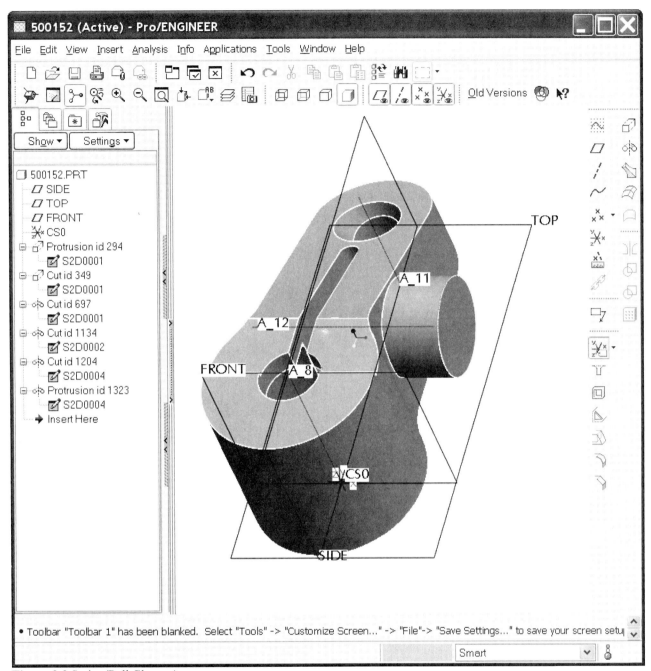

Figure 2.3 Swing/Pull Clamp Arm

Catalog Parts

In order to see and use Pro/ENGINEER's UI, we must have an active object (Fig. 2.3). Since you will not be modeling the part, you must download an existing model from the Catalog of parts from PTC. If you are using an Academic or Commercial version of Pro/ENGINEER Wildfire 3.0, you can download catalog parts directly. The Student Edition (SE) and Tryout Edition (TE) do not have access to these selections. *Note: if you do not have access to the catalog, go to www.cad-resources.com ⇒ Downloads ⇒ Lesson 2 Clamp ⇒ drag and drop the part into the Graphics Window as per the instructions.*

Throughout the text, a box surrounds all commands and menu selections.

START HERE: Open **Pro/ENGINEER Wildfire 3.0** using a shortcut icon on your Desktop *(or with WINDOWS, click: Start ⇒ Programs ⇒ proewildfire3.0)* ⇒ Pro/E will open on your computer ⇒ **Connections** tab ⇒ **Catalogs** (Fig. 2.4) ⇒ click: **SPX HYTEC**

Figure 2.4 Catalogs

NOTE: PTC will periodically change this page, so, it may look slightly different than what is displayed here.

Click: 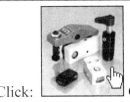 **Clamping** [Fig. 2.5(a)]

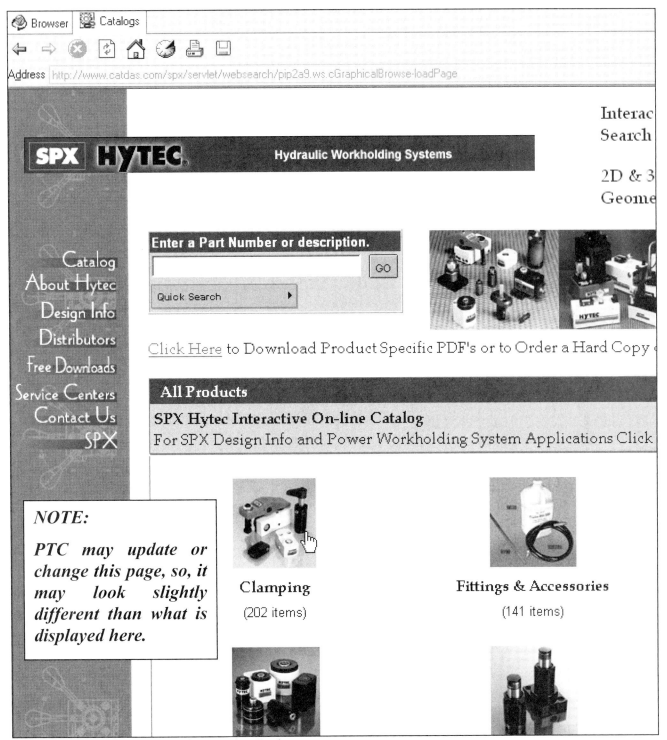

NOTE:

PTC may update or change this page, so, it may look slightly different than what is displayed here.

Figure 2.5(a) SPX Hytec Interactive On-Line Catalog

Click: 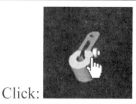 **Swing/Pull Clamp Arms** [Fig. 2.5(b)]

NOTE:

PTC may update or change this page, so, it may look slightly different than what is displayed here.

Figure 2.5(b) SPX Hytec Catalog

Double-click on the entry for Product Number **500152** (use the slide bar if needed) [Fig. 2.5(c)]:

| 500152 | 4500 Lbs. | 4500 Lbs. | 5000.0 PSI | 250 in.**3/min. | 0.5 sec. | 2.5 inches |

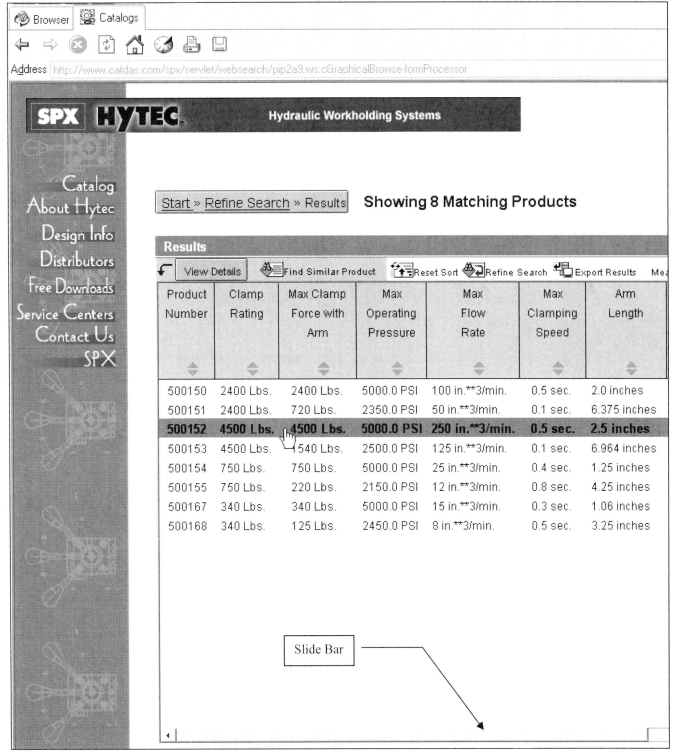

Figure 2.5(c) Swing/Pull Clamp Arms Part Number 500152

Click: **3D View** [Fig. 2.5(d)]

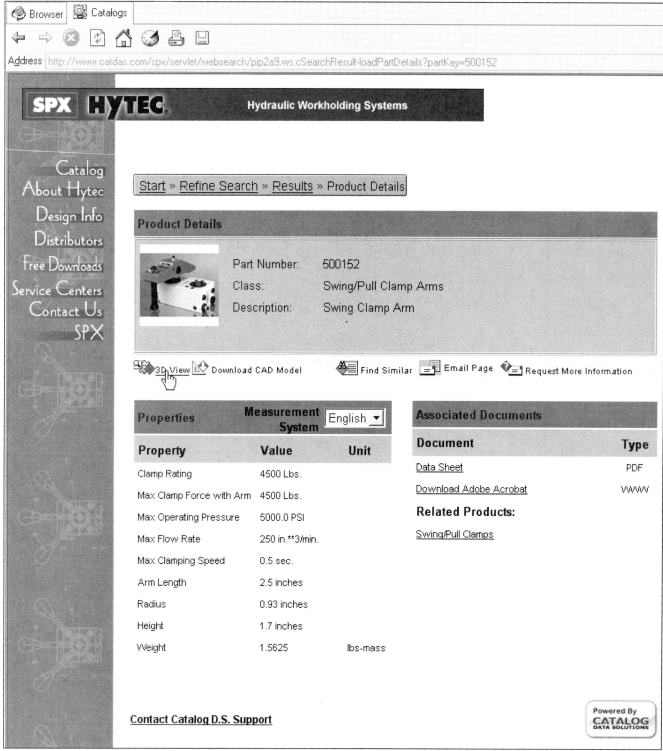

Figure 2.5(d) Swing/Pull Clamp Arms Part Number 500152 Properties

ProductView Lite

ProductView Lite is a data visualization tool that provides basic viewing capabilities for 3D models, including measurement and annotation. The Model Viewer lets you perform navigation tasks such as zoom, pan, rotate, fly-through, and exploding components of 3D models. You can also display models in wireframe, HLR (hidden lines removed), or shaded render modes. ProductView Lite provides the ability to view annotations that are associated with drawings or models. The annotations must be stored along with the file that you are viewing. Annotation sets appear along the left side of the ProductView Lite window as thumbnails. Click a thumbnail to view that annotation set.

You can use this tool to view annotations only; you cannot create annotations in ProductView Lite. In addition, 2D markups currently cannot be viewed with ProductView Lite; as a result, markups do not appear in the list of annotations.

The View Part window opens with the clamp displayed, click: Viewer Commands [Figs. 2.6(a-b)]

Model viewer commands: ProductView Lite for PTC PartsLink

The following toolbar buttons appear in the ProductView Lite Model viewer.

Shaded render mode
Causes the model to be rendered as a shaded solid.

HLR (hidden lines removed) render mode
Causes the model to be rendered as a solid line drawing. In this rendering mode, the lines that would be obscured by surfaces are not visible.

Wireframe render mode
Causes the model to be rendered as a wireframe structure.

End/Midpoint pick target
Snap to middle/end point causes coordinate selection to snap to the end point of the selected curve, line, or segment.

Center point pick target
Snap to center point causes coordinate selection to snap to the center of the chosen arc, line, or segment.

Curve pick target
Sets snapping to select a curve.

Surface pick target
Sets snapping to select a surface.

Component pick target
Snaps to a component in the assembly.

Figure 2.6(a) Model Viewer Commands

[🔍] Zoom all
Adjusts the image size in the display to show the entire drawing within the window.

Zoom window
Magnifies the window to fill the screen. To use the Zoom Window tool, click and drag with the mouse to define a window. When you release the mouse button, the window is magnified to the maximum allowable size, while still showing all the areas within the zoom window borders.

Restore location
Allows you to move all parts back to their original positions. This command restores the location of selected components. If no components are selected, it restores all components.

Select
The Select command allows you to select a component. The selected component is highlighted in red in the graphical display. The selected component(s) become the target of certain features. To select a component, click the Select button. Press **Ctrl** to select multiple objects. Press **Shift** to indicate when selection is available for your current pick target (such as point, curve, or surface).

Dimension measurement
Select **Dimension measurement** to display measurements for a selected pick target. To display dimensions of a component, click on that component. The dimensions appear in a label callout. For example, you can click on a cylinder to display the height, radius, and area. Or, you can click on an arc to display the length and radius. For more information, see Taking measurements.

Distance measurement
Select **Distance measurement** to click and measure the distance between two pick targets. To measure distance between items, specify the pick target (snap) for item 1, click on it to select it, then specify the pick target for item 2, and click to select that item. In other words, you can change pick targets between clicks for distance measurement. For more information, see Taking measurements.

Clear measurements
Causes all currently displayed measurements and annotations to be cleared from the view.

Figure 2.6(b) Model Viewer Commands (continued)

Close the Model viewer commands window, click: [X] ⇒ click and hold down on the model with **MMB** (**M**iddle **M**ouse **B**utton) and "fly the shuttle icon" 🚀 down to zoom out from the model until it is centered in the window and release the mouse button [Fig. 2.6(c)] ⇒ **RMB** to rotate the model (Fig. 2.7) ⇒ use both capabilities to position the model similar to that shown in Figure 2.7 ⇒ pick on the model with **LMB** to turn on the enclosing box and coordinate system (Fig. 2.8)

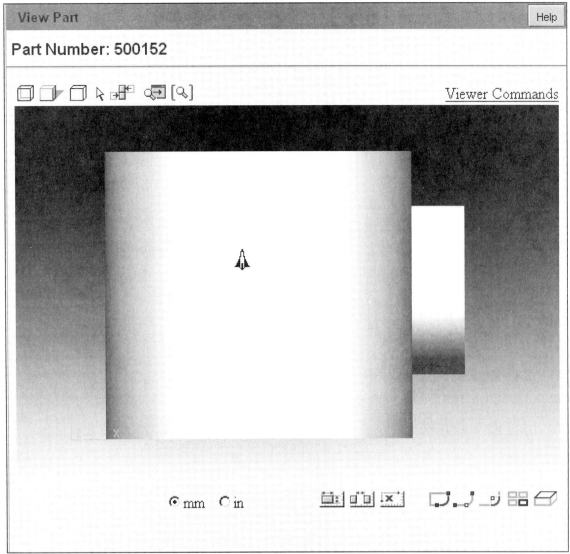

Figure 2.6(c) View Part Window with Clamp 500152 Displayed

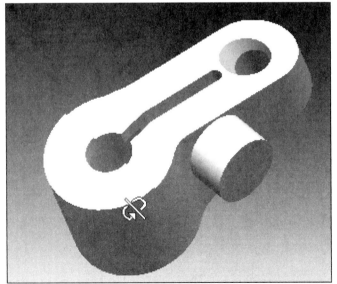

Figure 2.7 Fly and Rotate until Desired View is Obtained

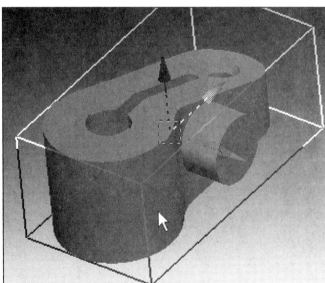

Figure 2.8 Pick on the Model with LMB

Click: ⊙in ⇒ 🔲 **Dimension Measurement** ⇒ 🔲 **Curve Snap** ⇒ pick the edge of the cylindrical

protrusion (Fig. 2.9) ⇒ 🔲 **Clear Measurements** ⇒ pick the edge of the cylindrical hole

(Fig. 2.10) ⇒ 🔲 **Clear Measurements** ⇒ select **ProE** as **CAD Model** format

CAD Model: Pro/E Part/Assembly (.prt) [Download] ⇒ **Download** ⇒ **Open** from File Download dialog box ⇒ **I Agree** from **WinZip** dialog box (if you do not have WinZip, you will have to download a free copy before this step at the prompt) (Fig. 2.11) ⇒ click on **500152.prt** ⇒ drag-and-drop into the Pro/E graphics window (Fig. 2.12) *(Do not drop it into the Browser window)* ⇒ close the WinZip dialog box ☒ (Fig. 2.13)

*(If you are using the **Tryout Edition** or the **Student Edition**, type **www.cad-resources.com** in the Browser Address Bar or in your Internet Browser's Address Bar ⇒ **Downloads** ⇒ **Clamp** ⇒ **TE & SE Version** ⇒ drag and drop the part from the WinZip dialog box into the Graphics Window)*

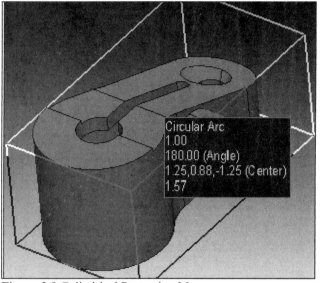

Figure 2.9 Cylindrical Protrusion Measurement

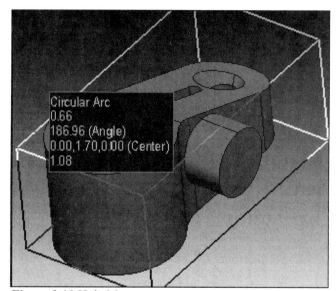

Figure 2.10 Hole Measurement

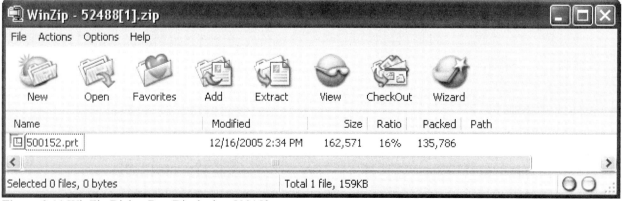

Figure 2.11 WinZip Dialog Box Displaying 500152.prt

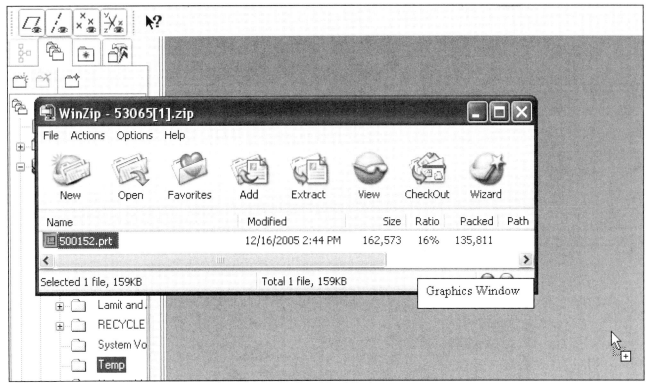

Figure 2.12 Drag-and-Drop 500152.prt into the Pro/E Graphics Window

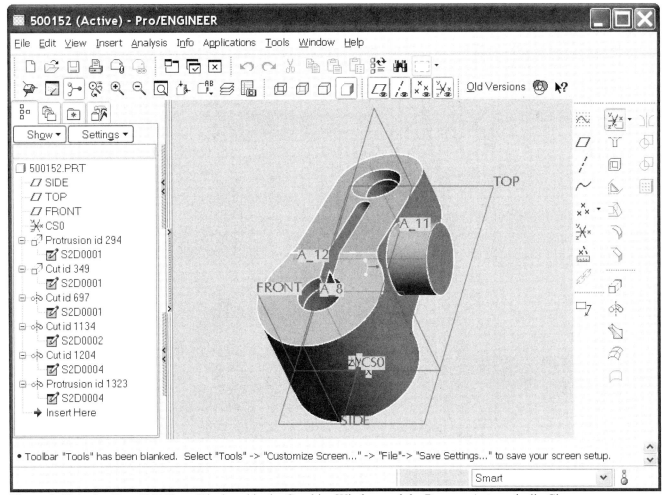

Figure 2.13 Active Part 500152.prt Displayed in the Graphics Window and the Browser Automatically Closes

77

File Functions

The **File** menu, from the menu bar, provides options for manipulating files (such as opening, creating, saving, renaming, backing up files, and printing). File functions include options for importing files from and exporting files to external formats, setting your working directory and performing operations on instances. Before using any File tool, make sure you have set your working directory to the folder where you wish to save objects for the project on which you are working. The Working Directory is a directory that you set up to contain Pro/E files.

You can save a Pro/E file using *Save* or *Save a Copy* from the File menu. The Save a Copy dialog box allows the exportation of Pro/E files to different formats, and to save files as images. Since the name already exists in session, you cannot save or Rename a file using the same name as the original file name, even if you save the file in a different directory. Pro/E forces you to enter a unique file name by displaying the message: *An object with that name exists in session. Choose a different name.*

Names for all Pro/E files are restricted to a maximum of 31 characters and must contain no spaces. A File can be a part, assembly, or drawing. Each is considered an "object".

Click: **File** ⇒ **Set Working Directory** (Fig. 2.14) ⇒ select the directory you wish all objects to be saved to for this Pro/E project ⇒ **OK** ⇒ **File** ⇒ **Save a Copy** ⇒ type in the New Name field: **CLAMP_500152** ⇒ **OK** (Fig. 2.15) ⇒ **Window** ⇒ **Close** ⇒ **File** ⇒ **Open** ⇒ **clamp_500152** ⇒ **Open**

[READ THIS- Depending on the computer you are using (yours or an organization's) the following may not work. If you have your mail server configured as the default then proceed, otherwise skip this step. **(Again- If you do not own the computer you are working on, then do not do this step)**]

Send email with object in active window ⇒ *email the object to yourself (or a friend with Pro/E software)* ⇒ check on **Create a ZIP file** ⇒ **OK** *(Fig. 2.16)* ⇒ *follow your email procedure as required*

Figure 2.14 Select the Working Directory

Figure 2.15 Save a Copy

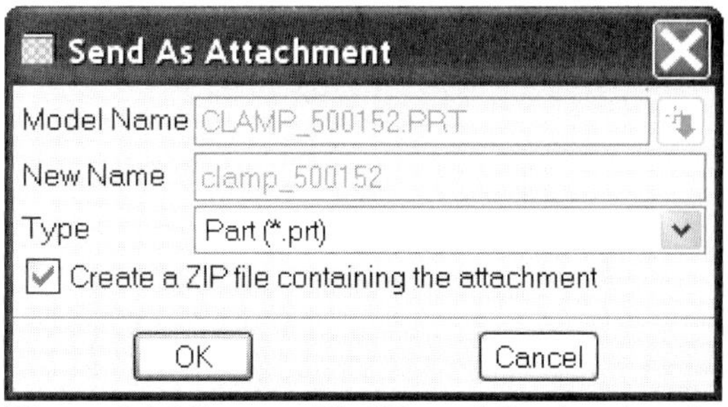

Figure 2.16 Create a Zip File and Send as Attachment

Help

Accessing the Help function is one of the best ways to learn CAD software. Use the Help tool as often as possible to understand the tool or command you are using at the time and to expand your knowledge of the other capabilities provided by Pro/ENGINEER. Use the **Help** menu to gain access to online information, Pro/E release information, and customer service information. The following commands are available on the Pro/E **Help** menu in standard Part and Assembly modes.

- **Help Center** Displays the context-sensitive online help system. When you select this, your supported network browser opens to display a navigation tree and search tools to aid you in finding specific help topics. You can also access these topics by clicking for context-sensitive Help from windows, menu commands, and dialog boxes.
- **What's This?** Enables context-sensitive Help mode.
- **Menu Mapper** To help you show where older release selections have been moved.
- **Online Resources** Useful tutorials, tools and various help guides.
- **What's New?** Displays what is new in this release of Pro/E software.
- **Technical Support Info** Displays product information, including the release level, license information, installation date, and customer support contact information.
- **About Pro/ENGINEER** Displays Pro/E copyright and release information.

Click: **Help** ⇒ **Help Center** [Fig. 2.17(a)] ⇒ **Fundamentals** ⇒ **Pro/ENGINEER Fundamentals**

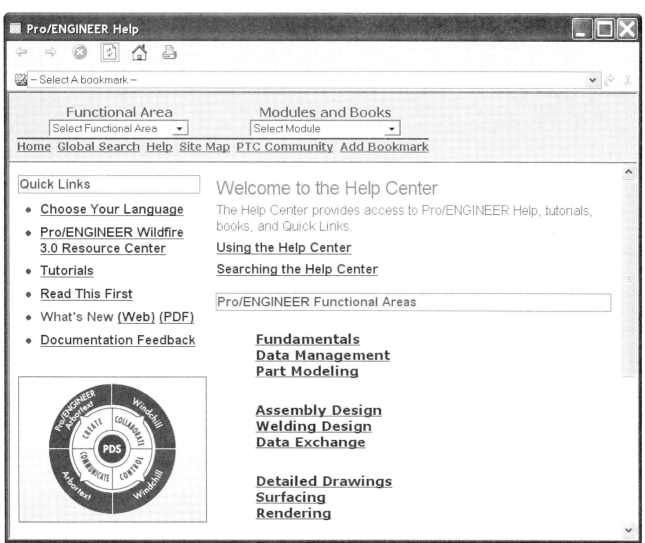

Figure 2.17(a) Pro/ENGINEER Help Center

Open the **Contents** choices (📖 to expand or 📕 to collapse) ⇒ explore the Help documentation before continuing ⇒ click: ❌ ⇒ ▶❓ **Context sensitive help** [Context sensitive help] from Top Toolchest ⇒ click on the **Orient Mode on/off** button from the Toolchest **Orient Mode on/off** ⇒ read the documentation [Fig. 2.17(b)] ⇒ close the window ❌ ⇒ repeat and choose some other buttons

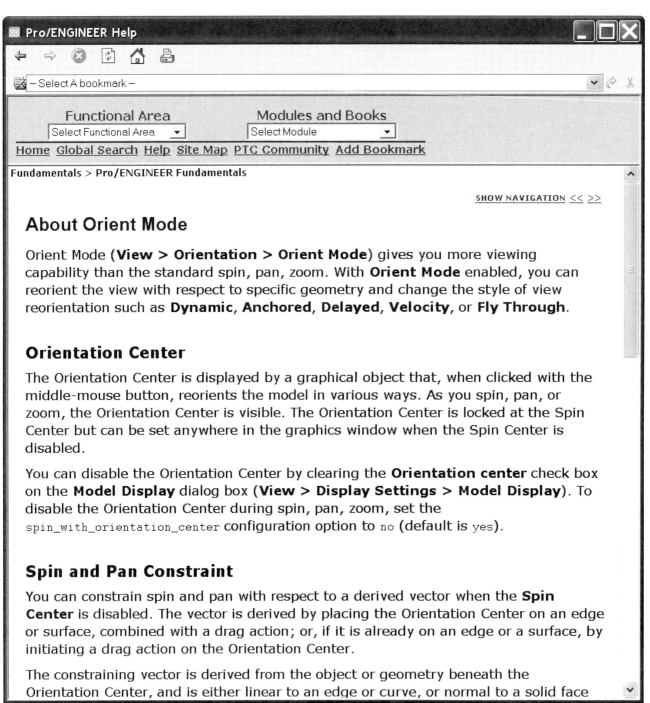

Figure 2.17(b) Pro/ENGINEER Help

View and Display Functions

Using the Pro/E **View** menu, you can adjust the model view, orient the view, hide and show entities, create and use advanced views, and set various model display options. The following list includes some of the View operations you can perform:

- Orient the model view in the following ways, using the **Orientation** dialog box: spin, pan, and zoom models and drawings, display the default orientation, revert to the previously displayed orientation, change the position or size of the model view, change the orientation (including changing the view angle in a drawing), and create new orientations
- Temporarily shade a model by using cosmetic shading
- Show, dim, or remove hidden lines
- Highlight items in the graphics window when you select them in the Model Tree
- Explode or unexplode an assembly view
- Repaint the Pro/E graphics window
- Refit the model to the Pro/E window after zooming in or out on the model
- Update drawings of model geometry
- Hide and show entities, and hide or show items during spin or animation
- Use advanced views
- Add perspective to the model view

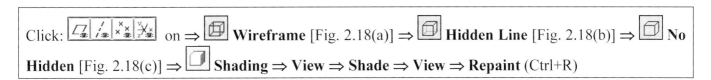
Click: [icons] on ⇒ [icon] **Wireframe** [Fig. 2.18(a)] ⇒ [icon] **Hidden Line** [Fig. 2.18(b)] ⇒ [icon] **No Hidden** [Fig. 2.18(c)] ⇒ [icon] **Shading** ⇒ **View** ⇒ **Shade** ⇒ **View** ⇒ **Repaint** (Ctrl+R)

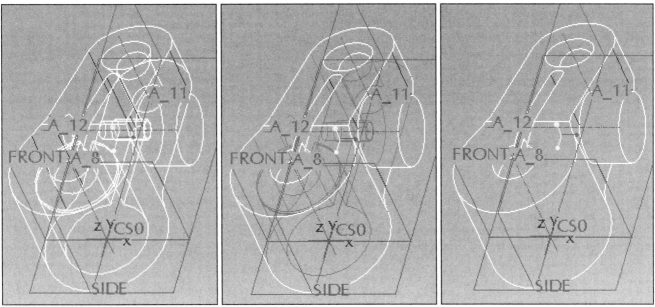

Figure 2.18(a) Wireframe   **Figure 2.18(b)** Hidden Line   **Figure 2.18(c)** No Hidden

To see the standard views provided, click: 🔲 ⇒ **Front** ⇒ 🔲 ⇒ **Right** [Figs. 2.19(a-b)] ⇒ try all the variations ⇒ 🔲 ⇒ **Standard Orientation**

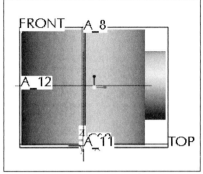

Figure 2.19(a) FRONT View

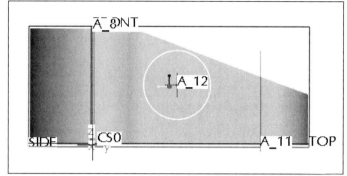

Figure 2.19(b) RIGHT View

View Tools

As with all CAD systems, Pro/E provides the typical view tools associated with CAD:

- 🔍 **Zoom In** Use this tool to zoom in on a specific portion of the model. Pick two positions of a rectangular zoom box.

- 🔍 **Zoom Out** Use this tool to reduce the view size of the model on the screen by 50%.

- 🔍 **Refit object to fully display it on the screen** Use this tool to refit the model to the screen so that you can view the entire model. A refitted model fills 80% of the graphics window.

Click: 🔍 **Zoom In** [Fig. 2.20(a)] ⇒ pick two positions about an area you wish to enlarge [Fig. 2.20(b)] ⇒ 🔍 **Zoom Out** ⇒ 🔍 **Refit** [Fig. 2.20(c)] ⇒ 🔳 **Redraw the current view**

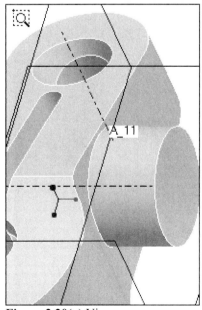

Figure 2.20(a) View

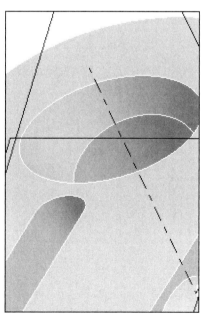

Figure 2.20(b) Zoom In

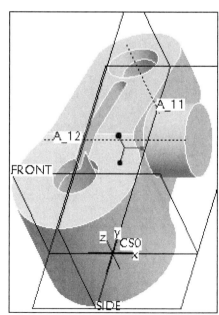

Figure 2.20(c) Refit

Using Mouse Buttons to Manipulate the Model

You can also dynamically reorient the model using the **MMB** by itself (**Spin**) or in conjunction with the **Shift** key (**Pan**) or **Ctrl** key (**Zoom, Turn**).

> Hold down **Ctrl** key and **MMB** in the graphics area near the model and move the cursor up (zoom out) [Fig. 2.21(a)] ⇒ hold down **Shift** key and **MMB** in the graphics area near the model and move the cursor about the screen (pan) [Fig. 2.21(b)] ⇒ hold down **MMB** in the graphics area near the model and move the cursor around (spin) [Figs. 2.22(a-b)] ⇒ [AB] ⇒ **Standard Orientation**

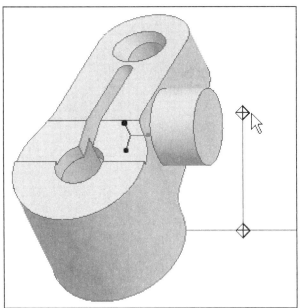

Figure 2.21(a) Zoom

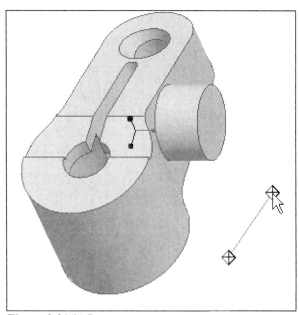

Figure 2.21(b) Pan

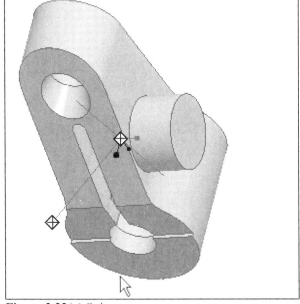

Figure 2.22(a) Spin

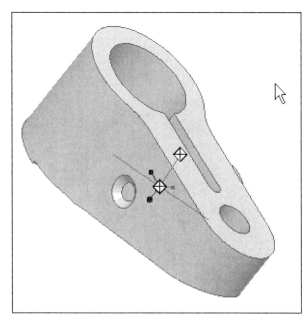

Figure 2.22(b) Spin Again

You may have noticed that the illustrations of the text have changed since Figure 2.19(b). The default for the background and geometry colors has been changed so that the illustrations will capture and print clearer. In the next section, you will learn how to change the system display settings.

System Display Settings

You can make a number of changes to the default colors furnished by Pro/E, customizing them for your own use:

- Define, save, and open color schemes
- Customize colors used in the user interface
- Change your entire color scheme to a predefined color scheme (such as black on white)
- Change the top or bottom background colors
- Redefine basic colors used in models
- Assign colors to be used by an entity
- Store a color scheme so you can reuse it
- Open a previously used color scheme

The **Scheme** menu includes the following color schemes (text uses the **Black on White** selection):

- **Black on White** Black entities shown on a white background
- **White on Black** White entities shown on a black background
- **White on Green** White entities shown on a dark-green background
- **Initial** Reset the color scheme to the one defined by the configuration file settings
- **Default** Reset the color scheme to the system default

Click: **View** ⇒ **Display Settings** ⇒ **System Colors** [Figs. 2.23(a-c)] ⇒ ☐ Blended Background ⇒ **Scheme** ⇒ **Black on White** ⇒ experiment with different color schemes and system colors ⇒ **Scheme** ⇒ **Default** ⇒ **OK**

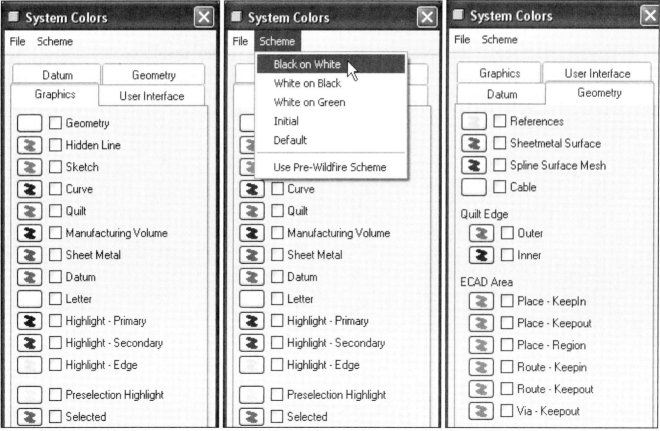

Figure 2.23(a) Graphics Tab   **Figure 2.23(b)** Scheme Tab   **Figure 2.23(c)** Geometry Tab

Information Tools

At any time during the design process you can request model, feature, or other information. Picking on a feature in the graphics window and then **RMB** ⇒ **Info** (Fig. 2.24) will provide information about that feature in the Browser [Figs. 2.25(a-b)]. This can also be accomplished by clicking on the feature name in the Model Tree and then **RMB**. Both feature and model information can be obtained using this method (Fig. 2.26). A variety of information can also be extracted using the **Info** tool from the menu bar.

Pick once on the revolved protrusion (Fig. 2.24) ⇒ **RMB** ⇒ **Info** ⇒ **Feature** [Figs. 2.25(a-b)] ⇒ click on the "quick sash" to collapse the Browser ⇒ in the graphics window, **RMB** (Fig. 2.26) ⇒ **Info** ⇒ **Model** (Fig. 2.27) ⇒ collapse the Browser

---

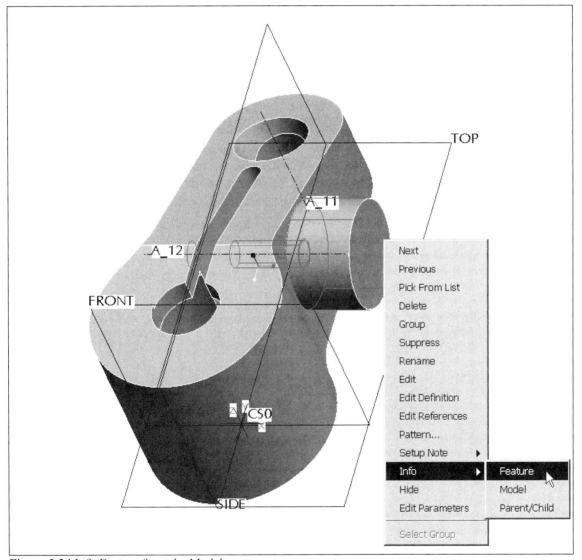

Figure 2.24 Info Feature from the Model

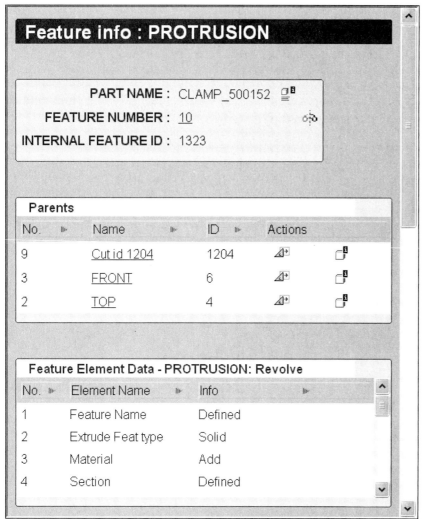

Figure 2.25(a) Feature Info Displayed in the Browser

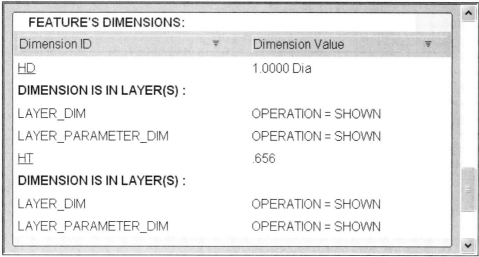

Figure 2.25(b) Feature Dimension Information

86

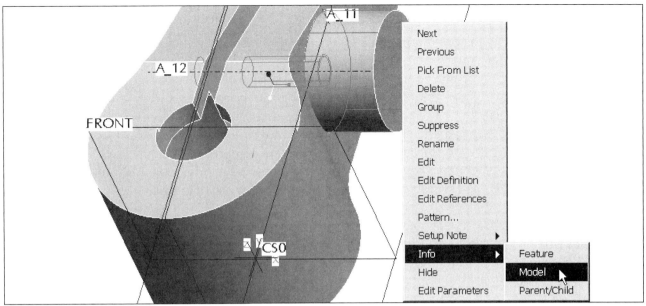

Figure 2.26 Info Model

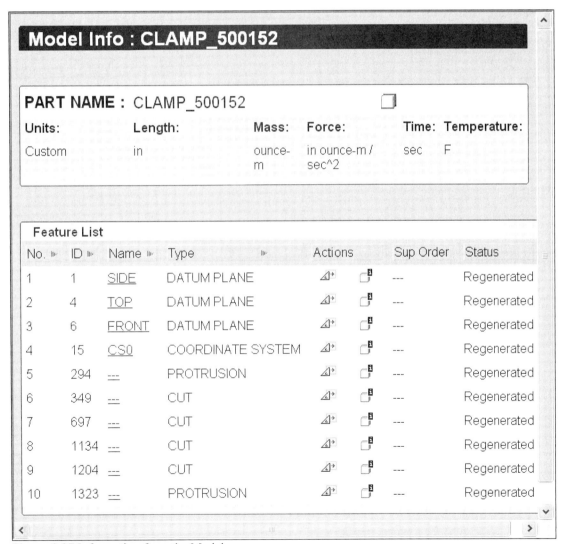

Figure 2.27 Information from the Model

The Model Tree

The **Model Tree** is a tabbed feature on the Pro/E navigator that displays a list of every feature or part in the current part, assembly, or drawing.

The model structure is displayed in hierarchical (tree) format with the root object (the current part or assembly) at the top of its tree and the subordinate objects (parts or features) below. If you have multiple Pro/E windows open, the Model Tree contents reflect the file in the current active window.

The Model Tree lists only the related feature and part level items in a current object and does not list the entities (such as edges, surfaces, curves, and so forth) that comprise the features.

Each Model Tree item contains an icon that reflects its object type, for example, hidden, assembly, part, feature, or datum plane (also a feature). The icon can also show the display status for a feature, part, or assembly, for example, suppressed.

Selection in the Model Tree is object-action oriented; you select objects in the Model Tree without first specifying what you intend to do with them. You can select components, parts, or features using the Model Tree. You cannot select the individual geometry that makes up a feature (entities). To select an entity, you must select it in the graphics window.

With the **Settings** tab you can control what is displayed in the Model Tree.

You can add informational columns to the Model Tree window, such as **Tree Columns** containing parameters and values, assigned layers, or feature name for each item. You can use the cells in the columns to perform context-sensitive edits and deletions. These options will be covered elsewhere in the text, as they are needed in the design process.

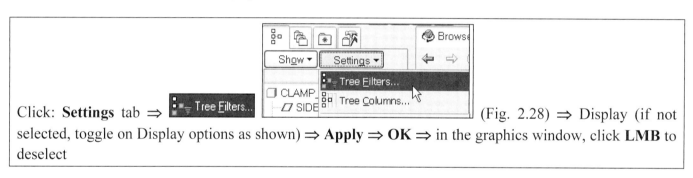

Click: **Settings** tab ⇒ Tree Filters... (Fig. 2.28) ⇒ Display (if not selected, toggle on Display options as shown) ⇒ **Apply** ⇒ **OK** ⇒ in the graphics window, click **LMB** to deselect

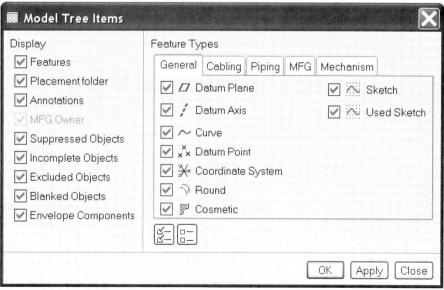

Figure 2.28 Model Tree Items Dialog Box

Working on the Model

In Pro/E, you can select objects to work on from within the graphics window or in the Model Tree by using the mouse or the keyboard. The object types that are available for selection vary depending on whether you select an object from within the graphics window or in the Model Tree. You can select any type of object, including features, 3-D notes, parts, datum objects (planes, axes, curves, points, and coordinate systems), and geometry (edges and surfaces) from within the graphics window. Additionally, since the Model Tree displays only parts, components, and features, you can select only those object types from within the Model Tree.

Selection in both the graphics window and the Model Tree can be action-object or object-action oriented, depending on the process you choose within Pro/E to build your model. You can specify the action you want to perform on an object before you select the object, or you can select the object before you specify the action.

You can *dynamically* modify certain features from within the graphics window as you work. Features can be edited by selecting them from the graphics window directly, or from the Model Tree. Dimensions of the following features can be modified dynamically:

- **Protrusions** Extruded protrusions with variable depth, and revolved protrusions of variable angle
- **Cuts** Extruded cuts with variable depth, and revolved cuts with variable angle
- **Surfaces** Extruded surfaces with variable depth, and revolved surfaces with variable angle
- **Rounds** Simple, constant, and edge chain

Pick on the cylindrical protrusion ⇒ **RMB** [Fig. 2.29(a)] ⇒ **Delete** ⇒ **OK** [Fig. 2.29(b)]

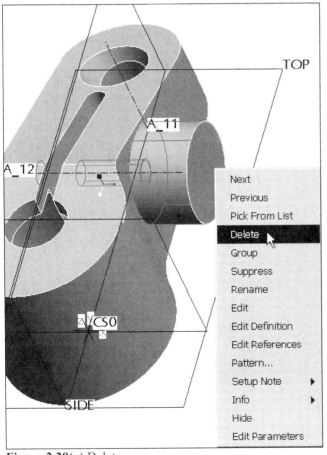

Figure 2.29(a) Delete

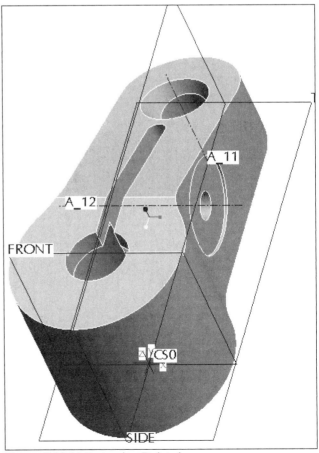

Figure 2.29(b) Protrusion Deleted

Click on the protrusion in the Model Tree (Fig. 2.30) ⇒ **RMB** ⇒ **Edit Definition** (Fig. 2.31)

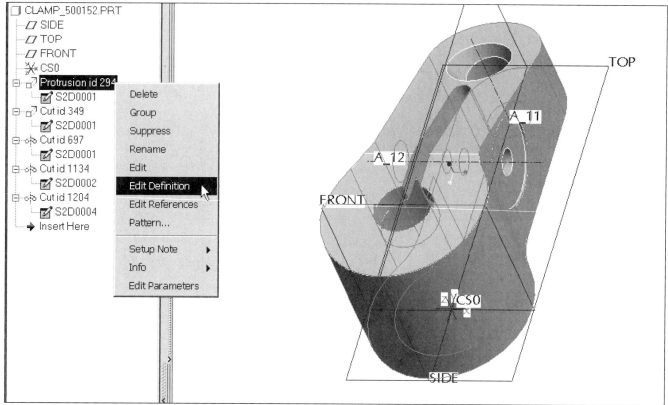

Figure 2.30 Redefining a Feature

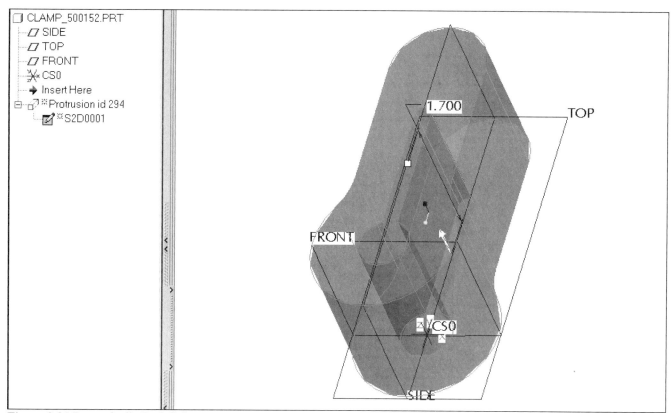

Figure 2.31 Dynamically Redefining a Feature

Pick on and drag the white nodal handle to change the protrusion's height from **1.700** (Fig. 2.32) to **2.000** (Fig. 2.33) ⇒ **MMB** to regenerate the model ⇒ **LMB** to deselect

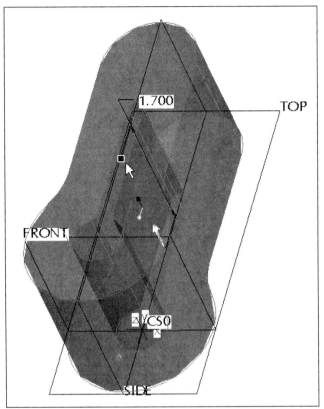

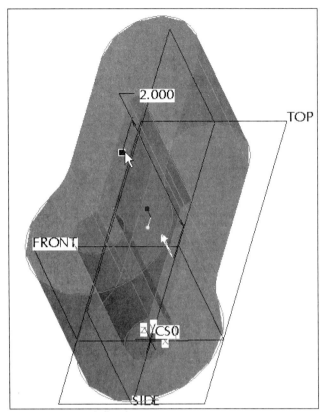

Figure 2.32 Dynamically Dragging a Nodal Handle

Figure 2.33 Redefined Feature

Although you may have not noticed it, there was other information and editing available during the redefine process. Repeat the command: Click on the protrusion in the Model Tree ⇒ **RMB** ⇒ **Edit Definition** ⇒ look at the "dashboard" on the lower part of the screen [Fig. 2.34(a)] ⇒ **Options** tab [Fig. 2.34(b)] ⇒ change the **2.000** to **3.000** ⇒ **Enter** ⇒ 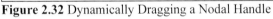 [Figs. 2.35(a-b)] ⇒ **Window** ⇒ **Close**

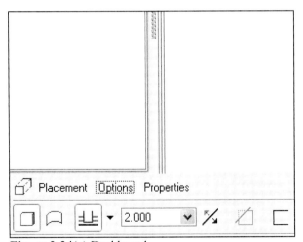

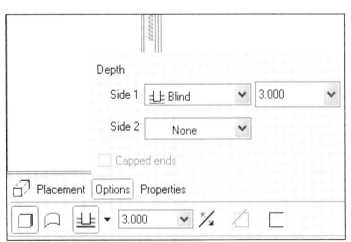

Figure 2.34(a) Dashboard

Figure 2.34(b) Dashboard Options Tab

91

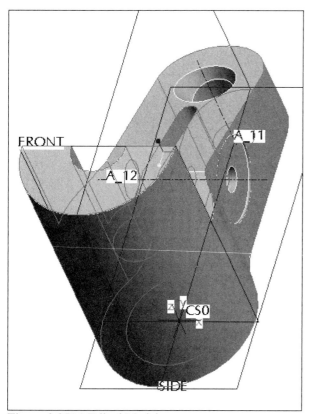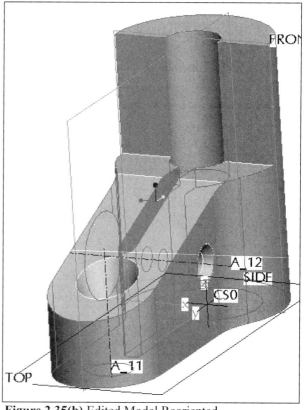

Figure 2.35(a) Edited Model   **Figure 2.35(b)** Edited Model Reoriented

About the Dashboard

As you create and modify your models using direct graphical manipulation in the graphics window, the **Dashboard** guides you throughout the modeling process. This context sensitive interface monitors your actions in the current tool and provides you with basic design requirements that need to be satisfied to complete your feature. As you select individual geometry, the Dashboard narrows the available options enabling you to make only targeted modeling decisions. For advanced modeling, separate slide-up panels provide all relevant advanced options for your current modeling action. These Dashboard panels remain hidden until needed. This enables you to remain focused on successfully capturing the design intent of your model. The Dashboard for a revolve feature is different from an extrude feature [Figs. 2.36(a-c)].

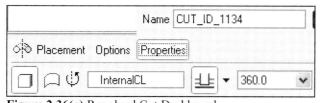

Figure 2.36(a) Revolved Cut Dashboard   **Figure 2.36(b)** Extruded Protrusion Dashboard

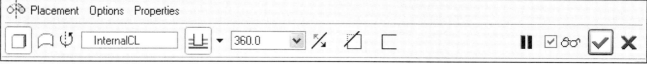

Figure 2.36(c) Revolved Protrusion Dashboard

This concludes the basic tour of Pro/E. Time permitting, download another model from the PTC Catalog and practice navigating the user interface and using commands and tools introduced previously. If you do not have Internet access, ask your instructor to provide another model. If you are using the TE or SE Edition, email *cad@cad-resources.com* or see the website *www.cad-resources.com*.

Lesson 3 Direct Modeling

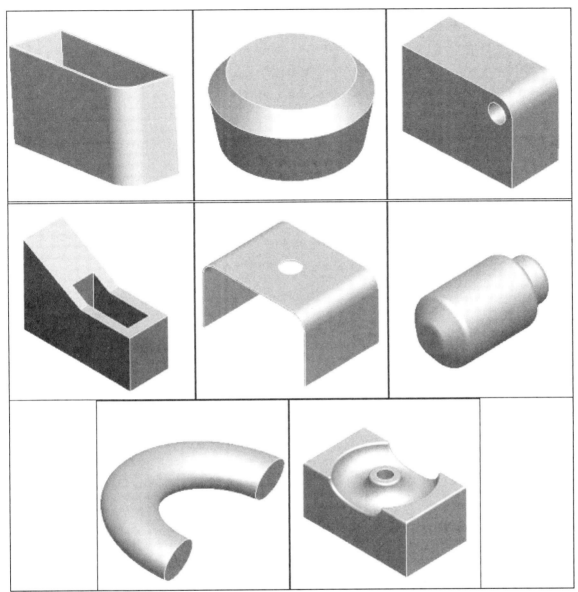

Figure 3.1 Eight Quick Modeling Parts

OBJECTIVES

- Modeling simple parts quickly using **default selections**
- Sample the **Extrude Tool** and the **Revolve Tool**
- Try out a variety of **Engineering Tools** including: **Hole**, **Shell**, **Round**, **Chamfer**, and **Draft**
- Sketch simple **sections**

Modeling

The purpose of this lesson is to quickly introduce you to a variety of **Feature Tools** and **Engineering Tools**. You will model a variety of very simple parts (Fig. 3.1). Little or no explanation of the methodology or theory of the Tool or process will accompany the instructions. By using almost all default selections; you will create models that will display the power and capability of Pro/E Wildfire 3.0.

In the next lesson, a detailed step-by-step systematic description will accompany all commands. Here, we hope to get you up and running on Pro/E without any belabored explanations.

Lesson 3 STEPS

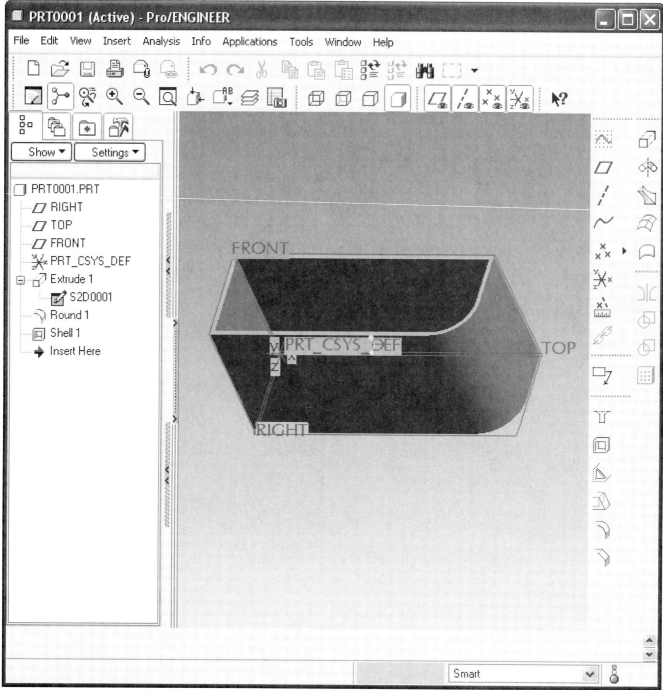

Figure 3.2(a) Part One

Part Model One (PRT0001.PRT) (Extrude)

This model [Fig. 3.2(a)] will introduce the **Extrude Tool** to create a simple box-shape, the **Round Tool** to add a round to one edge, and the **Shell Tool** to remove one surface and make the part walls a consistent thickness. In this direct modeling example, we will provide step-by-step illustrations. For subsequent parts, only important steps or illustrations displaying aspects of the command sequence that represent new material will be provided. The same applies to *Tool Tips* that appear as you pass your cursor over a button or icon. All parts have been created with out-of-the-box system settings for Pro/E Wildfire 3.0 including default templates, grid settings, and so forth.

Click: **Launch Pro/E** ⇒ **File** ⇒ **Set Working Directory** ⇒ select the working directory ⇒ **OK** ⇒ 🗋 **Create a new object** ⇒ ⊙ ▭ Part ⇒ **OK** ⇒ click the **FRONT** datum plane in the Model Tree [Fig. 3.2(b)]

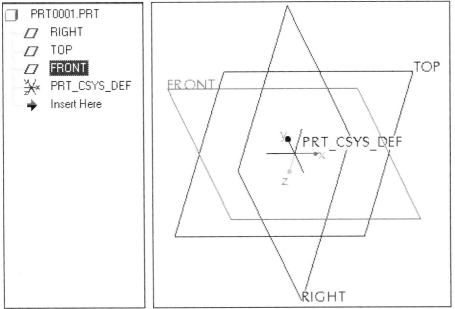

Figure 3.2(b) Pre-select the FRONT Datum Plane in the Model Tree

Click: 🗗 **Extrude Tool** ⇒ **RMB** ⇒ **Define Internal Sketch** [Fig. 3.2(c)] ⇒ Sketch dialog box opens. Pre-selected FRONT datum is the Sketch Plane; RIGHT datum is automatically selected as the Sketch Orientation Reference [Fig. 3.2(d)]

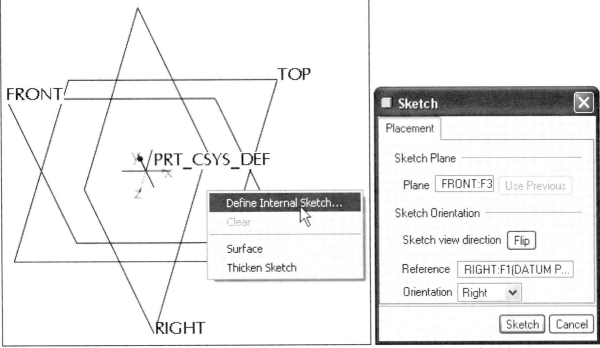

Figure 3.2(c) Define Internal Sketch   **Figure 3.2(d)** Sketch Dialog Box

With your mouse pointer in the Pro/E graphics window, click: **MMB** the Sketcher activates in order to define an independent section [Fig. 3.2(e)]

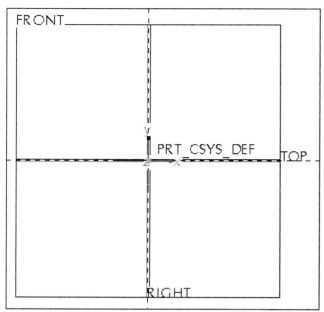

Figure 3.2(e) Active Sketcher Displays

Click: ☐ **Create rectangle** ⇒ sketch a rectangle by picking two corners [Fig. 3.2(f)] ⇒ ✓ **Continue with the current section**

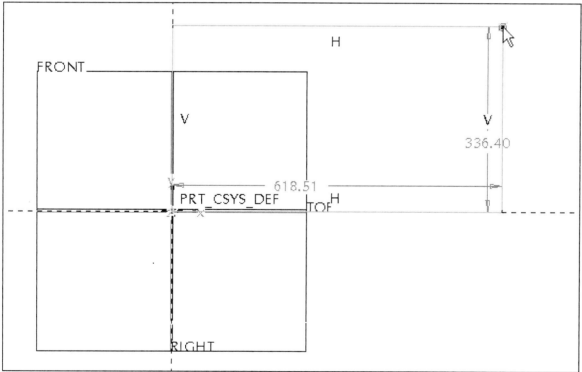

Figure 3.2(f) Sketched Rectangle

Click: [AB] ⇒ **Standard Orientation** [Fig. 3.2(g)] ⇒ [icon] **Shading** ⇒ **MMB** [Fig. 3.2(h)]

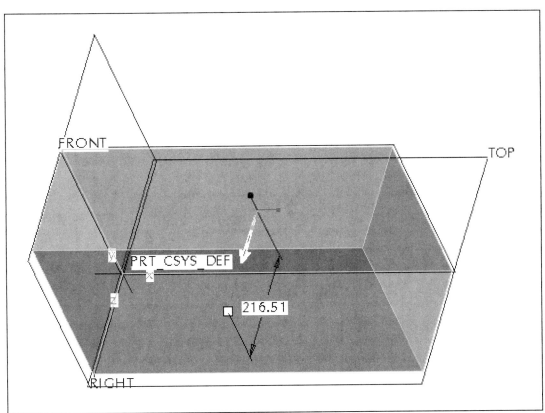

Figure 3.2(g) Trimetric View with Depth Handle Displayed

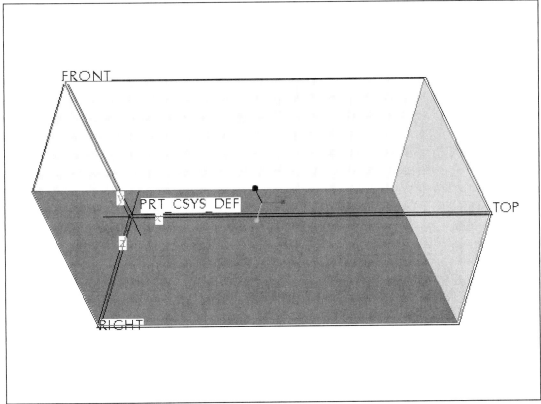

Figure 3.2(h) Completed Extruded Protrusion

Slowly pick on the right front edge of the part until it highlights [Fig. 3.2(i)] ⇒ **RMB** ⇒ **Round Edges** [Fig. 3.2(j)] ⇒ pick on and move a drag handle [Fig. 3.2(k)] ⇒ **MMB** [Fig. 3.2(l)]

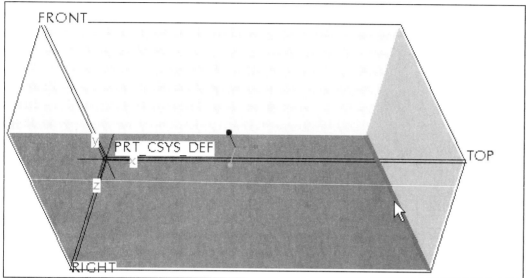

Figure 3.2(i) Pick on Edge to Highlight

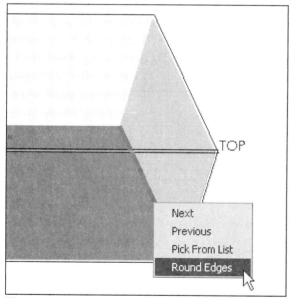

Figure 3.2(j) Round Edges

Figure 3.2(k) Move one of the Two Drag Handles

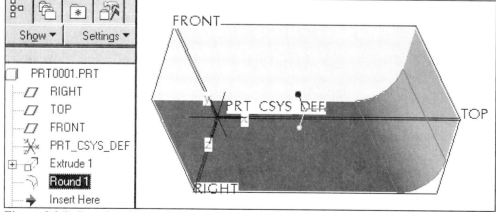

Figure 3.2(l) Completed Round

Slowly pick on the top surface (slightly move the mouse if needed) until it highlights [Fig. 3.2(m)] ⇒ 🔲
Shell Tool [Fig. 3.2(n)] ⇒ **MMB** [Fig. 3.2(o)] ⇒ 💾 ⇒ **MMB** ⇒ **File** ⇒ **Close Window**

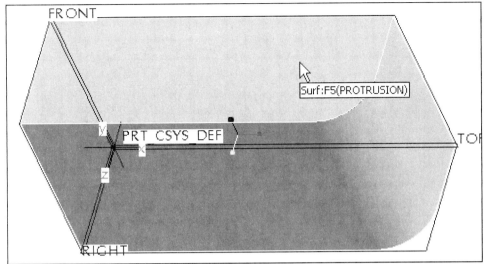

Figure 3.2(m) Selected Top Surface is Highlighted

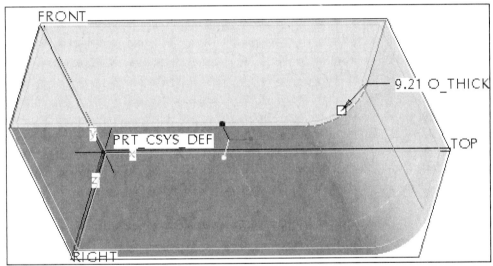

Figure 3.2(n) Shell Tool Applied

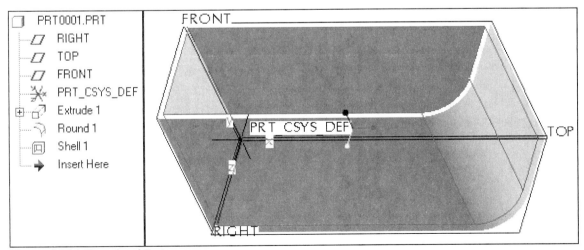

Figure 3.2(o) Completed Part

Part Model Two (PRT0002.PRT) (Draft)

Click: 🗋 ⇒ ◉ ▢ Part (PRT0002.PRT) ⇒ **MMB** ⇒ click on the **TOP** datum plane in the Model Tree ⇒ 🗗 **Extrude Tool** ⇒ **RMB** ⇒ **Define Internal Sketch** Sketch dialog box opens with Sketch Plane and Sketch Orientation selected [Fig. 3.3(a)] ⇒ **MMB** in the graphics window

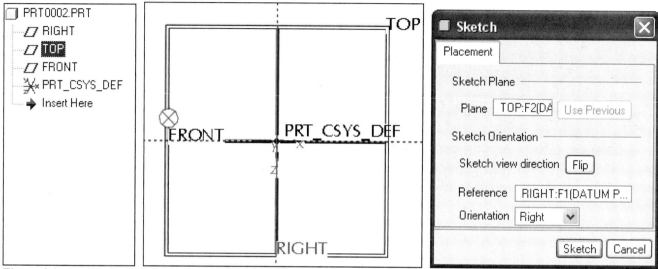

Figure 3.3(a) Pre-selected TOP Datum is the Sketch Plane, RIGHT Datum is Automatically Selected as the Sketch Orientation Reference

Click: **RMB** ⇒ **Circle** ⇒ sketch a circle by picking at the origin and then stretching the circle's diameter [Fig. 3.3(b)] to a convenient size ⇒ pick again to establish the circle's diameter [Fig. 3.3(c)]

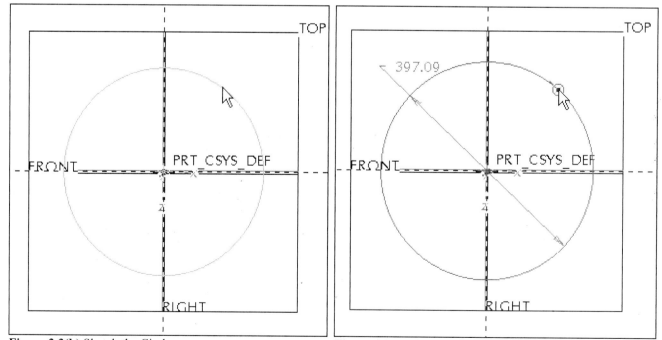

Figure 3.3(b) Sketch the Circle   **Figure 3.3(c)** Pick to Establish the Circle's Diameter

Click: [AB] ⇒ **Standard Orientation** [Fig. 3.3(d)] ⇒ [✓] [Fig. 3.3(e)] ⇒ **MMB** [Fig. 3.3(f)] ⇒ [cube] **Hidden Line** ⇒ pick on the curved surface until it is selected and highlights [Fig. 3.3(g)] ⇒ [icon] **Draft Tool**

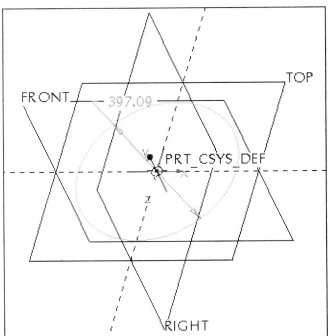

Figure 3.3(d) Sketch in Standard Orientation

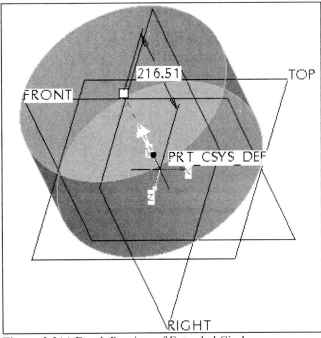

Figure 3.3(e) Depth Preview of Extruded Circle

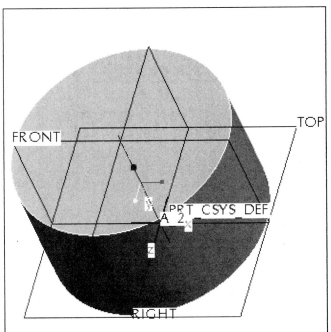

Figure 3.3(f) Completed Extruded Circle

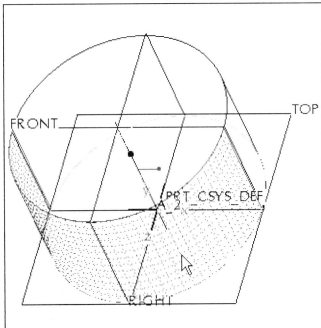

Figure 3.3(g) Select the Curved Surface

Click: References ⇒ pick the **TOP** datum plane as the Draft hinges [Fig. 3.3(h)] ⇒ click in the Angle dimension box and type **10** 10.00 ⇒ **Enter** ⇒ Shading ⇒ **MMB** [Fig. 3.3(i)] ⇒ pick on the top edge until it highlights [Fig. 3.3(j)] ⇒ **Chamfer Tool** ⇒ type **50** D 50.00 ⇒ **Enter** ⇒ **MMB** [Fig. 3.3(k)] ⇒ press and hold **MMB** and spin the part ⇒ pick on the bottom surface until it highlights ⇒ **Shell Tool** ⇒ **MMB** [Fig. 3.3(l)] ⇒ ⇒ **MMB** ⇒ **Window** ⇒ **Close**

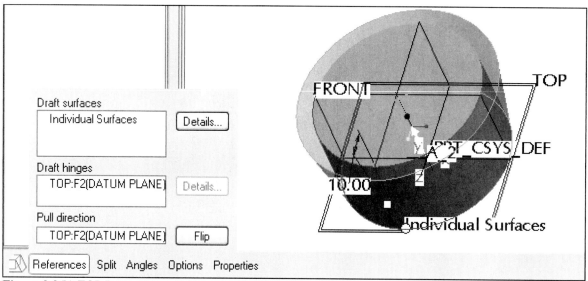

Figure 3.3(h) TOP Datum Plane Selected as the Draft Hinges

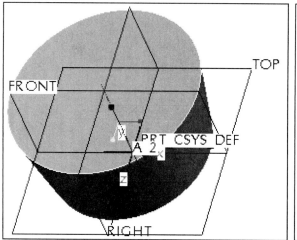

Figure 3.3(i) Completed Draft

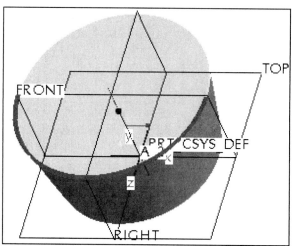

Figure 3.3(j) Select the Front Edge

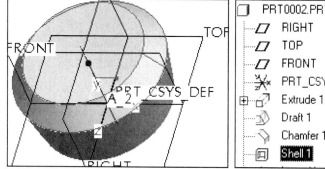

Figure 3.3(k) Completed Chamfer

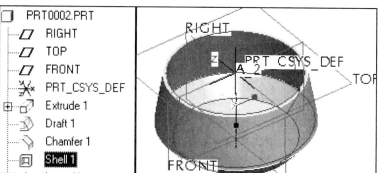

Figure 3.3(l) Completed Part

Part Model Three (PRT0003.PRT) (Hole)

Click: ▢ ⇒ ⊙ ▢ Part (PRT0003.PRT) ⇒ MMB ⇒ pick the FRONT datum ⇒ 🗔 Extrude Tool ⇒ RMB ⇒ Define Internal Sketch ⇒ MMB ⇒ ▢ Create rectangle ⇒ sketch a rectangle [Fig. 3.4(a)] ⇒ 📐 Create a circular fillet between two entities ⇒ pick the two lines that form the upper right-hand corner [Fig. 3.4(b)] ⇒ ✓ ⇒ 📑 ⇒ Standard Orientation [Fig. 3.4(c)] ⇒ MMB [Fig. 3.4(d)] ⇒ ∕ Datum Axis Tool ⇒ pick on the cylindrical surface [Fig. 3.4(e)] ⇒ MMB

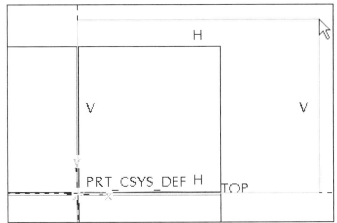

Figure 3.4(a) Sketch a Rectangle

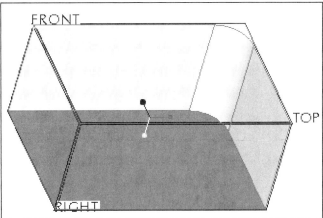

Figure 3.4(b) Create a Fillet

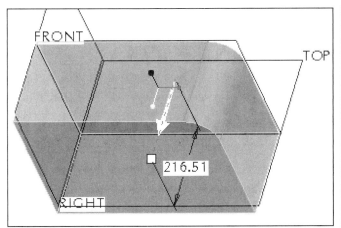

Figure 3.4(c) Depth Preview

Figure 3.4(d) Completed Protrusion

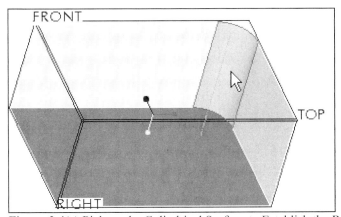

Figure 3.4(e) Pick on the Cylindrical Surface to Establish the References for the Datum Axis

With the datum axis still highlighted, click: **Hole Tool** [Fig. 3.4(f)] ⇒ Placement ⇒ click inside the Secondary references *collector* (turns yellow) [Fig. 3.4(g)] ⇒ pick on the front face [Fig. 3.4(h)] ⇒ expand depth options by opening slide-up panel ⇒ **Drill to intersect with all surfaces** [Fig. 3.4(i)] ⇒ MMB [Fig. 3.4(j)] ⇒ ⇒ MMB ⇒ Window ⇒ Close

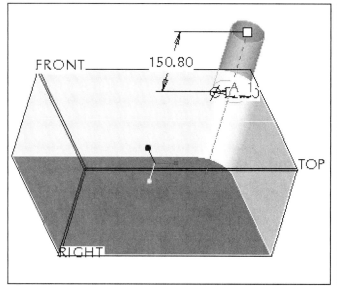

Figure 3.4(f) Hole Tool Display Preview

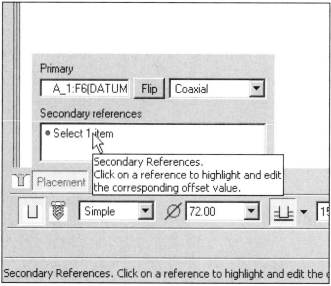

Figure 3.4(g) Click to Activate the Secondary References

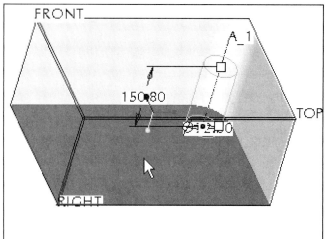

Figure 3.4(h) Pick on the Front Face

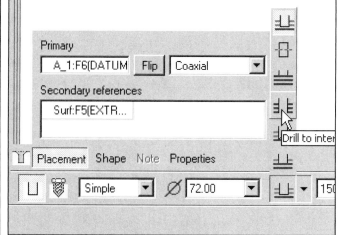

Figure 3.4(i) Select Drill Depth Option

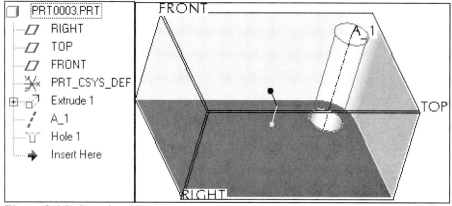

Figure 3.4(j) Completed Part

Part Model Four (PRT0004.PRT) (Cut)

Click: ▢ ⇒ ⦿ ▢ Part **(PRT0004.PRT)** ⇒ **MMB** ⇒ pick the **FRONT** datum ⇒ ⬚ ⇒ Placement from the dashboard ⇒ Define.. ⇒ **MMB** ⇒ ╲ **Create 2 point lines** ⇒ sketch the outline using five lines forming a closed section [Fig. 3.5(a)] ⇒ **MMB** to end the current tool ⇒ ✓ ⇒ AB ⇒ **Standard Orientation** [Fig. 3.5(b)] ⇒ **MMB** [Fig. 3.5(c)]

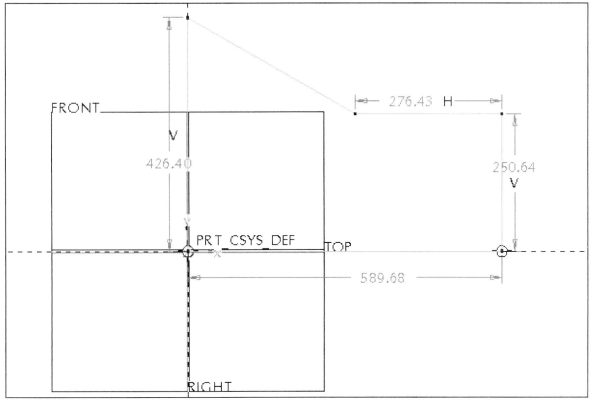

Figure 3.5(a) Sketch the Five Lines of the Enclosed Section (your weak dimensions will be different)

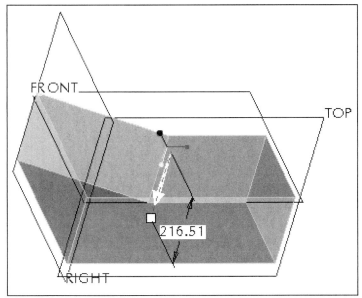

Figure 3.5(b) Depth Preview

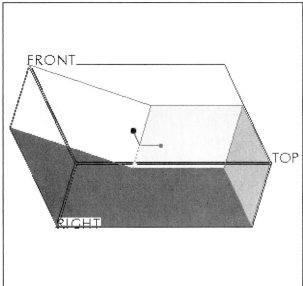

Figure 3.5(c) Completed Protrusion

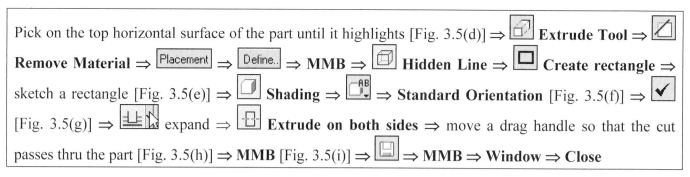

Pick on the top horizontal surface of the part until it highlights [Fig. 3.5(d)] ⇒ **Extrude Tool** ⇒ **Remove Material** ⇒ Placement ⇒ Define.. ⇒ **MMB** ⇒ **Hidden Line** ⇒ **Create rectangle** ⇒ sketch a rectangle [Fig. 3.5(e)] ⇒ **Shading** ⇒ ⇒ **Standard Orientation** [Fig. 3.5(f)] ⇒ [Fig. 3.5(g)] ⇒ expand ⇒ **Extrude on both sides** ⇒ move a drag handle so that the cut passes thru the part [Fig. 3.5(h)] ⇒ **MMB** [Fig. 3.5(i)] ⇒ ⇒ **MMB** ⇒ **Window** ⇒ **Close**

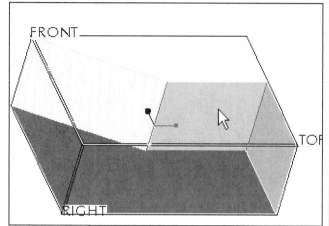

Figure 3.5(d) Select the Horizontal Surface

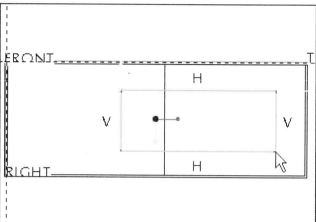

Figure 3.5(e) Sketch a Rectangle

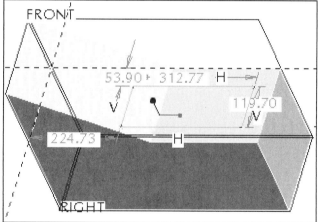

Figure 3.5(f) Sketch Displayed in Standard Orientation

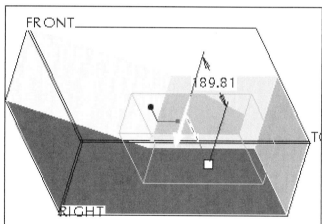

Figure 3.5(g) Previewed Cut

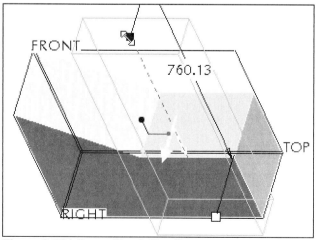

Figure 3.5(h) Drag a Depth Handle

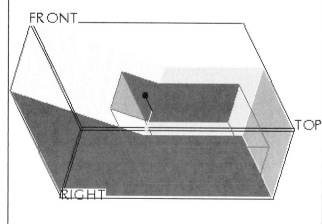

Figure 3.5(i) Completed Part

Part Model Five (PRT0005.PRT) (Mirror)

Click: ▢ ⇒ ◉ ▢ Part (PRT0005.PRT) ⇒ **MMB** ⇒ pick the **FRONT** datum ⇒ 🗗 ⇒ 🗏 **Thicken Sketch** from the dashboard ⇒ Placement ⇒ Define.. ⇒ **MMB** ⇒ **RMB** ⇒ **Centerline** ⇒ pick two points vertically on the edge of the RIGHT datum plane [Figs. 3.6(a-b)] ⇒ **MMB** to end the current tool ⇒ **RMB** ⇒ **Line** [Fig. 3.6(c)] ⇒ sketch the vertical and horizontal lines [Fig. 3.6(d)] ⇒ **MMB** to end the current line

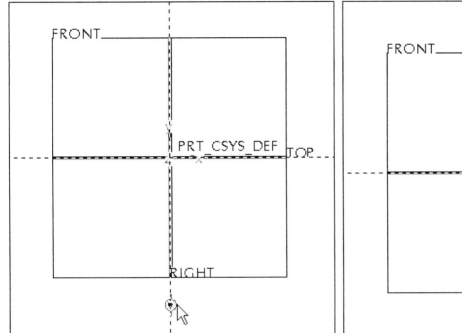

Figure 3.6(a) Pick the First Point of the Vertical Centerline

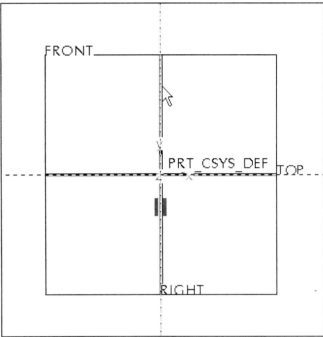

Figure 3.6(b) Pick the Second Point of the Vertical Centerline

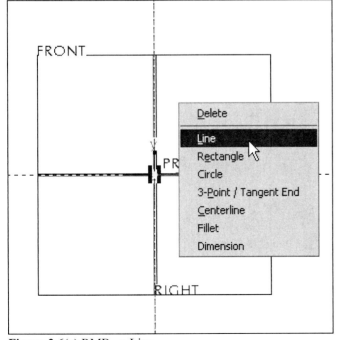

Figure 3.6(c) RMB ⇒ Line

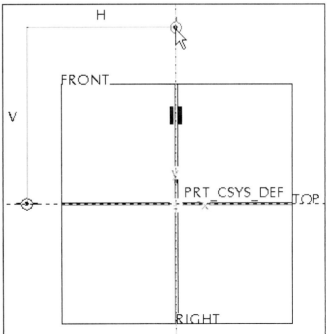

Figure 3.6(d) Sketch Two Lines

Click: **RMB** ⇒ **Fillet** [Fig. 3.6(e)] ⇒ pick the lines near the corner ⇒ **MMB** ⇒ press **LMB** and hold while dragging a window until it incorporates the two lines and fillet [Fig. 3.6(f)] ⇒ 🪞 **Mirror selected entities** ⇒ pick the centerline [Fig. 3.6(g)] ⇒ 🔲 ⇒ **Standard Orientation** ⇒ ✓ [Fig. 3.6(h)]

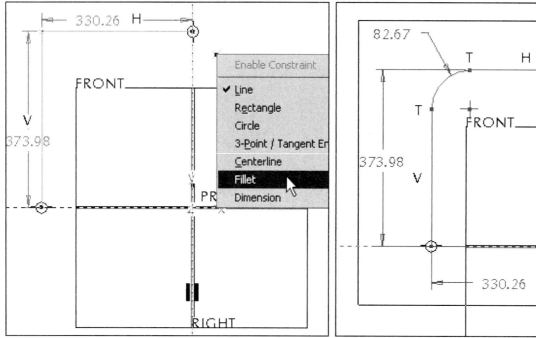

Figure 3.6(e) RMB ⇒ Fillet

Figure 3.6(f) Select Sketch Entities by Windowing

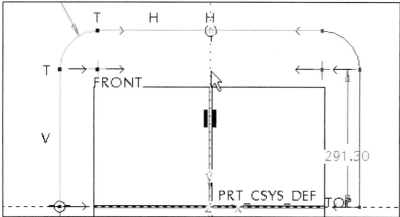

Figure 3.6(g) Mirrored Sketch

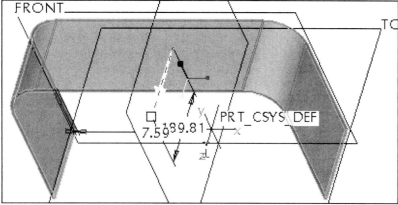

Figure 3.6(h) Depth Preview

Click: ⬚ ⇒ ⬚ **Extrude on both sides** ⇒ drag a depth handle [Fig. 3.6(i)] ⇒ **MMB** ⇒ **LMB** ⇒ press and hold the **Ctrl** key and select the **RIGHT** and **FRONT** datum planes in the Model Tree [Fig. 3.6(j)] ⇒ ⬚ **Datum Axis Tool** [Fig. 3.6(k)] ⇒ ⬚ **Hole Tool** ⇒ [Placement] ⇒ click inside the Secondary references *collector* (turns yellow) ⇒ pick on the top face ⇒ ⬚ ⬚ [Fig. 3.6(l)] ⇒ **MMB** ⇒ ⬚ ⇒ **MMB** ⇒ **Window** ⇒ **Close**

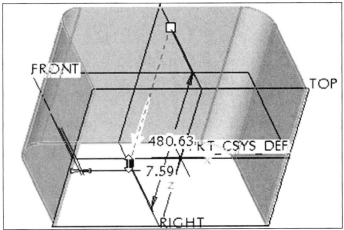

Figure 3.6(i) Drag Depth Handle   **Figure 3.6(j)** Select Datums

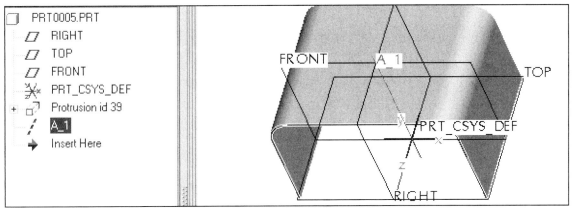

Figure 3.6(k) A_1 Axis Created

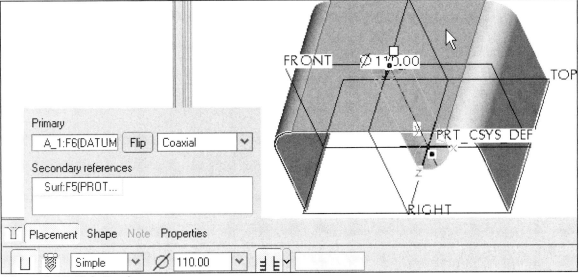

Figure 3.6(l) Hole Preview

Part Model Six (PRT0006.PRT) (Revolve)

Click: [icon] ⇒ [◉ ☐ Part] (PRT0006.PRT) ⇒ **MMB** ⇒ pick the **RIGHT** datum ⇒ [icon] **Revolve Tool** ⇒ **RMB** ⇒ **Define Internal Sketch** ⇒ Orientation **Top** [Fig. 3.7(a)] ⇒ [Sketch] ⇒ **RMB** ⇒ **Centerline** ⇒ create a horizontal centerline on the edge of the TOP datum ⇒ **MMB** ⇒ **RMB** ⇒ **Line** ⇒ sketch the outline of the closed section [Fig. 3.7(b)] ⇒ **MMB** to end the current line ⇒ [✓] ⇒ [icon] ⇒ **Standard Orientation** [Fig. 3.7(c)] ⇒ **MMB** ⇒ **LMB** [Fig. 3.7(d)]

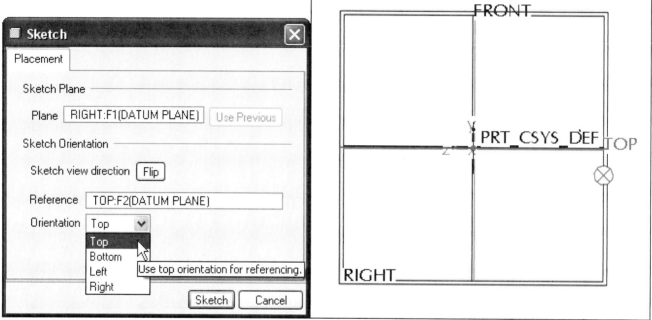

Figure 3.7(a) Sketch Dialog Box

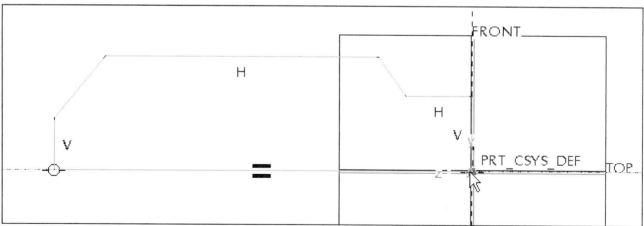

Figure 3.7(b) Sketch the Centerline and the Closed Section Outline

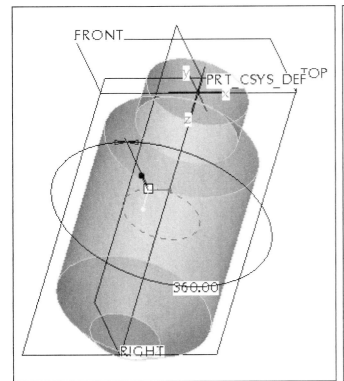

Figure 3.7(c) Revolve Preview

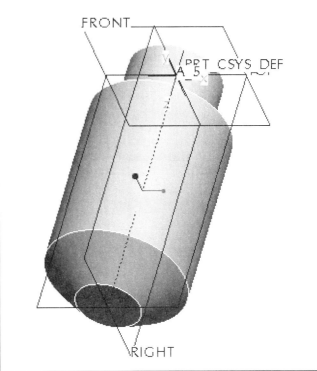

Figure 3.7(d) Completed Revolved Protrusion

Slowly pick on the front edge of the part until it highlights ⇒ press and hold the **Ctrl** key ⇒ pick the other visible edges ⇒ **RMB** ⇒ **Round Edges** [Fig. 3.7(e)] ⇒ drag a handle [Fig. 3.7(f)] ⇒ **MMB** ⇒ **LMB**

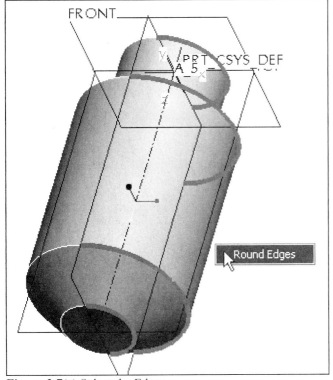

Figure 3.7(e) Select the Edges

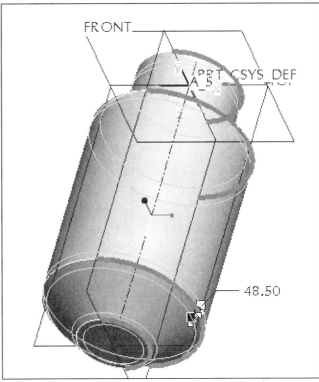

Figure 3.7(f) Move a Drag Handle to Adjust the Size

Click: **View** from the menu bar ⇒ [View Manager] [Fig. 3.7(g)] ⇒ [Xsec] [Fig. 3.7(h)] ⇒ [New] [Fig. 3.7(i)] ⇒ type **A** ⇒ **Enter** [Fig. 3.7(j)] ⇒ **MMB** ⇒ select **RIGHT** from the Model Tree [Fig. 3.7(k)] ⇒ [Display ▼] ⇒ [• Xhatching] [Fig. 3.7(l)] ⇒ [Close] ⇒ **LMB** ⇒ [💾] ⇒ **MMB** ⇒ **Window** ⇒ **Close**

Figure 3.7(g) View Manager

Figure 3.7(h) Select Xsec Tab

Figure 3.7(i) New Xsec

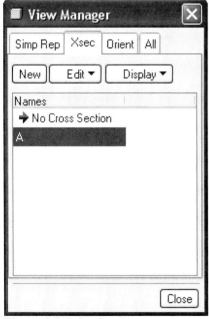

Figure 3.7(j) Section A

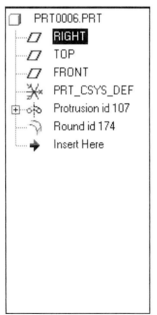

Figure 3.7(k) Select RIGHT

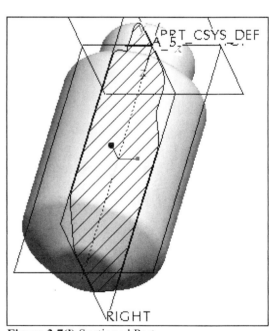

Figure 3.7(l) Sectioned Part

Part Model Seven (PRT0007.PRT) (Revolve Ellipse)

Click: ▢ ⇒ ◉ ▢ Part (PRT0007.PRT) ⇒ **MMB** ⇒ pick the **FRONT** datum ⇒ 🔄 **Revolve Tool** ⇒ **RMB** ⇒ **Define Internal Sketch** ⇒ **MMB** ⇒ **RMB** ⇒ **Centerline** ⇒ create a vertical centerline on the edge of the RIGHT datum ⇒ ⊙ ⇒ ⊙ **Create a full ellipse** ⇒ pick one point to locate the center and the second to determine the shape of the ellipse [Fig. 3.8(a)] ⇒ **MMB** to end the current tool ⇒ ✓ ⇒ 🔲 ⇒ **Standard Orientation** ⇒ Placement ⇒ 180.00 [Fig. 3.8(b)] ⇒ **MMB** [Fig. 3.8(c)] ⇒ **LMB** to deselect

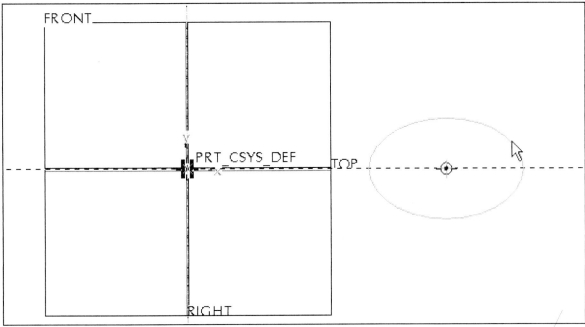

Figure 3.8(a) Sketch a Horizontal Centerline and an Ellipse

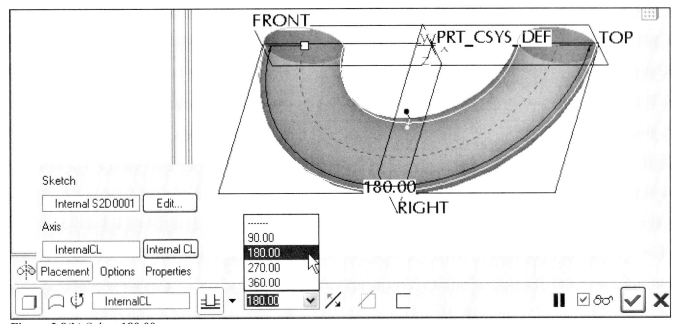

Figure 3.8(b) Select **180.00**

113

Click: **Info** from the menu bar ⇒ **Model** [Fig. 3.8(d)] ⇒ 🖫 ⇒ **MMB** ⇒ **Window** ⇒ **Close**

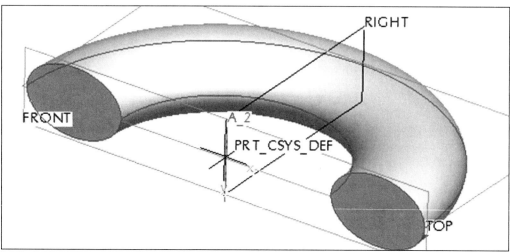

Figure 3.8(c) Completed Elliptical Torus

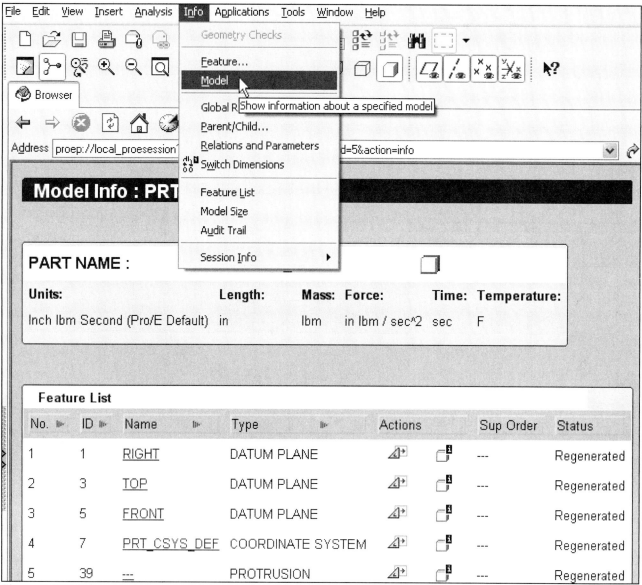

Figure 3.8(d) Information about the Model Displayed in the Embedded Web Browser

Part Model Eight (PRT0008.PRT) (Revolve Cut)

Click: ▢ ⇒ ◉ ▢ Part (**PRT0008.PRT**) ⇒ **MMB** ⇒ pick the **FRONT** datum ⇒ [icon] ⇒ [icon] ⇒ [icon] **Extrude on both sides** ⇒ Placement from the dashboard ⇒ Define.. ⇒ **MMB** ⇒ **RMB** ⇒ **Centerline** ⇒ create a vertical centerline on the edge of the RIGHT datum plane ⇒ **MMB** ⇒ **RMB** ⇒ **Rectangle** ⇒ sketch a rectangle [Fig. 3.9(a)] ⇒ **MMB** ⇒ ✓ ⇒ [icon] ⇒ **Standard Orientation** ⇒ **MMB** ⇒ **LMB**

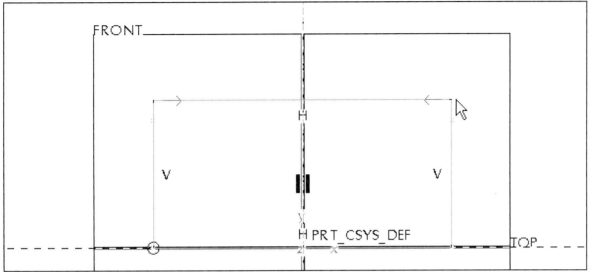

Figure 3.9(a) Sketch a Rectangle

Click: [icon] ⇒ [icon] **Remove Material** ⇒ Placement ⇒ Define.. ⇒ Use Previous ⇒ **Sketch** from the menu bar ⇒ **References** ⇒ [icon] **Select references** ⇒ pick the top edge of the part [Fig. 3.9(b)] ⇒ **MMB** ⇒ **MMB**

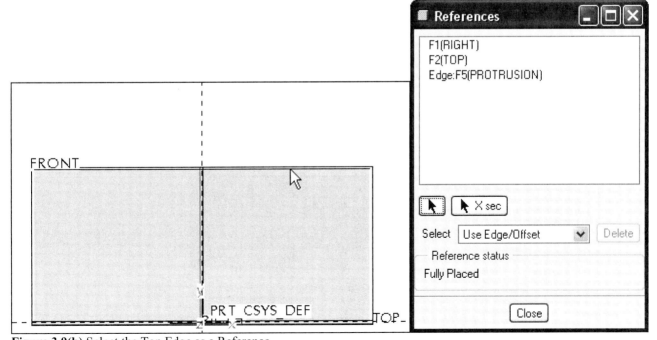

Figure 3.9(b) Select the Top Edge as a Reference

Click: **RMB** ⇒ **Centerline** ⇒ create a vertical centerline through the middle of the part ⇒ **RMB** ⇒ **Circle** ⇒ sketch a circle [Figs. 3.9(c-d)] ⇒ **MMB** ⇒ ✓ ⇒ **Ctrl+D** [Fig. 3.9(e)] ⇒ **MMB** [Fig. 3.9(f)] ⇒ **LMB**

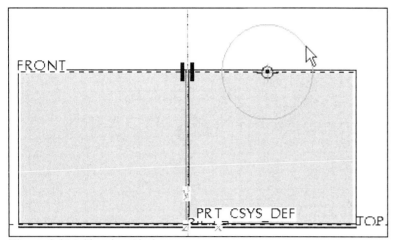

Figure 3.9(c) Sketch a Circle

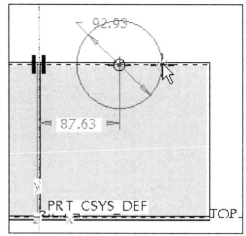

Figure 3.9(d) Completed Sketched Circle

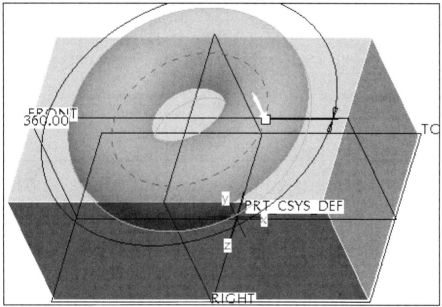

Figure 3.9(e) Revolved Cut Preview

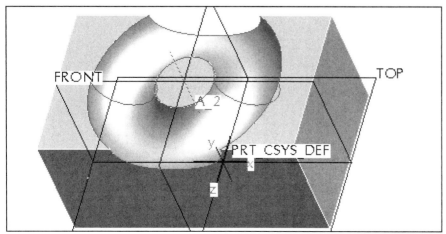

Figure 3.9(f) Completed Cut

Slowly pick on the edge of the part until it highlights ⇒ press and hold the **Ctrl** key ⇒ pick the other edges ⇒ **RMB** ⇒ **Round Edges** [Figs. 3.9(g-h)] ⇒ **MMB** [Fig. 3.9(i)] ⇒ **LMB**

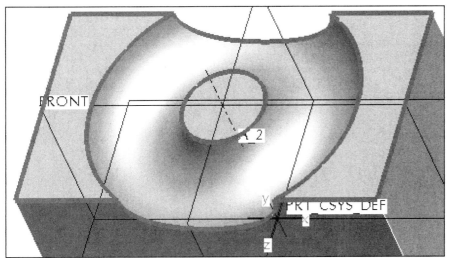

Figure 3.9(g) Select the Edges

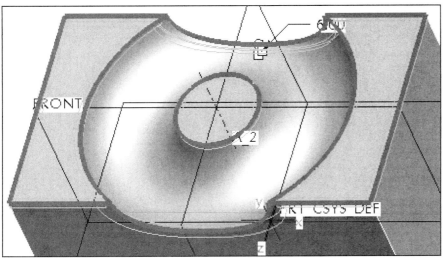

Figure 3.9(h) Round Preview

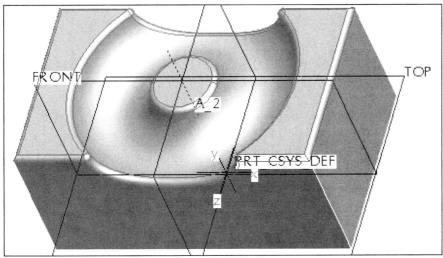

Figure 3.9(i) Completed Round

Place your cursor on Axis **A_2** ⇒ **LMB** until **A_2** is highlighted [Fig. 3.9(j)] ⇒ 🔲 **Hole Tool** ⇒ [Placement] ⇒ click inside the Secondary references *collector* (turns yellow) ⇒ pick on the top face of the part [Fig. 3.9(k)] ⇒ 🔲 ⇒ 🔲 ⇒ **MMB** [Fig. 3.9(l)] ⇒ **LMB** ⇒ 🔲 ⇒ **MMB** ⇒ **Window** ⇒ **Close**

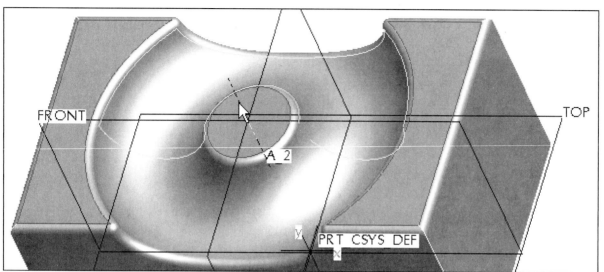

Figure 3.9(j) Select Axis **A_2**

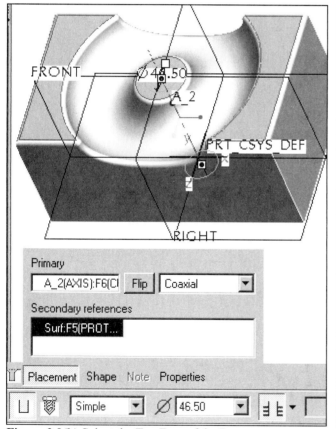

Figure 3.9(k) Select the Top Face of the Part

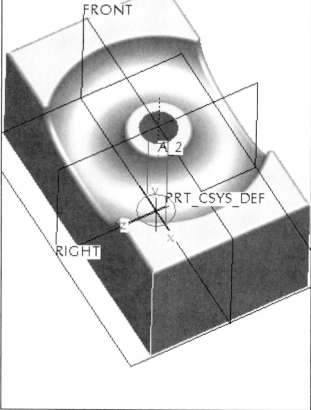

Figure 3.9(l) Completed Part

See **www.cad-resources.com** ⇒ **Downloads**, for extra Lessons and projects.

Lesson 4 Extrusions

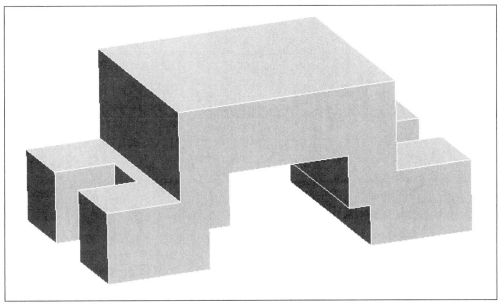

Figure 4.1 Clamp

OBJECTIVES

- Create a feature using an **Extruded** protrusion
- Understand **Setup** and **Environment** settings
- Define and set a **Material** type
- Create and use **Datum** features
- Sketch protrusion and cut feature geometry using the **Sketcher**
- Understand the feature **Dashboard**
- **Copy** a feature
- **Save** and **Delete Old Versions** of an object

Extrusions

The design of a part using Pro/E starts with the creation of base features (normally datum planes), and a solid protrusion. Other protrusions and cuts are then added in sequence as required by the design. You can use various types of Pro/E features as building blocks in the progressive creation of solid parts (Fig. 4.1). Certain features, by necessity, precede other more dependent features in the design process. Those dependent features rely on the previously defined features for dimensional and geometric references.

The progressive design of features creates these dependent feature relationships known as *parent-child relationships*. The actual sequential history of the design is displayed in the Model Tree. The parent-child relationship is one of the most powerful aspects of Pro/E and parametric modeling in general. It is also very important after you modify a part. After a parent feature in a part is modified, all children are automatically modified to reflect the changes in the parent feature. It is therefore essential to reference feature dimensions so that Pro/E can correctly propagate design modifications throughout the model.

An **extrusion** is a part feature that adds or removes material. A protrusion is *always the first solid feature created*. This is usually the first feature created after a base feature of datum planes. The **Extrude Tool** is used to create both protrusions and cuts. A toolchest button is available for this command or it can be initiated using Insert ⇒ Extrude from the menu bar. Figure 4.2 shows four different types of basic protrusions.

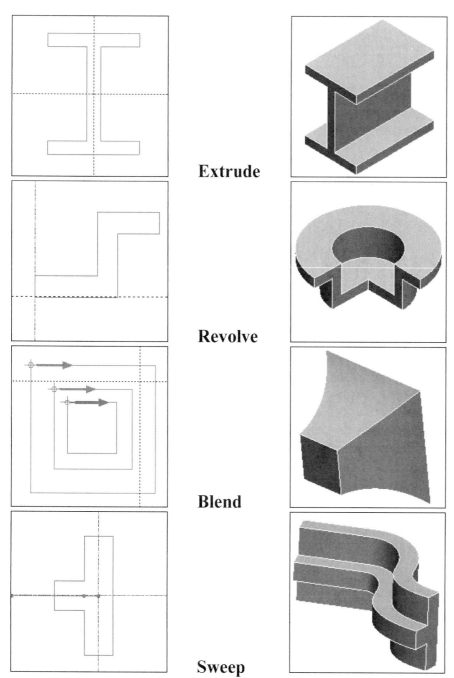

Figure 4.2 Basic Protrusions

The Design Process

It is tempting to directly start creating models. Nevertheless, in order to build value into a design, you need to create a product that can keep up with the constant design changes associated with the design-through-manufacturing process. Flexibility must be integral to the design. Flexibility is the key to a friendly and robust product design while maintaining design intent, and you can accomplish it through planning. To plan a design, you need to understand the overall function, form, and fit of the product. This understanding includes the following points:

- Overall size of the part
- Basic part characteristics
- The way in which the part can be assembled
- Approximate number of assembly components
- The manufacturing processes required to produce the part

Lesson 4 STEPS

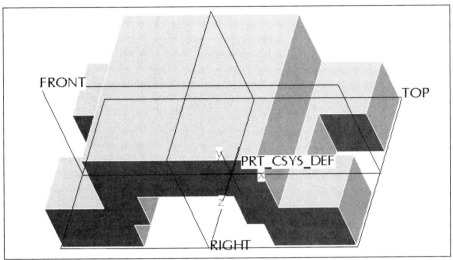

Figure 4.3 Clamp and Datum Planes

Clamp

The clamp in Figure 4.3 is composed of a protrusion and two cuts. A number of things need to be established before you actually start modeling. These include setting up the *environment*, selecting the *units*, and establishing the *material* for the part.

Before you begin any part using Pro/E, you must plan the design. The **design intent** will depend on a number of things that are out of your control and on a number that you can establish. Asking yourself a few questions will clear up the design intent you will follow: Is the part a component of an assembly? If so, what surfaces or features are used to connect one part to another? Will geometric tolerancing be used on the part and assembly? What units are being used in the design, SI or decimal inch? What is the part's material? What is the primary part feature? How should I model the part, and what features are best used for the primary protrusion (the first solid mass)? On what datum plane should I sketch to model the first protrusion? These and many other questions will be answered as you follow the systematic lesson part. However, you must answer many of the questions on your own when completing the *lesson project*, which does not come with systematic instructions.

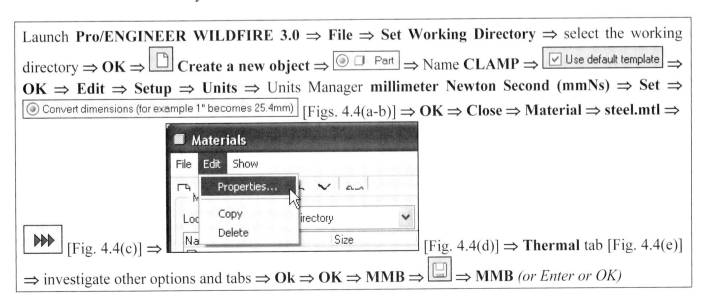

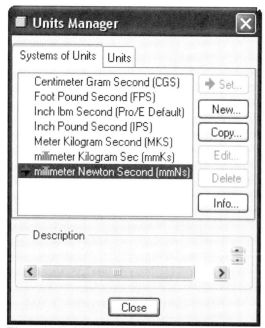
Figure 4.4(a) Units Manager Dialog Box

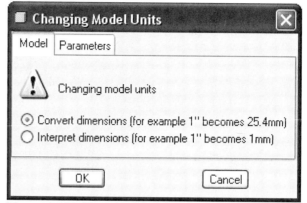
Figure 4.4(b) Changing Model Units Dialog Box

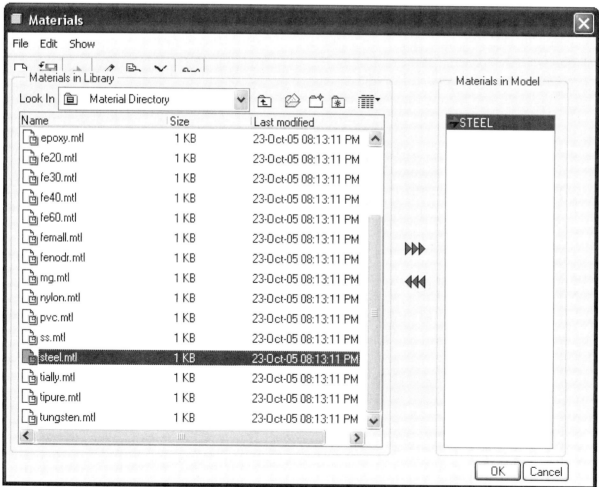
Figure 4.4(c) Material File

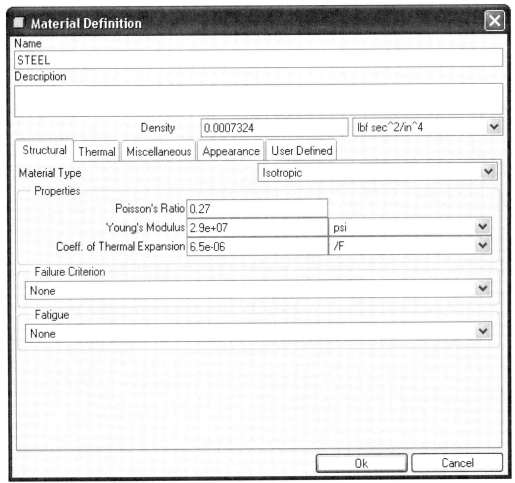

Figure 4.4(d) Material Definition, Structural Tab

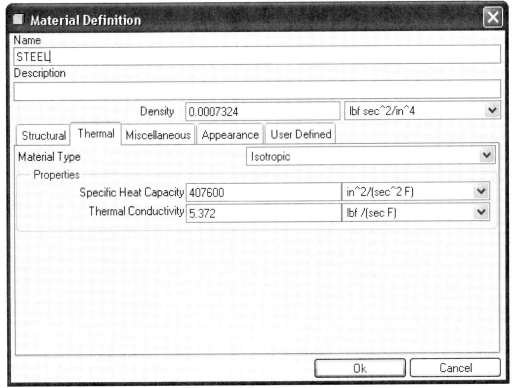

Figure 4.4(e) Material Definition, Thermal Tab

Since ☑ Use default template was selected, the default datum planes and the default coordinate system are displayed in the graphics window and in the Model Tree (Fig. 4.5). *The **default datum planes** and the **default coordinate system** will be the first features on all parts and assemblies.* The datum planes are used to sketch on and to orient the part's features. Having datum planes as the first features of a part, instead of the first protrusion, gives the designer more flexibility during the design process. Picking on an item in the Model Tree will highlight that item on the model (Fig. 4.5).

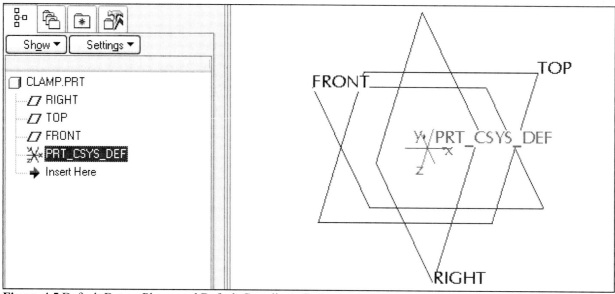

Figure 4.5 Default Datum Planes and Default Coordinate System

Click on the **FRONT** datum plane in the Model Tree ⇒ **Sketch Tool** from Right Toolchest ⇒ Sketch dialog box opens [Fig. 4.6(a)] ⇒ *accept the default selections*, click: Sketch

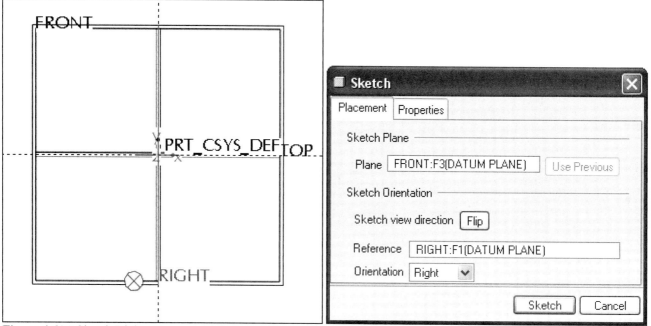

Figure 4.6(a) Sketch Dialog Box

From menu bar, click: **Sketch** ⇒ **References** [Fig. 4.6(b)] *(the RIGHT and TOP datum planes are the positional/dimensional references)* ⇒ **Close** ⇒ **Toggle the grid** on from Top Toolchest [Fig. 4.6(c)]

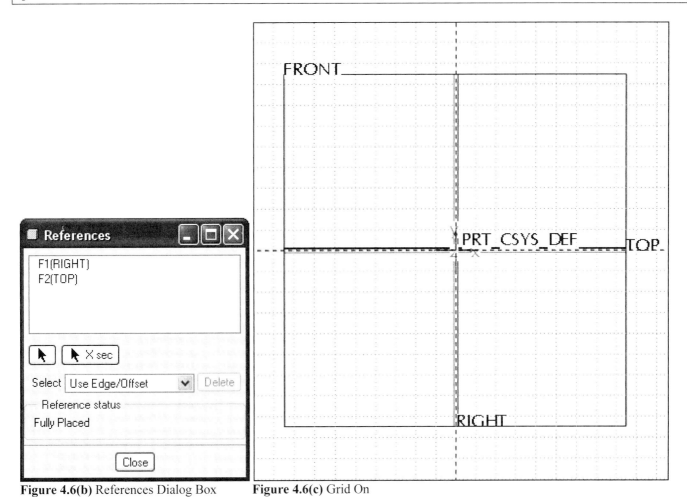

Figure 4.6(b) References Dialog Box   **Figure 4.6(c)** Grid On

The sketch is now displayed and oriented in 2D [Fig. 4.6(c)]. The coordinate system is at the middle of the sketch, where datum RIGHT and datum TOP intersect. The X coordinate arrow points to the right and the Y coordinate arrow points up. The Z arrow is pointing toward you (out from the screen). The square box you see is the limited display of datum FRONT. This is similar to sketching on a piece of graph paper. Pro/E is not coordinate-based software, so you need not enter geometry with X, Y, and Z coordinates.

Use **Shift+MMB** and **Ctrl+MMB** to reposition and resize the sketch as needed. Since you now have a visible grid, it is a good idea to have your sketch picks snap to the grid. Click: **Tools** from the menu bar ⇒ **Environment** ⇒ **Snap to Grid** [Fig. 4.6(d)] ⇒ **Apply** ⇒ **OK**

You can control many aspects of the environment in which Pro/E runs with the Environment dialog box. To open the Environment dialog box, click Tools ⇒ Environment on the menu bar or click the appropriate icon in the toolbar. When you make a change in the Environment dialog box, it takes effect for the current session only. When you start Pro/E, the environment settings are defined by Pro/E configuration defaults.

Depending on which Pro/E Mode is active (here it is the Part Mode), some or all of the following options may be available in the **Environment** dialog box:

Display:

Dimension Tolerances Display model dimensions with tolerances
Datum Planes Display the datum planes and their names
Datum Axes Display the datum axes and their names
Point Symbols Display the datum points and their names
Coordinate Systems Display the coordinate systems and their names
Spin Center Display the spin center for the model
3D Notes Display model notes
Notes as Names Display the note as a name, not the full note
Reference Designators Display reference designation of cabling, ECAD, and Piping components
Thick Cables Display a cable with 3-D thickness

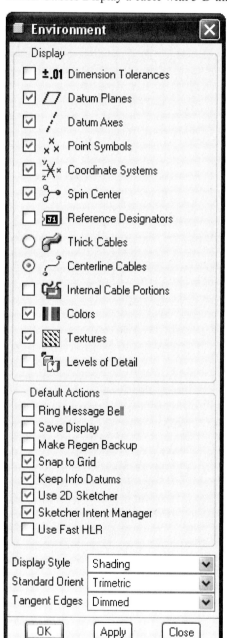

Centerline Cables Display the centerline of a cable with location points
Internal Cable Portions Display cable portions that are hidden from view
Colors Display colors assigned to model surfaces
Textures Display textures on shaded models
Levels of Detail Controls levels of detail available in a shaded model during dynamic orientation

Default Actions:

Ring Message Bell Ring bell (beep) after each prompt or system message
Save Display Save objects with their most recent screen display
Make Regen Backup Backs up the current model before every regeneration
Snap to Grid Make points you select on the Sketcher screen snap to a grid
Keep Info Datums Control how Pro/E treats datum planes, datum points, datum axes, and coordinate systems created on the fly under the Info functionality
Use 2D Sketcher Control the initial model orientation in Sketcher mode
Sketcher Intent Manager Use the Intent Manager when in Sketcher
Use Fast HLR Make possible the hardware acceleration of dynamic spinning with hidden lines, datums, and axes

Display Style:

- **Wireframe** Model is displayed with no distinction between visible and hidden lines
- **Hidden Line** Hidden lines are shown in gray
- **No Hidden** Hidden lines are not shown
- **Shading** All surfaces and solids are displayed as shaded

Standard Orient:

- **Isometric** Standard isometric orientation
- **Trimetric** Standard trimetric orientation
- **User Defined** User-defined orientation

Tangent Edges:

- **Solid** Display tangent edges as solid lines
- **No Display** Blank tangent edges
- **Phantom** Display tangent edges in phantom font
- **Centerline** Display tangent edges in centerline font
- **Dimmed** Display tangent edges in the Dimmed Menu system

Figure 4.6(d) Environment Dialog Box

Because you checked ☑ Snap to Grid, you can now sketch by simply picking grid points representing the part's geometry (outline). Because this is a sketch in the true sense of the word, you need only create geometry that *approximates* the shape of the feature; the sketch does not have to be accurate as far as size or dimensions are concerned. No two sketches will be the same between those using these steps, unless you count grid spaces (which is not necessary). Even with the grid snap off Pro/E, constrains the geometry according to rules, which include but are not limited to the following:

- **RULE:** Symmetry
  DESCRIPTION: Entities sketched symmetrically about a centerline are assigned equal values with respect to the centerline
- **RULE:** Horizontal and vertical lines
  DESCRIPTION: Lines that are approximately horizontal or vertical are considered exactly horizontal or vertical
- **RULE:** Parallel and perpendicular lines
  DESCRIPTION: Lines that are sketched approximately parallel or perpendicular are considered exactly parallel or perpendicular
- **RULE:** Tangency
  DESCRIPTION: Entities sketched approximately tangent to arcs or circles are assumed to be exactly tangent

The outline of the part's primary feature is sketched using a set of connected lines. The part's dimensions and general shape are provided in Figure 4.6(e). The cut on the front and sides will be the created with separate sketched features. Sketch only one series of lines (8 lines in this sketch). ***Do not sketch lines on top of lines.***

It is important not to create any unintended constraints while sketching. Therefore, remember to exaggerate the sketch geometry and not to align geometric items that have no relationship. Pro/E is very smart: if you draw two lines at the same horizontal level, Pro/E assumes they are horizontally aligned. Two lines the same length will be constrained as so.

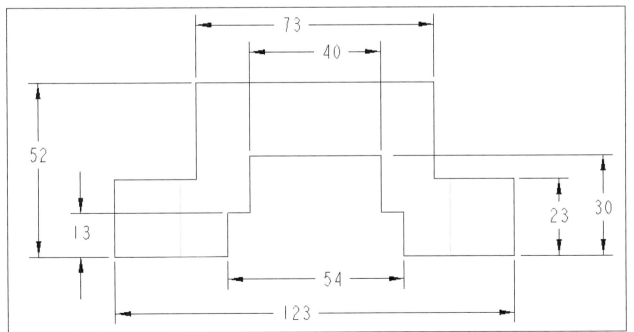

Figure 4.6(e) Front View of Drawing Showing Dimensions for the Clamp

With your cursor anywhere in the graphics window, but not on an object, click: **RMB** [Fig. 4.6(f)] ⇒ **Centerline** ⇒ pick two vertical positions on the RIGHT datum plane to create the centerline [Fig. 4.6(g)]

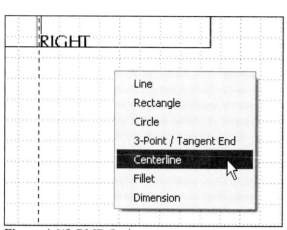

Figure 4.6(f) RMB Options

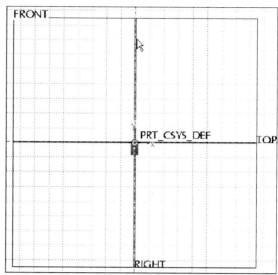

Figure 4.6(g) Create the Centerline

Click: **RMB** ⇒ **Line** ⇒ sketch the eight lines of the outline [Fig. 4.6(h)] ⇒ **MMB** to end the line sequence [Fig. 4.6(i)] ⇒ **MMB**

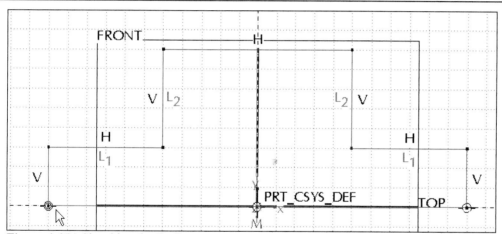

Figure 4.6(h) Sketching the Outline

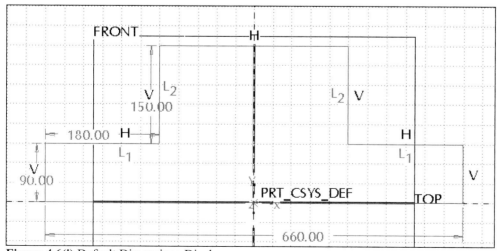

Figure 4.6(i) Default Dimensions Display

A sketcher *constraint symbol* [L₂ V] appears next to the entity that is controlled by that constraint. Sketcher constraints can be turned on or off (enabled or disabled) while sketching. Simply click your RMB as you sketch- before clicking the position- and the constraint that is displaying will have a slash imposed over it. This will disable it for that entity. An **H** next to a line means horizontal; a **T** means tangent. Dimensions display, as they are needed according to the references selected and the constraints. Seldom are they the same as the required dimensioning scheme needed to manufacture the part. You can add, delete, and move dimensions as required. *The dimensioning scheme is important, **not** the dimension value, which can be modified now or later.*

Place and create the dimensions as required [Fig. 4.6(j)]. Do not be concerned with the perfect positioning of the dimensions, but in general, follow the spacing and positioning standards found in the **ASME Geometric Tolerancing and Dimensioning** standards. This saves you time when you create a drawing of the part. Dimensions placed at this stage of the design process are displayed on the drawing document by simply showing all the dimensions.

To dimension between two lines, simply pick the lines with the left mouse button (LMB) and place the dimension value with the middle mouse button (MMB). To dimension a single line, pick on the line (LMB), and then place the dimension with MMB.

Click: **Tools** ⇒ [icon] **Environment** ⇒ [☐ Snap to Grid] *(it is easier to position the dimensions with Snap to Grid off)* ⇒ **Apply** ⇒ **OK** ⇒ **RMB** ⇒ **Dimension** ⇒ add and reposition dimensions [Fig. 4.6(j)]

(To move a dimension—click: [cursor icon] *⇒ pick a dimension ⇒ hold down the LMB ⇒ move it to a new position ⇒ release the LMB)*

If any of the dimension values are *light gray* in color, they are called *weak* dimensions. If a weak dimension matches your dimensioning scheme, make them *strong* ⇒ pick on a weak dimension value *(will highlight in Red)* ⇒ **RMB** ⇒ **Strong** [Fig. 4.6(j)]

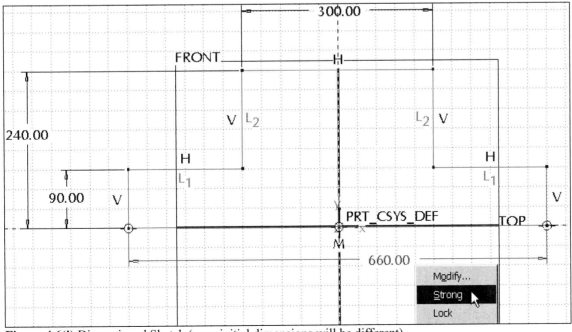

Figure 4.6(j) Dimensioned Sketch (your initial dimensions will be different)

Next, control the sketch by adding symmetry constraints, click: ⬚ **Impose sketcher constraints on the section** ⇒ ⊹⊦ **Make two points or vertices symmetric about a centerline** [Fig. 4.6(k)] ⇒ pick the centerline and then pick two vertices (endpoints) to be symmetric ⊦← [Fig. 4.6(l)] ⇒ repeat the process and make the sketch symmetrical [Fig. 4.6(m)] ⇒ **Close**

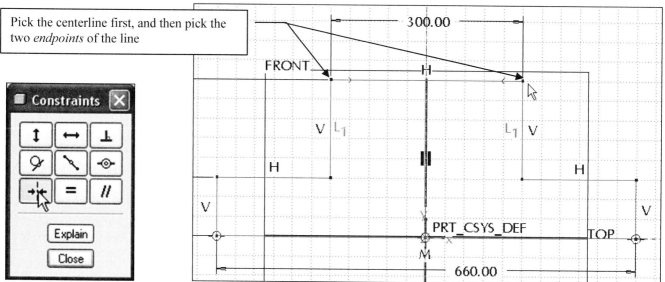

Pick the centerline first, and then pick the two *endpoints* of the line

Figure 4.6(k) Constraints Dialog   **Figure 4.6(l)** Adding Symmetry Constraint

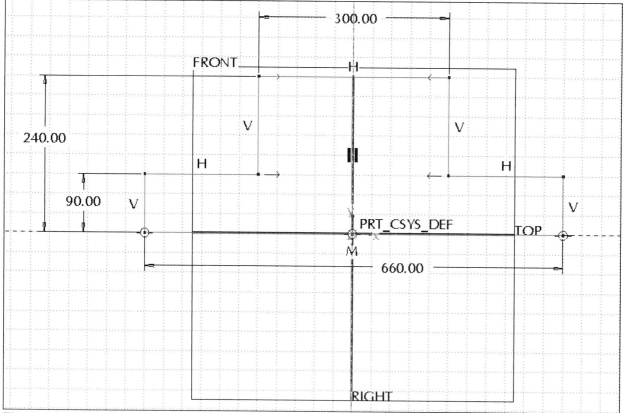

Figure 4.6(m) Sketch is Symmetrical (your values will be different!)

Your original sketch values will be different from the example, but the final design values will be the same. **DO NOT CHANGE YOUR SKETCH DIMENSION VALUES TO THOSE IN FIGURE 4.6(m).**

Click: [arrow] ⇒ Window-in the sketch (place the cursor at one corner of the window with the **LMB** depressed, drag the cursor to the opposite corner of the window and release the **LMB**) to capture all four dimensions. They will turn red. ⇒ **RMB** ⇒ **Modify** ⇒ [✓ Lock Scale] ⇒ [✓ Regenerate] [Fig. 4.6(n)] ⇒ double-click on length dimension (*here it is 660, but your dimension may be different*) in the Modify Dimensions dialog box and type the design value at the prompt (**123**) [Fig. 4.6(o)] ⇒ **Enter** ⇒ [✓] **Regenerate the section and close the dialog** ⇒ double-click on another dimension and modify the value [Fig. 4.6(p)] ⇒ **Enter** ⇒ continue until all of the values are changed to the design sizes [Fig. 4.6(q)]

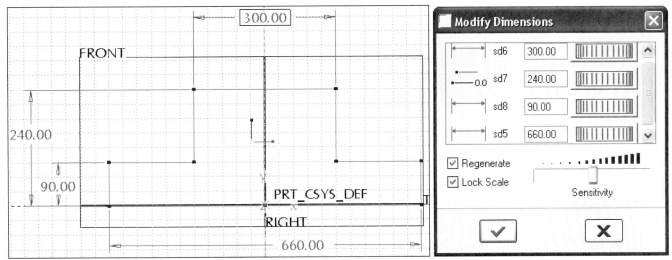

Figure 4.6(n) Modify Dimensions

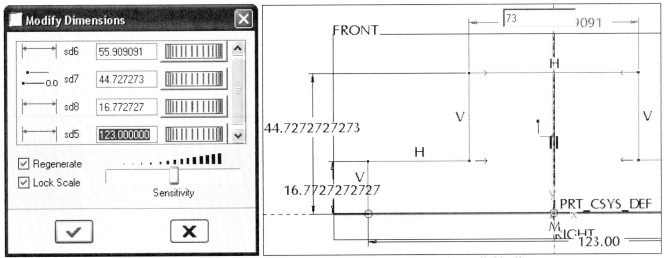

Figure 4.6(o) Modify the Dimension   **Figure 4.6(p)** Modify each Dimension Individually

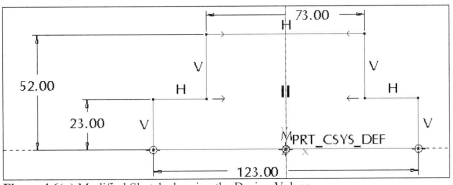

Figure 4.6(q) Modified Sketch showing the Design Values

Click: 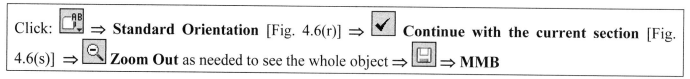 ⇒ **Standard Orientation** [Fig. 4.6(r)] ⇒ ✓ **Continue with the current section** [Fig. 4.6(s)] ⇒ 🔍 **Zoom Out** as needed to see the whole object ⇒ 💾 ⇒ **MMB**

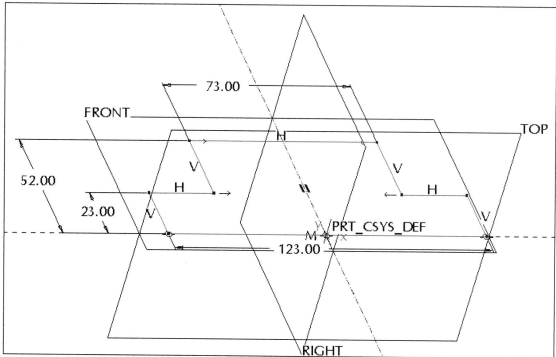

Figure 4.6(r) Regenerated Dimensions

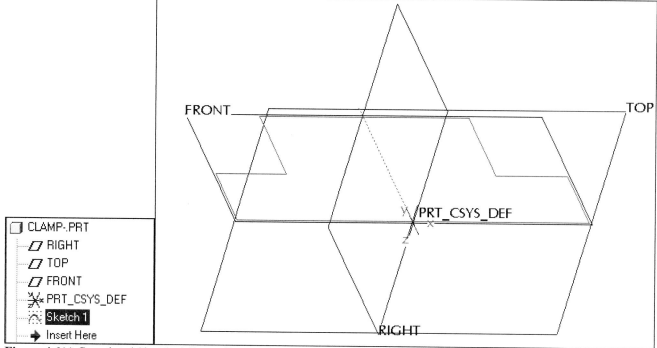

Figure 4.6(s) Completed Sketched Curve (Datum Curve)

The datum curve (Sketch1) will remain highlighted, active and therefore selected.

Click: **Extrude Tool** [Fig. 4.7(a)] ⇒ double-click on the depth value on the model ⇒ type **70** [Fig. 4.7(b)] ⇒ **Enter** ⇒ place your pointer over the square drag handle (it will turn black) ⇒ **RMB** ⇒ **Symmetric** [Fig. 4.7(c)] ⇒ **MMB** [Fig. 4.7(d)] ⇒ ⇒ **MMB**

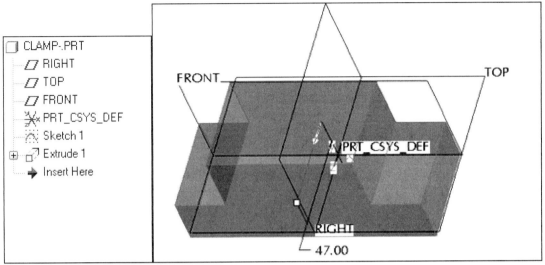

Figure 4.7(a) Depth of Extrusion Previewed

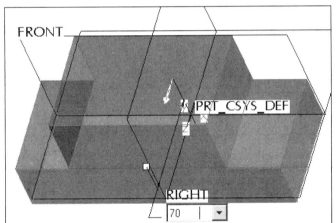

Figure 4.7(b) Modify the Depth Value

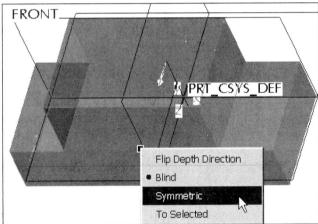

Figure 4.7(c) Symmetric

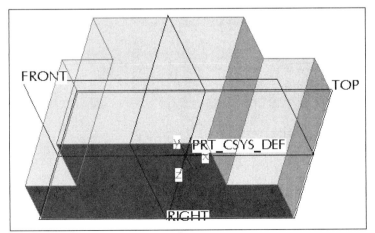

Figure 4.7(d) Completed Extrusion

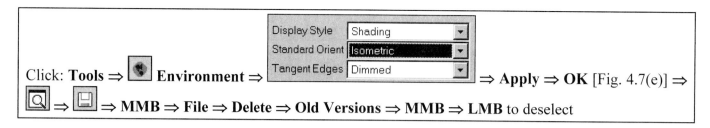

Storing an object on the disk does not overwrite an existing object file. To preserve earlier versions, Pro/E saves the object to a new file with the same object name but with an updated version number. Every time you store an object using Save, you create a new version of the object in memory, and write the previous version to disk. Pro/E numbers each version of an object storage file consecutively (for example, box.sec.1, box.sec.2, box.sec.3). If you save 25 times, you have 25 versions of the object, all at different stages of completion. You can use *File ⇒ Delete ⇒ Old Versions* after the *Save* command to eliminate previous versions of the object that may have been stored.

When opening an existing object file, you can open any version that is saved. Although Pro/E automatically retrieves the latest saved version of an object, you can retrieve any previous version by entering the full file name with extension and version number (for example, **partname.prt.5**). If you do not know the specific version number, you can enter a number relative to the latest version. For example, to retrieve a part from two versions ago, enter **partname.prt.3** *(or partname.prt.-2)*.

You use *File ⇒ Erase* to remove the object and its associated objects from memory. If you close a window before erasing it, the object is still in memory. In this case, you use *File ⇒ Erase ⇒ Not Displayed* to remove the object and its associated objects from memory. This does not delete the object. It just removes it from active memory. *File ⇒ Delete ⇒ All Versions* removes the file from memory and from disk completely. You are prompted with a Delete All Confirm dialog box when choosing this command. Be careful not to delete needed files.

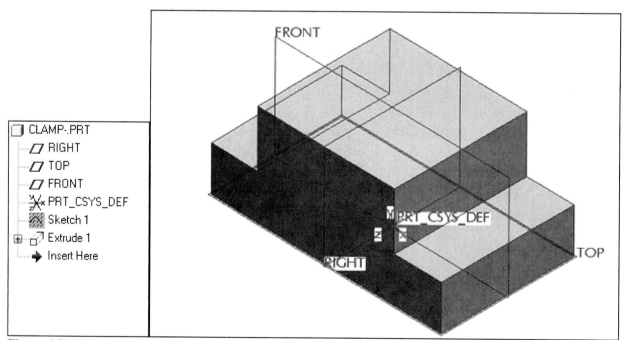

Figure 4.7(e) Isometric Orientation

Next, the cut through the middle of the part will be modeled.

Click: ⬚ **Extrude Tool** ⇒ **RMB** ⇒ **Remove Material** ⇒ **RMB** ⇒ **Define Internal Sketch** [Fig. 4.8(a)] ⇒ **Use Previous** from the Sketch dialog box [Plane | Use Previous] [Fig. 4.8(b)] ⇒ **RMB** ⇒ **Centerline** [Fig. 4.8(c)] ⇒ create a vertical centerline ⇒ **RMB** ⇒ **Line** ⇒ sketch the seven lines of the open outline [Fig. 4.8(d)] ⇒ **MMB** ⇒ ⬚ **Impose sketcher constraints on the section** ⇒ ⬚ [Fig. 4.6(k)] ⇒ pick the centerline and then pick two vertices to be symmetric ⬚ ⇒ repeat the process and make the sketch symmetrical ⇒ If you attempt to create too many constraints, Pro/E will open the Resolve Sketch dialog box [Fig. 4.8(e)]. **Delete** the extra symmetric constraint if this happens. ⇒ **Close**

Figure 4.8(a) RMB Options

Figure 4.8(b) Sketch Dialog Box

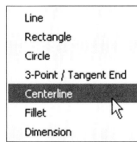
Figure 4.8(c) Centerline

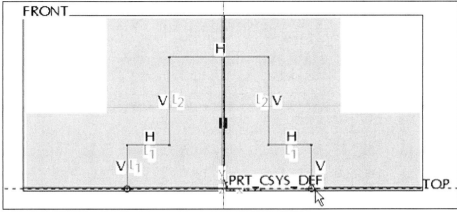
Figure 4.8(d) Sketch

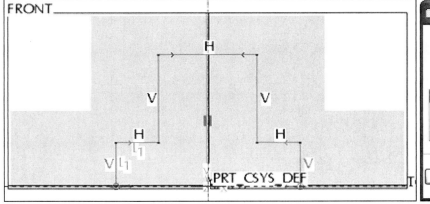

Figure 4.8(e) Resolve Sketch Dialog Box

Click: **RMB** ⇒ **Dimension** ⇒ add and reposition dimensions, *your values will be different* ⇒ 🔲
Hidden Line [Fig. 4.8(f)] ⇒ **MMB** to deselect dimension tool and activate 🔺 **Select items**

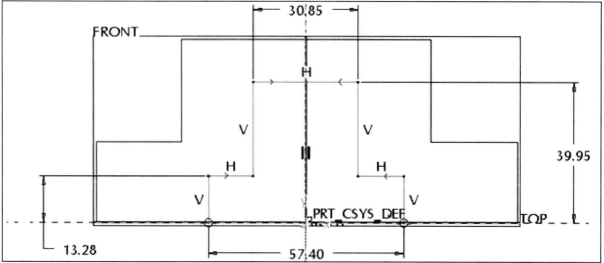

Figure 4.8(f) Dimensioned Sketch

Window-in the sketch to capture all four dimensions. They will turn red. ⇒ **RMB** ⇒ **Modify** ⇒ ☐ Lock Scale ⇒ ☑ Regenerate ⇒ modify the values [Fig. 4.8(g)] ⇒ ✓ [Fig. 4.8(h)]

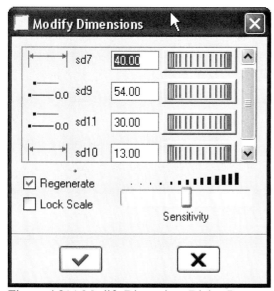

Figure 4.8(g) Modify Dimensions Dialog Box

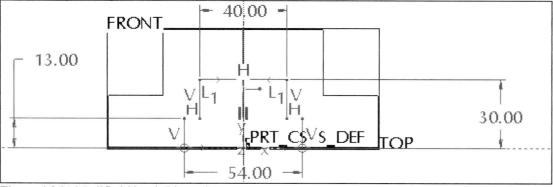

Figure 4.8(h) Modified Sketch Dimensions

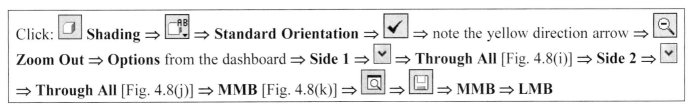

Figure 4.8(i) Options Side 1   **Figure 4.8(j)** Options Side 2

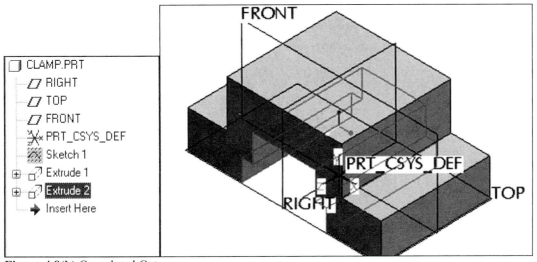

Figure 4.8(k) Completed Cut

The next feature will be a **20 X 20** centered cut (Fig. 4.9). Because the cut feature is identical on both sides of the part, you can mirror and copy the cut after it has been created.

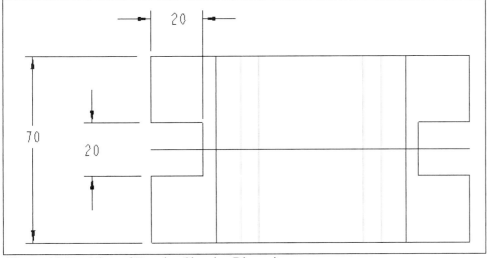

Figure 4.9 Top View of Drawing Showing Dimensions

Click: [icon] **Extrude Tool** ⇒ [Placement] from the dashboard ⇒ [Define...] Sketch dialog box opens ⇒ Sketch Plane--- Plane: select **TOP** datum from the model as the sketch plane [Fig. 4.10(a)] ⇒ [Sketch] ⇒ **Sketch** from the menu bar ⇒ **References** ⇒ pick the left edge of the part to add it to the References dialog box [Fig. 4.10(b)] ⇒ **Close** ⇒ **Tools** ⇒ **Environment** ⇒ [☐ Snap to Grid] ⇒ **OK**

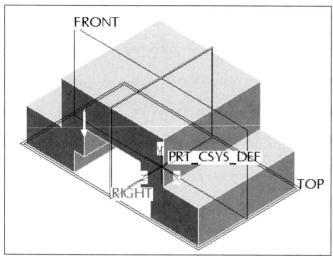 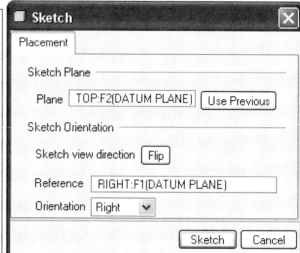

Figure 4.10(a) Sketch Plane Selection and Sketch Dialog Box

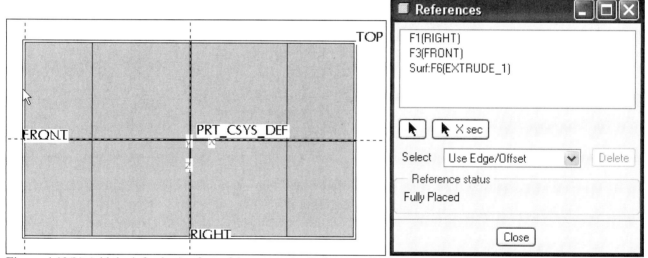

Figure 4.10(b) Add the left edge/surface of the part to the References Dialog Box

Click: [icon] **Hidden Line** ⇒ **RMB** ⇒ **Centerline** ⇒ create a horizontal centerline through the center of the part ⇒ **RMB** ⇒ **Line** ⇒ place the mouse on the left edge and create three lines [Fig. 4.10(c)] ⇒ **MMB** to end the line sequence [Fig. 4.10(d)] ⇒ [icon] **Impose sketcher constraints on the section** ⇒ [icon] ⇒ pick the centerline and then pick two vertices to be symmetric, add the required dimensions ⇒ **MMB** ⇒ **MMB** ⇒ move and modify the values for the two dimensions (**20 X 20**) [Fig. 4.10(e)] ⇒ [✓] ⇒ [icon] ⇒ **Standard Orientation** ⇒ from the dashboard [icon] **Remove Material** ⇒ **Options** tab ⇒ [▼] ⇒ [icon] **Through All** [Fig. 4.10(f)] ⇒ [✓] ⇒ [icon] ⇒ [icon] **Shading** ⇒ rotate your model using **MMB** to see the cut clearly [Fig. 4.10(g)] ⇒ [icon] ⇒ **MMB**

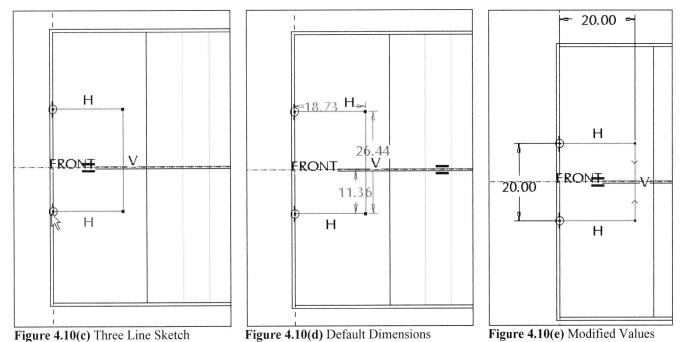

Figure 4.10(c) Three Line Sketch   **Figure 4.10(d)** Default Dimensions   **Figure 4.10(e)** Modified Values

Figure 4.10(f) Options Through All

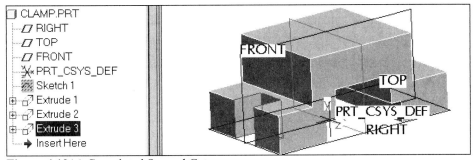

Figure 4.10(g) Completed Second Cut

Click: [icon] ⇒ **Standard Orientation** ⇒ with the new cut still highlighted [Fig. 4.11(a)], click: [icon] **Mirror Tool** ⇒ select the **RIGHT** datum plane from the Model Tree [Fig. 4.11(b)] ⇒ **MMB** [Fig. 4.11(c)] ⇒ **File** ⇒ **Save** ⇒ **MMB** ⇒ **MMB** rotate part [Fig. 4.11(d)] ⇒ **File** ⇒ **Close Window**

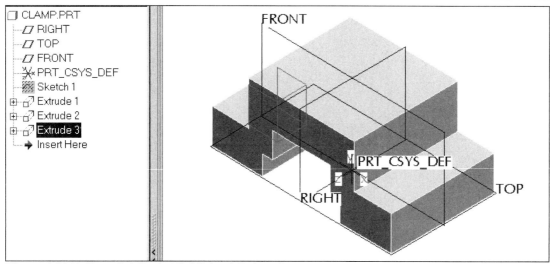

Figure 4.11(a) Extruded Cut is Highlighted (Selected)

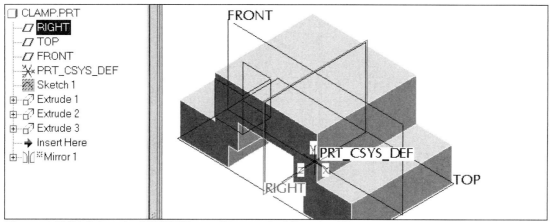

Figure 4.11(b) Select the RIGHT Datum Plane as the Mirroring Plane

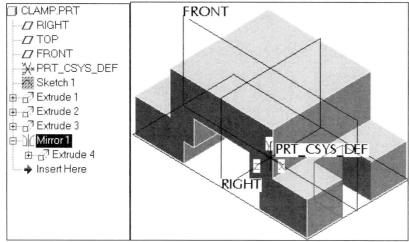

Figure 4.11(c) Mirror Cut

Figure 4.11(d) Completed Part

The lesson is now complete. If you wish to model a project without instructions, a complete set of projects and illustrations are available at ***www.cad-resources.com*** ⇒ ***Downloads***.

Lesson 5 Datums, Layers, and Sections

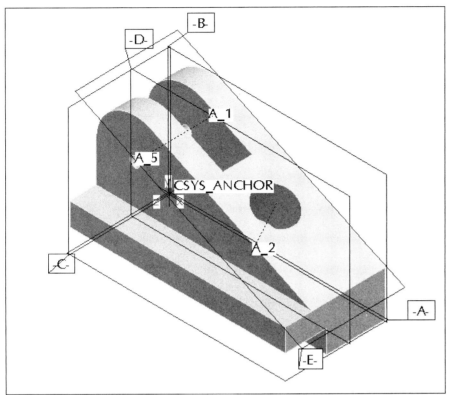

Figure 5.1 Anchor Model with Datum Features

OBJECTIVES

- Create **Datums** to locate features
- Set datum planes for **geometric tolerancing**
- Learn how to change the **Color** and **Shading** of models
- Use **Layers** to organize features
- Use datum planes to establish **Model Sectioning**
- Add a simple **Relation** to control a feature
- Use **Info** command to extract **Relations** information

DATUMS AND LAYERS

Datums and **layers** are two of the most useful mechanisms for creating and organizing your design (Fig. 5.1). Features such as *datum planes* and *datum axes* are essential for the creation of all parts, assemblies, and drawings using Pro/E.

Layers are an essential tool for grouping items and performing operations on them, such as selecting, hide/unhide, plotting, and suppressing. Any number of layers can be created. User-defined names are available, so layer names can be easily recognized.

Most companies have a layering scheme that serves as a *default standard* so that all projects follow the same naming conventions and objects/items are easily located by anyone with access. Layer information, such as display status, is stored with each individual part, assembly, or drawing.

Lesson 5 STEPS

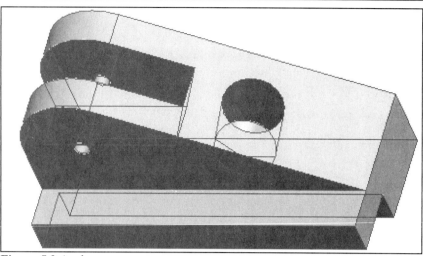

Figure 5.2 Anchor

Anchor

Though default datum planes have been sufficient in previous lessons, the Anchor (Fig. 5.2) incorporates the creation of user-defined datums and the assignment of datums to layers. The datum planes will be set as geometric tolerance features and put on a separate layer.

Launch **Pro/ENGINEER WILDFIRE 3.0** ⇒ **File** ⇒ **Set Working Directory** ⇒ navigate to your directory if necessary ⇒ **OK** ⇒ Create a new object ⇒ **Part** ⇒ Name **ANCHOR** ⇒ Use default template ⇒ **OK** ⇒ **Edit** ⇒ **Setup** ⇒ **Units** ⇒ Units Manager **Inch lbm Second** ⇒ **Close** ⇒ ⇒ **Material** ⇒ **steel.mtl** ⇒ ⇒ **OK** ⇒ **Done** ⇒ click on the coordinate system-- **PRT_CSYS_DEF** ⇒ **RMB** [Fig. 5.3(a)] ⇒ **Rename** ⇒ type **CSYS_ANCHOR** in the Model Tree [Fig. 5.3(b)] ⇒ **Enter**

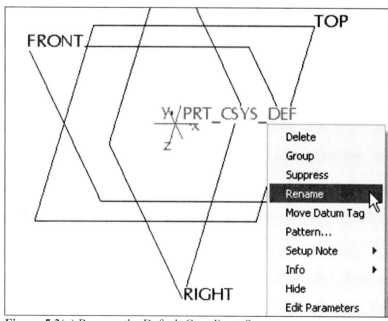

Figure 5.3(a) Rename the Default Coordinate System

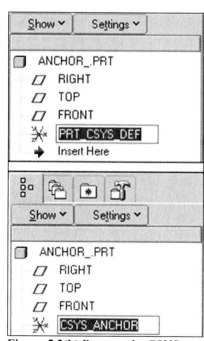

Figure 5.3(b) Rename the CSYS

It is considered good practice to rename the default coordinate system to something similar to the components name. If an assembly has 25 components, it will have 25 default coordinate systems. Renaming the coordinate system to the component name will make it easier to identify.

In the next set of steps, you will show the Layer Tree and see the default layering that automatically takes place as you model. The default datum planes and the default coordinate system are automatically placed on two layers each. For the datum planes; the *part default datum plane layer* and the *part all datum plane layer*.

The coordinate system will also be layered in a similar fashion. In Figure 5.3(d), the *part all datum layer* for the coordinate system displays the new name created for the coordinate system.

Click: **Show** in the Navigator ⇒ **Layer Tree** Layer Tree displays in place of the Model Tree [Figs. 5.3(c-d)] *(or click* ⬛ *)* ⇒ 💾 **Save** ⇒ **OK** ⇒ **Show** in the Navigator ⇒ **Model Tree**

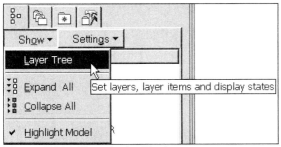

Figure 5.3(c) Show in the Navigator

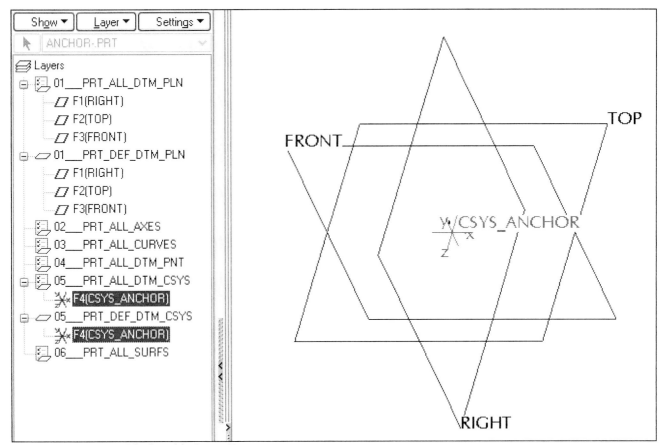

Figure 5.3(d) Layers Displayed in Layer Tree

As in previous lessons, the first protrusion will be sketched on **FRONT**. Use Figure 5.4 for the protrusion dimensions. Use only the **5.50, R1.00, 1.125,** and **25°** dimensions.

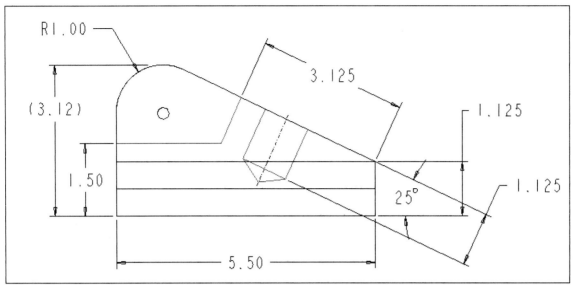

Figure 5.4 Anchor Drawing Front View

Create the first protrusion, click: **Extrude Tool** ⇒ **Placement** ⇒ **Define** [Fig. 5.5(a)] ⇒ Sketch Plane--- Plane: select **FRONT** datum from the model or Model Tree ⇒ **Sketch** from Sketch dialog box ⇒ **Tools** ⇒ **Environment** ⇒ Snap to Grid ⇒ **OK** ⇒ **Toggle the grid** on ⇒ **RMB** anywhere in the graphics window ⇒ **Line** [Fig. 5.5(b)]

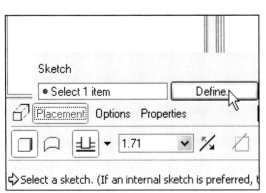

Figure 5.5(a) Dashboard

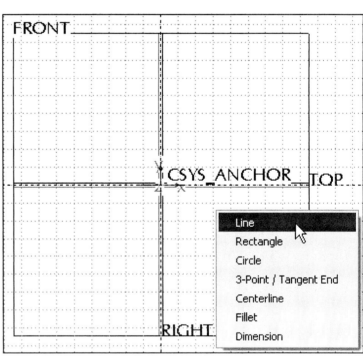

Figure 5.5(b) RMB ⇒ Line

144

Sketch the four lines ⇒ **MMB** [Fig. 5.5(c)] ⇒ [icon] **Create a circular fillet between two entities** ⇒ pick the left vertical line and the angled line to create a fillet [Fig. 5.5(d)] ⇒ [icon] ⇒ pick on the horizontal line and then the angled line and use **MMB** to position the dimension value [Fig. 5.5(e)] ⇒ **Tools** from menu bar ⇒ [icon] **Environment** ⇒ [Snap to Grid] ⇒ **OK** ⇒ [icon] **Toggle the grid** off

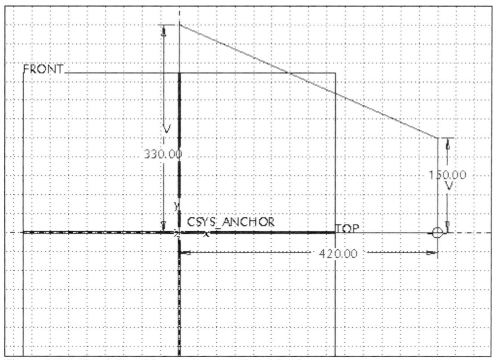

Figure 5.5(c) Sketch Four Lines to Create a Closed Section

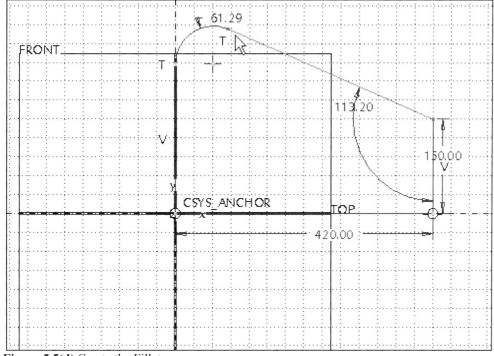

Figure 5.5(d) Create the Fillet

Click: ▶ **Select items** ⇒ reposition the default dimensions [Fig. 5.5(e)] ⇒ 📝 **Modify the values of dimensions** [Fig. 5.5(f)] ⇒ ✓ **Regenerate the section and close the dialog** ⇒ ✓ **Continue** ⇒ 📋 ⇒ **Standard Orientation** ⇒ in the dashboard type: **2.5625** [Fig. 5.5(g)] ⇒ **Enter** ⇒ **Coordinate systems** off [Fig. 5.5(h)] ⇒ ☑👓 [Fig. 5.5(i)] ⇒ ▶ ⇒ ✓ ⇒ 💾 ⇒ **MMB** ⇒ **Ctrl+D**

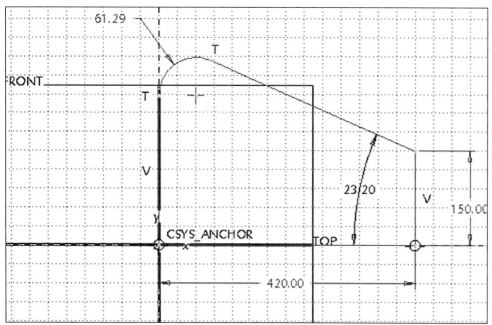

Figure 5.5(e) Create the Angle Dimension and Reposition the other Dimensions (your dimension values will be different)

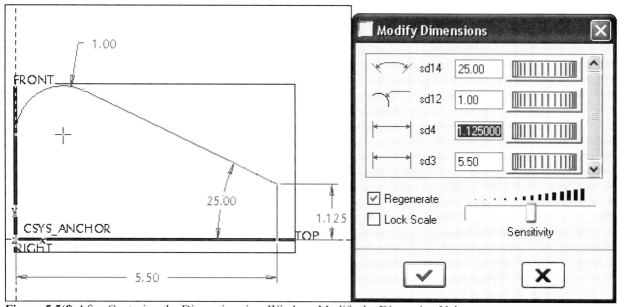

Figure 5.5(f) After Capturing the Dimensions in a Window, Modify the Dimension Values

Figure 5.5(g) Modify the Depth to **2.5625** in the Dashboard

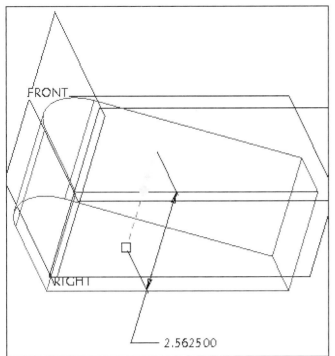
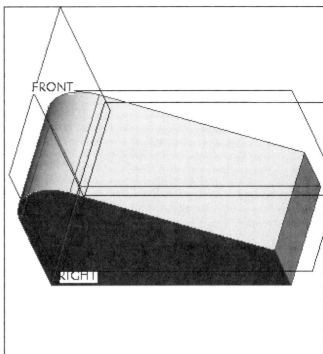

Figure 5.5(h) Feature Depth   **Figure 5.5(i)** Preview (with Shading on)

Colors

Colors are used to define material and light properties. Use the **Color Editor** to create and modify colors and to customize the display of geometric items. You can access the Color Editor by clicking a color swatch to set the **Appearance Color** or the **Highlight Color**. Avoid the colors that are used as Pro/E defaults. Because feature and entity highlighting is defaulted to *red*, colors similar to *red* should be avoided. Datum planes have *tan* and *gray* sides; therefore, they should be avoided. It is up to you to select colors that work with the type of project you are modeling. If the parts will be used in an assembly, each component should have a unique color scheme.

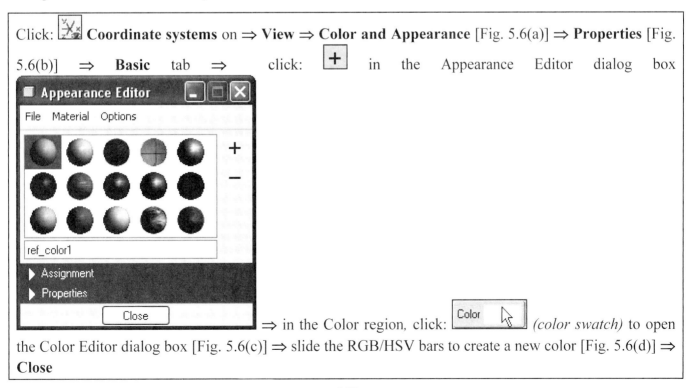

Click: **Coordinate systems** on ⇒ **View** ⇒ **Color and Appearance** [Fig. 5.6(a)] ⇒ **Properties** [Fig. 5.6(b)] ⇒ **Basic** tab ⇒ click: **+** in the Appearance Editor dialog box ⇒ in the Color region, click: **Color** *(color swatch)* to open the Color Editor dialog box [Fig. 5.6(c)] ⇒ slide the RGB/HSV bars to create a new color [Fig. 5.6(d)] ⇒ **Close**

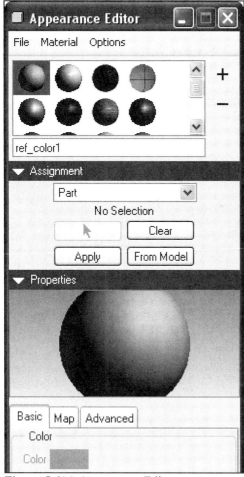

Figure 5.6(a) Appearance Editor

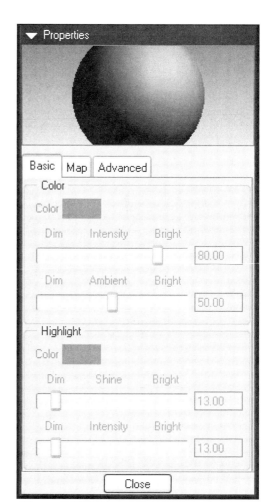

Figure 5.6(b) Properties

Repeat to make a few more colors, adding each to the palette [Fig. 5.6(h)] ⇒ also, add some Highlights to the colors using the same basic method [Fig. 5.6(e)] ⇒ also, try creating colors with the **Color Wheel** or the **Blending Palette** [Figs. 5.6(f-g)] ⇒ **File** ⇒ **Save** ⇒ type a name for your color file *(you can open and use this file for other projects instead of recreating new colors)* ⇒ **OK** [Fig. 5.6(h)] ⇒ **Apply** ⇒ **Close** from the Appearance Editor dialog box

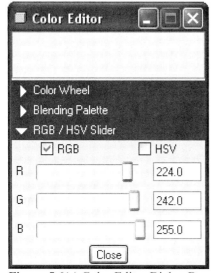

Figure 5.6(c) Color Editor Dialog Box

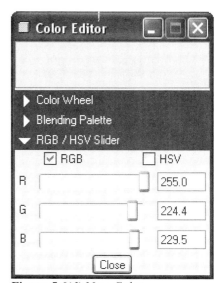

Figure 5.6(d) New Color

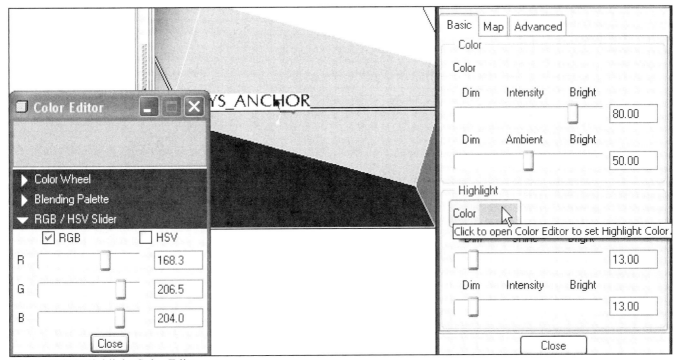

Figure 5.6(e) Highlight Color Editor

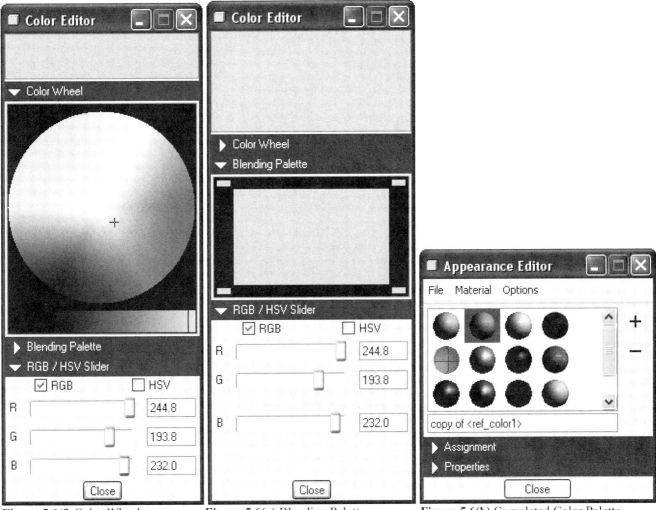

Figure 5.6(f) Color Wheel   **Figure 5.6(g)** Blending Palette   **Figure 5.6(h)** Completed Color Palette

The next two features will remove material. For both of the cuts, use the RIGHT datum plane as the sketching plane, and TOP datum plane as the reference plane. Each cut requires just two lines, and two dimensions. Use the first cut's sketching/placement plane and reference/orientation (**Use Previous**) for the second cut. Create the cuts separately, as open sections. Figure 5.7 shows the cuts dimensions.

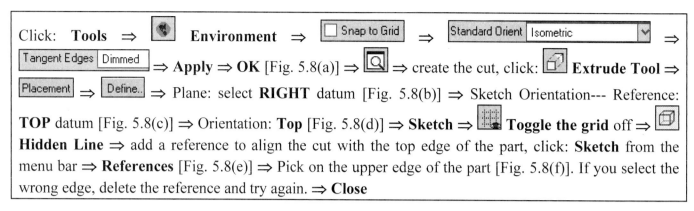

Click: **Tools** ⇒ [icon] **Environment** ⇒ ☐ Snap to Grid ⇒ Standard Orient |Isometric| ⇒ |Tangent Edges| |Dimmed| ⇒ **Apply** ⇒ **OK** [Fig. 5.8(a)] ⇒ [icon] ⇒ create the cut, click: [icon] **Extrude Tool** ⇒ |Placement| ⇒ |Define..| ⇒ Plane: select **RIGHT** datum [Fig. 5.8(b)] ⇒ Sketch Orientation--- Reference: **TOP** datum [Fig. 5.8(c)] ⇒ Orientation: **Top** [Fig. 5.8(d)] ⇒ **Sketch** ⇒ [icon] **Toggle the grid** off ⇒ [icon] **Hidden Line** ⇒ add a reference to align the cut with the top edge of the part, click: **Sketch** from the menu bar ⇒ **References** [Fig. 5.8(e)] ⇒ Pick on the upper edge of the part [Fig. 5.8(f)]. If you select the wrong edge, delete the reference and try again. ⇒ **Close**

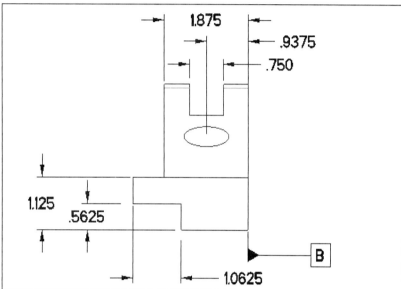

Figure 5.7 Side View showing Dimensions for the Cuts

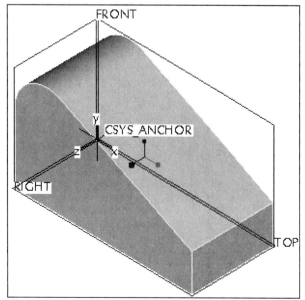

Figure 5.8(a) Standard Orientation Isometric

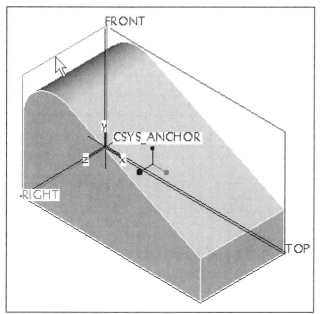

Figure 5.8(b) Select the Sketch Plane (RIGHT)

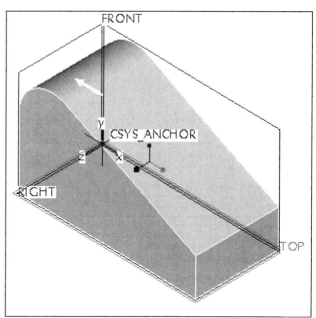

Figure 5.8(c) Select the Orientation Plane (TOP)

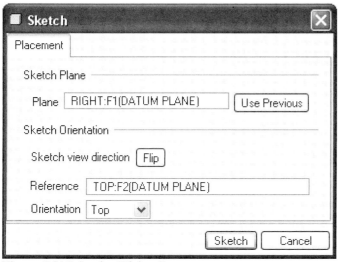

Figure 5.8(d) Section Dialog Box

Figure 5.8(e) References Dialog Box

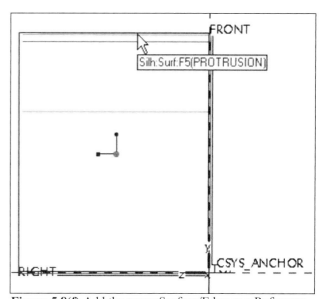

Figure 5.8(f) Add the upper Surface/Edge as a Reference

Click: ![line icon] **Create 2 point lines** ⇒ sketch the lines- start from the top with a vertical line [Fig. 5.8(g)] ⇒ **MMB** to end the line sequence ⇒ ![arrow icon] ⇒ ![dim icon] **Toggle display of dimensions** on [Fig. 5.8(h)] ⇒ ![redraw icon] **Redraw the current view** ⇒ reposition the default dimensions [Fig. 5.8(i)]

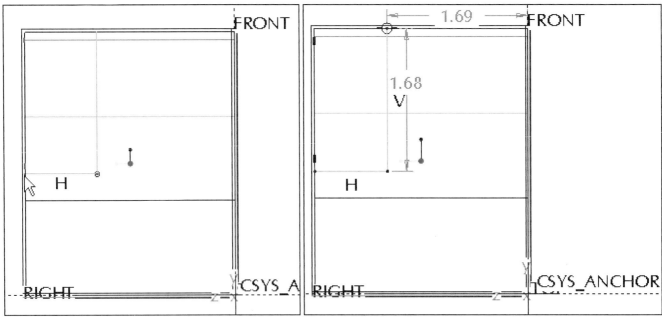

Figure 5.8(g) Sketch a Vertical and a Horizontal Line   **Figure 5.8(h)** Default Dimensions (Weak) Displayed

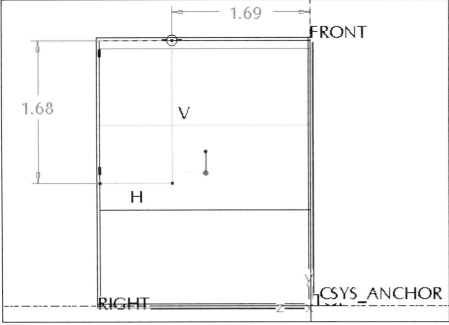

Figure 5.8(i) Repositioned Default Dimensions

Click: ![arrow icon] ⇒ pick on the horizontal dimension ⇒ **RMB** ⇒ **Strong** [Fig. 5.8(j)] ⇒ **RMB** ⇒ **Dimension** ⇒ add the new vertical dimension [Fig. 5.8(k)] *(the weak vertical dimension will disappear)* ⇒ **MMB** ⇒ double-click on each dimension and change to the design values [Figs. 5.8(l-m)] ⇒ ![check] ⇒ ![AB icon] ⇒ **Standard Orientation** ⇒ ![shading icon] **Shading** [Fig. 5.8(n)]

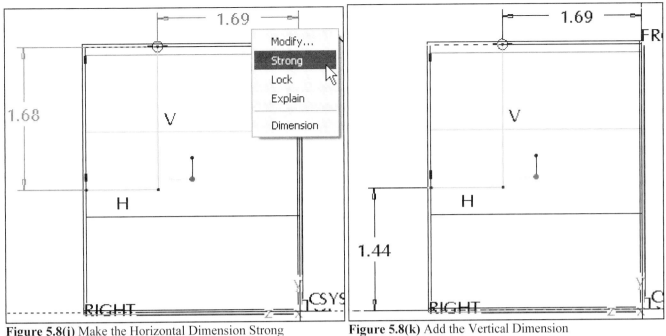

Figure 5.8(j) Make the Horizontal Dimension Strong

Figure 5.8(k) Add the Vertical Dimension

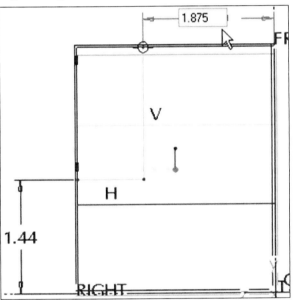

Figure 5.8(l) Modify Dimension to **1.875**

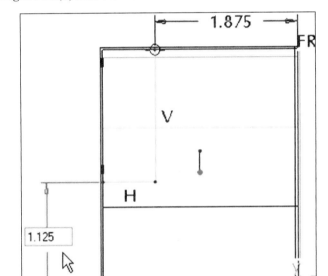

Figure 5.8(m) Modify Dimension to **1.125**

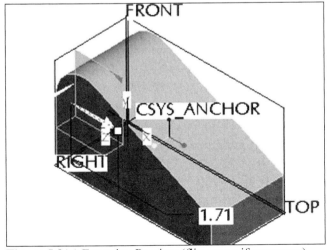

Figure 5.8(n) Extrusion Preview (flip arrow if necessary)

From the dashboard click: ⬜ **Remove Material** ⇒ **Options** tab ⇒ ▤ **Through All** [Fig. 5.8(o)] ⇒ ☑👓 [Fig. 5.8(p)] ⇒ ▶ ⇒ ✓ ⇒ 💾 ⇒ **MMB**

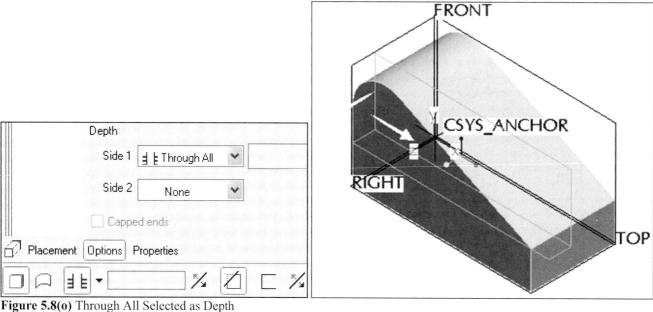

Figure 5.8(o) Through All Selected as Depth

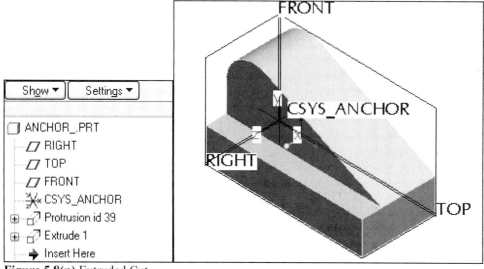

Figure 5.8(p) Extruded Cut

Complete the second cut, click: ⬜ ⇒ **RMB** ⇒ **Define Internal Sketch** ⇒ **Use Previous** ⇒ ⬜ **No Hidden** ⇒ **Sketch** ⇒ **References** ⇒ add the left vertical edge/surface as a reference [Fig. 5.9(a)] ⇒ **Close** ⇒ **RMB** ⇒ **Line** ⇒ sketch the vertical and horizontal lines [Fig. 5.9(b)] ⇒ **MMB** ⇒ **MMB** ⇒ ⬜ **Toggle display of dimensions** on ⇒ reposition the default dimensions [Fig. 5.9(c)] ⇒ double-click on the vertical dimension and type **.5625** [Fig. 5.9(d)] ⇒ **Enter** ⇒ double-click on the horizontal dimension and type **1.0625** [Fig. 5.9(e)] ⇒ **Enter** ⇒ ✓ ⇒ ⬜ ⇒ **Standard Orientation** ⇒ ⬜ ⇒ ⬜ **Remove Material** ⇒ **Options** tab ⇒ ▤ **Through All** [Fig. 5.9(f)] ⇒ **Ctrl+MMB** to zoom in if necessary ⇒ ☑👓 [Fig. 5.9(g)] ⇒ ▶ ⇒ ✓ ⇒ 🔍 ⇒ 💾 ⇒ **MMB** [Fig. 5.9(h)] ⇒ **LMB**

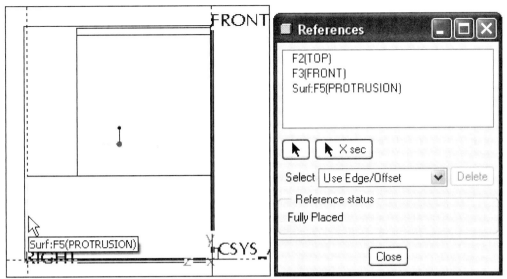

Figure 5.9(a) Add the Left Vertical Edge/Surface as a Reference

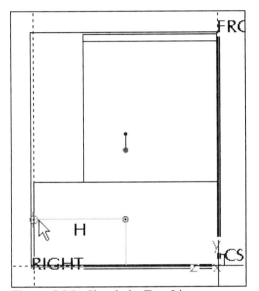

Figure 5.9(b) Sketch the Two Lines

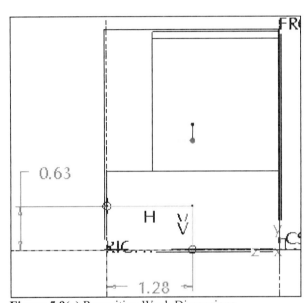

Figure 5.9(c) Reposition Weak Dimensions

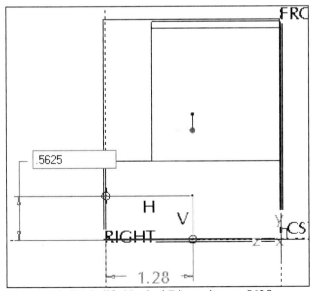

Figure 5.9(d) Modify Vertical Dimension to **.5625**

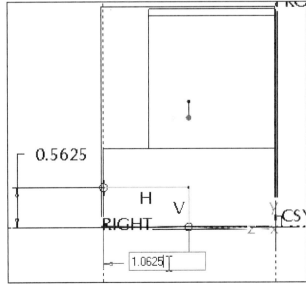

Figure 5.9(e) Modify Horizontal Dimension to **1.0625**

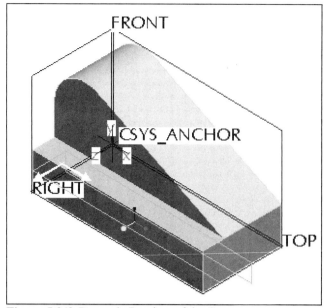

Figure 5.9(f) Dynamic Preview of the Cut

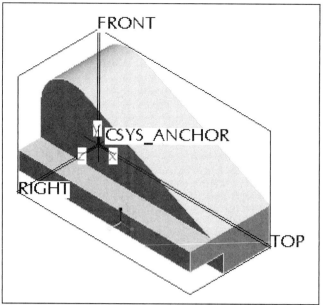

Figure 5.9(g) Verify Mode to View Attached Geometry

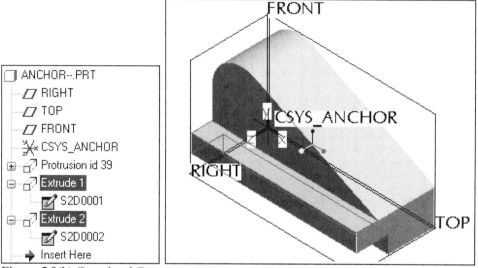

Figure 5.9(h) Completed Cuts

For the next feature, you will need to create a new datum plane on which to sketch. New part datum planes are by default numbered sequentially, DTM1, DTM2, and so on. Also, a datum axis (A-1) will be created through the curved top of the part and used later for the axial location of a small hole. Holes and circular features automatically have axes when they are created. Features created with arcs, fillets; and so on need to have axes added, unless set in the *configuration file* as the default.

Click: **Coordinate systems** off ⇒ **Datum Plane Tool** ⇒ References: pick datum **FRONT** as the plane to offset from [Fig. 5.10(a)] ⇒ in the DATUM PLANE dialog box, Offset: Translation, type the distance **1.875/2** *(1.875/2 = .9375)* ⇒ **Enter** ⇒ **OK** ⇒ **LMB** to deselect ⇒ **Datum Axis Tool** ⇒ References: pick the curved surface [Fig. 5.10(b)] ⇒ **OK** ⇒ **LMB** to deselect ⇒ **Datum Plane Tool** ⇒ References: pick the angled surface [Fig. 5.10(c)] ⇒ click **Offset** [Fig. 5.10(d)] ⇒ click **Offset** again to see options ⇒ click **Through** [Figs. 5.10(e-f)] ⇒ **OK** Fig. 5.10(g) ⇒ ⇒ **OK** ⇒ **LMB** to deselect ⇒ **Coordinate systems** on

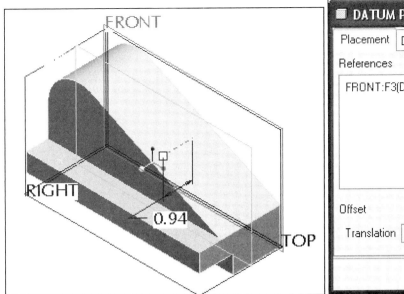

Figure 5.10(a) Offset Datum

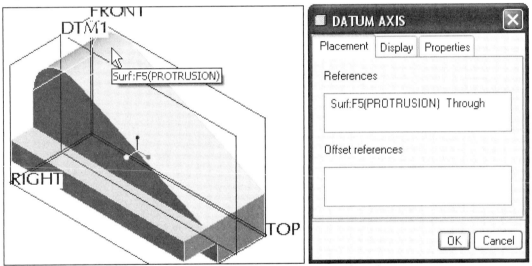

Figure 5.10(b) Datum Axis A_1 Through the Center of the Curved Surface

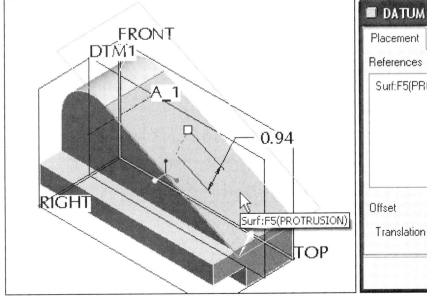

Figure 5.10(c) Pick on the Angled Surface

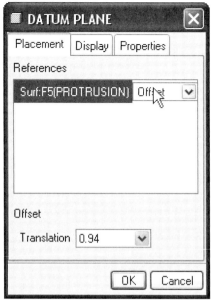
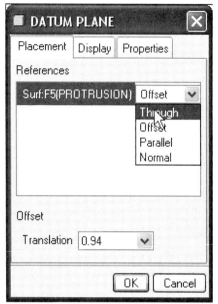

Figure 5.10(d) Click on Offset **Figure 5.10(e)** Click Through **Figure 5.10(f)** Through

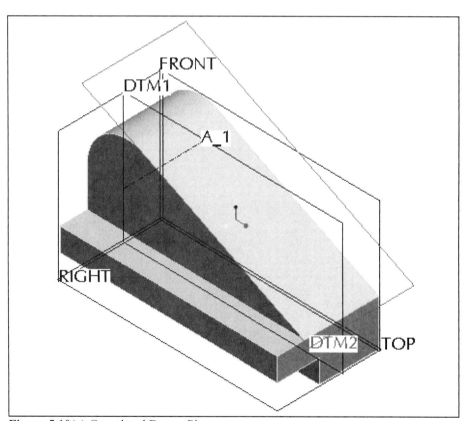

Figure 5.10(g) Completed Datum Plane

Create a new layer and add the new datum planes and axis to it, click: **Show** in the Navigator ⇒ **Layer Tree** Layer Tree displays in place of the Model Tree [Fig. 5.11(a)] ⇒ click on and highlight **Layers** in the Layer Tree ⇒ **RMB** ⇒ **New Layer** [Fig. 5.11(b)]

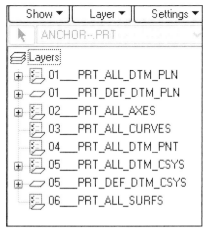

Figure 5.11(a) Layer Tree

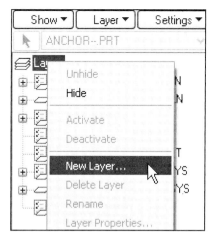

Figure 5.11(b) New Layer

In the Layer Properties dialog box [Fig. 5.11(c)]-Name: type **DATUM_FEATURES** *(do not press Enter)* as the name for new layer [Fig. 5.11(d)] ⇒ select the two new datum planes from the model ⇒ select the axis from the model, use **RMB** to toggle to the correct selection ⇒ select the axis with **LMB** [Fig. 5.11(e)] ⇒ **OK** [Fig. 5.11(f)] ⇒ expand layer to see items ![DATUM_FEATURES] [Fig. 5.11(g)] ⇒ **Show** in the Navigator ⇒ **Model Tree** ⇒ **Ctrl+R**

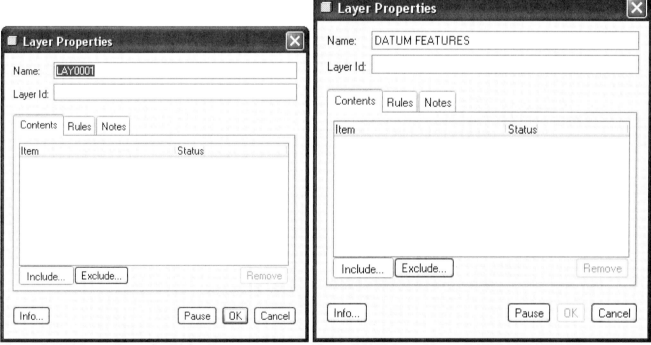

Figure 5.11(c) Layer Properties Dialog Box   **Figure 5.11(d)** Layer Name- DATUM_FEATURES

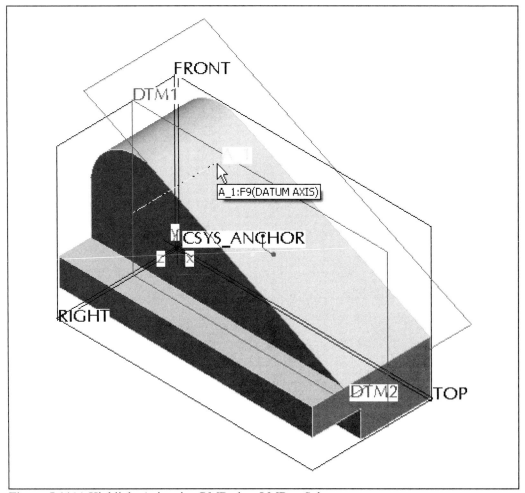

Figure 5.11(e) Highlight Axis using RMB, then LMB to Select

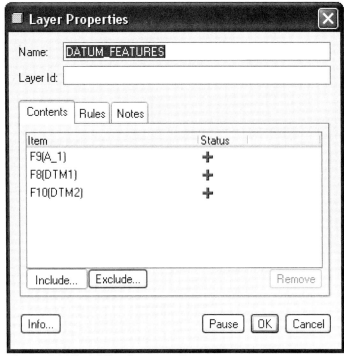

Figure 5.11(f) Adding Items to a Layer

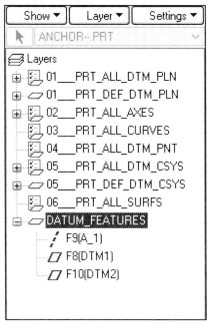

Figure 5.11(g) DATUM_FEATURES Layer

Geometric Tolerances

Before continuing with the modeling of the part, the datums used and created thus far will be "set" for geometric tolerancing. Geometric tolerances (GTOLs) provide a comprehensive method of specifying where on a part the critical surfaces are, how they relate to one another, and how the part must be inspected to determine if it is acceptable. They provide a method for controlling the location, form, profile, orientation, and run out of features. When you store a Pro/E GTOL in a solid model, it contains parametric references to the geometry or feature it controls—its reference entity—and parametric references to referenced datums and axes. In Assembly mode, you can create a GTOL in a subassembly or a part. A GTOL that you create in Part or Assembly mode automatically belongs to the part or assembly that occupies the window; however, it can refer only to set datums belonging to that model itself, or to components within it. It cannot refer to datums outside of its model in some encompassing assembly, unlike assembly created features. You can add GTOLs in Part or Drawing mode, but they are reflected in all other modes. Pro/E treats them as annotations, and they are always associated with the model. Unlike dimensional tolerances, though, GTOLs do not affect part geometry. Before you can reference a datum plane or axis in a GTOL, you must set it as a reference. Pro/E encloses its name using the set datum symbol. After you have set a datum, you can use it in the usual way to create features and assemble parts. You enter the set datum command by clicking Edit ⇒ Setup ⇒ Geom Tol ⇒ Set Datum, and then select the datum plane or axis. Pro/E encloses the datum name in a feature control frame. If needed, type a new name in the Name field of the Datum dialog box. Most datums will follow the alphabet, A, B, C, D, and so on. You can hide (not display) a reference datum plane by placing it on a layer and then hiding the layer *(or using View ⇒ Visibility ⇒ Hide)*.

Click: **Edit** ⇒ **Setup** ⇒ **Geom Tol** ⇒ **Set Datum** ⇒ pick **TOP** from the model [Fig. 5.12(a)] ⇒ double-click in the Name field- type **A** ⇒ **OK** ⇒ pick **FRONT** ⇒ Name- type **B** ⇒ **OK** ⇒ pick **RIGHT** ⇒ Name- type **C** ⇒ **OK** ⇒ pick **DTM1** ⇒ Name- type **D** ⇒ **OK** ⇒ pick **DTM2** ⇒ Name- type **E** ⇒ **OK** ⇒ **Done/Return** ⇒ **Done** [Fig. 5.12(b)] ⇒ 🔍 ⇒ 💾 ⇒ **MMB** ⇒ **LMB** to deselect

---

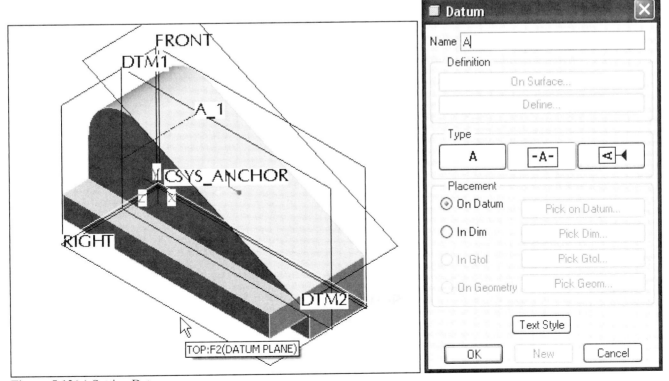

Figure 5.12(a) Setting Datums

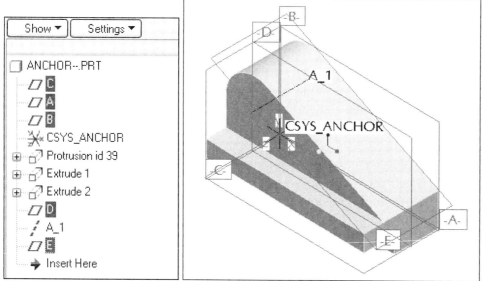

Figure 5.12(b) Set Datums

Create the slot on the top of the part by sketching on datum D and projecting the cut toward both sides. Use the Model Tree to select the appropriate datum planes.

Click: [icon] ⇒ **RMB** ⇒ **Define Internal Sketch** ⇒ Plane: select datum **D** [Fig. 5.13(a)] ⇒ Sketch Orientation--- Reference: datum **C** ⇒ Orientation: **Right** ⇒ **Sketch** ⇒ [icon] ⇒ **Standard Orientation** ⇒ add a reference to align the cut with the angled edge of the part, click: **Sketch** from the menu bar ⇒ **References** ⇒ pick on datum **E** ⇒ pick the small vertical surface on the right side of the part as the fourth reference [Fig. 5.13(b)] ⇒ **Close**

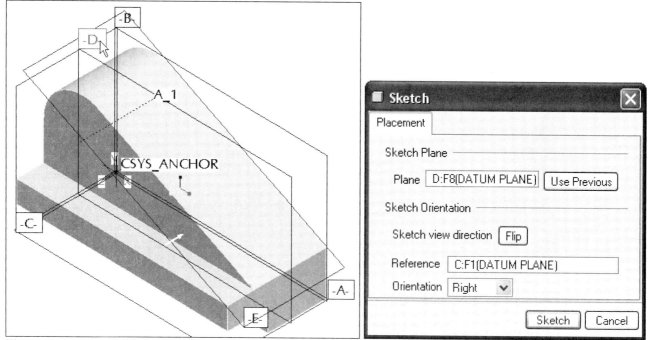

Figure 5.13(a) Sketch Plane and Sketch Orientation Reference

162

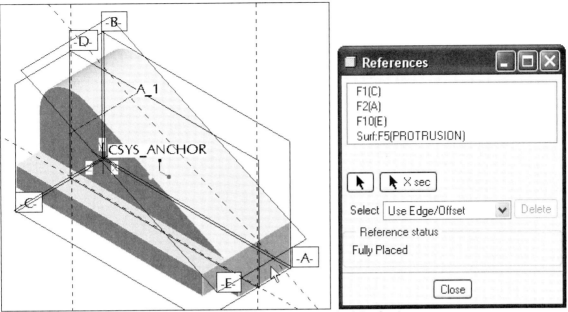

Figure 5.13(b) New References (Datum E and Small Right Vertical Surface)

Click: 🟦 **Hidden Line** ⇒ 🗗 **Orient the sketching plane parallel to the screen** ⇒ ✳ **Create points** pick at the corner of the angled datum **E** and the right vertical reference [Fig. 5.13(c)] ⇒ **MMB**

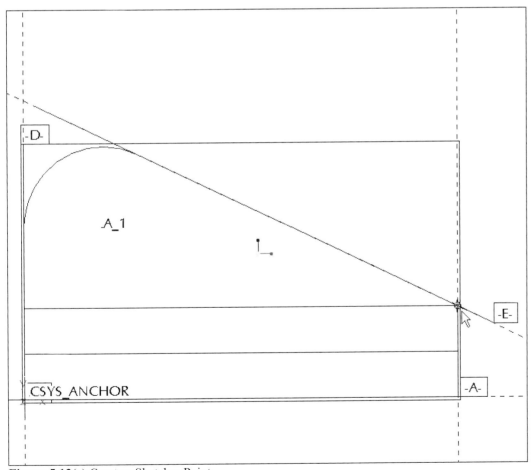

Figure 5.13(c) Create a Sketcher Point

Click: **Create 2 point lines** ⇒ sketch the first line from and perpendicular to the angled edge [Fig. 5.13(d)] ⇒ continue to sketch the second line horizontal [Fig. 5.13(e)] ⇒ **MMB** [Fig. 5.13(f)] ⇒ **MMB**

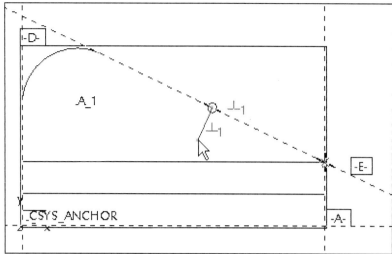

Figure 5.13(d) Create the first Line Perpendicular to Datum E

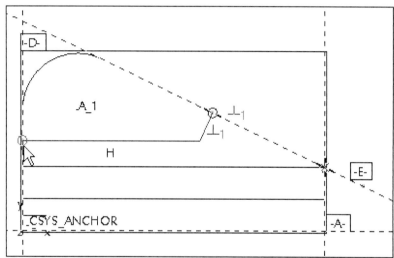

Figure 5.13(e) Create the Second Line Horizontal

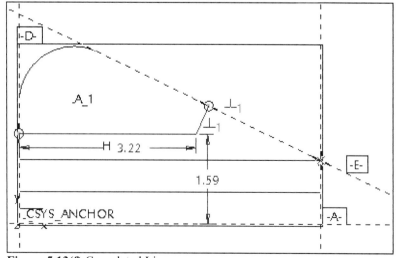

Figure 5.13(f) Completed Lines

Click: |↔| ⇒ add a dimension by picking the first sketched line [Fig. 5.13(g)] and then the point *(do not pick endpoint to point)* ⇒ **MMB** to place the dimension [Fig. 5.13(h)] ⇒ add a vertical dimension [Fig. 5.13(i)] ⇒ **Ctrl+R**

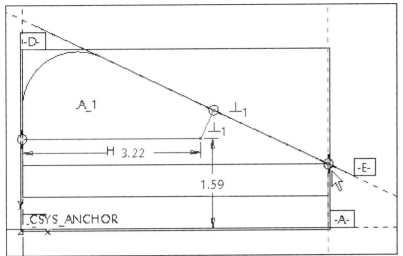

Figure 5.13(g) Create Dimension from First Line to Point (not point to point!)

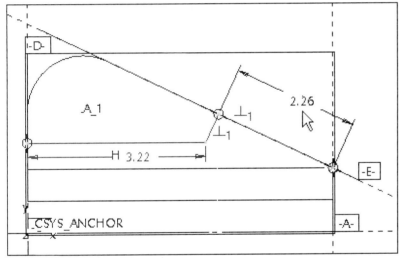

Figure 5.13(h) Place Dimension

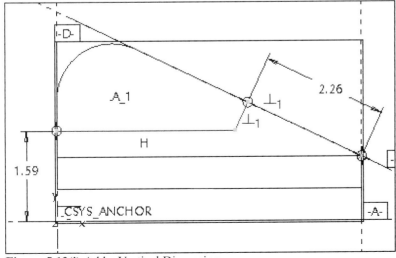

Figure 5.13(i) Add a Vertical Dimension

Click: [icon] **Modify the values of dimensions** ⇒ select both dimensions, modify the values, angled dimension is **3.125** and vertical dimension is **1.50** [Fig. 5.13(j)] ⇒ [✓] ⇒ [✓] ⇒ [icon] ⇒ **Standard Orientation** ⇒ [icon] **Shading** ⇒ [icon] **Remove Material** ⇒ depth value **.75** ⇒ **Enter** ⇒ **Options** tab ⇒ [icon] **Symmetric** [Fig. 5.13(k)] ⇒ *if necessary* [icon] **Change depth direction** ⇒ [icon] ⇒ [▶] ⇒ [✓] ⇒ [icon] ⇒ **MMB** [Fig. 5.13(l)] ⇒ **LMB** to deselect ⇒ [icon] **Refit**

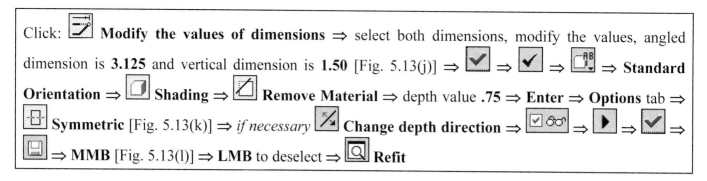

Figure 5.13(j) Modify Dimensions to **3.125** and **1.50**

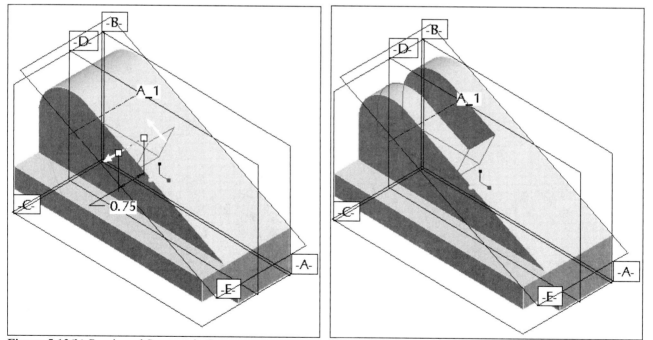

Figure 5.13(k) Previewed Symmetric Cut   **Figure 5.13(l)** Completed Cut

The hole drilled in the angled surface appears to be aligned with datum **D**. Upon closer inspection (Fig. 5.7), it can be seen that the hole is at a different distance *(.875 from B)* and is not in line with the slot and datum plane *(D was offset .9375 from B)*. Create the feature using a sketched hole. The drill tip (**118** degrees) at the bottom of the hole needs to be modeled so the hole is created as a sketched hole.

Click: **Hole Tool** ⇒ pick on the angled face and a hole will display with handles ⇒ drag one handle to datum **B** and the other to the edge between the angled surface and the right vertical face ⇒ double-click on the dimension from the hole's center to datum **B** and modify the value to **.875** ⇒ **Enter** ⇒ double-click on the dimension from the hole's center to the edge and modify the value to **2.0625** ⇒ **Enter** ⇒ Simple ⇒ Sketched ⇒ **Placement** tab [Figs. 5.14(a-b)] ⇒ **Activates Sketcher to create a section** *(a new window will open)* ⇒ **Tools** ⇒ **Environment** ⇒ Snap to Grid ⇒ **OK** ⇒ **Toggle the grid** on ⇒ **RMB** ⇒ **Centerline** ⇒ sketch a vertical centerline ⇒ sketch the four lines required to describe half of the hole's shape *(closed section)* [Figs. 5.14(c-d)] ⇒ **Toggle display of dimensions** on ⇒ **MMB** ⇒ **RMB** ⇒ **Dimension** ⇒ Create a diameter dimension by picking the centerline with the **LMB**, then the edge to be dimensioned with the **LMB**, and then the centerline a second time with the **LMB**. Place the dimension by picking a position with the **MMB**. ⇒ add the angle and the depth dimension [Fig. 5.14(e)] ⇒ add centerlines, point, and symmetric constraint [Fig. 5.14(f)] ⇒ add the full angle dimension ⇒ delete the half-angle dimension ⇒ **Modify** to the design values [Fig. 5.14(g)] ⇒ ⇒ ⇒ ⇒ **Standard Orientation** ⇒ [Fig. 5.14(h)] ⇒ ⇒ **MMB** ⇒ **LMB** to deselect

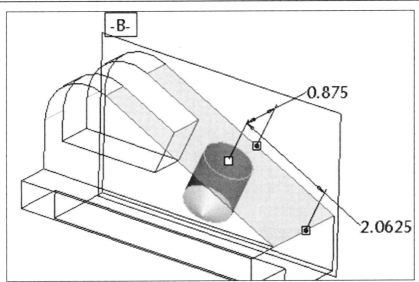

Figure 5.14(a) Hole Placement (view display has been altered- Hidden Line ⇒ MMB held down)

Figure 5.14(b) Hole Placement

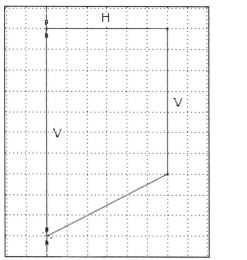

Figure 5.14(c) Sketch One Half of Hole Geometry

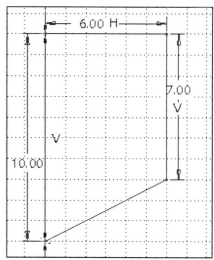

Figure 5.14(d) Default Dimensions

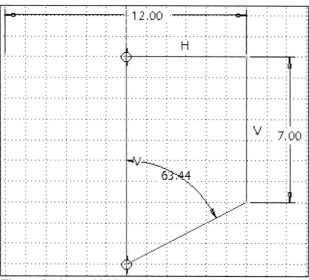

Figure 5.14(e) Add Dimensions

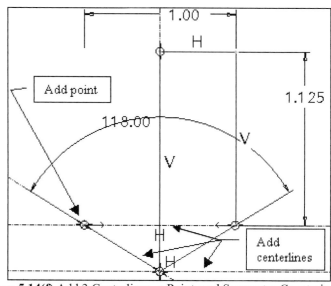

Figure 5.14(f) Add 3 Centerlines, a Point, and Symmetry Constraint

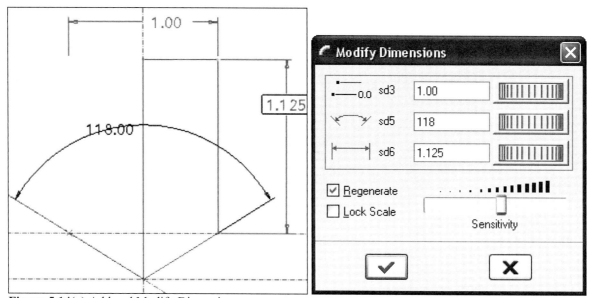

Figure 5.14(g) Add and Modify Dimensions

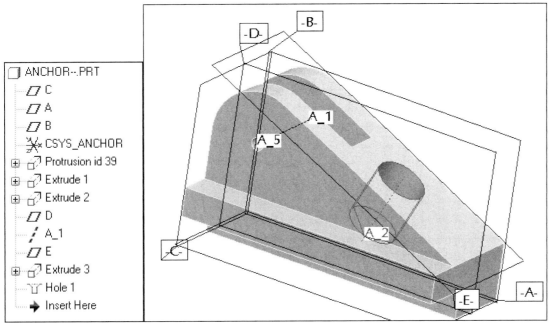

Figure 5.14(h) Completed Sketched Hole

The last feature to create is a ⌀**.250** hole. Click: **Hole Tool** ⇒ pick **A_1** from the model or the Model Tree ⇒ **RMB** ⇒ **Secondary References Collector** ⇒ pick datum **D** from the model or the Model Tree ⇒ **Shape** tab ⇒ diameter **.250** ⇒ **Through All** (both sides) [Fig. 5.15(a)] ⇒ ⇒ [Fig. 5.15(b)] ⇒ ⇒ **MMB** ⇒ **LMB** to deselect

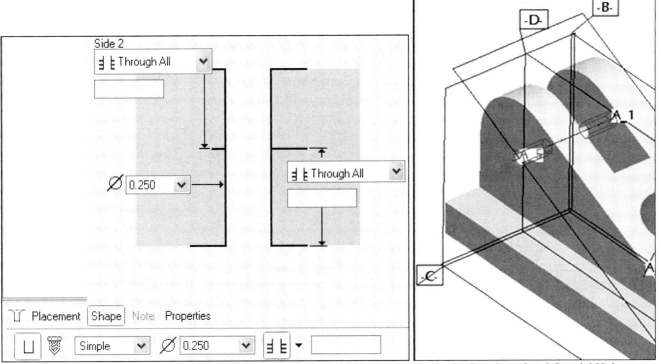

Figure 5.15(a) Shape

Figure 5.15(b) Completed Coaxial Hole

Suppressing and Resuming Features

Next, you will create a new layer and add the two holes to it. The holes will then be selected in the Layer Tree and suppressed. Suppressed features are temporarily removed from the model along with their children (if any). In the example, you will notice that the holes and one axis (A_4) will be suppressed. Since axis A_1 is the parent of the small hole, it will not be suppressed. Suppressing features is like removing them from regeneration temporarily. However, you can "unsuppress" (resume) suppressed features at any time.

You can suppress features on a part to simplify the part model and decrease regeneration time. For example, while you work on one end of a shaft, it may be desirable to suppress features on the other end of the shaft. Similarly, while working on a complex assembly, you can suppress some of the features and components for which the detail is not essential to the current assembly process. Suppress features to do the following:

- Concentrate on the current working area by suppressing other areas
- Speed up a modification process because there is less to update
- Speed up the display process because there is less to display
- Temporarily remove features to try different design iterations

Unlike other features, the base feature cannot be suppressed. If you are not satisfied with your base feature, you can redefine the section of the feature, or start another part.

Click: **Set Layers** (Layer Tree displays in place of the Model Tree) ⇒ click on and highlight **Layers** in the Layer Tree ⇒ **RMB** ⇒ **New Layer** ⇒ Name: type **HOLES** as the name for the new layer *(do not press Enter)* ⇒ Selection Filter (bottom right-hand side below graphics window) click: **Feature** ⇒ select the two holes from the model [Fig. 5.16(a)] ⇒ **OK** ⇒ expand the **HOLES** layer ⇒ press and hold down the **Ctrl** key and pick on the two holes in the **HOLES** layer [Fig. 5.16(b)] ⇒ **Edit** from the menu bar ⇒ **Suppress** ⇒ **Suppress** ⇒ **OK** from the Suppress dialog box ⇒ **LMB** to deselect *(look at the Layer Tree)* [Fig. 5.16(c)] ⇒ **Edit** from the menu bar ⇒ **Resume** ⇒ **Resume All** ⇒ **Show** ⇒ **Model Tree** ⇒ ⇒ **OK** ⇒ **LMB** to deselect

---

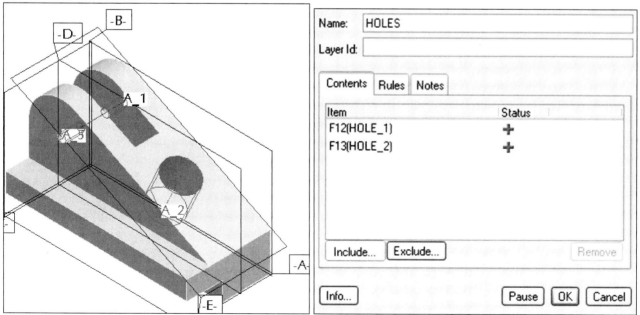

Figure 5.16(a) Two Holes on Layer HOLES

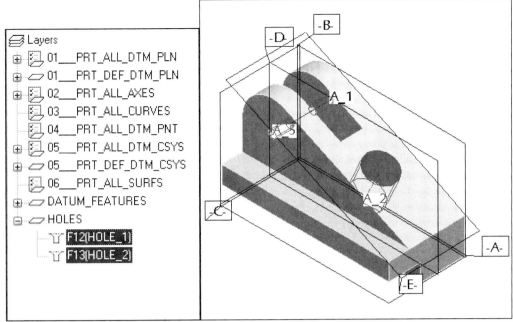

Figure 5.16(b) HOLES Layer

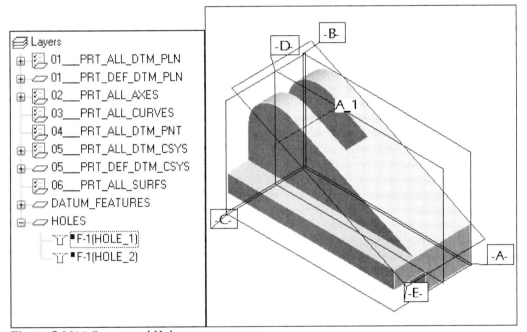

Figure 5.16(c) Suppressed Holes

Cross Sections

There are two types of cross sections: **planar** and **offset**. Planar cross sections can be crosshatched or filled, while offset cross sections can be crosshatched but not filled. You will be creating a planar cross section. Pro/E can create standard planar cross sections of models (parts or assemblies), offset cross sections of models (parts or assemblies), planar cross sections of datum surfaces or quilts (Part mode only) and planar cross sections that automatically intersect all quilts and all geometry in the current model.

Cross section cut planes do not intersect cosmetic features in a model.

Click: 🗔 **Start the view manager** from Top Toolchest ⇒ View Manager dialog box displays [Fig. 5.17(a)] ⇒ **Xsec** tab [Fig. 5.17(b)] ⇒ **New** [Fig. 5.17(c)] ⇒ type name **A** [Fig. 5.17(d)] ⇒ **Enter** ⇒ **Planar** [Fig. 5.17(e)] ⇒ **Single** [Fig. 5.17(f)] ⇒ **MMB** ⇒ **Plane** [Fig. 5.17(g)] ⇒ pick datum **D** ⇒ **Display** ⇒ **Xhatching** ⇒ **Edit** ⇒ **Redefine** [Fig. 5.17(h)] ⇒ **Hatching** ⇒ **Fill** [Fig. 5.17(i)] ⇒ **Hatch** ⇒ **Spacing** ⇒ **Half** [Fig. 5.17(j)] ⇒ **MMB** ⇒ **MMB** ⇒ **Display** ⇒ **Normal** ⇒ **Close** ⇒ 💾 ⇒ **MMB**

Figure 5.17(a) View Manager

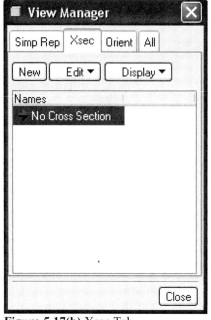
Figure 5.17(b) Xsec Tab

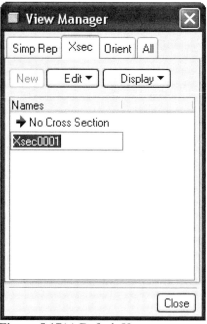
Figure 5.17(c) Default Xsec

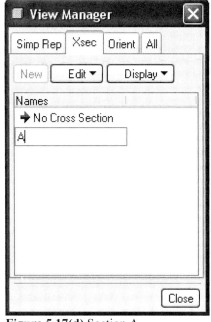
Figure 5.17(d) Section A

Figure 5.17(e) Planar

Figure 5.17(f) Done

Figure 5.17(g) Plane

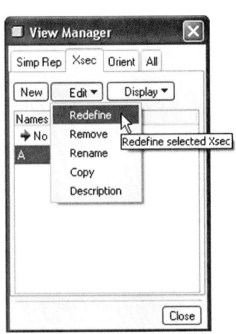
Figure 5.17(h) Redefine Xsec

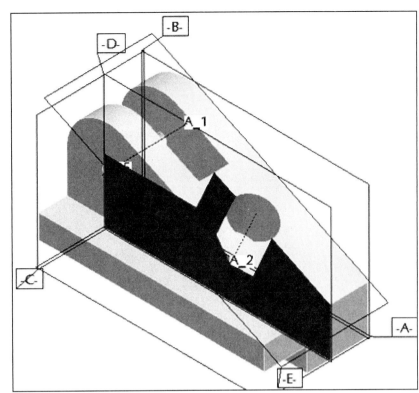
Figure 5.17(i) Model X-Section Fill

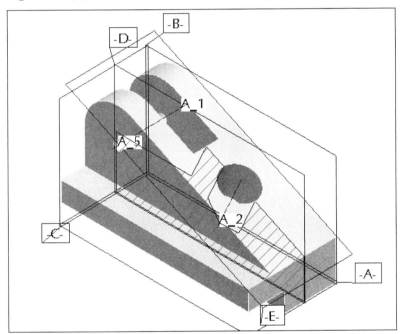
Figure 5.17(j) Model X-Section Hatch Half Spacing

The section passes through datum D. The slot is the *child* of datum D. If datum D moves, so will the slot and the small hole (and X-Section A). In order to ensure that the datum D stays centered on the upper portion of the part, you will need to create a relation to control the location of datum D. Relations will be covered in-depth in a later lesson. The first cut used a dimension from datum B for location (**1.875**). The relation should state that the distance from datum B to datum D will be one-half the value of the distance from datum B to the first cut. If the thickness of the upper portion of the part (**1.875**) changes datum D will remain centered as will the slot and the X-Section. To start, you must first find out the feature dimension symbols (**d#**) required for the relation. (Almost everyone will have a different d# than the ones displayed here)

Select the first cut from the model or the Model Tree ⇒ **RMB** ⇒ **Edit** [Figs. 5.18(a-b)] ⇒ **Info** from the menu bar ⇒ [Switch Dimensions] note the **d** symbol **d10** *(Note: your "d" values may be different)* [Fig. 5.18(c)] ⇒ click on datum **D** in the Model Tree ⇒ **RMB** ⇒ **Edit** ⇒ note the **d** symbol **d13** *(Note: your "d" values may be different)* [Fig. 5.18(d)]

(If you have the incorrect symbol, your relation will not work. Use your d symbols not the ones in the text!)

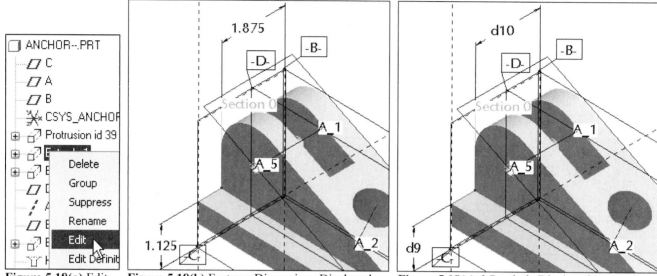

Figure 5.18(a) Edit   **Figure 5.18(b)** Features Dimensions Displayed   **Figure 5.18(c) d** Symbols Displayed

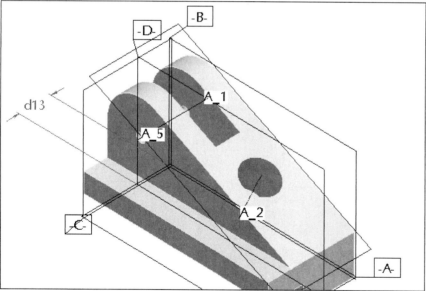

Figure 5.18(d) Using Edit to Display Datum D Dimension Symbol

Click: **Tools** from the menu bar ⇒ **Relations** Relations dialog box displays ⇒ **Local Parameters** to see the parameters of the part ⇒ type **d13=d10/2** in the Relations field [Fig. 5.18(e)] *(your "d" values may be different)* ⇒ ☑ **Execute/Verify** ⇒ **OK** ⇒ **Ok** ⇒ **Info** ⇒ **Switch Dimensions** ⇒ 💾 ⇒ **MMB**

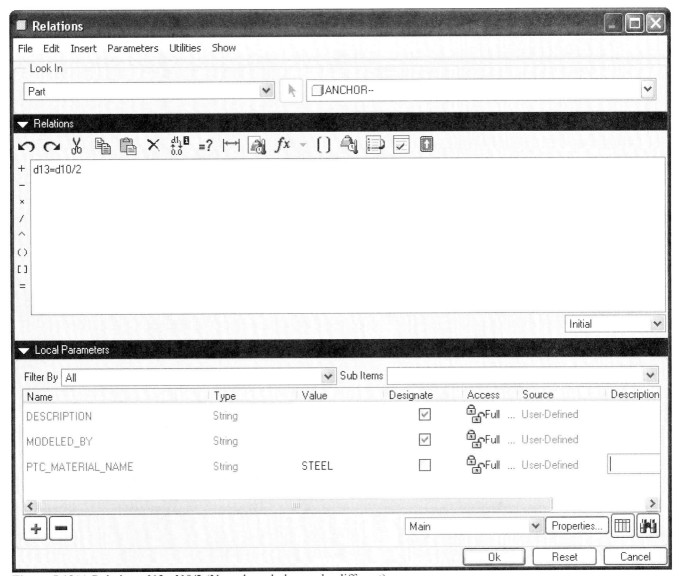

Figure 5.18(e) Relations **d13=d10/2** (Your d symbols may be different)

In the real world, you will seldom encounter a situation where the project is designed and modeled without a "design change" or **ECO** (Engineering Change Order). Therefore, let us assume that an ECO has been "issued" that states: *the location of the hole on the angled surface must be aligned with the center of the slot.*

Double-click on the large hole [Fig. 5.18(f)] ⇒ **Info** ⇒ **Switch Dimensions**- note the **d** symbol **d24** *(your "d" values may be different)* [Fig. 5.18(g)] (hole dimension from datum B) ⇒ **Tools** ⇒ **Relations** ⇒ below the first relation type **d24=d10/2** *(your "d" values may be different [Fig. 5.18(j)])* ⇒ **Ok** ⇒ **Info** ⇒ **Switch Dimensions** ⇒ double-click on the first cut on the model ⇒ double-click on **1.875** [Fig. 5.18(h)] and modify to **2.00** [Fig. 5.18(i)] ⇒ **Enter** ⇒ [icon] ⇒ [icon] ⇒ **TOP** [Fig. 5.18(k)]

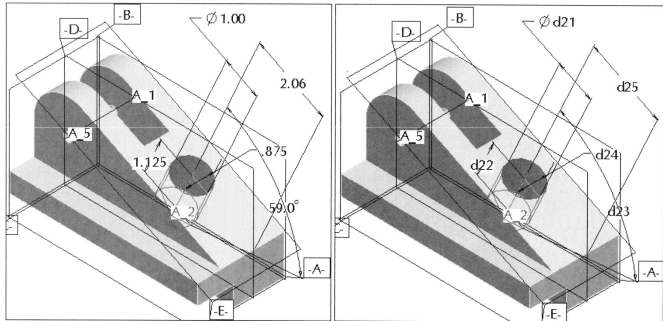

Figure 5.18(f) Hole Dimensions

Figure 5.18(g) Hole Dimension Symbols

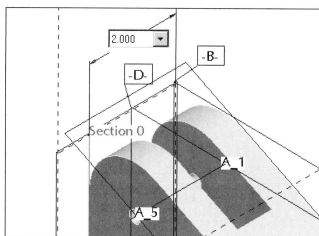

Figure 5.18(h) Cut Dimensions

Figure 5.18(i) Edited Cut

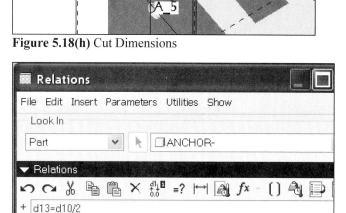

Figure 5.18(j) Relations

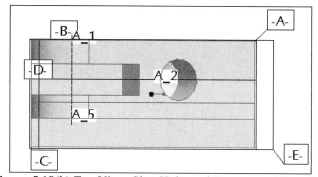

Figure 5.18(k) Top View, Slot, Hole, and Datum are Aligned

Click: [AB] ⇒ **Standard Orientation** ⇒ **Info** ⇒ **Switch Dimensions** ⇒ [Undo] **Undo: Edit Value** ⇒ **Info** ⇒ **Relations and Parameters** ⇒ click on **d10**, **d13**, and **d24** in the Browser in order to display them in the graphics window [Fig. 5.18(l)] ⇒ close the Browser with the quick sash ⇒ **Info** ⇒ **Switch Dimensions** ⇒ [save] ⇒ **MMB** ⇒ **File** ⇒ **Delete** ⇒ **Old Versions** ⇒ **MMB** ⇒ **Info** ⇒ **Feature List** [Fig. 5.18(m)] ⇒ **File** ⇒ **Close Window**

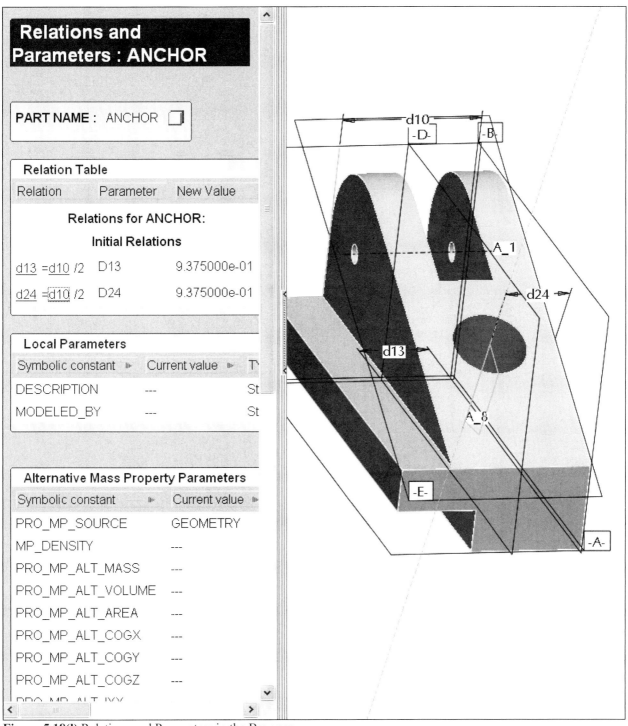

Figure 5.18(l) Relations and Parameters in the Browser

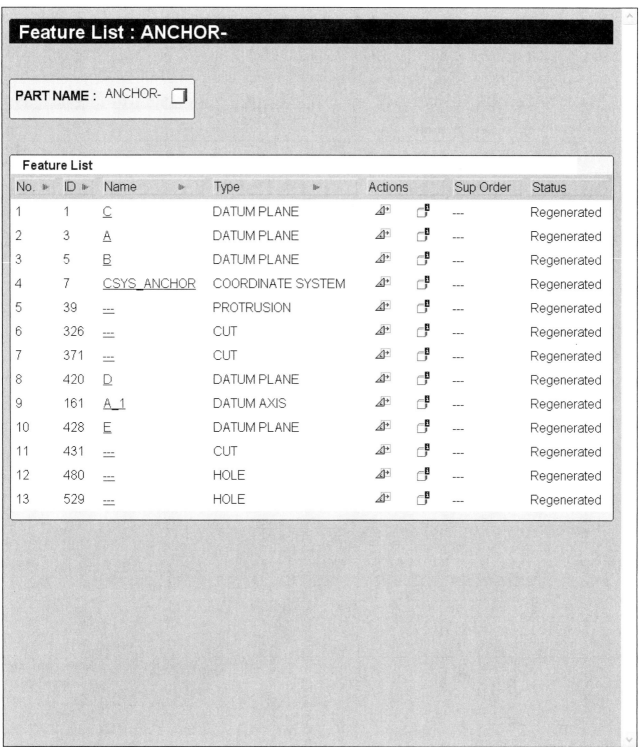

Figure 5.18(m) Feature List

The lesson is now complete. If you wish to model a project without instructions, a complete set of projects and illustrations are available at ***www.cad-resources.com*** ⇒ ***Downloads***.

Lesson 6 Revolved Features

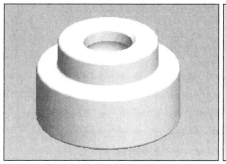
Figure 6.1(a) Clamp Foot

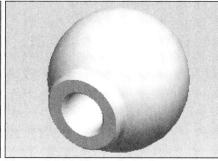
Figure 6.1(b) Clamp Swivel

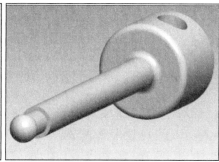
Figure 6.1(c) Clamp Ball

OBJECTIVES

- Master the **Revolve Tool** (Figs. 6.1(a-c)]
- Create **Chamfers** along part edges
- Learn how to **Sketch in 3D**
- Understand and use the **Navigation browser**
- Alter and set the **Items** and **Columns** displayed in the **Model Tree**
- Create standard **Tapped Holes**
- Create **Cosmetic Threads** and complete **tabular information** for threads
- Edit **Dimension Properties**
- Use the **Model Player** to extract information and dimensions
- Get a hard copy using the **Print** command

REVOLVED FEATURES

The **Revolve Tool** creates a *revolved solid* or a *revolved cut* by revolving a sketched section around a centerline from the sketching plane (Fig. 6.2). You can have any number of centerlines in your sketch/section, but only one will be used to rotate your section geometry. Rules for sketching a revolved feature include:

- The revolved section must have a centerline
- By default Pro/E uses the first centerline sketched as the *axis of revolution* (you may select a different axis of revolution)
- The geometry must be sketched on only one side of the *axis of revolution*
- The section must be closed for a solid (Fig. 6.2) but can be open for a cut

Figure 6.2 Revolved Protrusion (CADTRAIN, COAch for Pro/ENGINEER)

A variety of geometric shapes and constructions are used on revolved features. For instance, **chamfers** are created at selected edges of the part. Chamfers are *pick-and-place* features.

Threads can be a *cosmetic feature* representing the *nominal diameter* or the *root diameter* of the thread. Information can be embedded in the feature. Threads show as a unique color. By putting cosmetic threads on a separate layer, you can hide and unhide them.

Chamfers

Chamfers are created between abutting edges of two surfaces on the solid model. An edge chamfer removes a flat section of material from a selected edge to create a beveled surface between the two original surfaces common to that edge. Multiple edges can be selected.

There are four basic dimensioning schemes for edge chamfers: (Fig. 6.3).

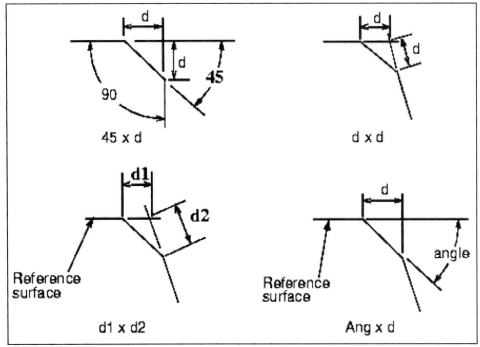

Figure 6.3 Chamfer Options

- **45 x d** Creates a chamfer that is at an angle of **45°** to both surfaces and a distance **d** from the edge along each surface. The distance is the only dimension to appear when edited. **45 x d** chamfers can be created only on an edge formed by the intersection of two *perpendicular* surfaces.
- **d x d** Creates a chamfer that is a distance **d** from the edge along each surface. The distance is the only dimension to appear when edited.
- **d1 x d2** Creates a chamfer at a distance **d1** from the selected edge along one surface and a distance **d2** from the selected edge along the other surface. Both distances appear along their respective surfaces.
- **Ang x d** Creates a chamfer at a distance **d** from the selected edge along one adjacent surface at an **Angle** to that surface.

Threads

Cosmetic threads are displayed with *magenta/purple* lines and circles. Cosmetic threads can be external or internal, blind or through. A cosmetic thread has a set of embedded parameters that can be defined at its creation or later, when the thread is added.

Standard Holes

Standard holes are a combination of sketched and extruded geometry. It is based on industry-standard fastener tables. You can calculate either the tapped or clearance diameter appropriate to the selected fastener. You can use Pro/E supplied standard lookup tables for these diameters or create your own. Besides threads, standard holes can be created with chamfers.

Navigation Window

Besides using the File command and corresponding options (File ⇒ Set Working Directory), the *Navigation window* can be used to directly access similar functions.

As previously mentioned, the working directory is a directory that you set up to contain Pro/E files. You must have read/write access to this directory. You usually start Pro/E from your working directory. A new working directory setting is not saved when you exit the current Pro/E session. By default, if you retrieve a file from a non-working directory, rename the file and then save it, the renamed file is saved to the directory from which it was originally retrieved, if you have read/write access to that directory. It is not saved in the current working directory, unless the config.pro option *save_object_in_current* is set to *yes*.

The navigation area is located on the left side of the Pro/E main window. It includes tabs for the Model Tree and Layer Tree, Folder Browser, Favorites, and Connections:

- **Model Tree** (default)
- **Layer Tree** (Show ⇒ Layer Tree)
- **Folder Browser**
- **Favorites**
- **Connections**

Folder Browser

The Folder browser (**Folder Browser**) is an expandable tree that lets you browse the file systems and other locations that are accessible from your computer. As you navigate the folders, the contents of the selected folder appear in the Pro/E browser as a Contents page. The Folder browser contains top-level nodes for accessing file systems and other locations that are known to Pro/E:

- **In Session** Pro/E objects that have been retrieved into local memory.
- **Shared Spaces** This is a shared file location accessed through the PTC Conference Center.
- **All registered servers** The browser lists all servers that you have registered with the Server Registry dialog box. These registered servers may include a Windchill server, a Pro/INTRALINK server, and an FTP server.
- **Node for the local file system** When you open the Folder browser, the local file system appears in the browser with the startup directory node expanded and highlighted.
- **Network Neighborhood** *(only for Windows)* The navigator shows computers on the networks to which you have access. The operations you can perform depend on your permissions to the remote computers.

Manipulating Folders

To work with folders, you can use the Folder Browser toolbar or the shortcut menu. You can perform the following tasks with the toolbar:

- **Create a new folder**
- **Delete selected folders**
- **Open the Windchill Cabinets** This icon is available only when you are working with an active Windchill server
- **Open the Windchill Workspace** This icon is available only when you are working with an active Windchill server
- **Working directory**

Using the Shortcut Menu in the Folder Browser

To open a shortcut menu, click **RMB** on an item in the Folder Browser. The commands on the shortcut menu vary depending on the task and your permissions. The shortcut menu lists the following commands:

- **New Folder** Add a subfolder to the selected folder
- **Open** Open a folder in the Pro/ENGINEER browser
- **Expand** Expand a node (if not open), **Collapse** a node (if open)
- **Server Registry** Access the Server Registry dialog box
- **Set Working Directory** Designate the selected directory as the working directory
- **Rename** Rename a selected folder
- **Delete** Delete a selected folder and all subfolders

Click: **Folder Browser** in the Navigator ⇒ click on the directory you wish to set as the working directory ⇒ **RMB** ⇒ **Set Working Directory** [Fig. 6.4(a)] *(yours will be different). Do not change your directory from your default Workfolder unless you are instructed to do so by your teacher or you are on your own computer.*

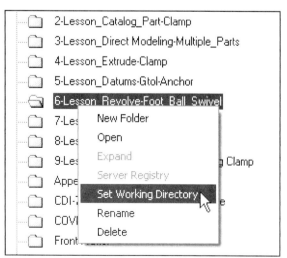

Figure 6.4(a) Folder Browser (**yours will be different**). Do not change your directory from your default Workfolder unless you are instructed to do so by your teacher or you are on your own computer.

As an example, you can set your working directory and then click on the Pro/E object that you wish to preview and open. Figure 6.4(b) is an example of what the Navigator and Browser would look like. You can use the MMB to rotate the part or assembly in the preview window as shown in Figure 6.4(c). Double-clicking on the file name or clicking [icon] **Open file in Pro/E** will open the part. After you complete this lesson you will have this same part in your set directory.

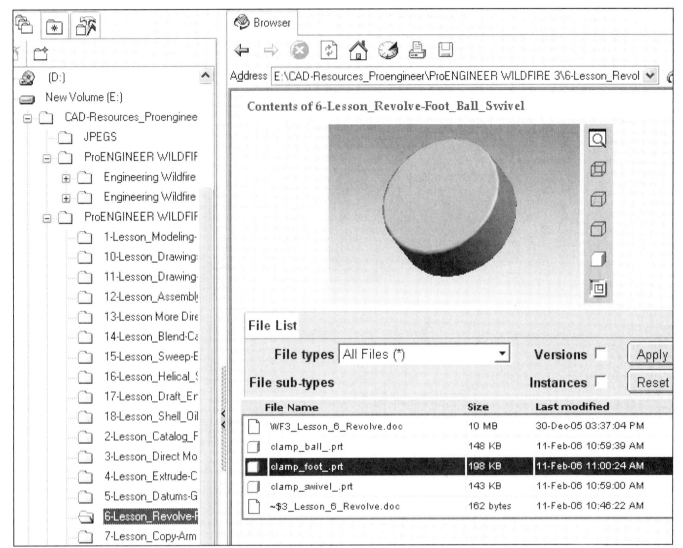

Figure 6.4(b) Folder Browser with File List and Previewed Part

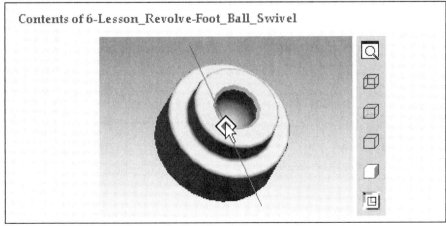

Figure 6.4(c) MMB to Rotate the Part Preview

Lesson 6 STEPS

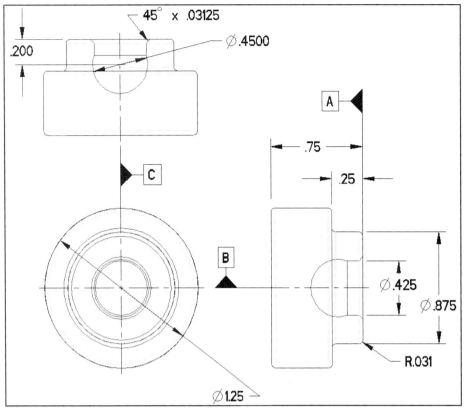

Figure 6.5 Clamp Foot Detail

Clamp Foot

The Clamp Foot (Fig. 6.5) is the first of three revolved parts created for Lesson 6. The Clamp Foot, Clamp Ball, and Clamp Swivel are three revolved parts needed for the Clamp assembly and drawings later in the text. This part requires a revolved extrusion (protrusion), a revolved cut, a chamfer and rounds.

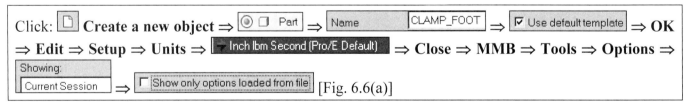

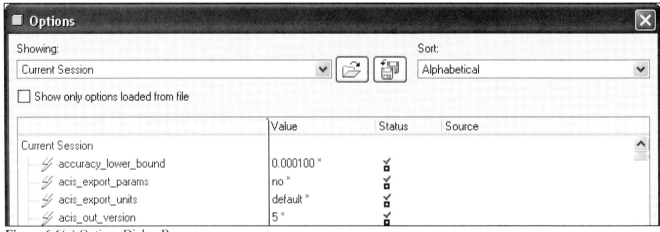

Figure 6.6(a) Options Dialog Box

Slide the bar down to the option or type the option. Option: ***default_dec_places*** ⇒ Value: **3** [Fig. 6.6(b)] ⇒ **Enter** [Fig. 6.6(c)]

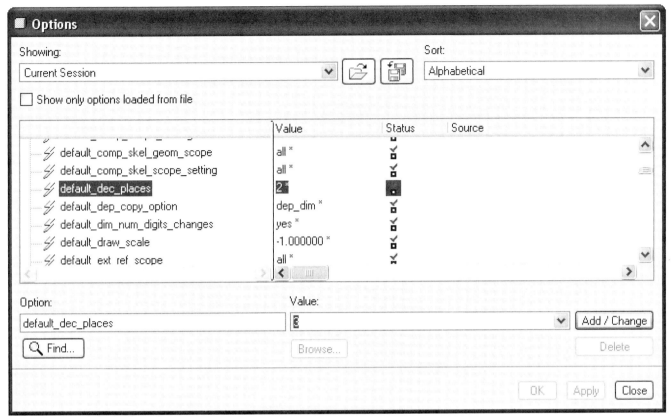

Figure 6.6(b) default_dec_places

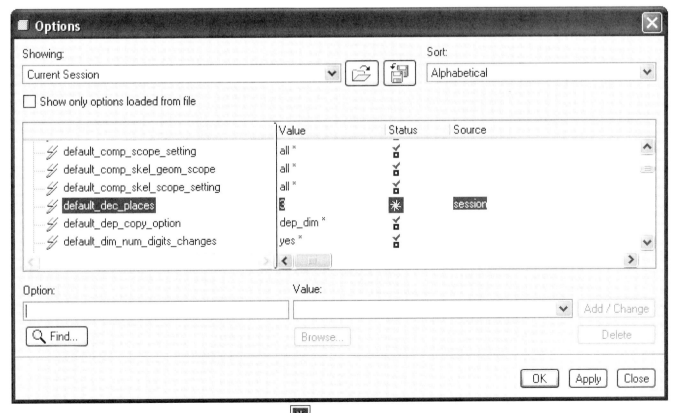

Figure 6.6(c) default_dec_places   Value: 3   Status: [✻]   Source: session

Slide the bar down to the option or type. Option: **sketcher_dec_places** ⇒ Value: **2** ⇒ **Enter** [Fig. 6.6(d)] ⇒ Option: **sketcher_starts_in_2d** ⇒ Value: **no** [Fig. 6.6(e)] ⇒ [Add / Change] *(same as Enter)* ⇒ Option: (type) *(you must type this)* **def_layer** ⇒ Value: ▼ ⇒ **layer_axis** *(space here)* **datum_axes** *(after you select layer_axis, type datum_axes)* [Fig. 6.6(f)] ⇒ **Add/Change** ⇒ **Apply** [Fig. 6.6(g)]

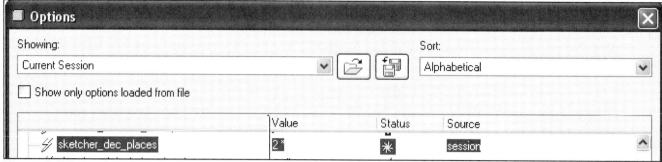

Figure 6.6(d) sketcher_dec_places

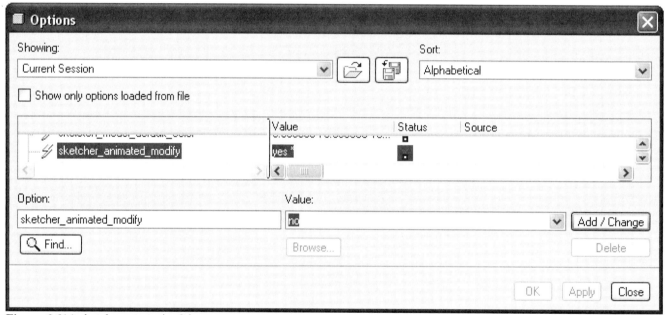

Figure 6.6(e) sketcher_starts_in_2d   no

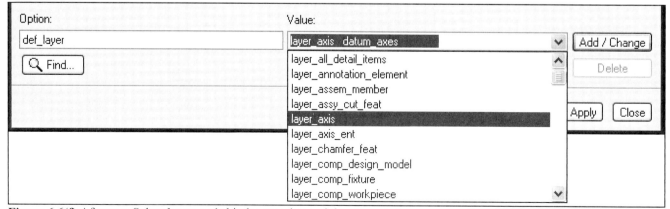

Figure 6.6(f) After you Select **layer_axis** hit the spacebar and then type **datum_axes**   *(layer_axis space datum_axes)*

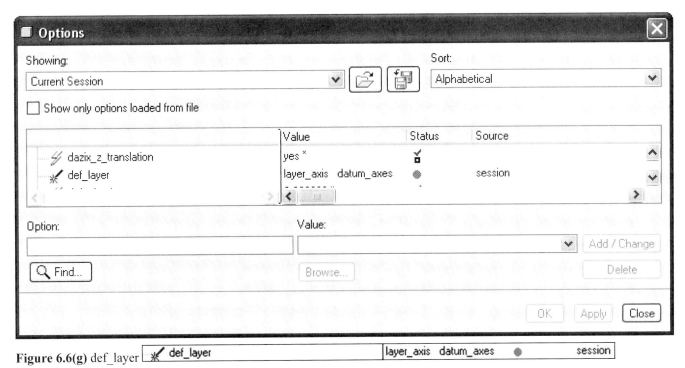

Figure 6.6(g) def_layer

☑ Show only options loaded from file [Fig. 6.6(h)] ⇒ 💾 **Save a copy of the currently displayed configuration file** *(you will use this in other sessions)* ⇒ Name **clamp.pro** [Fig. 6.6(i)] ⇒ **Ok** ⇒ **Close**

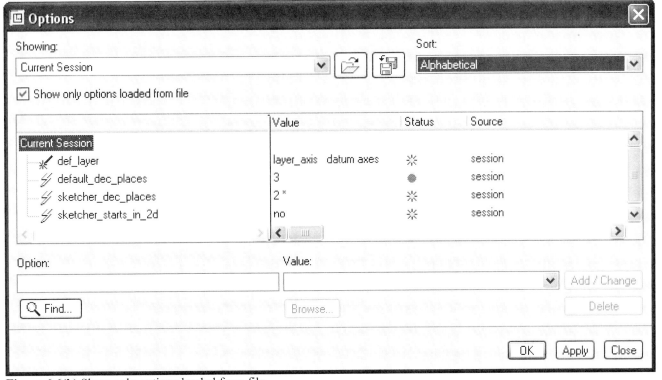

Figure 6.6(h) Show only options loaded from file

Figure 6.6(i) clamp.pro Options File

The first protrusion is a revolved protrusion created with the Revolve Tool, you will be sketching the section in 3D since the configuration option; *sketcher_starts_in_2d*, was previously set to *no*. In the Environment dialog box, you can see that Use 2D Sketcher is deactivated.

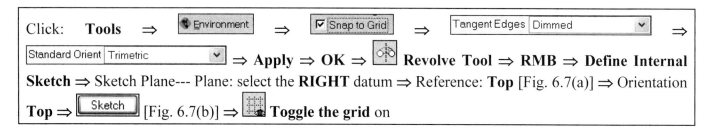

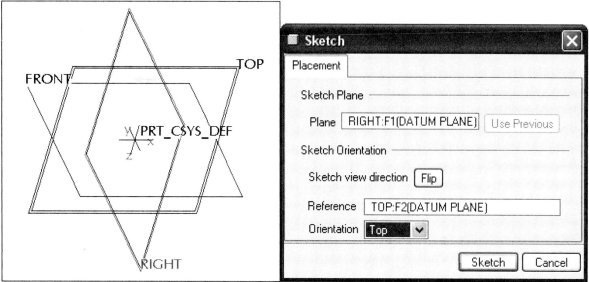

Figure 6.7(a) Sketch Plane and Reference

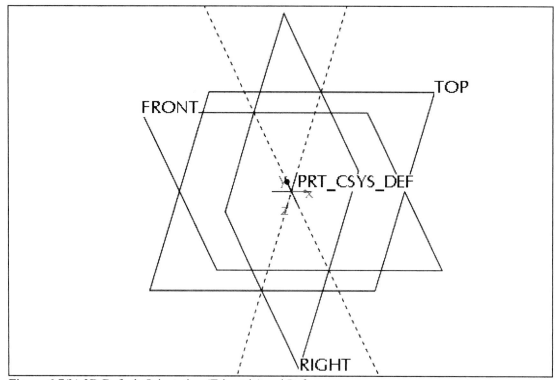

Figure 6.7(b) 3D Default Orientation (Trimetric) and References

Click: **RMB** ⇒ **Centerline** sketch a *vertical* centerline through the default coordinate system to be used as the axis of revolution [Fig. 6.7(c)] ⇒ **MMB** rotate the part to see the vertical plane clearly [Fig. 6.7(d)] ⇒ **RMB** ⇒ **Line** sketch the six lines on the RIGHT datum [Fig. 6.7(e)] ⇒ **MMB** ⇒ **MMB**

(Note: if you have difficulty, you can delete the lines and start again after clicking  ***Orient the sketching plane parallel to the screen*** *and sketch the lines in 2D as in previous lessons)*

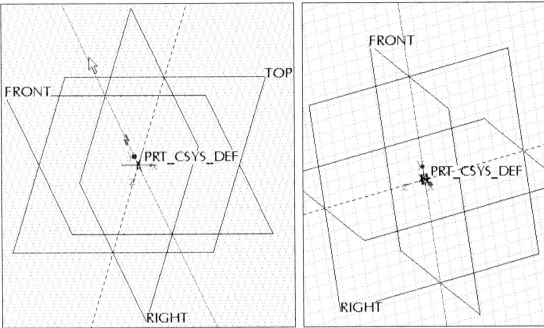

Figure 6.7(c) Sketch a Vertical Centerline.   **Figure 6.7(d)** MMB Rotate the sketch

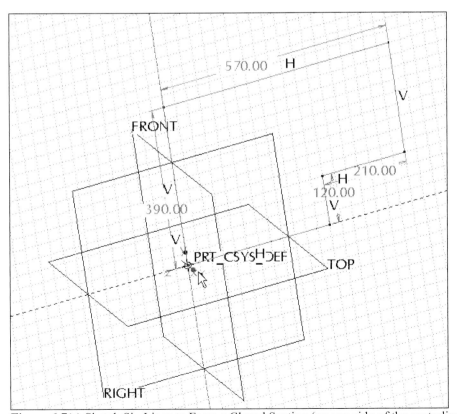

Figure 6.7(e) Sketch Six Lines to Form a Closed Section (on one side of the centerline)

Click: **Sketch** from menu bar ⇒ **Options** ⇒ [Snap To Grid] [Fig. 6.7(f)] ⇒ ✓ ⇒ off ⇒ **Ctrl+R** ⇒ **RMB** ⇒ **Dimension** add the two vertical dimensions [Fig. 6.7(g)] ⇒ add a diameter dimension by picking the centerline, then pick the outer vertical edge line, and then pick the centerline again ⇒ **MMB** to place the dimension ⇒ repeat to dimension the other diameter [Fig. 6.7(h)] ⇒ **MMB** ⇒ **Ctrl+R**

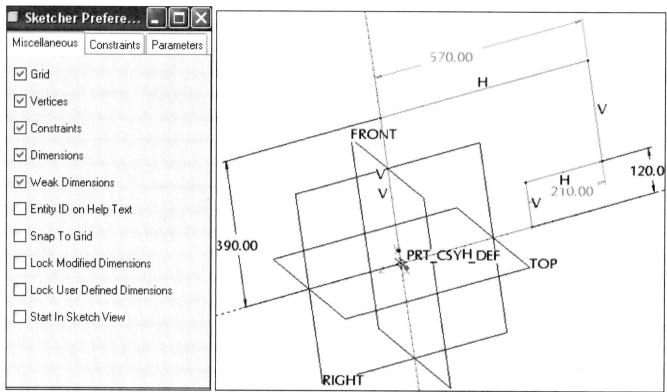

Figure 6.7(f) Sketcher Preferences   **Figure 6.7(g)** Add the Height (Vertical) Dimensions (your sketch values will be different)

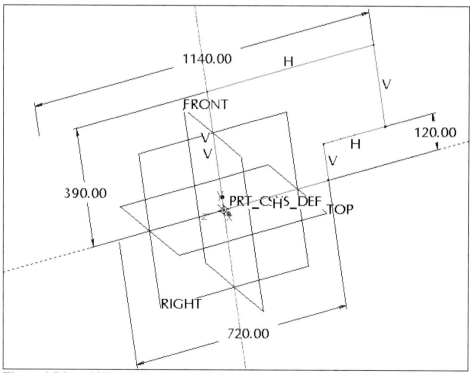

Figure 6.7(h) Add Two Diameter Dimensions

Modify the dimensions, click: [arrow] ⇒ window-in the sketch to capture all four dimensions ⇒ **RMB** ⇒ **Modify** ⇒ [Lock Scale] ⇒ [Regenerate] [Fig. 6.7(i)] ⇒ modify the overall height to the design value (see Fig. 6.5) of **.750** ⇒ **Enter** ⇒ [✓] [Fig. 6.7(j)] ⇒ **Ctrl+R** repaint

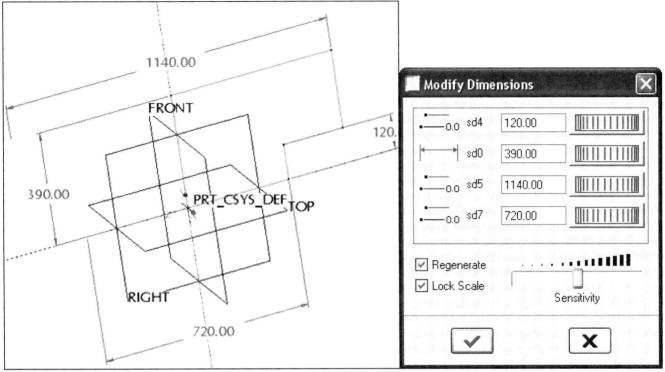

Figure 6.7(i) Capture the Dimensions (your values will be different)

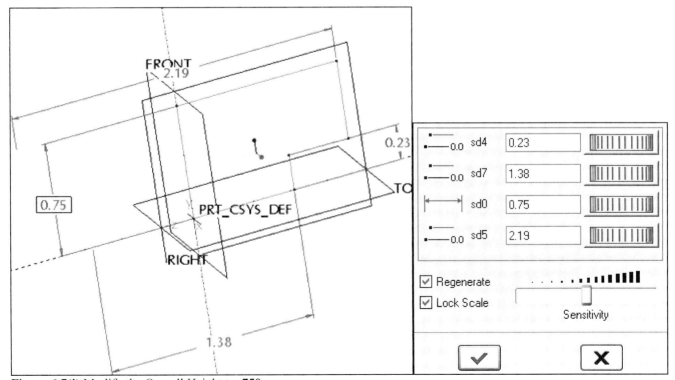

Figure 6.7(j) Modify the Overall Height to **.750**

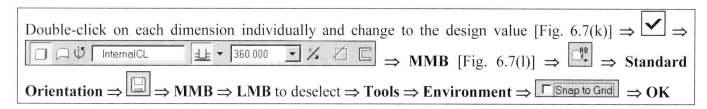

Double-click on each dimension individually and change to the design value [Fig. 6.7(k)] ⇒ ✓ ⇒ [toolbar: InternalCL, 360.000] ⇒ **MMB** [Fig. 6.7(l)] ⇒ ⇒ **Standard Orientation** ⇒ ⇒ **MMB** ⇒ **LMB** to deselect ⇒ **Tools** ⇒ **Environment** ⇒ Snap to Grid ⇒ **OK**

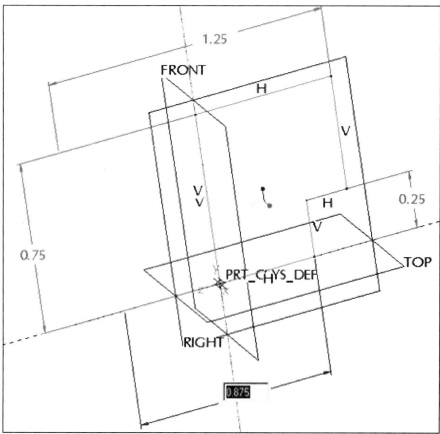

Figure 6.7(k) Modify the Remaining Dimensions (.75, 1.25, .25, .875)

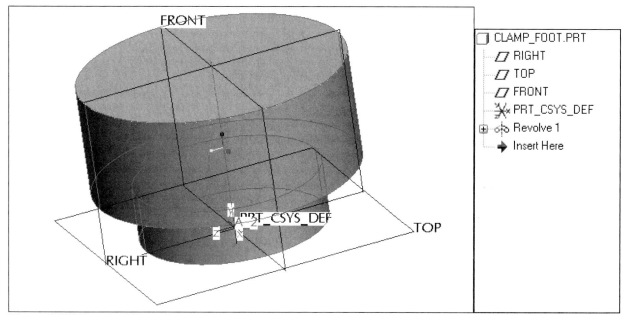

Figure 6.7(l) Revolved Extrusion

Click: ⊕ **Revolve Tool** ⇒ △ **Remove Material** ⇒ **RMB** ⇒ **Define Internal Sketch** ⇒ [Use Previous] [Figs. 6.8(a-b)] ⇒ 🗒 **Orient the sketching plane parallel to the screen** [Fig. 6.8(c)]

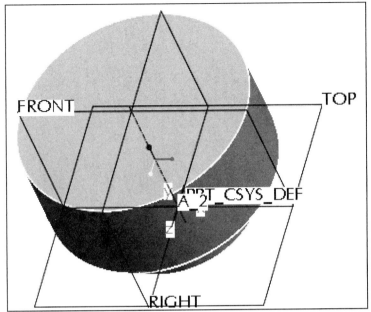

Figure 6.8(a) Sketch Plane and Reference

Figure 6.8(b) Sketch Dialog Box

Click: **RMB** ⇒ **Centerline** sketch a *vertical* centerline through the default coordinate system to be used as the axis of revolution [Fig. 6.8(d)] ⇒ **MMB** ⇒ 🗆 [Fig. 6.8(e)]

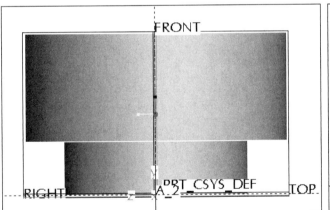

Figure 6.8(c) Sketch Plane and Reference   **Figure 6.8(d)** Sketch a Vertical Centerline

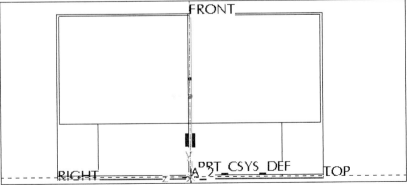

Figure 6.8(e) Hidden Line Display

From the menu bar click: **Sketch** ⇒ **Arc** ⇒ **Center and Ends** ⇒ sketch the arc [Fig. 6.8(f)] ⇒ **RMB** ⇒ **Line** sketch the three lines [Fig. 6.8(g)] *(it must be a closed section, so three lines are required)* ⇒ **MMB** ⇒ **MMB** ⇒ [icons] off ⇒ pick on each dimension and move to a better location ⇒ [icon] **Dynamically trim section entities** ⇒ press and hold **LMB** and draw a curve through the two elements you wish to remove [Fig. 6.8(h)] ⇒ release the **LMB** to complete the trim [Fig. 6.8(i)] ⇒ **Ctrl+R** *(Note: if you delete the wrong items, click:* [icon] *Undo sketcher operations and try again)*

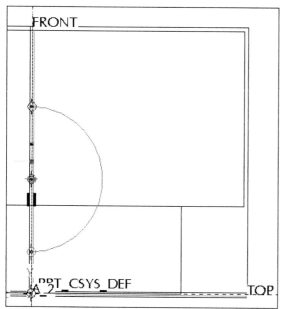

Figure 6.8(f) Sketch the Arc

Figure 6.8(g) Sketch the Lines

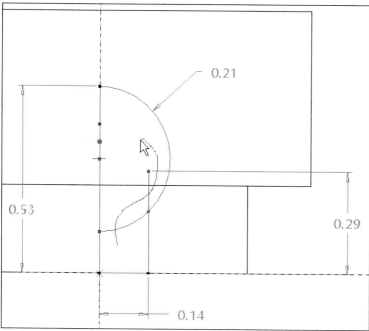

Figure 6.8(h) Dynamically Trim the Arc and Line

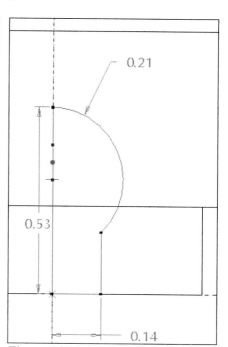

Figure 6.8(i) Sketch

Click: **RMB** ⇒ **Dimension** ⇒ double-click on the arc edge ⇒ **MMB** to place the diameter dimension ⇒ add a dimension to the center of the arc from the bottom of the part ⇒ pick the centerline, then pick the vertical line, and then pick the centerline again ⇒ **MMB** to place the diameter dimension [Fig. 6.8(j)] ⇒ **MMB** to deselect ⇒ [icon] ⇒ window–in all three dimensions ⇒ **RMB** ⇒ **Modify** modify the dimensions to the design values (Fig. 6.5) [Fig. 6.8(k)] ⇒ [✓] ⇒ [icons] on ⇒ [icon] ⇒ [icon] ⇒ **Standard Orientation** ⇒ [✓] ⇒ [icons InternalCL] [360.000] [icons] ⇒ **MMB** ⇒ [icon] ⇒ **MMB** rotate the part to view the hole/cut [Fig. 6.8(l)] ⇒ [icon] ⇒ **MMB** ⇒ **LMB**

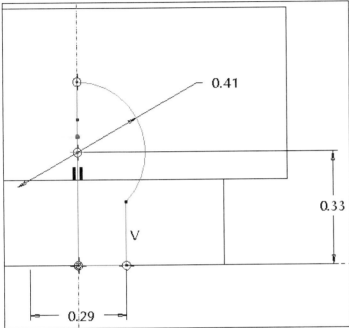

Figure 6.8(j) Correct Dimensioning Scheme (your dimension values will be different)

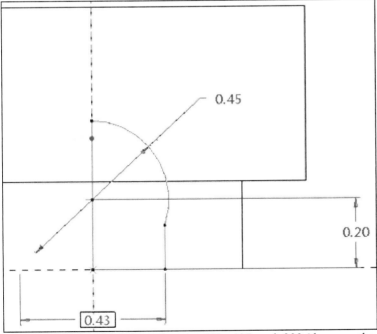

Figure 6.8(k) Modify Dimensions to **.450, .425,** and **.200** (these are the design values)

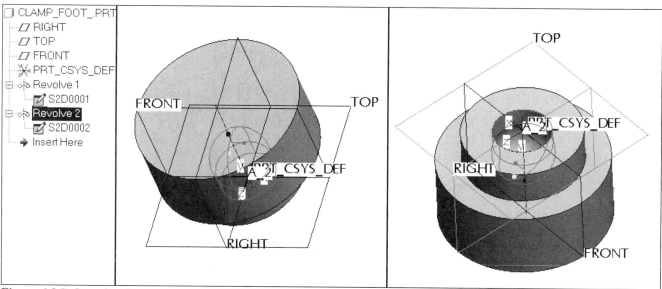

Figure 6.8(l) Completed Cut

Slowly click on the edge of the cut until selected (highlights) [Fig. 6.8(m)] ⇒ 🗒 **Chamfer Tool** [Fig. 6.8(n)] ⇒ double-click on the dimension and modify to the design value of **.03125** [Fig. 6.8(o)] ⇒ **Enter** ⇒ **MMB** [Fig. 6.8(p)] ⇒ **LMB** to deselect ⇒ 💾 ⇒ **MMB** ⇒ **File** ⇒ **Delete** ⇒ **Old Versions** ⇒ **MMB**

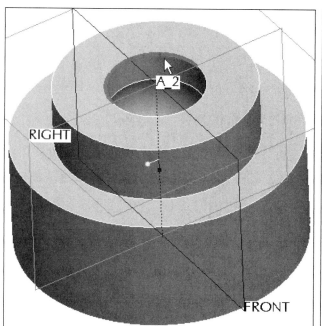

Figure 6.8(m) Pick on Edge

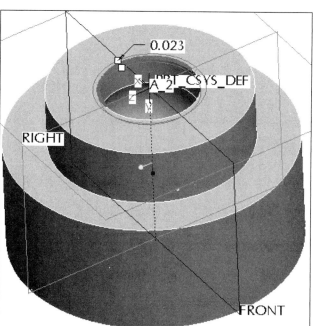

Figure 6.8(n) Chamfer Preview

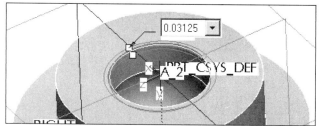

Figure 6.8(o) Chamfer Dimension

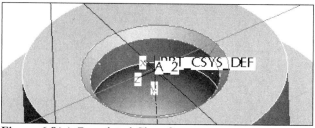

Figure 6.8(p) Completed Chamfer

Select one edge, then press and hold the **Ctrl** key and pick on the remaining three edges ⇒ **RMB** ⇒ **Round Edges** [Fig. 6.8(q)] ⇒ double-click on the dimension and modify to the design value of **.03125** [Fig. 6.8(r)] ⇒ **Enter** ⇒ **MMB** ⇒ 🖫 ⇒ **MMB** ⇒ **LMB**

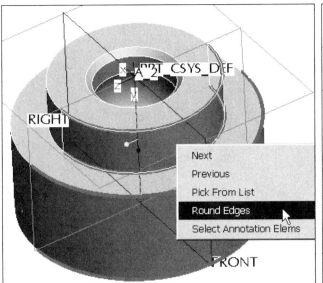
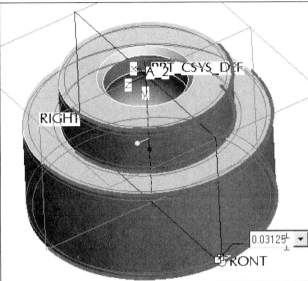

Figure 6.8(q) Round Edges　　　　　　　　　　**Figure 6.8(r) .03125** Round

Click: **Edit** ⇒ **Setup** ⇒ **Geom Tol** ⇒ **Set Datum** ⇒ pick **TOP** from the model ⇒ Name- type **A** ⇒ **OK** ⇒ pick **RIGHT** ⇒ Name- type **B** ⇒ **OK** ⇒ pick **FRONT** ⇒ Name- type **C** ⇒ **OK** [Fig. 6.8(s)] ⇒ **Done/Return** ⇒ **Material** ⇒ 📄 steel.mtl ⇒ ▶▶▶ ⇒ **OK** ⇒ **Done** ⇒ 🖫 ⇒ **OK** ⇒ **File** ⇒ **Delete** ⇒ **Old Versions** ⇒ **MMB** ⇒ **Window** ⇒ **Close**

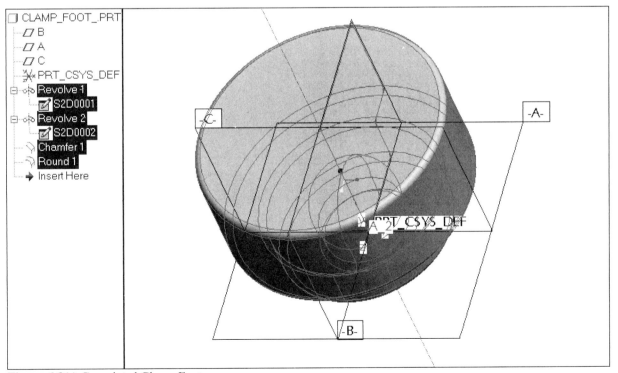

Figure 6.8(s) Completed Clamp Foot

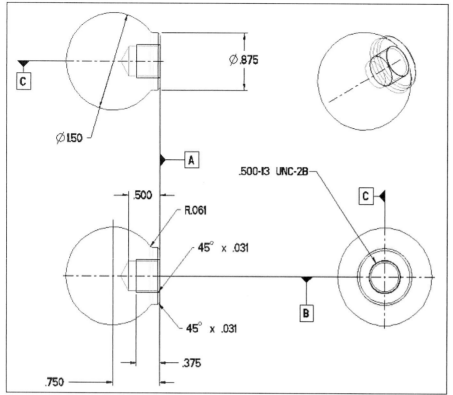

Figure 6.9 Clamp Ball Detail

Clamp Ball

The second part for this lesson is the Clamp Ball (Fig. 6.9). Much of the sketching is the same for this part as was done for the first revolved feature of the Clamp Foot. Instead of an internal revolved cut, you will create a standard hole. The Clamp Ball is made of *nylon*.

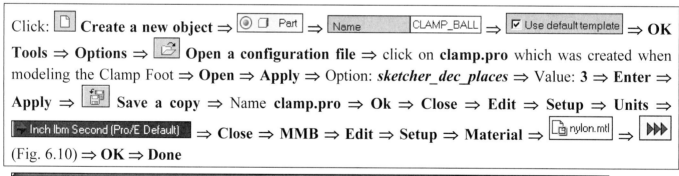

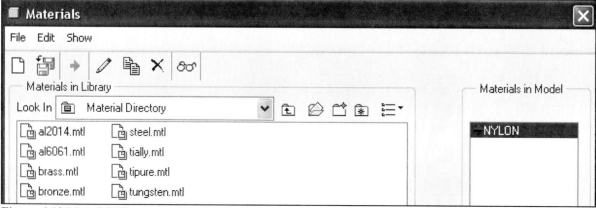

Figure 6.10 Material File

Model Tree

The Model Tree is a tabbed feature on the Pro/E navigator that contains a list of every feature (or component) in the current part, assembly, or drawing. The model structure is displayed in hierarchical (tree) format with the root object (the current feature, part or assembly) at the top of its tree and the subordinate objects (features, parts, or assemblies) below. If you have multiple Pro/E windows open, the Model Tree contents reflect the file in the current "active" window. The Model Tree lists only the related feature- and part-level objects in a current file and does not list the entities (such as edges, surfaces, curves, and so forth) that comprise the features.

Each Model Tree item displays an icon that reflects its object type, for example, assembly, part, feature, or datum plane (also a feature). The icon can also show the display status for a feature, part, or assembly, for example, suppressed or hidden. The information in the Model Tree can be saved as a text (.txt) file. Selection in the Model Tree is object-action oriented; you select objects in the Model Tree without first specifying what you intend to do with them. You can select components, parts, or features using the Model Tree. You cannot select the individual geometry that makes up a feature (entities). To select an entity, you must select it in the main graphics window. Items can be added or removed from the Model Tree column display using Settings in the Navigator:

- Select features, parts, or assemblies, and perform object-specific operations on them using the shortcut menu.
- Filter the display by item type, for example, hiding or unhiding datum features.
- Open a part within an assembly file by right-clicking the part in the Model Tree.
- Create or modify features and perform other operations such as deleting or redefining parts or features, and rerouting parts or features using the shortcut menu.
- Search the Model Tree for model properties or other feature information.
- Show the display status for an object, for example, suppressed or hidden.

Click: **Settings** in the Navigator [Fig. 6.11(a)] ⇒ **Tree Filters** ⇒ check all Display options on ⇒ [Fig. 6.11(b)] ⇒ **Apply** ⇒ **OK**

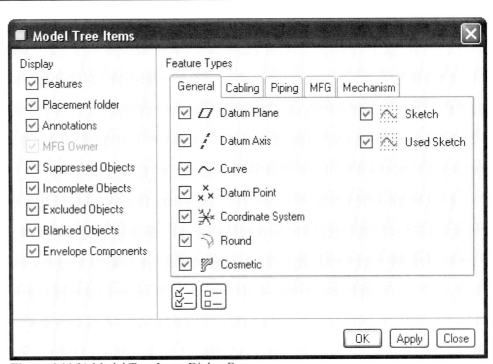

Figure 6.11(a) Tree Filters   **Figure 6.11(b)** Model Tree Items Dialog Box

Click: **Settings** in the Navigator ⇒ [Tree Columns...] [Fig. 6.12(a)] ⇒ Model Tree Columns dialog box opens: Type **Info** [Fig. 6.12(b)] ⇒ add types to the Displayed list, click: **Feat #** ⇒ [»] ⇒ **Feat ID** ⇒ [»] ⇒ Type [v] ⇒ select **Layer** [Fig. 6.12(c)] ⇒ **Layer Names** ⇒ [»] ⇒ **Layer Status** ⇒ [»] [Fig. 6.12(d)] ⇒ **Apply** ⇒ **OK**

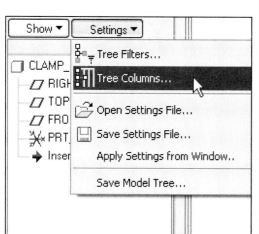

Figure 6.12(a) Model Tree Columns

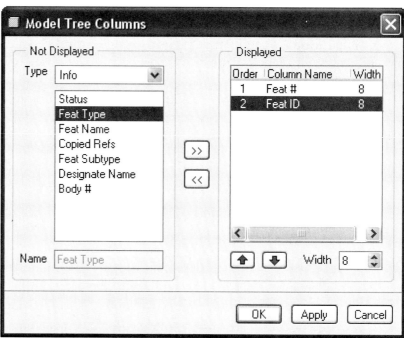

Figure 6.12(b) Model Tree Columns Dialog Box, Type: Info

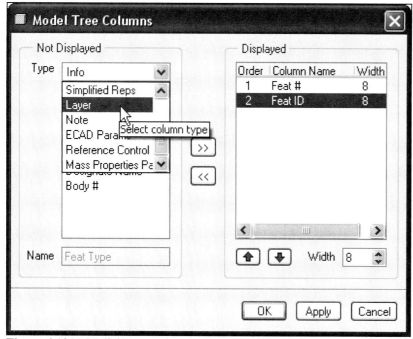

Figure 6.12(c) Model Tree Columns Dialog Box, Type: Layer

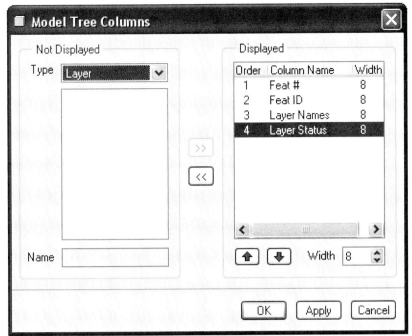

Figure 6.12(d) Model Tree Columns Dialog Box, Displayed List

Click on the sash ⇒ drag sash to expand Model Tree [Fig. 6.12(e)] ⇒ adjust the width of each column by dragging the column divider ⇒ drag the sash to the left to decrease the Model Tree size [Fig. 6.12(f)]

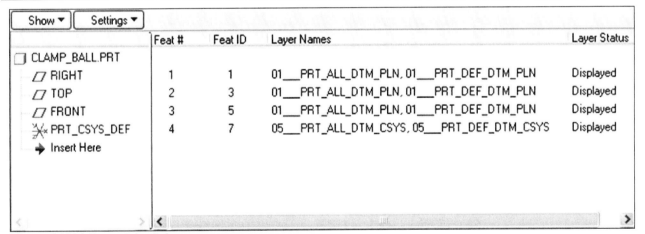

Figure 6.12(e) Expand the Columns

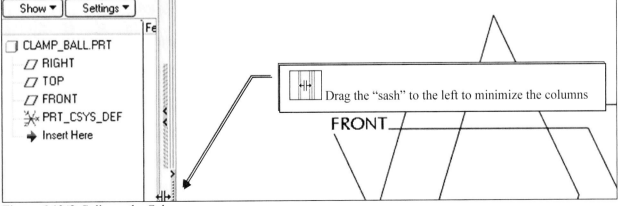

Figure 6.12(f) Collapse the Columns

201

Click: **Tools** ⇒ [Environment] ⇒ [☑ Snap to Grid] ⇒ [☑ Use 2D Sketcher] ⇒ **Apply** ⇒ **OK** ⇒ pick on the **RIGHT** datum plane in the graphics window or the Model Tree (it will highlight) ⇒ [icon] **Revolve Tool** ⇒ **RMB** ⇒ **Define Internal Sketch** ⇒ Orientation: **Top** ⇒ **MMB** ⇒ [icon] **Toggle the grid on** ⇒ [icon] ⇒ [icon] ⇒ **RMB** ⇒ **Centerline** sketch a *vertical* centerline through the default coordinate system to be used as the axis of revolution ⇒ [icons] ⇒ [icon] ⇒ sketch a half-circle [Fig. 6.13(a)] ⇒ **RMB** ⇒ **Line** sketch the three lines [Fig. 6.13(b)] ⇒ **MMB** ⇒ [icon] **Dynamically trim section entities** ⇒ press and hold **LMB** and draw a curve through the two elements you wish to remove [Fig. 6.13(c)] ⇒ **RMB** ⇒ **Dimension** ⇒ dimension as per design requirements (Fig. 6.9) *(Your dimension values will be different. Your dimensioning "scheme" will be the same)* ⇒ [icons] off ⇒ [icon] **Toggle the grid off** ⇒ [icon] **Spin Center** off [Fig. 6.13(d)] ⇒ **Ctrl+R**

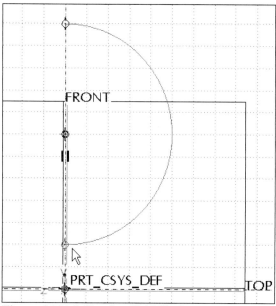

Figure 6.13(a) Sketch a Half-Circle

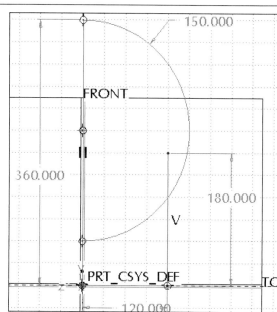

Figure 6.13(b) Sketch Three Lines

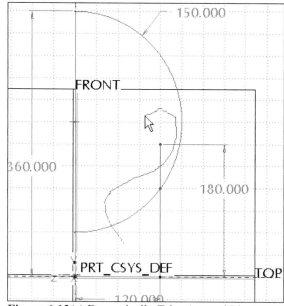

Figure 6.13(c) Dynamically Trim Arc and Line

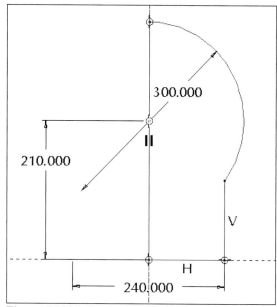

Figure 6.13(d) Correct Dimensioning Scheme

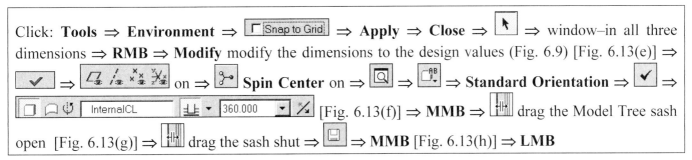

Click: **Tools** ⇒ **Environment** ⇒ ☐ Snap to Grid ⇒ **Apply** ⇒ **Close** ⇒ [arrow] ⇒ window–in all three dimensions ⇒ **RMB** ⇒ **Modify** modify the dimensions to the design values (Fig. 6.9) [Fig. 6.13(e)] ⇒ ✓ ⇒ [icons] on ⇒ **Spin Center** on ⇒ [icon] ⇒ [icon] ⇒ **Standard Orientation** ⇒ ✓ ⇒ [toolbar: InternalCL | 360.000] [Fig. 6.13(f)] ⇒ **MMB** ⇒ [icon] drag the Model Tree sash open [Fig. 6.13(g)] ⇒ [icon] drag the sash shut ⇒ [icon] ⇒ **MMB** [Fig. 6.13(h)] ⇒ **LMB**

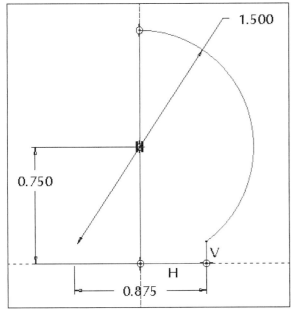

Figure 6.13(e) Modified Dimensions

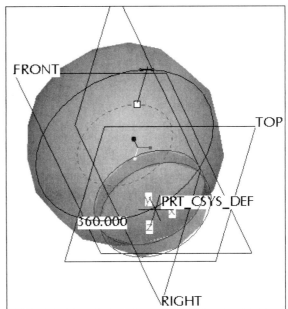

Figure 6.13(f) Feature Preview

	Feat #	Feat ID	Layer Names	Layer Status
☐ CLAMP_BALL_.PRT				
⧠ C	1	1	01___PRT_ALL_DTM_PLN, 01___PRT_DEF_DTM_PLN	Displayed
⧠ A	2	3	01___PRT_ALL_DTM_PLN, 01___PRT_DEF_DTM_PLN	Displayed
⧠ B	3	5	01___PRT_ALL_DTM_PLN, 01___PRT_DEF_DTM_PLN	Displayed
✹ CLAMP_BALL_CSYS	4	7	05___PRT_ALL_DTM_CSYS, 05___PRT_DEF_DTM_CSYS	Displayed
⊞ ⧫ Revolve 1	5	39	02___PRT_ALL_AXES	Displayed

Figure 6.13(g) Revolve Feature Shown in Model Tree

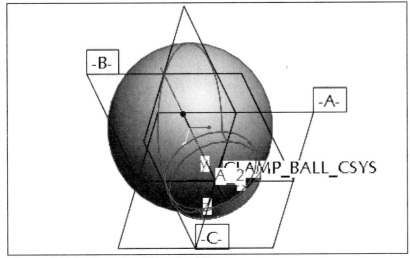

Figure 6.13(h) Completed Revolved Feature

Holes

Hole charts are used to lookup diameters for a given fastener size. You can create custom hole charts and specify their directory location with the configuration file option *hole_parameter_file_path*. UNC, UNF and ISO hole charts are supplied with Pro/E. Create a standard **.500-13 UNC-2B** hole, **.375** thread depth, **.50** tap drill. Include a standard chamfer.

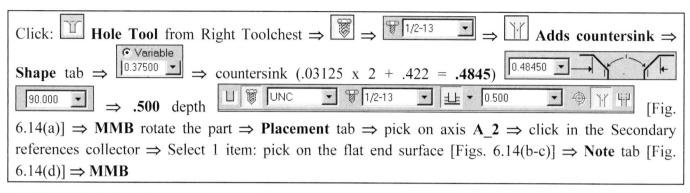

Click: **Hole Tool** from Right Toolchest ⇒ ⇒ 1/2-13 ⇒ **Adds countersink** ⇒ **Shape** tab ⇒ 0.37500 ⇒ countersink (.03125 x 2 + .422 = **.4845**) 0.48450 ⇒ 90.000 ⇒ **.500** depth UNC 1/2-13 0.500 [Fig. 6.14(a)] ⇒ **MMB** rotate the part ⇒ **Placement** tab ⇒ pick on axis **A_2** ⇒ click in the Secondary references collector ⇒ Select 1 item: pick on the flat end surface [Figs. 6.14(b-c)] ⇒ **Note** tab [Fig. 6.14(d)] ⇒ **MMB**

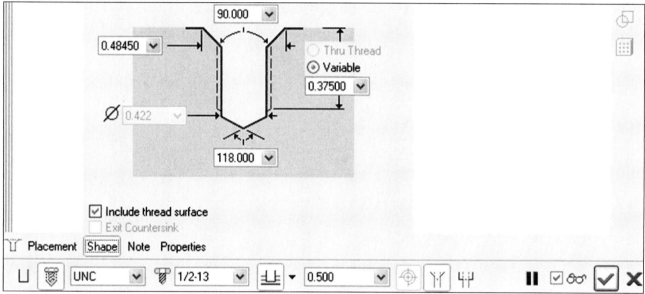

Figure 6.14(a) Standard Hole Shape

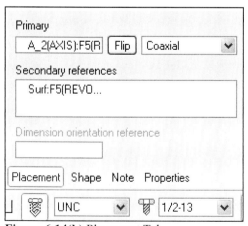

Figure 6.14(b) Placement Tab

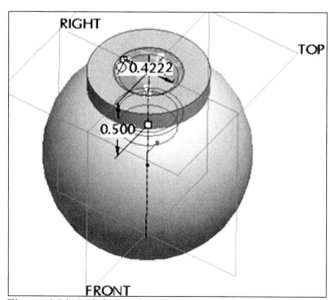

Figure 6.14(c) Hole Preview

204

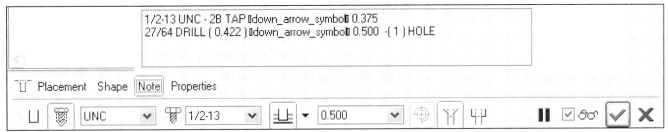

Figure 6.14(d) Hole Note

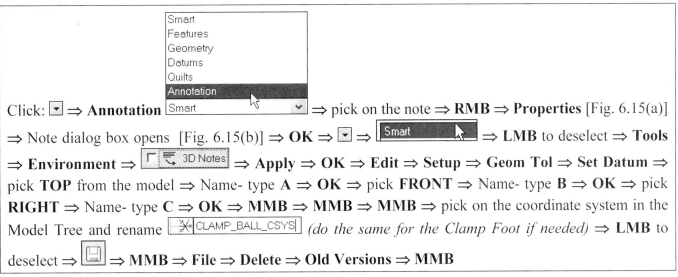

Click: ▼ ⇒ **Annotation** [Smart] ⇒ pick on the note ⇒ **RMB** ⇒ **Properties** [Fig. 6.15(a)] ⇒ Note dialog box opens [Fig. 6.15(b)] ⇒ **OK** ⇒ ▼ ⇒ [Smart] ⇒ **LMB** to deselect ⇒ **Tools** ⇒ **Environment** ⇒ [3D Notes] ⇒ **Apply** ⇒ **OK** ⇒ **Edit** ⇒ **Setup** ⇒ **Geom Tol** ⇒ **Set Datum** ⇒ pick **TOP** from the model ⇒ Name- type **A** ⇒ **OK** ⇒ pick **FRONT** ⇒ Name- type **B** ⇒ **OK** ⇒ pick **RIGHT** ⇒ Name- type **C** ⇒ **OK** ⇒ **MMB** ⇒ **MMB** ⇒ **MMB** ⇒ pick on the coordinate system in the Model Tree and rename [CLAMP_BALL_CSYS] *(do the same for the Clamp Foot if needed)* ⇒ **LMB** to deselect ⇒ 💾 ⇒ **MMB** ⇒ **File** ⇒ **Delete** ⇒ **Old Versions** ⇒ **MMB**

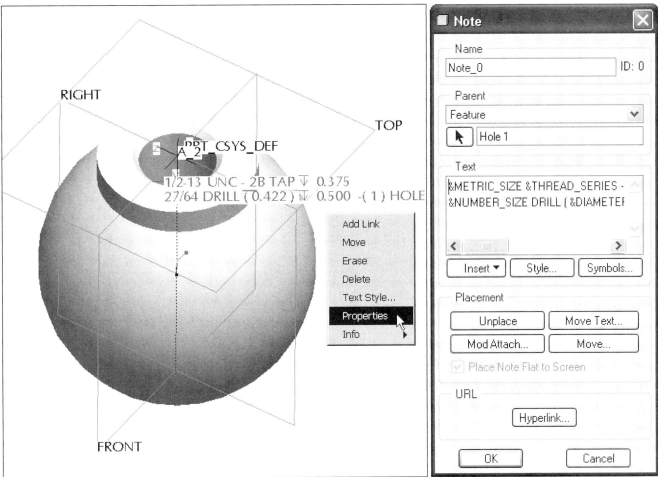

Figure 6.15(a) Note Properties   **Figure 6.15(b)** Note Dialog Box

Dimension Properties

Besides the value of a dimension and its position, there are several attributes that you can change using the Dimension Properties dialog box. The Properties tab provides options for *Value and tolerance* modification, *Format* options, and *Display* variables including *Flip Arrows*, which allows the toggling of dimension arrows inside or outside of its extension lines.

Move, which is available on all three tabs, moves the dimension itself, and the associated leader lines, to a new location. *Move Text* only moves the text associated with the dimension to a new location. *Text Symbol* provides a dialog box with symbology.

Double-click on the model to display the dimensions for the protrusion ⇒ pick on the ⌀.875 dimension ⇒ **RMB** [Fig. 6.16(a)] ⇒ **Properties** Dimension Properties dialog box displays [Fig. 6.16(b)]

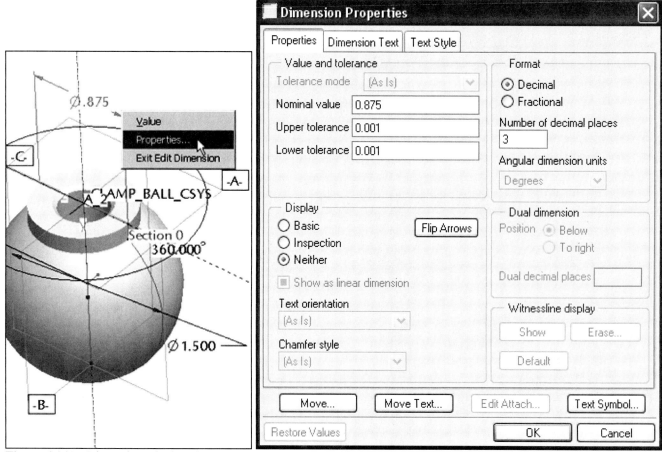

Figure 6.16(a) Dimension Displayed   **Figure 6.16(b)** Dimension Properties Dialog Box

The Dimension Text tab [Fig. 6.16(c)] shows the parametric dimension symbol. The Text Style tab [Fig. 6.16(d)] provides options for Character, and Note/Dimension variations. Take some time and explore the options provided in the Dimension Properties dialog box.

Click: **Move** ⇒ select a new position for the ⌀.875 dimension [Fig. 6.16(e)] ⇒ **OK** ⇒ repeat to move the other dimensions as required ⇒ **LMB** to deselect ⇒ **Ctrl+S** ⇒ **MMB** ⇒ **File** ⇒ **Delete** ⇒ **Old Versions** ⇒ **MMB**

Figure 6.16(c) Dimension Properties, Dimension Text Tab

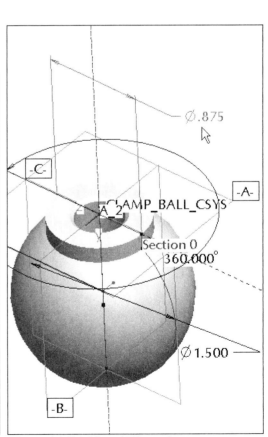

Figure 6.16(d) Dimension Properties, Text Style Tab

Figure 6.16(e) Moved Dimension

Click: **LMB** to clear dimensions from the screen ⇒ [Smart] (lower right corner of below graphics window) ⇒ [▼] ⇒ [Geometry] ⇒ pick on the edge [Fig. 6.17(a)] ⇒ **RMB** ⇒ **Round Edges** ⇒ double-click on the dimension and modify to the design value of **.06125** [Fig. 6.17(b)] ⇒ **Enter** ⇒ **MMB** ⇒ **Ctrl+S** ⇒ **Enter** ⇒ **MMB** ⇒ **LMB**

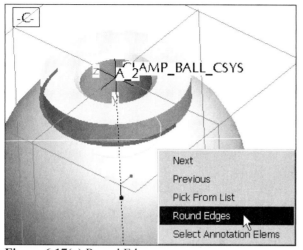
Figure 6.17(a) Round Edges

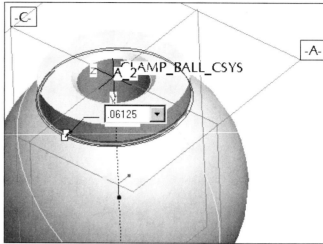
Figure 6.17(b) Radius .06125

Pick on the edge of the protrusion [Fig. 6.18(a)] ⇒ **Chamfer Tool** [Fig. 6.18(b)] ⇒ [Sets Transitions Pieces Options Properties / DxD D 0.031250] ⇒ **MMB** ⇒ **MMB** ⇒ **LMB** to deselect ⇒ [▼] ⇒ [Smart] ⇒ **File** ⇒ **Save** ⇒ **Enter** ⇒ **File** ⇒ **Delete** ⇒ **Old Versions** ⇒ **MMB** ⇒ **File** ⇒ **Close Window**

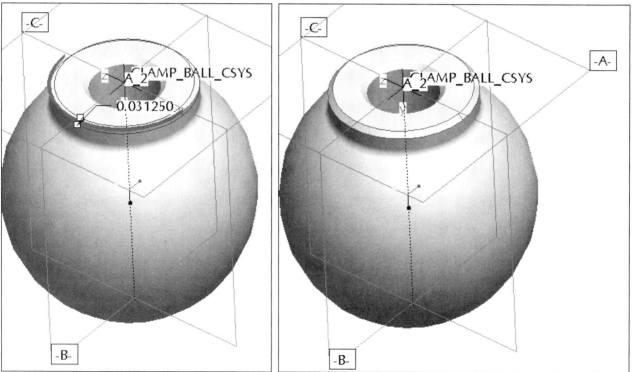

Figure 6.18(a) Chamfer .03125   **Figure 6.18(b)** Completed Chamfer

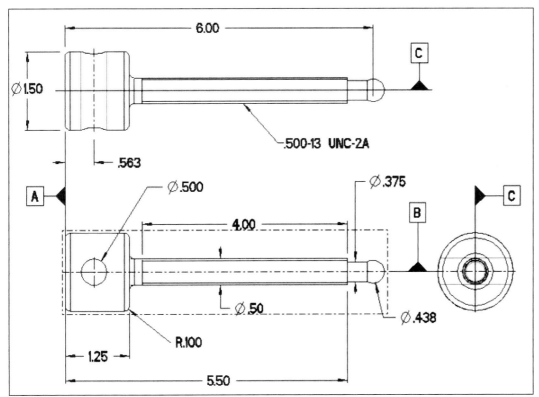

Figure 6.19(a) Clamp Swivel Detail

Clamp Swivel

The Clamp Swivel is the third part created by revolving one section about a centerline [Fig. 6.19(a)]. The Clamp Swivel is a component of the Clamp Assembly [Fig. 6.19(b)].

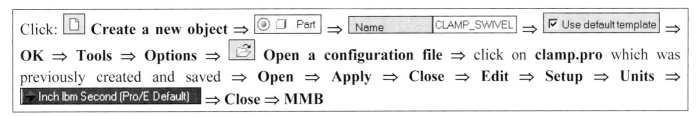

Click: **Create a new object** ⇒ **Part** ⇒ Name **CLAMP_SWIVEL** ⇒ **Use default template** ⇒ **OK** ⇒ **Tools** ⇒ **Options** ⇒ **Open a configuration file** ⇒ click on **clamp.pro** which was previously created and saved ⇒ **Open** ⇒ **Apply** ⇒ **Close** ⇒ **Edit** ⇒ **Setup** ⇒ **Units** ⇒ **Inch lbm Second (Pro/E Default)** ⇒ **Close** ⇒ **MMB**

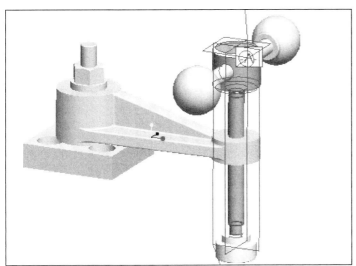

Figure 6.19(b) Clamp Swivel Highlighted in Clamp Assembly

Open the **Browser** by clicking on the quick sash ⇒ type Address http://www.matweb.com ⇒ **Enter** ⇒ type **Steel** in the upper right-hand corner box Steel SEARCH [Fig. 6.20(a)] ⇒ click: **SEARCH** ⇒ pick ☑ 5 AISI 1006 Steel, cold drawn bar, 19-32 mm (0.75-1.25 in) round [Fig. 6.20(b)] ⇒ click the quick sash to close the **Browser**

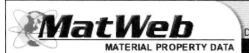

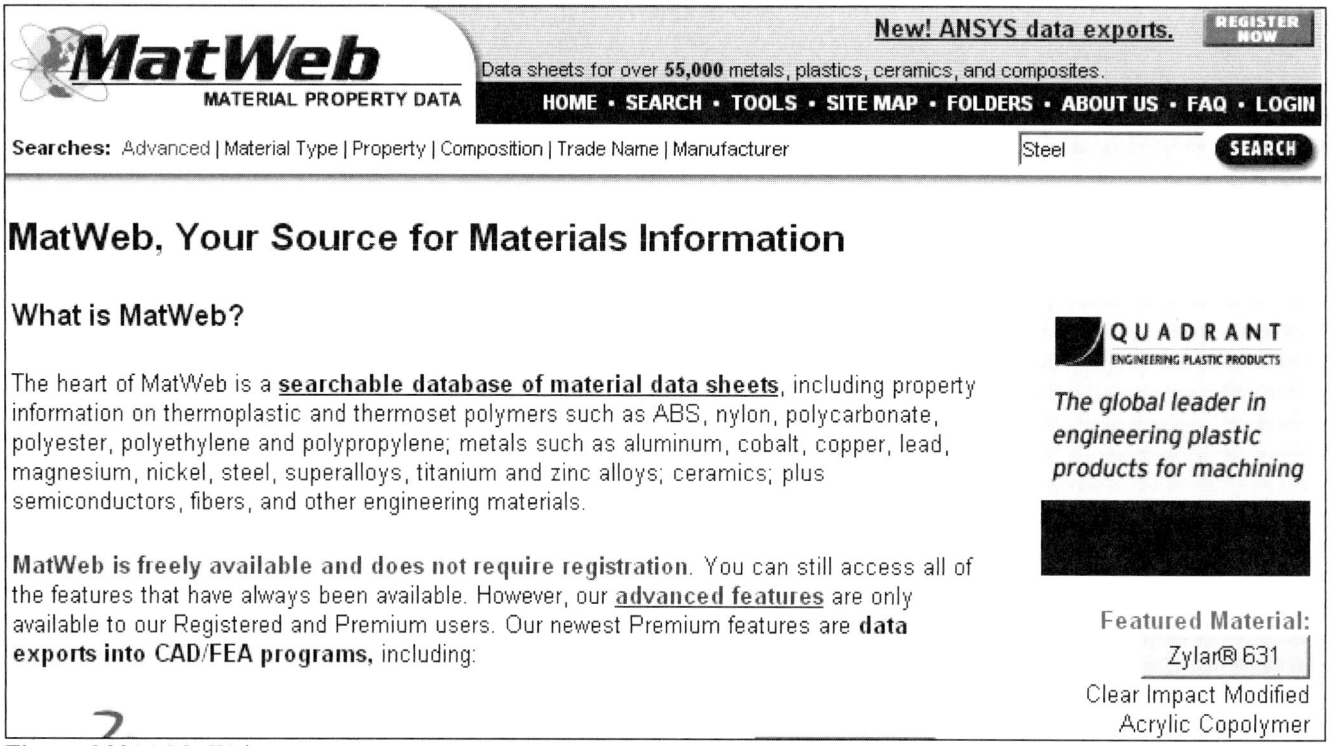

Figure 6.20(a) MatWeb.com

Mechanical Properties			
Hardness, Brinell	95	95	
Hardness, Knoop	113	113	Converted from Brinell hardness.
Hardness, Rockwell B	55	55	Converted from Brinell hardness.
Hardness, Vickers	98	98	Converted from Brinell hardness.
Tensile Strength, Ultimate	330 MPa	47900 psi	
Tensile Strength, Yield	285 MPa	41300 psi	
Elongation at Break	20 %	20 %	in 50 mm
Reduction of Area	45 %	45 %	
Modulus of Elasticity	200 GPa	29000 ksi	Typical for steel
Bulk Modulus	140 GPa	20300 ksi	Typical for steel
Poisson's Ratio	0.29	0.29	Typical For Steel
Machinability	50 %	50 %	Based on AISI 1212 steel. as 100% machinability. The machinability of Group I bar, rod, and wire products can be improved by cold drawing.
Shear Modulus	80 GPa	11600 ksi	Typical for steel

Figure 6.20(b) Steel Properties

Use the material specifications from MatWeb [Fig. 6.20(b)] to complete as much of the material table as required ⇒ **Edit** ⇒ **Setup** ⇒ **Material** ⇒ ⬜ **Create new material** [Fig. 6.21(a)] ⇒ type **STEEL_1006** ⇒ fill in the form [Fig. 6.21(b)] ⇒ **Save To Model** [Fig. 6.21(c)] ⇒ **OK** ⇒ **Done**

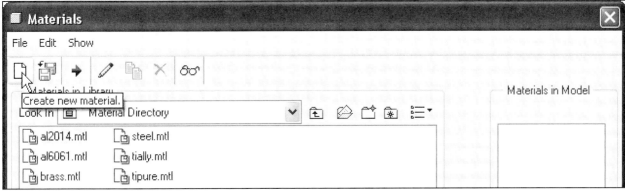

Figure 6.21(a) Materials Dialog Box

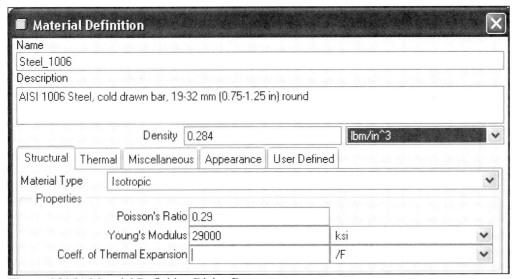

Figure 6.21(b) Material Definition Dialog Box

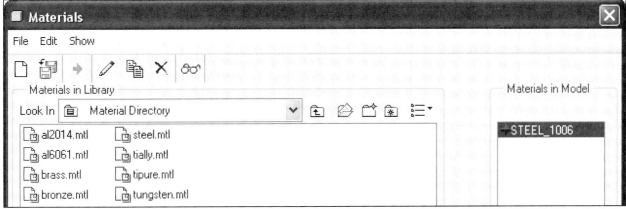

Figure 6.21(c) STEEL_1006

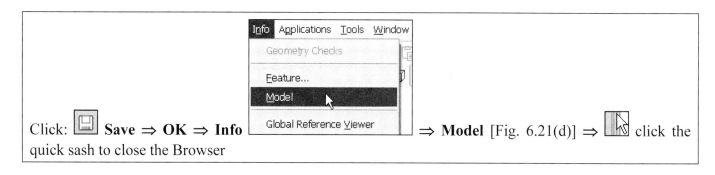

Click: 🖫 **Save** ⇒ **OK** ⇒ **Info** ⇒ **Model** [Fig. 6.21(d)] ⇒ click the quick sash to close the Browser

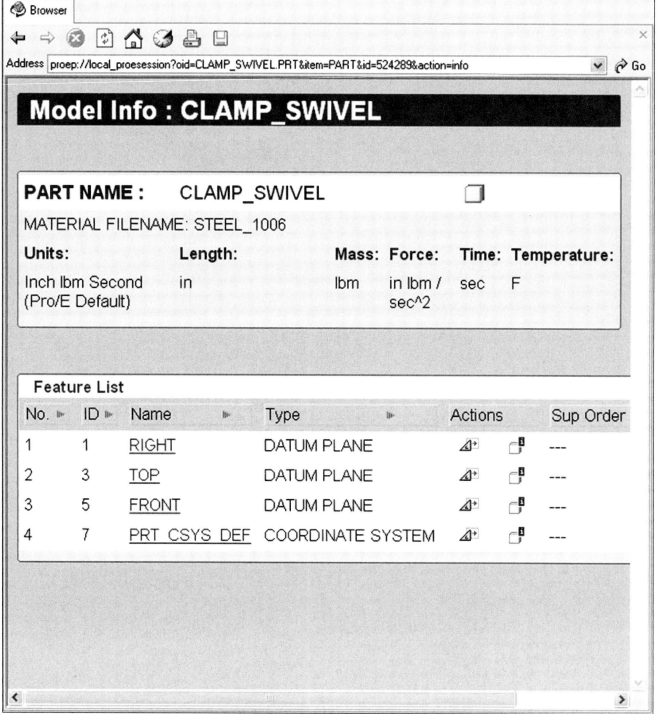

Figure 6.21(d) Model Info: Material Steel_1006

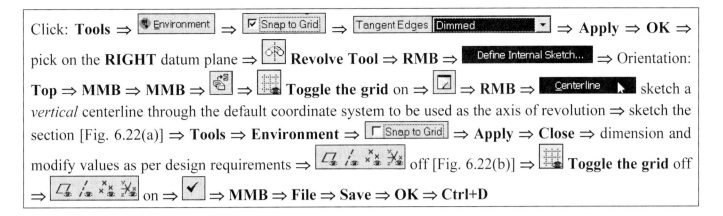

Click: **Tools** ⇒ [Environment] ⇒ [Snap to Grid] ⇒ [Tangent Edges Dimmed] ⇒ **Apply** ⇒ **OK** ⇒ pick on the **RIGHT** datum plane ⇒ [Revolve Tool] ⇒ **RMB** ⇒ [Define Internal Sketch...] ⇒ Orientation: **Top** ⇒ **MMB** ⇒ **MMB** ⇒ [icon] ⇒ [Toggle the grid on] ⇒ [icon] ⇒ **RMB** ⇒ [Centerline] sketch a *vertical* centerline through the default coordinate system to be used as the axis of revolution ⇒ sketch the section [Fig. 6.22(a)] ⇒ **Tools** ⇒ **Environment** ⇒ [Snap to Grid] ⇒ **Apply** ⇒ **Close** ⇒ dimension and modify values as per design requirements ⇒ [icons] off [Fig. 6.22(b)] ⇒ [Toggle the grid] off ⇒ [icons] on ⇒ [✓] ⇒ **MMB** ⇒ **File** ⇒ **Save** ⇒ **OK** ⇒ **Ctrl+D**

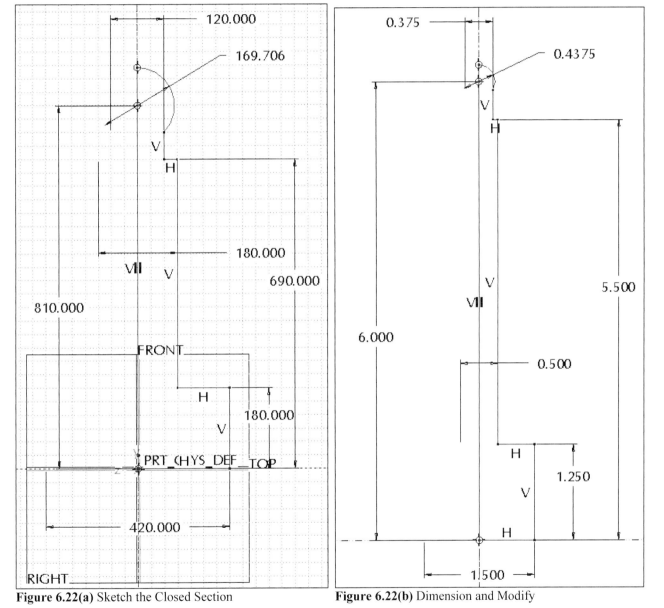

Figure 6.22(a) Sketch the Closed Section   **Figure 6.22(b)** Dimension and Modify

Click: **Edit** ⇒ **Setup** ⇒ **Geom Tol** ⇒ **Set Datum** ⇒ pick **TOP** from the model ⇒ Name- type **A** ⇒ **MMB** ⇒ pick **RIGHT** ⇒ Name- type **B** ⇒ **MMB** ⇒ pick **FRONT** ⇒ Name- type **C** ⇒ **MMB** ⇒ **MMB** ⇒ **MMB** ⇒ **MMB** ⇒ rename the coordinate system [CLAMP_SWIVEL_CSYS] ⇒ [💾] ⇒ **MMB** ⇒ **MMB** (rotate the model as needed) (Fig. 6.23)

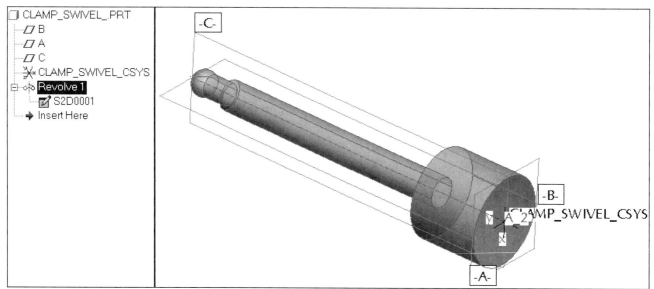

Figure 6.23 Completed Revolved Extrusion

Pick on datum plane **B** ⇒ [Hole icon] **Hole Tool** ⇒ [⌀ 0.500] ⇒ **Enter** ⇒ [icon] ⇒ **RMB** ⇒ [Secondary References Collector] ⇒ **Placement** tab ⇒ pick datum **A** ⇒ Offset **.5625** [A:F2(DATUM... Offset 0.5625] ⇒ press and hold the **Ctrl** key and pick on datum **C** ⇒ Offset **.000** [C:F3(DATUM... Offset 0.000] [Fig. 6.24(a)] ⇒ **Shape** tab [Fig. 6.24(b)] ⇒ Side 2: [Through All] ⇒ **MMB** rotate the model [Fig. 6.24(c)] ⇒ **MMB** [Fig. 6.24(d)] ⇒ **LMB** to deselect

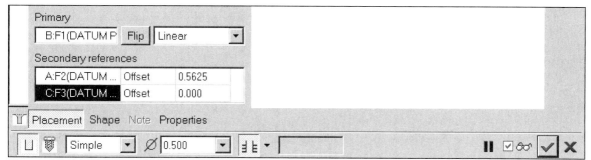

Figure 6.24(a) Placement Tab

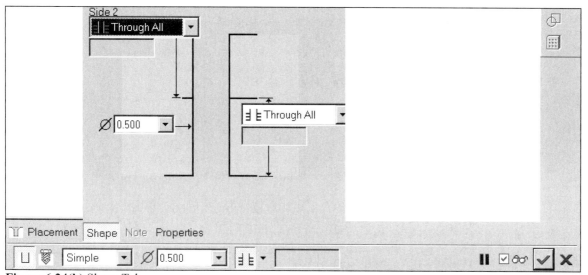

Figure 6.24(b) Shape Tab

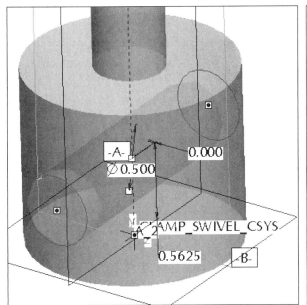
Figure 6.24(c) Hole Preview

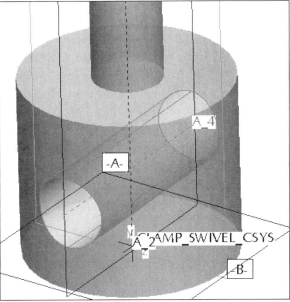
Figure 6.24(d) Completed Hole

Pick an edge, press and hold the **Ctrl** key and pick two more edges ⇒ **RMB** [Fig. 6.25(a)] ⇒ **Round Edges** ⇒ double-click on the dimension and modify to the design value of **.100** ⇒ **Enter** ⇒ **MMB** [Fig. 6.25(b)] ⇒ ⇒ **MMB** ⇒ **LMB** to deselect [Fig. 6.25(c)]

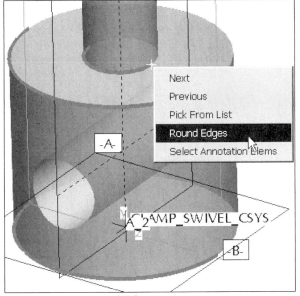
Figure 6.25(a) Round Edges

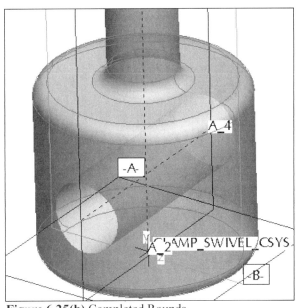
Figure 6.25(b) Completed Rounds

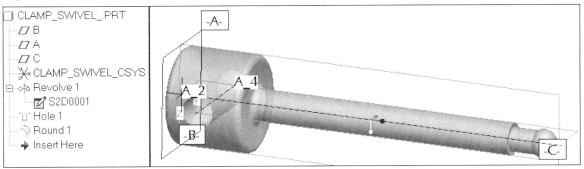
Figure 6.25(c) Completed Clamp_Swivel

Cosmetic Threads

A **cosmetic thread** is a feature that "represents" the diameter of a thread without having to show the actual threaded feature. Since a threaded feature is memory intensive, using cosmetic threads can save an enormous amount of visual memory on your computer. It is displayed in a unique default color. Unlike other cosmetic features, you cannot modify the line style of a cosmetic thread, nor are cosmetic threads affected by hidden line display settings in the ENVIRONMENT menu.

Threads are created with the default tolerance setting of limits. Cosmetic threads can be external or internal, and blind or through. You create cosmetic threads by specifying the minor or major diameter (for external and internal threads, respectively), starting surface, and thread length or ending edge.

For a starting surface, you can select a quilt surface, regular Pro/E surface, or split surface (such as a surface that belongs to a revolved feature, chamfer, round, or swept feature). For an "up to" surface, you can select any solid surface or a datum plane. A cosmetic thread that uses a depth parameter (a blind thread) cannot be defined from a non-planar surface.

The following table lists the parameters that can be defined for a cosmetic thread at its creation or later after the cosmetic thread has been added. In this table, "pitch" is the distance between two threads.

PARAMETER NAME	PARAMETER TYPE	PARAMETER DESCRIPTION
MAJOR_DIAMETER	Real Number	Thread major diameter
THREADS_PER_INCH	Real Number	Threads per inch (1/pitch)
THREAD FORM	String	Thread form
CLASS	String	Thread class
PLACEMENT	Character	Thread placement (A-external, B-internal)
METRIC	YesNo (True/False)	Thread is metric

Here, the external cosmetic thread is specified by the minor diameter (external threads), starting surface, and ending edge.

- **Internal cosmetic threads** are created automatically when holes are created using the Hole Tool (Standard- Tapped). In situations where the internal thread is unique and the Hole Tool cannot be used, internal cosmetic threads can be added to the hole.
- **External cosmetic threads** represent the *root diameter*. For threaded shafts you must create the external cosmetic threads.

After creating the cosmetic thread, edit the thread table. The thread size of an external thread must be changed to the nominal size from the root diameter defaulted on the thread table. Create an external cosmetic thread (**.500-13 UNC-2A**) using the Ø**.500** surface [Fig. 6.26(a)]. The thread starts at the "neck" and extends **4.00** along the swivel's shaft [Fig. 6.26(b)].

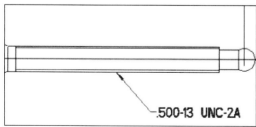

Figure 6.26(a) External Thread

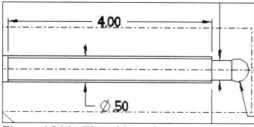

Figure 6.26(b) Thread Length

Click: **Insert** ⇒ **Cosmetic** ⇒ **Thread** ⇒ pick the cylindrical surface [Fig. 6.27(a)] ⇒ pick the thread start surface--the edge lip surface [Fig. 6.27(b)] ⇒ **Okay** [Fig. 6.27(c)] ⇒ **Blind** ⇒ **Done** ⇒ **4.00** ⇨ Enter depth 4.00 ⇒ **Enter** ⇒ type the value of the cosmetic thread root diameter: **.4485** ⇨ Enter DIAMETER .4485 ⇒ **Enter** [Fig. 6.27(d)] ⇒ **Mod Params** ⇒ edit the table

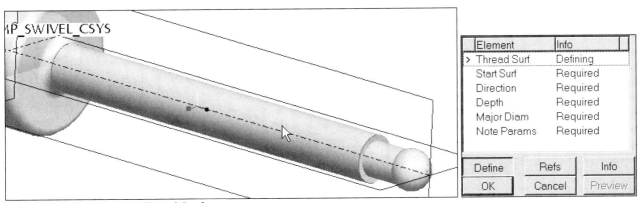

Figure 6.27(a) Select the Thread Surface

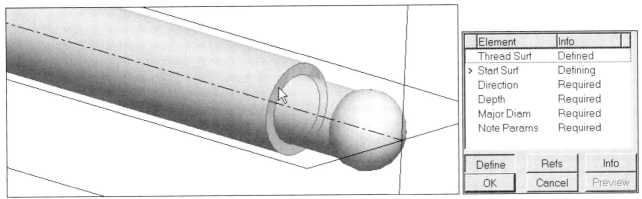

Figure 6.27(b) Select the Thread Start Surface

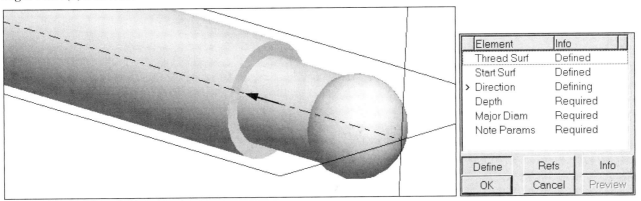

Figure 6.27(c) Thread Direction Arrow

Figure 6.27(d) COSEMETIC: Thread Dialog Box

The Pro/TABLE thread parameter [Fig. 6.27(e)] shows the diameter as **0.4485**. Since you are cosmetically representing the *root diameter (.4485)* of the external thread on the model, the *thread diameter* is *smaller* than the *nominal (.50) thread size*.

Click in each table field and change the values [Fig. 6.27(f)] ⇒ MAJOR_DIAMETER **.5** ⇒ THREADS_PER_INCH **13** ⇒ FORM **UNC** ⇒ CLASS **2** ⇒ PLACEMENT **A** ⇒ METRIC **FALSE** ⇒ from Pro/TABLE, click: **File** ⇒ **Save** ⇒ **File** ⇒ **Exit** ⇒ **Show** [Fig. 6.27(g)] ⇒ **Close** ⇒ **MMB** [Fig. 6.27(h)] ⇒ **OK** [Fig. 6.27(i)] ⇒ **LMB** ⇒ **Hidden Line** (to see the cosmetic surface) ⇒ **Shade**

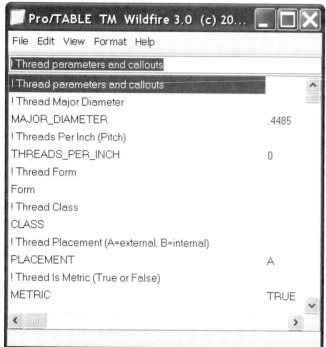

Figure 6.27(e) Pro/Table Thread Parameters

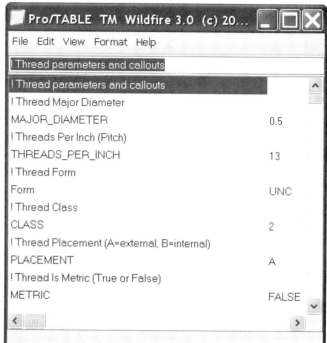

Figure 6.27(f) Edited Table Parameters

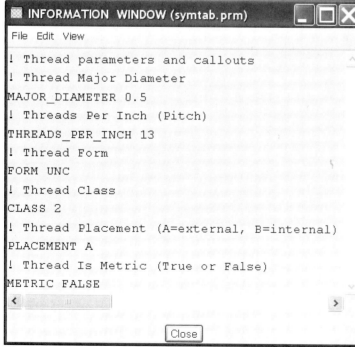

Figure 6.27(g) Thread Information Window

Figure 6.27(h) Completed Thread Elements

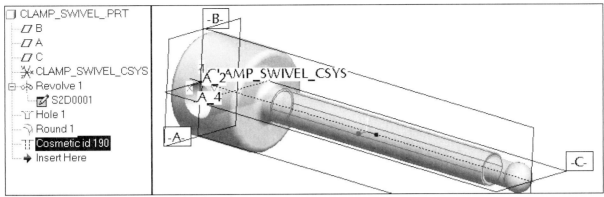

Figure 6.27(i) Completed Cosmetic Thread

Click on **Cosmetic id** in the Model Tree (Fig. 6.28) ⇒ **RMB** ⇒ **Info** ⇒ **Feature** ⇒ click on the quick sash to close the Browser ⇒ **Ctrl+S** ⇒ **Enter** ⇒ **LMB** to deselect

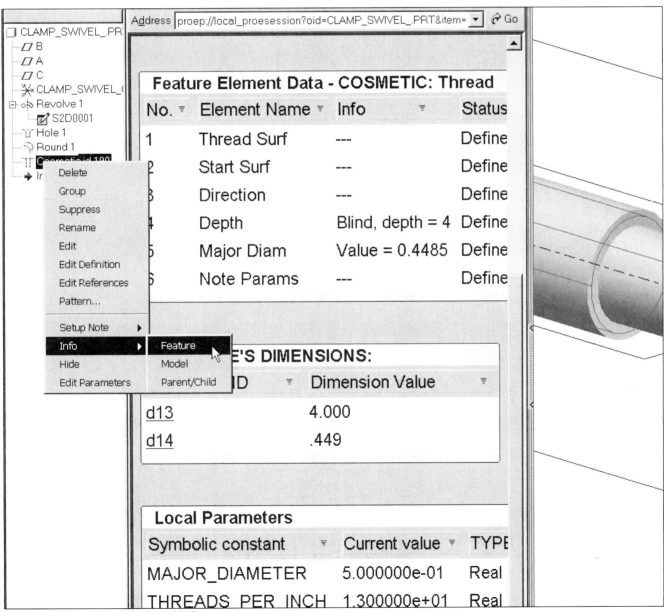

Figure 6.28 COSMETIC Thread Info

Using the Model Player

The **Model Player** (Fig. 6.29) option on the **Tools** menu lets you observe how a part is built. You can:

- Move backward or forward through the feature-creation history of the model in order to observe how the model was created. You can start the model playback at any point in its creation history
- Regenerate each feature in sequence, starting from the specified feature
- Display each feature as it is regenerated or rolled forward
- Update (regenerate all the features in) the entire display when you reach the desired feature
- Obtain information about the current feature (you can show dimensions, obtain regular feature information, investigate geometry errors, and enter Fix Model mode)

To open the **Model Player** dialog box, click Tools ⇒ Model Player. You can select one of the following:

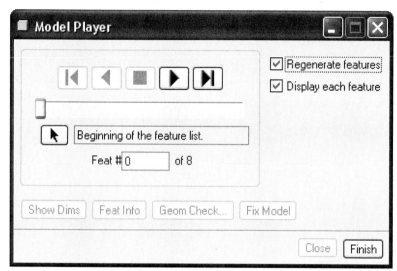

- **Regenerate features** Regenerates each feature in sequence, starting from the specified feature
- **Display each feature** Displays each feature in the graphics window as it is being regenerated
- **Compute CL** (Available in Manufacturing mode only) When selected, the CL data is recalculated for each NC sequence during regeneration

Figure 6.29 Model Player

Select one of the following commands:

Go to the beginning of the model moves immediately to the beginning of the model

Step backward through the model one feature at-a-time and regenerates the preceding feature

Stop play

Step forward through the model one feature at-a-time and regenerates the next feature

Go to the last feature in the model moves immediately to the end of the model (resume all)

Slider Bar Drag the slider handle to the feature at which you want model playback to begin. The features are highlighted in the graphics window as you move through their position with the slider handle. The feature number and type are displayed in the selection panel [such as #4 (COORDINATE SYSTEM)] [Fig. 6.30(b)], and the feature number is displayed in the **Feat #** box.

Select feature from screen or model tree Lets you select a starting feature from the graphics window or the Model Tree. Opens the **SELECT FEAT** and **SELECT** menus. After you select a starting feature, its number and ID are displayed in the selection panel, and the feature number is displayed in the **Feat #** box.

Feat # 5 of 25 Lets you specify a starting feature by typing the feature number in the box. After you enter the feature number, the model immediately rolls or regenerates to that feature.

To stop playback, click the **Stop play** button. Use the following commands for information:

- **Show Dims** Displays the dimensions of the current feature
- **Feat Info** Provides regular feature information about the current feature in an Information window
- **Geom Check** Investigates the geometry error for the current feature
- **Fix Model** Activates Resolve mode by forcing the current feature to abort regeneration
- **Close** Closes the Model Player and enters Insert mode at the current feature
- **Finish** Closes the Model Player and returns to the last feature in the model

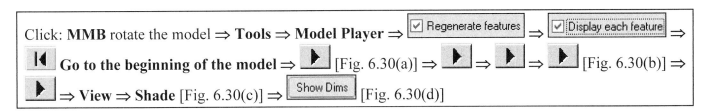

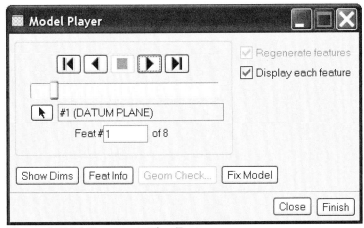
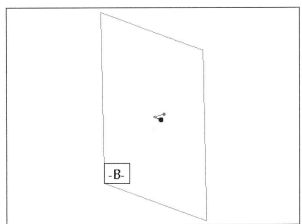

Figure 6.30(a) Regenerate First Feature

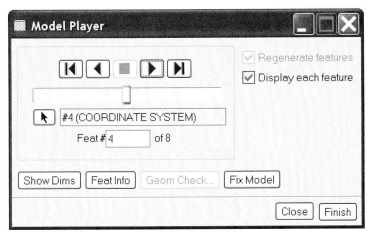
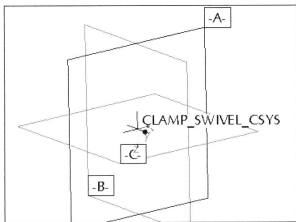

Figure 6.30(b) Regenerate the Coordinate System

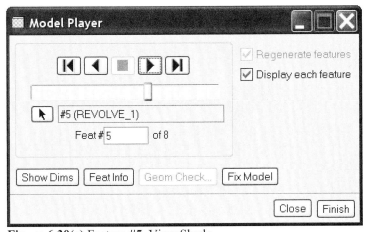
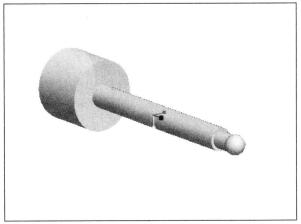

Figure 6.30(c) Feature #5, View Shade

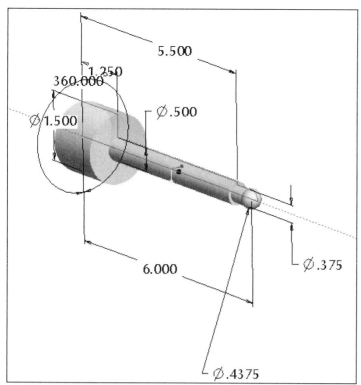

Figure 6.30(d) Feature #5, Revolve Dimensions

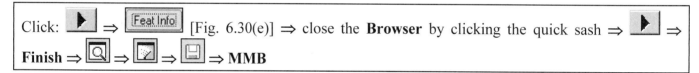

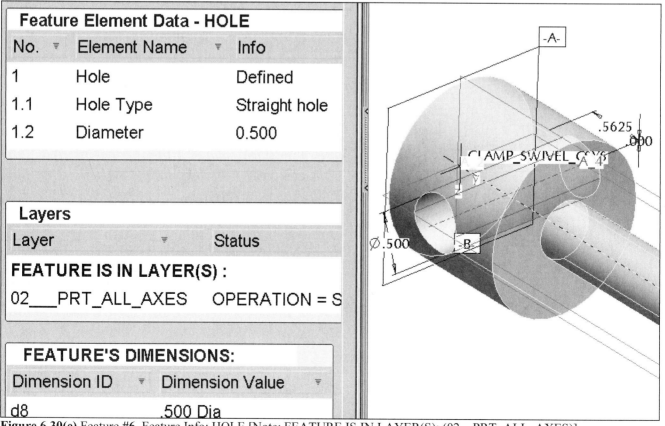

Figure 6.30(e) Feature #6, Feature Info: HOLE [Note: FEATURE IS IN LAYER(S): (02__PRT_ALL_AXES)]

Change the names of two datums. Double-click on datum **B** and change it to datum **D** ⇒ double-click on datum **C** and change it to datum **B** ⇒ double-click on datum **D** and change it to datum **C** [Figs. 6.31(a-f)]
[Q] ⇒ [/] ⇒ [🖫] ⇒ **MMB** ⇒ **RMB** ⇒ **Info** ⇒ **Model** (Fig. 6.32) ⇒ close the Browser

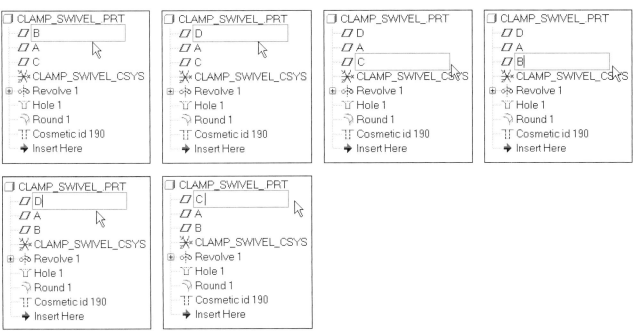

Figures 6.31(a-f) Change the names of the Datums

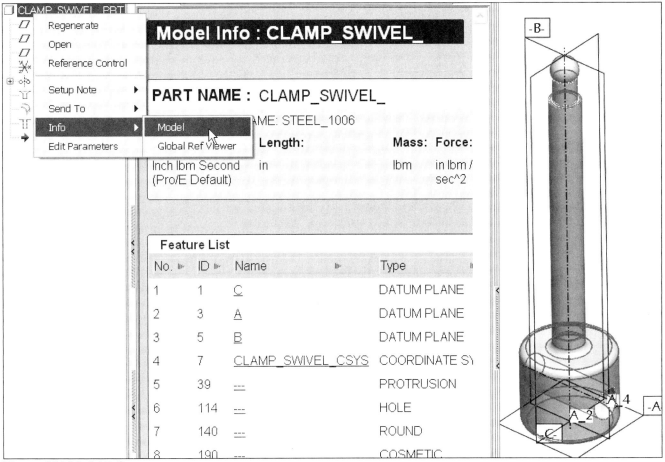

Figure 6.32 Model Info

Printing and Plotting

From the File menu, you can print with the following options: scaling, clipping, displaying the plot on the screen, or sending the plot directly to the printer. Shaded images can also be printed from this menu. You can create plot files of the current object (sketch, part, assembly, drawing, or layout) and send them to the print queue of a plotter. The plotting interface to HPGL and PostScript formats are standard.

You can configure your printer using the MS Printer Manager Print dialog box, available from the Pro/E Print dialog box. If you are printing a shaded image, the Printer Configuration dialog box opens.

The following applies for plotting:

- Hidden lines appear as gray for a screen plot, but as dashed lines on paper.
- When Pro/E plots Pro/E line fonts, it scales them to the size of a sheet. It does not scale the user-defined line fonts, which do not plot as defined.
- You can use the configuration file option *use_software_linefonts* to make sure that the plotter plots a user-defined font exactly as it appears in Pro/E.
- You can plot a cross section from Part or Assembly mode.
- With a Pro/PLOT license, you can write plot files in a variety of formats.

Print your object, click: **Print the active object** ⇒ **OK** [Fig. 6.33(a)] ⇒ **OK** [Fig. 6.33(b)] ⇒ ⇒ **MMB** ⇒ **File** ⇒ **Delete** ⇒ **Old Versions** ⇒ **MMB** ⇒ **Window** ⇒ **Close**

Figure 6.33(a) Print Dialog Box

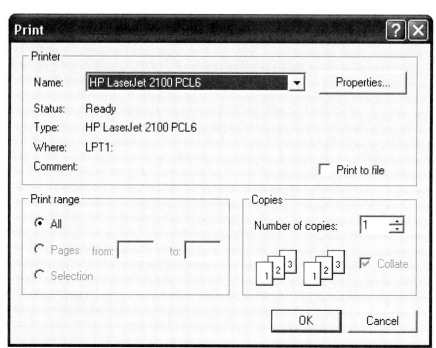

Figure 6.33(b) MS Printer Manager Print Dialog Box

The lesson is now complete. If you wish to model a project without instructions, a complete set of projects and illustrations are available at ***www.cad-resources.com*** ⇒ ***Downloads.***

Lesson 7 Feature Operations

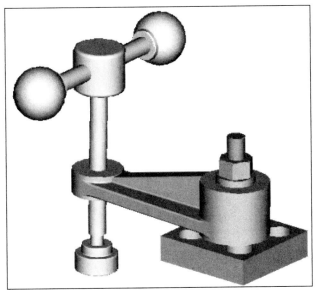

Figure 7.1 Swing Clamp Assembly

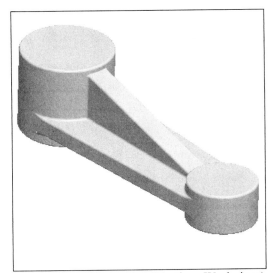

Figure 7.2(a) Clamp Arm (Casting- Workpiece)

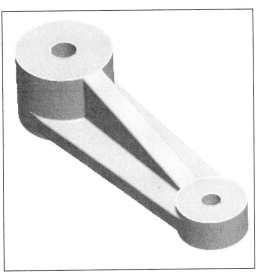

Figure 7.2(b) Clamp Arm (Machined- Design Part)

OBJECTIVES

- Use **Copy** and **Paste Special**
- Create **Ribs**
- Write **Relations** to control features
- Understand **Parameters**
- Understand and solve **Failures**
- Create a workpiece using a **Family Table**

FEATURE OPERATIONS

The Clamp Arm is used in the Swing Clamp Assembly (Fig. 7.1). This lesson will cover a wide range of Pro/E capabilities including: **Feature Operations**, **Copy** and **Paste Special**, **Relations**, **Parameters**, **Failures**, **Family Tables**, and the **Rib Tool**. You will create two versions of the Clamp Arm; one with all cast surfaces [Fig. 7.2(a)] and the other with machined ends [Fig. 7.2(b)] (using a Family Table).

Ribs

A rib is a special type of protrusion designed to create a thin fin or web that is attached to a part. You always sketch a rib from a side view, and it grows about the sketching plane symmetrically or to either side. Because of the way ribs are attached to the parent geometry, they are always sketched as open sections.

When sketching an open section, Pro/E may be uncertain about the side to which to add the rib. Pro/E adds all material in the direction of the arrow. If the incorrect choice is made, toggle the arrow direction by clicking on the direction arrow on the screen.

A rib must "see" material everywhere it attaches to the part; otherwise, it becomes an unattached feature. There are two types of ribs: straight [Fig. 7.3(a)] and rotational [Fig. 7.3(b)]. The type is automatically set according to the attaching geometry (planar or curved).

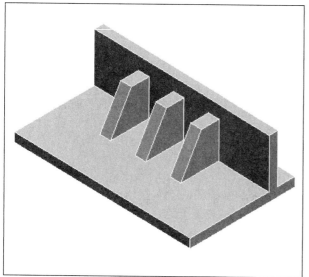

Figure 7.3(a) Straight Rib

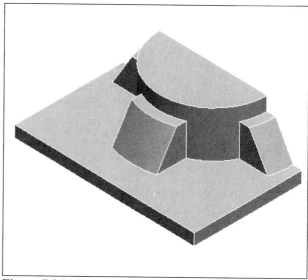

Figure 7.3(b) Rotational Rib

Relations

Relations (also known as parametric relations) are user-defined equations written between symbolic dimensions and parameters. Relations capture design relationships within features or parts, or among assembly components, thereby allowing users to control the effects of modifications on models.

Relations can be used to control the effects of modifications on models, to define values for dimensions in parts and assemblies, and to act as constraints for design conditions (for example, specifying the location of a hole in relation to the edge of a part). They are used in the design process to describe conditional relationships between different features of a part or an assembly.

Relations can be used to provide a value for a dimension. However, they can also be used to notify you when a condition has been violated, such as when a dimension exceeds a certain value. There are two basic types of relations, equality and comparison.

An equality relation equates a parameter on the left side of the equation to an expression on the right side. This type of relation is used for assigning values to dimensions and parameters. The following are a few examples of equality relations:

$$d2 = 25.500 \qquad d8 = d4/2 \qquad d7 = d1+d6/2 \qquad d6 = d2*(sqrt(d7/3.0+d4))$$

A comparison relation compares an expression on the left side of the equation to an expression on the right side. This type of relation is commonly used as a constraint or as a conditional statement for logical branching. The following are examples of comparison relations:

    d1 + d2 > (d3 + 5.5)   Used as a constraint
    IF (d1 + 5.5) > = d7   Used in a conditional statement

Parameter Symbols

Four types of parameter symbols are used in relations:

- **Dimensions** These are dimension symbols, such as **d8, d12**.
- **Tolerances** These are parameters associated with ± symmetrical and plus-minus tolerance formats. These symbols appear when dimensions are switched from numeric to symbolic.
- **Number of Instances** These are integer parameters for the number of instances in a direction of a pattern.
- **User Parameter** These can be parameters defined by adding a parameter or a relation (e.g., **Volume = d3 * d4 * d5**).

Operators and Functions

The following operators and functions can be used in equations and conditional statements:

Arithmetic Operators

+	Addition
−	Subtraction
/	Division
*	Multiplication
^	Exponentiation
( )	Parentheses for grouping [for example, (d0 = (d1–d2)*d3)]

Assignment Operators

=	Equal to

    The = (equals) sign is an assignment operator that equates the two sides of an equation or relation. When it is used, the equation can have only a single parameter on the left side.

Comparison Operators

Comparison operators are used whenever a TRUE/FALSE value can be returned. For example, the relation **d1 >= 3.5** returns TRUE whenever d1 is greater than or equal to **3.5**. It returns FALSE whenever **d1** is less than **3.5**. The following comparison operators are supported:

==	Equal to
>	Greater than
>=	Greater than or equal to
!=, <>, ~=	Not equal to
<	Less than

<=	Less than or equal to
|	Or
&	And
~, !	Not

Mathematical Functions

The following operators can be used in relations, both in equations and in conditional statements. Relations may include the following mathematical functions:

cos ( )	cosine
tan ( )	tangent
sin ( )	sine
sqrt ( )	square root
asin ( )	arc sine
acos ( )	arc cosine
atan ( )	arc tangent
sinh ( )	hyperbolic sine
cosh ( )	hyperbolic cosine
tanh ( )	hyperbolic tangent

Failures

Sometimes model geometry cannot be constructed because features that have been modified or created conflict with or invalidate other features. This can happen when the following occurs:

- A protrusion is created that is unattached and has a one-sided edge.
- New features are created that are unattached and have one-sided edges.
- A feature is resumed that now conflicts with another feature (i.e. two chamfers on the same edge).
- The intersection of features is no longer valid because dimensional changes have moved the intersecting surfaces.
- A relation constraint has been violated.

Resolve Feature

After a feature fails, Pro/E enters Resolve Feature mode. Use the commands in the RESOLVE FEAT menu to fix the failed feature:

- **Undo Changes** Undo the changes that caused the failed regeneration attempt, and return to the last successfully regenerated model.
- **Investigate** Investigate the cause of the regeneration failure using the Investigate submenu.
- **Fix Model** Roll the model back to the state before failure and select commands to fix the problem.
- **Quick Fix** Choose an option from the **QUICK FIX** menu, the options are as follows:
    - **Redefine** Redefine the failed feature.
    - **Reroute** Reroute the failed feature.
    - **Suppress** Suppress the failed feature and its children.
    - **Clip Supp** Suppress the failed feature and all the features after it.
    - **Delete** Delete the failed feature.

Failed Features

If a feature fails during creation and it does not use the dialog box interface, Pro/E displays the FEAT FAILED menu with the following options:

- **Redefine** Redefine the feature
- **Show Ref** Display the SHOW REF menu so you can see the references of the failed feature. Pro/E displays the reference number in the Message Window.
- **Geom Check** Check for problems with overlapping geometry, misalignment, and so on
- **Feat Info** Get information about the feature

If a feature fails, you can redisplay the part with all failed geometry highlighted in different colors. Pro/E displays the corresponding error messages in an Information Window. Features can fail during creation for the following reasons:

- **Overlapping geometry** A surface intersects itself. If Pro/E finds a self-intersecting surface, it does not perform any further surface checks.
- **Surface has edges that coincide** The surface has no area. Pro/E highlights the surface in red.
- **Inverted geometry** Pro/E highlights the inverted geometry in purple and displays an error message.
- **Bad edges** Pro/E highlights bad edges in blue and displays an error message.

Family Tables

Family Tables are effective for two main reasons: they provide a beneficial tool, and they are easy to use. You need to understand the functionality of Family Tables, and you must understand when a Family Table is required and what circumstances should promote its use.

To determine whether a model is a candidate for a Family Table: establish whether the original and the variation would ever have to co-exist at the same time (both in the same assembly, both shown in the same drawing, both with an independent Bill of Materials) and whether they should be tied together (most of the same dimensions, features, and parameters). If so, the component is a candidate for the creation of a Family Table [Figs. 7.4(a-b)], otherwise, the model may be a candidate for copying to an independent model.

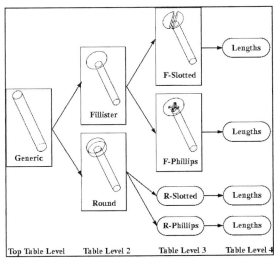

Figure 7.4(a) Screw Family (CADTRAIN)

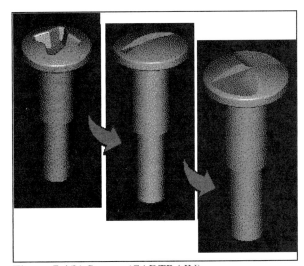

Figure 7.4(b) Screws (CADTRAIN)

Lesson 7 STEPS

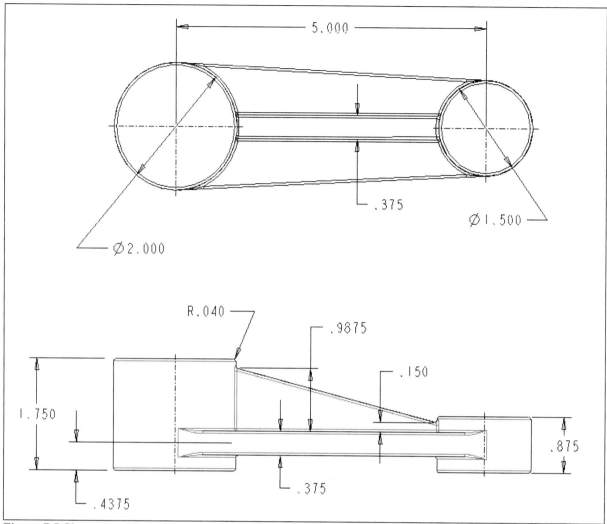

Figure 7.5 Clamp Arm Workpiece Drawing

Clamp Arm

The Clamp_Arm is modeled in two stages. Model the casting (Fig. 7.5) using the casting detail, and then use the machine detail to complete the last (machined) features. The last step will be to create a Family Table with an instance that suppresses the machined features. By having a *casting part* (which is called a ***workpiece*** in **Pro/NC**) and a separate but almost identical *machined part* (which is called a ***design part*** in **Pro/NC**), you can create an operation for machining and an NC sequence. During the manufacturing process, *you merge the workpiece into the design part* and create a ***manufacturing model***.

The difference between the two files is the difference between the volume of the workpiece/casting part and the volume of the design part/machined part. The removed volume can be seen as *material removal* when you are performing an **NC Check** operation on the manufacturing model. If the machining process gouges the part, the gouge will display the interference. The cutter location can also be displayed as an animated machining process.

For additional tutorials on Pro/MANUFACTURING: Pro/NC, and Expert Machinist see: www.cad-resources.com and click ***CDI72 Pro/ENGINEER Advanced***.

Start a new part, click: **File** ⇒ **Set Working Directory** select your working directory ⇒ **OK** ⇒ **File** ⇒ **New** ⇒ `● □ Part` ⇒ `Name  CLAMP_ARM` ⇒ `☑ Use default template` ⇒ **OK** ⇒ **Tools** ⇒ **Options** ⇒ Showing: **Current Session** ⇒ `📂` ⇒ *if available* open your previously created option file ⇒ **Apply** ⇒ **Close** ⇒ **Tools** ⇒ `Environment` ⇒ `☐ Snap to Grid` ⇒ `☑ Use 2D Sketcher` ⇒ `Tangent Edges | Dimmed` ⇒ **OK** ⇒ **Edit** ⇒ **Setup** ⇒ **Material** ⇒ `steel.mtl` ⇒ `▶▶▶` ⇒ **OK** ⇒ **MMB** ⇒ change the coordinate system name to `CLAMP_ARM_CSYS` ⇒ `💾` ⇒ **MMB** ⇒ pick on datum **FRONT** to pre-select it ⇒ `🗂` ⇒ **RMB** ⇒ `Define Internal Sketch...` [Fig. 7.6(a)] ⇒ **MMB** ⇒ **RMB** ⇒ `Circle` sketch a circle ⇒ **MMB** ⇒ **MMB** ⇒ **Ctrl+D** ⇒ double-click on the diameter dimension and change to **2.00** [Fig. 7.6(b)] ⇒ **Enter** ⇒ `✔` [Fig. 7.6(c)] ⇒ double-click on the height dimension and change to **1.75** [Fig. 7.6(c)] ⇒ **Enter** ⇒ **MMB** ⇒ **Ctrl+S** ⇒ **MMB**

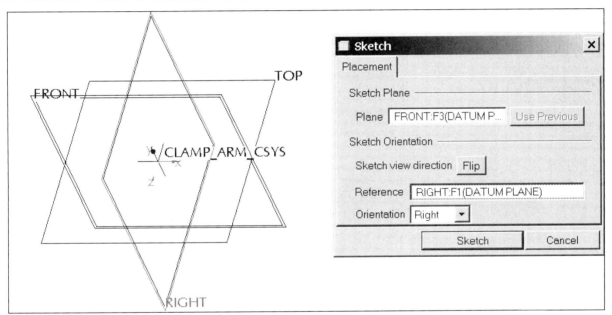

Figure 7.6(a) Sketch Plane and Orientation

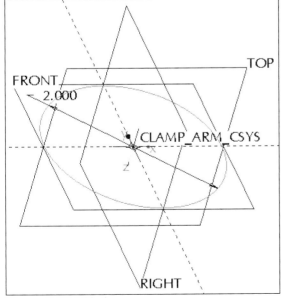

Figure 7.6(b) Circle

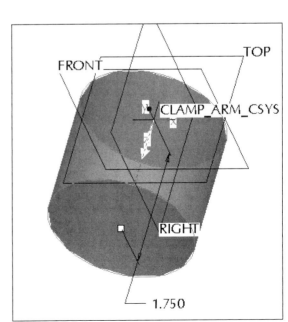

Figure 7.6(c) Extrusion Preview

With the extrusion selected/highlighted, click: **Ctrl+C** ⇒ **Edit** ⇒ [Paste Special...] [Fig. 7.7(a)] ⇒ [☐ Dependent copy] ⇒ [☑ Apply Move/Rotate transformations to copies] ⇒ **OK** ⇒ [Transformations] ⇒ pick datum **Right** as the **Direction reference** ⇒ move the drag handle to the right until **5.00** [Fig. 7.7(b)] ⇒ **MMB** ⇒ **Ctrl+S** ⇒ **Enter** ⇒ **LMB** to deselect

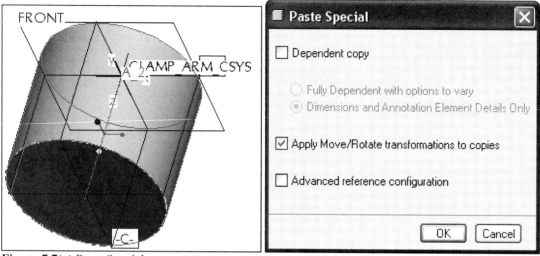

Figure 7.7(a) Paste Special

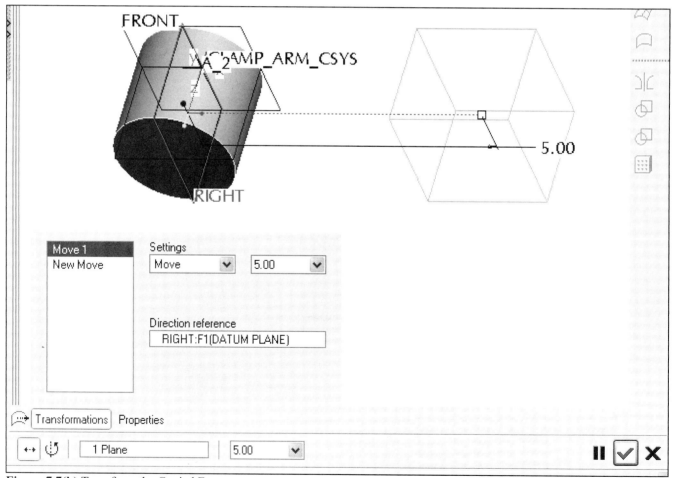

Figure 7.7(b) Transform the Copied Feature

Double-click on the copied feature and modify the diameter dimension to **1.50** and the height dimension to **.875** [Fig. 7.7(c)] ⇒ **Regenerates Model** [Fig. 7.7(d)] ⇒ **Ctrl+S** ⇒ **MMB**

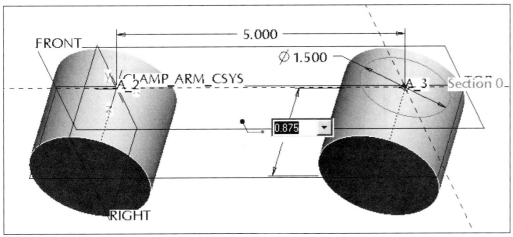

Figure 7.7(c) Modify the Dimensions

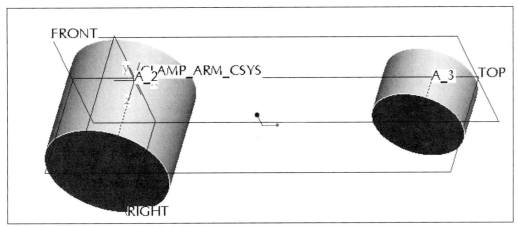

Figure 7.7(d) Regenerated Model

Pick on the **FRONT** datum plane ⇒ **Datum Plane Tool** ⇒ Translation 0.4375 (Fig. 7.8) ⇒ **OK**

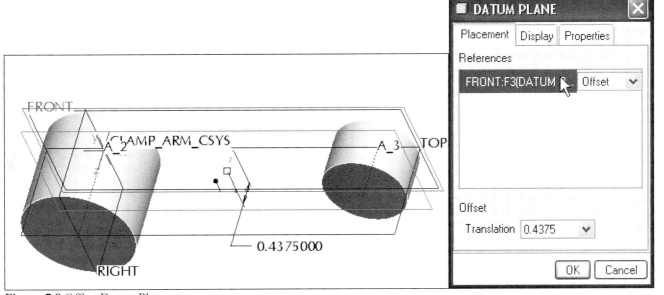

Figure 7.8 Offset Datum Plane

233

With datum **DTM1** selected, click: ▢ **Extrude Tool** ⇒ **RMB** ⇒ [Define Internal Sketch...] ⇒ **MMB** ⇒ **Sketch** ⇒ **References** [Figs. 7.9(a-b)] ⇒ add one arc from both circles [Figs. 7.9(c-d)] ⇒ **Close**

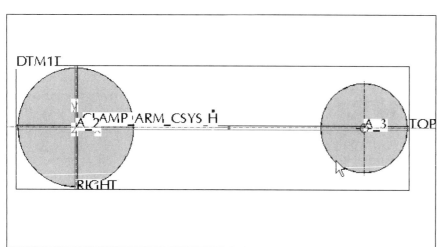

Figure 7.9(a) References **Figure 7.9(b)** References Dialog Box

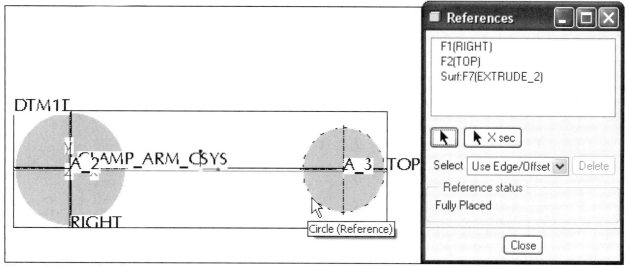

Figure 7.9(c) Add Reference

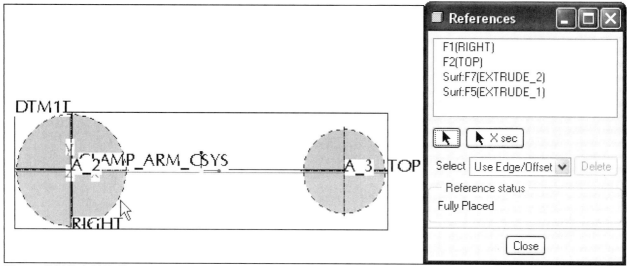

Figure 7.9(d) Add Reference

Click: **RMB** ⇒ **Centerline** ⇒ create a horizontal centerline [Fig. 7.9(e)] ⇒ **RMB** ⇒ **Line** ⇒ create a line that is tangent to both circular extrusions, pick a tangent starting position [Figs. 7.9(f-g)] ⇒ complete the line by picking on the opposite circular reference [Figs. 7.9(h-i)] ⇒ **MMB** ⇒ create the bottom tangent line ⇒ **MMB** ⇒ **Hidden Line** [Fig. 7.9(j)] ⇒ **MMB**

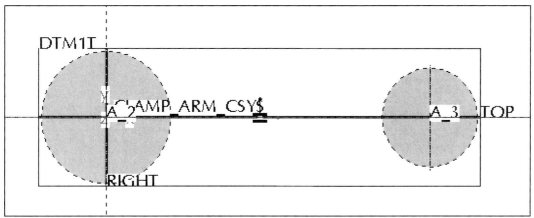

Figure 7.9(e) Add Centerline

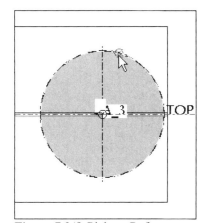 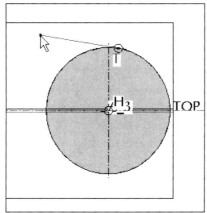 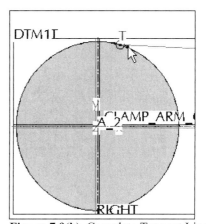

Figure 7.9(f) Pick on Reference   **Figure 7.9(g)** Make Sure T Displays   **Figure 7.9(h)** Complete Tangent Line

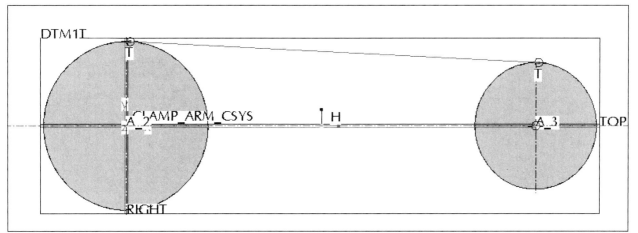

Figure 7.9(i) Tangent Line

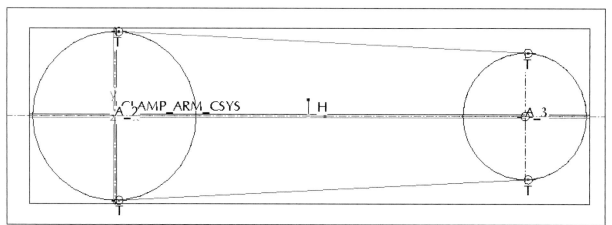

Figure 7.9(j) Tangent Lines

Click: [icon] **Create an arc** ⇒ pick the center and ends (from each tangent end of the line) [Fig. 7.9(k)] ⇒ repeat on the opposite end [Fig. 7.9(l)] ⇒ **MMB** ⇒ **LMB** to deselect ⇒ [icon] [Fig. 7.9(m)]

Figure 7.9(k) Arc **Figure 7.9(l)** Arc Opposite End

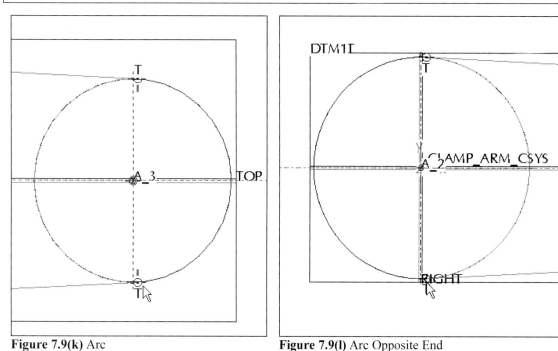

Figure 7.9(m) Completed Sketch

Click: 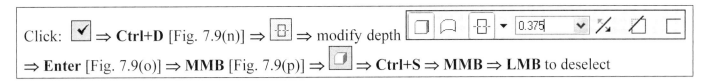 ⇒ **Ctrl+D** [Fig. 7.9(n)] ⇒ ⊞ ⇒ modify depth
⇒ **Enter** [Fig. 7.9(o)] ⇒ **MMB** [Fig. 7.9(p)] ⇒ ⊡ ⇒ **Ctrl+S** ⇒ **MMB** ⇒ **LMB** to deselect

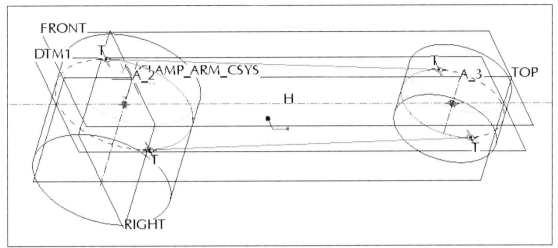

Figure 7.9(n) 3D Sketch

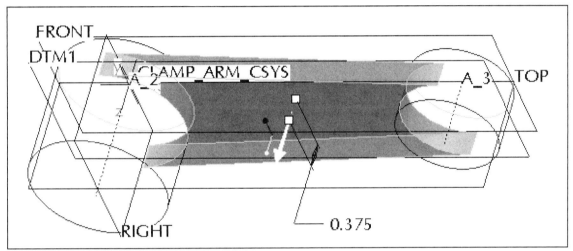

Figure 7.9(o) Preview

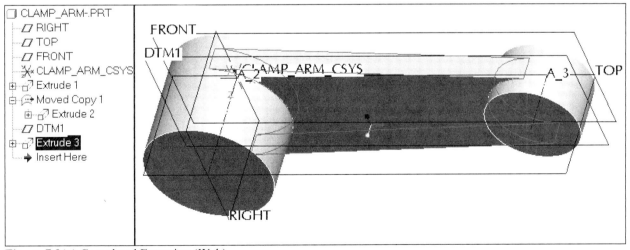

Figure 7.9(p) Completed Extrusion (Web)

Click: **Rib Tool** ⇒ **RMB** ⇒ Define Internal Sketch... ⇒ Sketch Plane- Plane: pick **TOP** [Fig. 7.10(a)] ⇒ **Flip** Sketch view direction [Flip] [Fig. 7.10(b)] ⇒ **Sketch** from dialog box ⇒ off ⇒ **Sketch** from Toolchest ⇒ **References** ⇒ delete the two existing references [Fig. 7.10(c)] ⇒ [Fig. 7.10(d)]

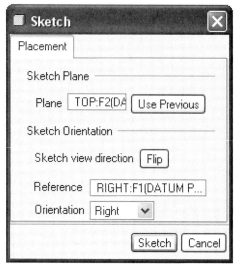

Figure 7.10(a) Sketch Dialog Box

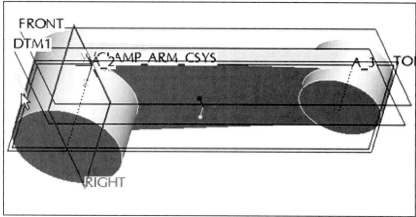

Figure 7.10(b) Flip the Viewing Direction Arrow

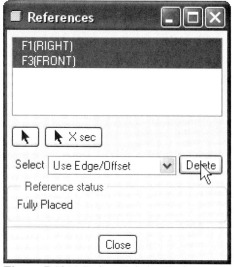

Figure 7.10(c) Delete Existing References

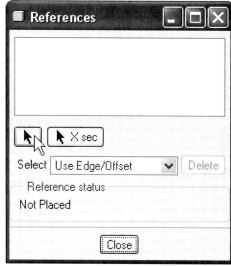

Figure 7.10(d) Select Button

Add three new references, pick on the inside vertical edge of both cylindrical extrusions and the top edge of the web extrusion [Fig. 7.10(e-f)] ⇒ **Close** ⇒ [icon] on ⇒ **RMB** ⇒ **Line** ⇒ **Ctrl+MMB** to zoom in ⇒ draw one angled line from one vertical reference to the other [Fig. 7.10(g)] (zoom in if necessary to pick the vertical surface/edge, not the endpoint) ⇒ **MMB** ⇒ **RMB** ⇒ **Dimension** add the vertical dimensions [Fig. 7.10(h)] ⇒ **MMB** ⇒ double-click on a dimension ⇒ change the dimension value [Fig. 7.10(i)] ⇒ **Enter** ⇒ repeat for the other dimension [(**.9875** and **.15** are the design values)]

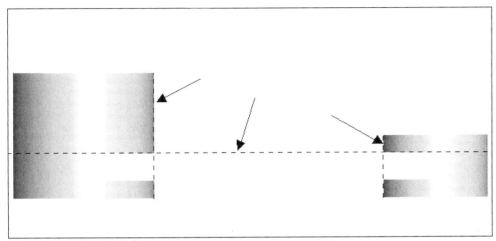

Figure 7.10(e) New References

Figure 7.10(f) References Dialog Box

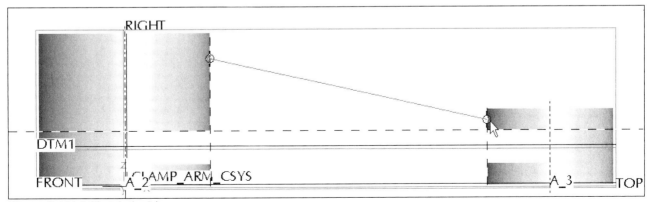

Figure 7.10(g) Sketch the Line

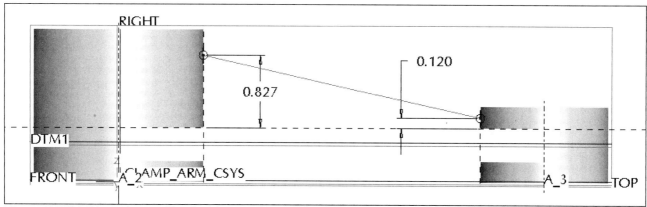

Figure 7.10(h) Dimensions as Sketched (your values will be different)

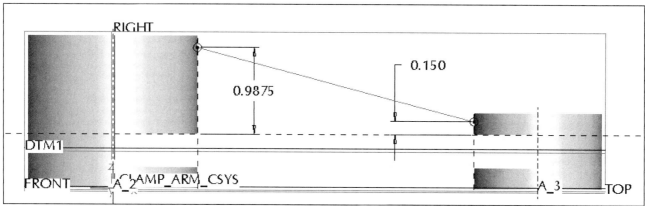

Figure 7.10(i) Modify the Dimensions to the Design Values

Click: **Ctrl+D** ⇒ ☑ ⇒ if necessary, pick on the yellow arrow to flip the direction of the rib creation towards the part ⇒ move the drag handle until a rib thickness of **.375** is displayed [Fig. 7.10(j)] ⌐ 0.375 ⌐ (or type the value and hit Enter) ⇒ **MMB** ⇒ **Ctrl+S** ⇒ **MMB** ⇒ **LMB** in graphics area to deselect

(Note: clicking ⧄ in the **dashboard** of the **Rib Tool** toggles the rib from centered, to the right side, or to the left side of the sketch plane).

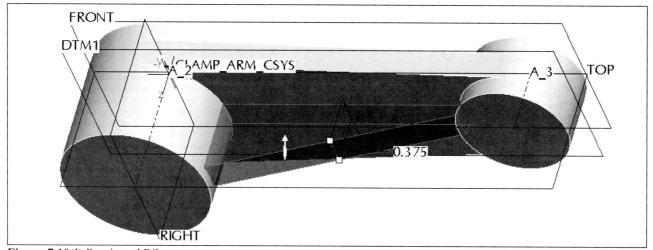

Figure 7.10(j) Previewed Rib

Create the rounds, click *slowly* on an edge until it highlights in red ⇒ press and hold the **Ctrl** key and pick on the other edges until they highlight ⇒ **RMB** ⇒ **Round Edges** [Fig. 7.11(a)] ⇒ modify the radius to **.04** [.04] ⇒ **Enter** ⇒ **MMB** [Fig. 7.11(b)] ⇒ rotate the part ⇒ **View** ⇒ **Shade** ⇒ select the six edges [Fig. 7.11(c)] ⇒ **RMB** ⇒ **Round Edges** ⇒ **MMB** ⇒ **Ctrl+S** ⇒ **MMB** ⇒ **LMB** to deselect

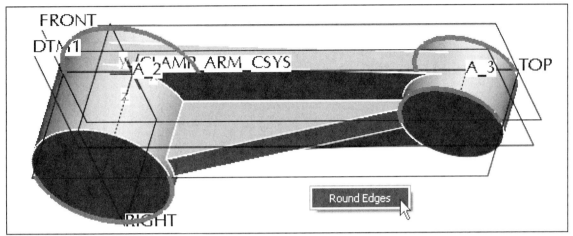

Figure 7.11(a) Previewed Rounds

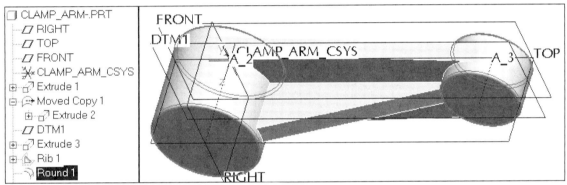

Figure 7.11(b) Rounds Created

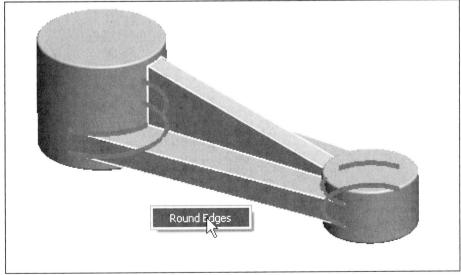

Figure 7.11(c) Select the Six Edges to be Rounded

Click: [icon] off ⇒ pick on the six edges [Fig. 7.11(d)] ⇒ **RMB** ⇒ **Round Edges** ⇒ [icon 0.040] ⇒ **Enter** ⇒ **MMB** ⇒ **LMB** to deselect ⇒ pick on the edge [Fig. 7.11(e)] ⇒ **RMB** ⇒ **Round Edges** ⇒ **MMB** [Fig. 7.11(f)] ⇒ [icon] on ⇒ **Ctrl+S** ⇒ **MMB** ⇒ **LMB** to deselect ⇒ **Ctrl+D**

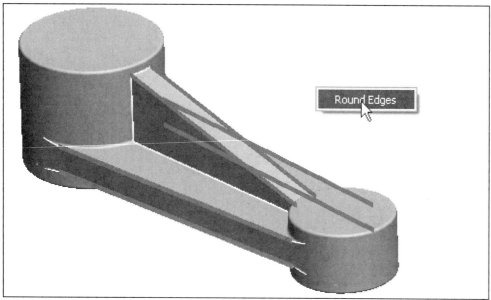

Figure 7.11(d) Round Edges

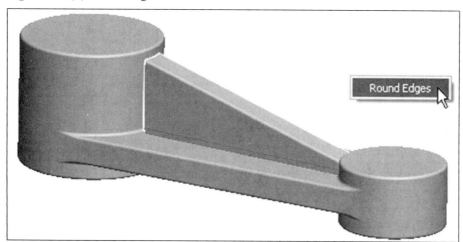

Figure 7.11(e) Edge to be Rounded

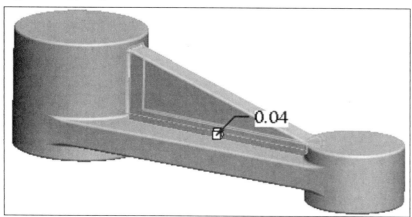

Figure 7.11(f) Round is Automatically Generated around the Rib

The "machined" features of the part can be created with cuts to "face" the ends of the cylindrical protrusions. A standard tapped hole, and a thru hole with two countersinks are also added as "machined" features. You will be using the dimensions from the machine drawing (Fig. 7.12).

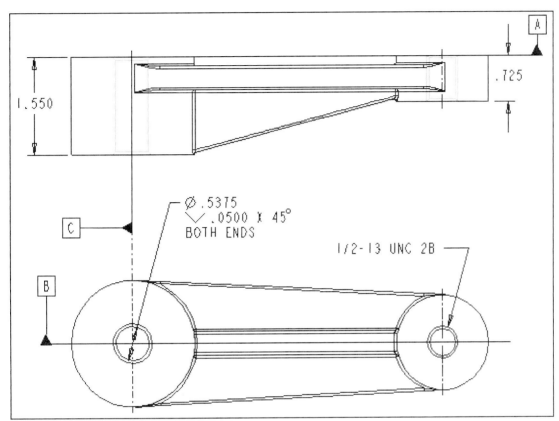

Figure 7.12 Clamp Arm Machining Drawing

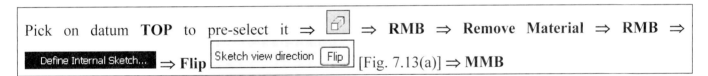

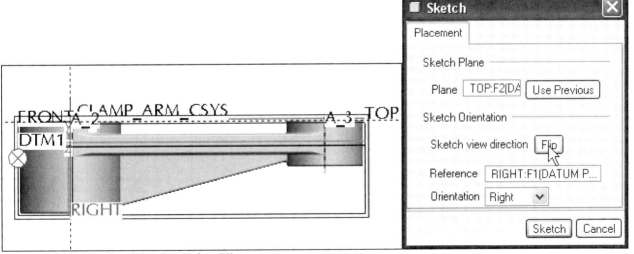

Figure 7.13(a) Viewing Direction before Flip

Click: [icons] off ⇒ **Sketch** ⇒ **References** ⇒ [icon] ⇒ pick on the outside vertical edge of both cylindrical extrusions as references [Fig. 7.13(b)] ⇒ **RMB** ⇒ **Line** ⇒ **Ctrl+MMB** to zoom in ⇒ draw a horizontal line from one vertical reference to the other [Fig. 7.13(c)] ⇒ **MMB** ⇒ **MMB** ⇒ double-click the dimension ⇒ **.100** ⇒ **Enter** ⇒ [icons] on ⇒ [✓] ⇒ **Ctrl+D** ⇒ [Options] ⇒ [Side 1 Through All] ⇒ [Side 2 Through All] ⇒ if necessary, flip material direction arrow [Fig. 7.13(d)]

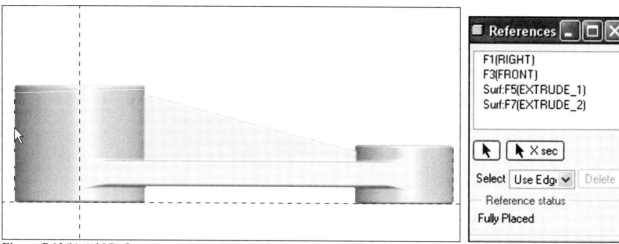

Figure 7.13(b) Add References

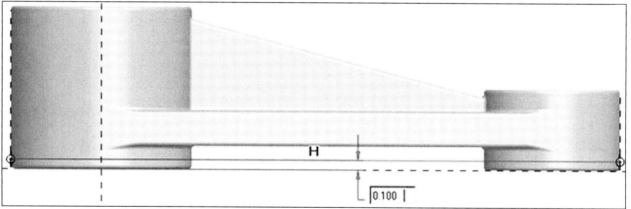

Figure 7.13(c) Draw a Line and Modify the Dimension

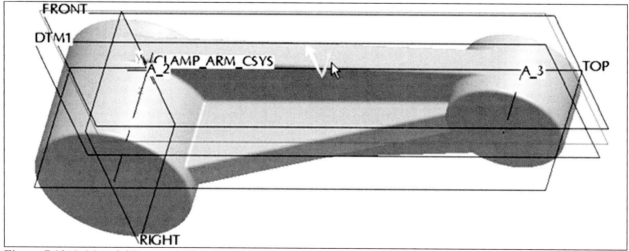

Figure 7.13(d) Material Removal Direction

Click: [icon] ⇒ [icon] ⇒ **MMB** ⇒ **Ctrl+D** ⇒ **RMB** ⇒ **Edit** [Figs. 7.13(e-f)] ⇒ **LMB**

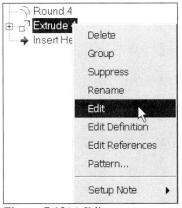

Figure 7.13(e) Edit

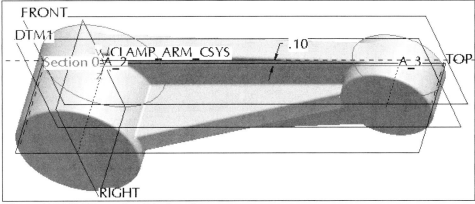

Figure 7.13(f) Dimension Displayed

Click: **Ctrl+S** ⇒ **MMB** ⇒ [icon] ⇒ **RMB** ⇒ **Remove Material** ⇒ **RMB** ⇒ **Define Internal Sketch...** ⇒ pick the top surface of the large circular protrusion [Fig. 7.14(a)] ⇒ **Sketch** ⇒ **Sketch** from menu bar ⇒ **References** ⇒ **MMB** rotate the part ⇒ pick on the circular surface of the large protrusion to add as a reference [Fig. 7.14(b)] ⇒ **Close** ⇒ [icon] ⇒ **RMB** ⇒ **Circle** sketch a circle by picking the center (at references) and one edge (on surface reference) [Fig. 7.14(c)] *(since the circle is locked into references, no dimensions will be displayed, if you get dimensions, redo the circle)* ⇒ **Ctrl+D** ⇒ [✓]

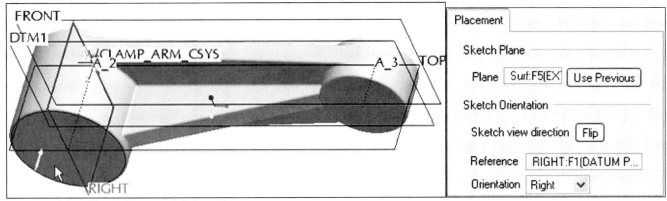

Figure 7.14(a) Sketch Plane and Orientation

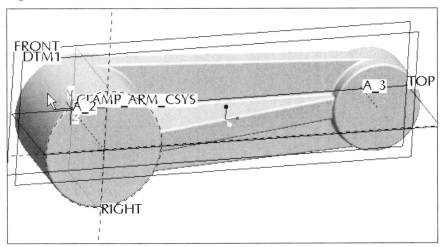

Figure 7.14(b) Add a Surface Reference

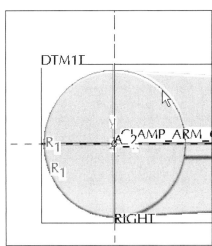

Figure 7.14(c) Sketch a Circle

Type depth `0.10` [Fig. 7.14(d)] ⇒ **Enter** ⇒ **MMB** [Fig. 7.14(e)] ⇒ **LMB** ⇒ repeat the process and cut **.050** from the top of the small end [Fig. 7.14(f)] ⇒ **MMB** ⇒ **Ctrl+S** ⇒ **MMB**

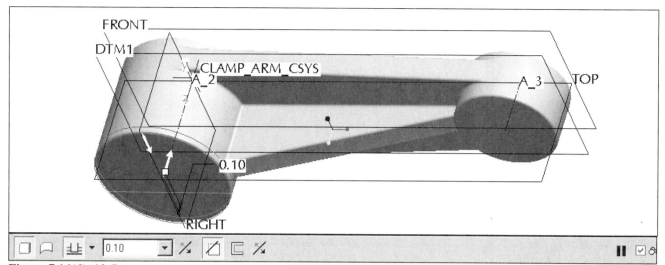

Figure 7.14(d) .10 Cut

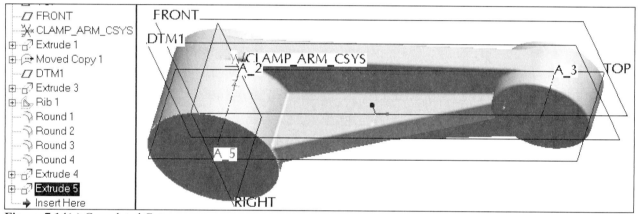

Figure 7.14(e) Completed Cut

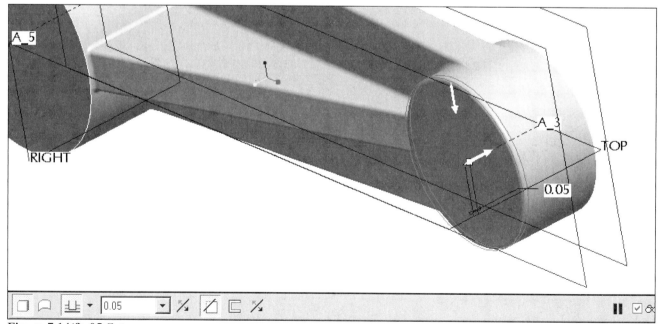

Figure 7.14(f) .05 Cut

Flexing the Model

During the design of a typical part there are many modifications made to the design. The ability to make changes without causing failures is important. "Flexing" the model, changing and editing dimension values to see if the model integrity withstands these modifications, establishes how robust your design is. Modify the part's geometry and observe the change, undo after each edit. If you get a failure, **Undo Changes**.

MMB rotate the part ⇒ pick on the rib ⇒ **RMB** ⇒ **Edit** ⇒ double-click on the **.375** rib width dimension and modify to **1.00** ⇒ **Enter** ⇒ [icon] [Fig. 7.15(a)] ⇒ pick on the first circular extrusion ⇒ **RMB** ⇒ **Edit** ⇒ double-click on the **1.75** dimension and change to **2.50** ⇒ **Enter** ⇒ [icon] [Fig. 7.15(b)] ⇒ pick on the second circular extrusion ⇒ **RMB** ⇒ **Edit** ⇒ double-click on the **1.50** dimension and change to **2.75** ⇒ **Enter** ⇒ [icon] [Fig. 7.15(b)] ⇒ [icon] ⇒ [icon] ⇒ [icon] ⇒ **Ctrl+S** ⇒ **MMB**

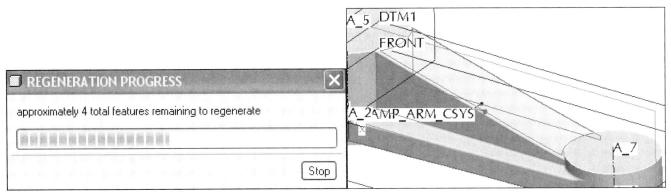

Figure 7.15(a) Regenerate the Rib

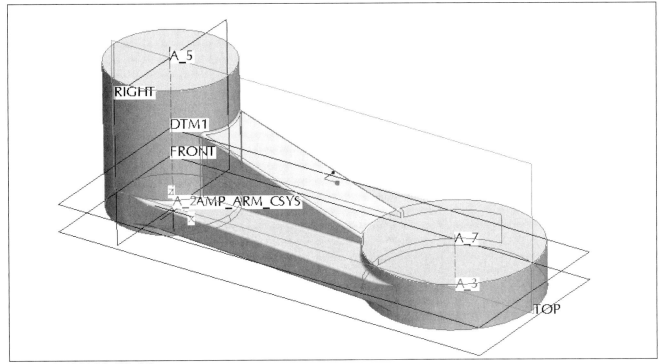

Figure 7.15(b) Modified Part

Measuring Geometry

Using analysis measure, you can measure model geometry with one of the following commands:

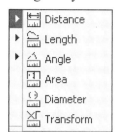

- **Distance** Displays the distance between two entities
- **Length** Displays the length of the curve or edge
- **Angle** Displays the angle between two entities
- **Area** Displays the area of the selected surface, quilt, facets, or an entire model
- **Diameter** Displays the diameter of the surface
- **Transform** Displays a note showing the transformation matrix values between two coordinate systems

Click: **Analysis** ⇒ **Measure** ⇒ [Distance] ⇒ pick the top of the large circular protrusion [Fig. 7.16(a)] ⇒ place your cursor over the bottom of the protrusion and click **RMB** until it highlights [Fig. 7.16(b)] ⇒ **LMB** to select *(measures the height)* [Figs. 7.16(c-d)] ⇒ ✓

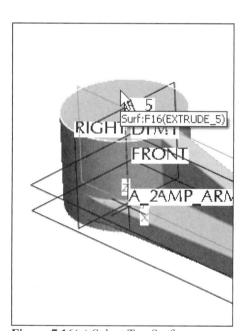
Figure 7.16(a) Select Top Surface

Figure 7.16(b) Select Bottom Surface

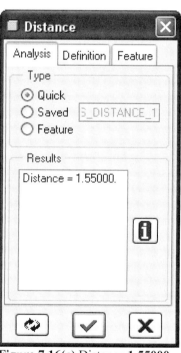
Figure 7.16(c) Distance **1.55000**

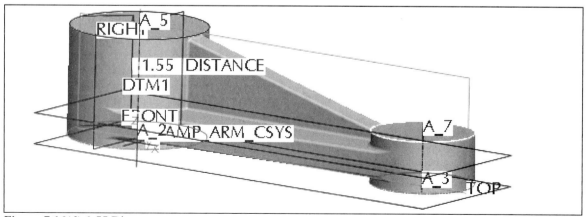
Figure 7.16(d) 1.55 Distance

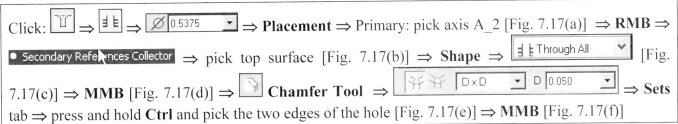

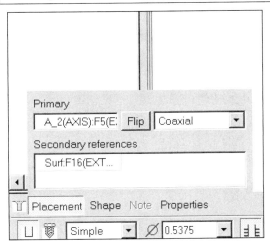

Figure 7.17(a) Placement Tab

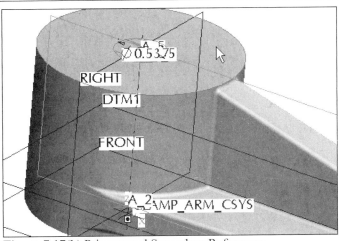

Figure 7.17(b) Primary and Secondary References

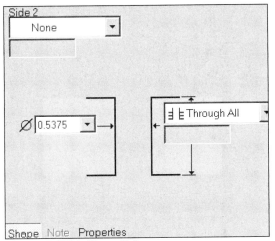

Figure 7.17(c) Shape Tab

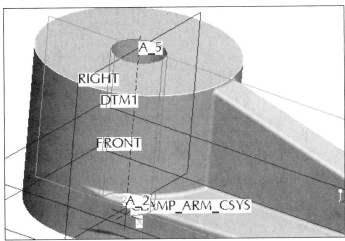

Figure 7.17(d) Hole

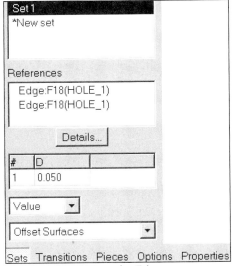

Figure 7.17(e) Chamfer Edges

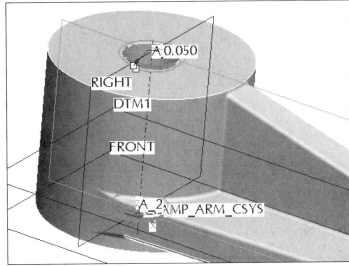

Figure 7.17(f) Chamfers

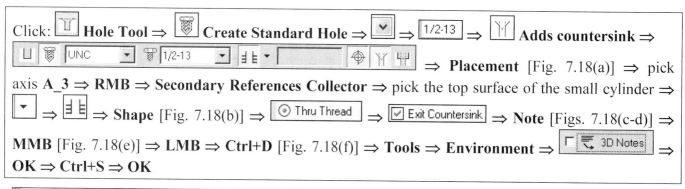

Figure 7.18(a) Placement Tab   **Figure 7.18(b)** Shape Tab   **Figure 7.18(c)** Note Tab

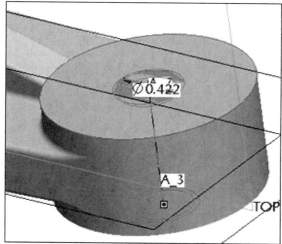 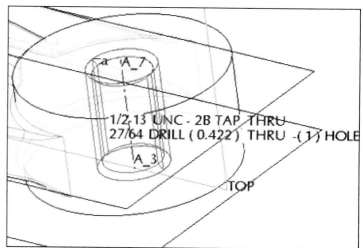

Figure 7.18(d) Previewed Hole   **Figure 7.18(e)** Standard Hole

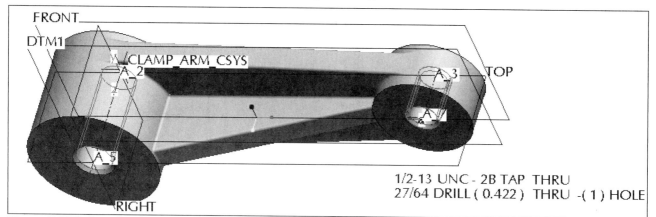

Figure 7.18(f) Completed Holes

Family Tables

Family Tables are used any time a part or assembly has several unique iterations developed from the original model. The iterations must be considered as separate models, not just iterations of the original model. In this lesson, you will add instances of a family table to create a machined part and a casting of the Clamp Arm (a version without machined cuts or holes).

You will be creating a Family Table from the generic model. The (base) model is the **Generic**. Each variation is referred to as an **Instance**. When you create a Family Table, Pro/E allows you to *select dimensions,* which can vary between instances. You can also *select features* to add to the Family Table. Features can vary by being suppressed or resumed in an instance. When you are finished selecting items (e.g., dimensions, features, and parameters), the Family Table is automatically generated.

When adding features to the table, enter an **N** to suppress the feature, or a **Y** to resume the feature. Each instance must have a unique name.

Family tables are spreadsheets, consisting of columns and rows. *Rows* contain instances and their corresponding values; *columns* are used for items. The column headings include the *instance name* and the names of all of the *dimensions, parameters, features, members,* and *groups* that were selected to be in the table. The Family Table dialog box is used to create and modify family tables.

Family tables include:

- The base object (generic object or *generic*) on which all members of the family are based.
- Dimensions, parameters, feature numbers, user-defined feature names, and assembly member names that are selected to be table-driven (*items*).
    - **Dimensions** are listed by name (for example, **d125**) with the associated symbol name (if any) on the line below it (for example, depth).
    - **Parameters** are listed by name (dim symbol).
    - **Features** are listed by feature number with the associated feature type (for example, [cut]) or feature name on the line below it. The generic model is the first row in the table. Only modifying the actual part, suppressing, or resuming features can change the table entries belonging to the generic; *you cannot change the generic model by editing its row entries in the family table.*
- Names of all family members (*instances*) created in the table and the corresponding values for each of the table-driven items

Use the following commands to create a family table, click: **Tools** ⇒ **Family Table**--the Family Table: dialog box opens [Fig. 7.19(a)]

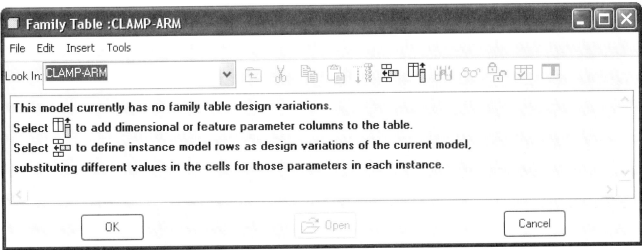

Figure 7.19(a) Family Table Dialog Box

Click: ▦ **Add/delete the table columns** ⇒ ⊙ Feature (from the Add Item options) ⇒ select the cuts, holes, and chamfers from the model or the Model Tree [Fig. 7.19(b)] ⇒ **OK** from Select dialog box ⇒ **MMB** ⇒ **MMB** [Fig. 7.19(c)] ⇒ ▦ **Insert a new instance at the selected row** ⇒ click on the name of the new instance **CLAMP_ARM_INST** ⇒ type **CLAMP_ARM_DESIGN** ⇒ **Enter** [Fig. 7.19(d)] *(adds a new instance)*

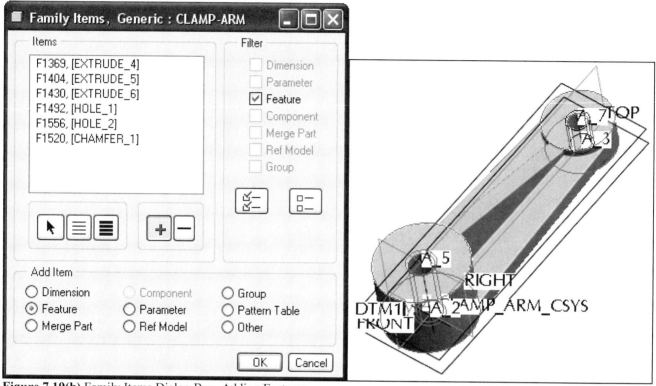

Figure 7.19(b) Family Items Dialog Box, Adding Features

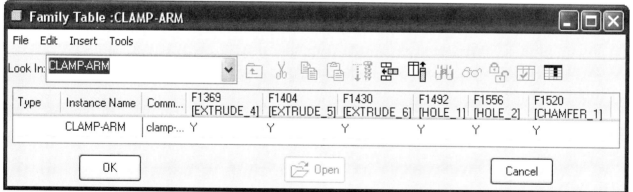

Figure 7.19(c) New Family Table

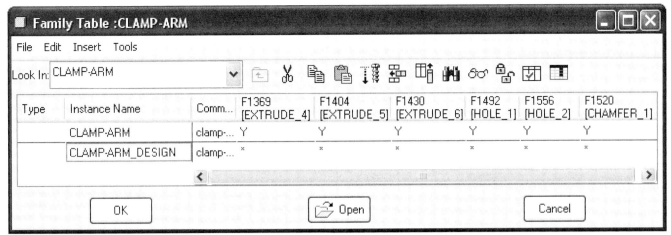

Figure 7.19(d) Add an Instance

The second instance **CLAMP_ARM_INST** should be highlighted ⇒ type **CLAMP_ARM_WORKPIECE** [Fig. 7.19(e)] ⇒ click in the cell of the first feature and change to **N** (not used) [Fig. 7.19(f)] ⇒ change all cells for the **CLAMP_ARM_WORKPIECE** to **N** [Fig. 7.19(g)]

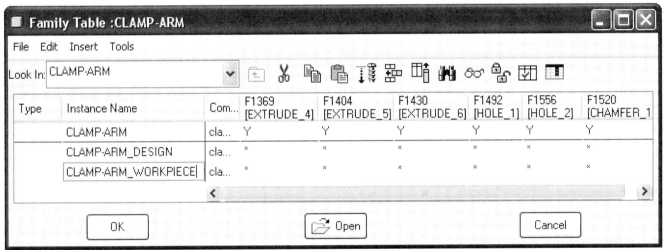

Figure 7.19(e) Add a Second Instance

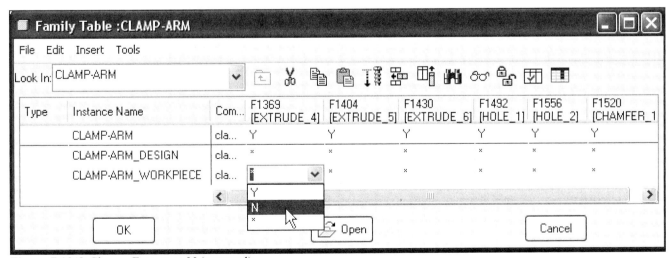

Figure 7.19(f) Change Feature to N (not used)

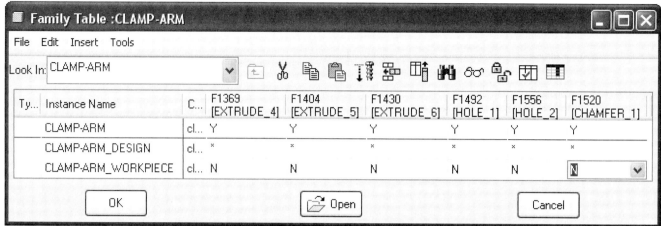

Figure 7.19(g) N for all Machined Features

Click: 🔲 **Verify instances of the family** [Fig. 7.19(h)] ⇒ **Verify** [Fig. 7.19(i)] ⇒ **Close** [Fig. 7.19(j)] ⇒ **OK** from Family Table dialog box ⇒ **Ctrl+S** ⇒ **Enter**

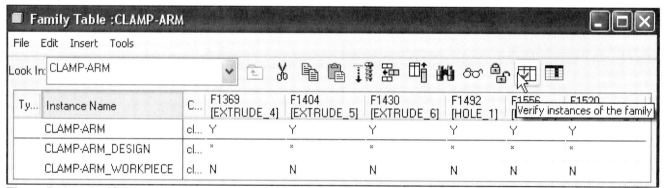

Figure 7.19(h) Verify Instances of the Family

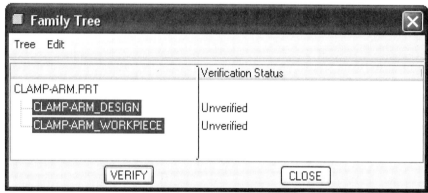

Figure 7.19(i) Verify

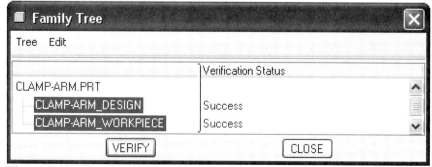

Figure 7.19(j) Verification Status

A Family Table controls whether a feature is present or not for a given design instance, not whether a feature is displayed. The Generic is the base model [Fig. 7.19(k)].

Click: **Tools** ⇒ **Family Table** ⇒ click on **CLAMP_ARM_DESIGN** ⇒ Open [Fig. 7.19(l)] ⇒ adjust your windows to view both models ⇒ from Generic part window, click: **Window** ⇒ **Activate** ⇒ **Tools** ⇒ **Family Table** ⇒ click on **CLAMP_ARM_WORKPIECE** ⇒ **RMB** ⇒ Open [Fig. 7.19(m)] ⇒ **Window** ⇒ **Close** ⇒ **Window** from the menu bar of **CLAMP_ARM_DESIGN** ⇒ **Activate** ⇒ **Window** ⇒ **Close** ⇒ **Window** ⇒ **Activate** ⇒ **Ctrl+S** ⇒ **MMB**

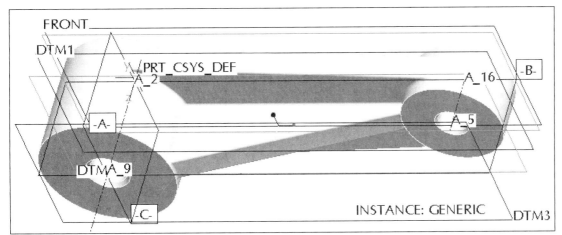

Figure 7.19(k) Instance: GENERIC

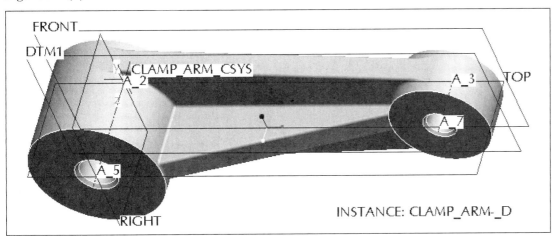

Figure 7.19(l) Instance: CLAMP_ARM_DESIGN

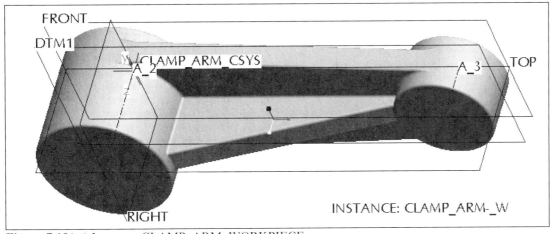

Figure 7.19(m) Instance: CLAMP_ARM_WORKPIECE

Write a relation to keep the thickness of the rib the same as that for the "web".

Click: [icons] off ⇒ with the **Ctrl** key pressed, pick on the "web" extrusion and the rib ⇒ **RMB** ⇒ **Edit** [Fig. 7.20(a)] ⇒ **Info** ⇒ [Switch Dimensions] [Fig. 7.20(b)]

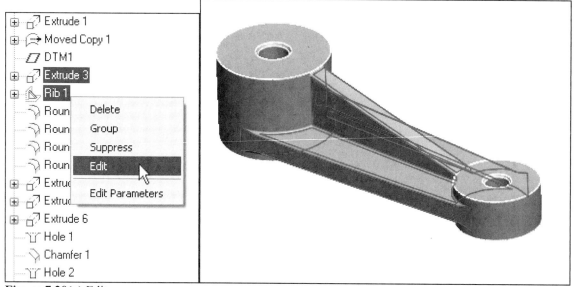

Figure 7.20(a) Edit

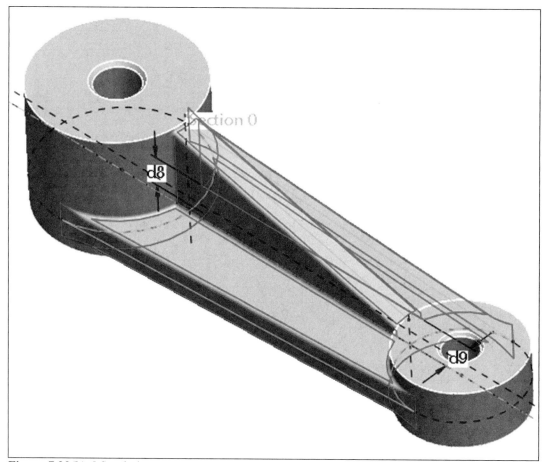

Figure 7.20(b) d Symbols

Pick on a **"d"** symbol *(your "d" values may be different)* [Fig. 7.20(c)] ⇒ **RMB** ⇒ **Properties** ⇒ **Move** [Fig. 7.20(d)] ⇒ pick a new location for the dimension ⇒ [Flip Arrows] ⇒ **OK** ⇒ repeat for the other dimension [Fig. 7.20(e)]

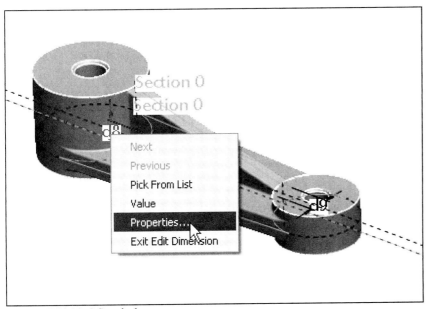

Figure 7.20(c) d Symbols

Figure 7.20(d) Dimension Properties, Move

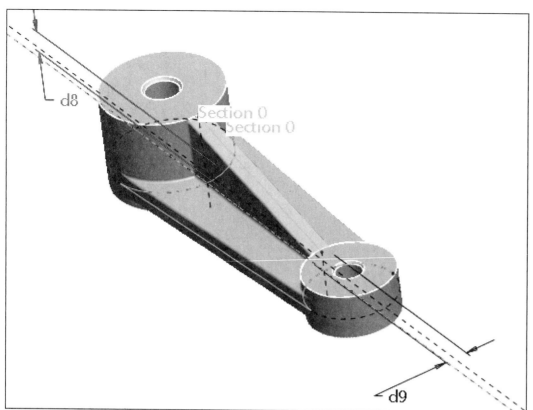

Figure 7.20(e) Repositioned **d** Symbols

Click: **Tools** ⇒ **Relations** ⇒ type **d9=d8** [Fig. 7.20(f)] *(your "d" values may be different)* ⇒ ☑
Execute/Verify ⇒ **OK** ⇒ **Ok**

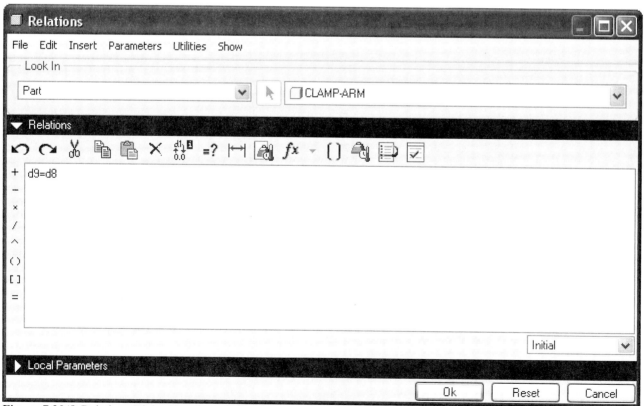

Figure 7.20(f) Relations Dialog Box (your "d" values may be different)

Click: **Info** ⇒ [Switch Dimensions] ⇒ **Ctrl+R** ⇒ pick on the "web" ⇒ **RMB** ⇒ **Edit** ⇒ double-click on the .375 dimension [Fig. 7.21(a)] and change to **.60** [Fig. 7.21(b)] ⇒ **Enter** ⇒ [Regenerate] [Fig. 7.21(c)] ⇒ **Undo Changes** [Fig. 7.21(d)] ⇒ **Confirm**

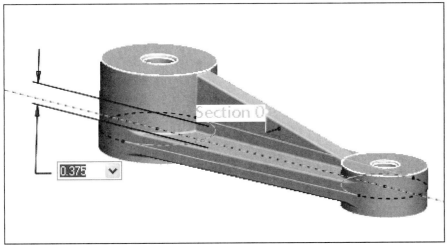

Figure 7.21(a) Edit Dimension

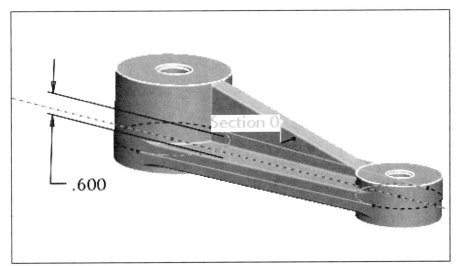

Figure 7.21(b) Dimension .600

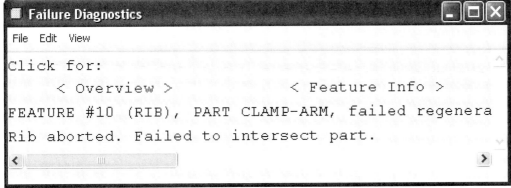 

Figure 7.21(c) Failure Diagnostics Dialog Box   **Figure 7.21(d)** Resolve

Modify the value to **.20** [Fig. 7.21(e)] ⇒ **Enter** ⇒ **Ctrl+G** regenerates model ⇒ [↶] ⇒ [icons] on ⇒ [icon] ⇒ **Standard Orientation** ⇒ [icon] ⇒ **OK** [Fig. 7.21(f)]

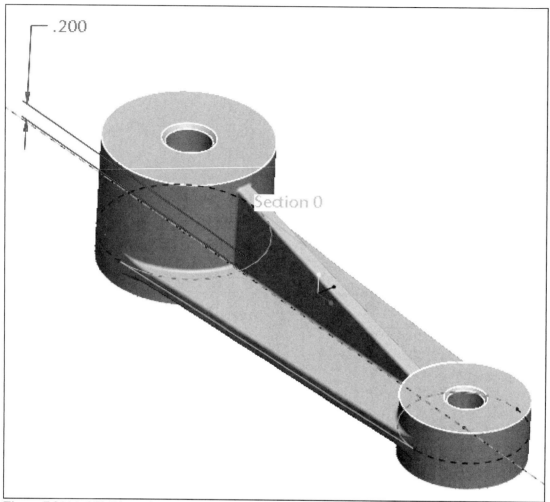

Figure 7.21(e) .200 Dimension

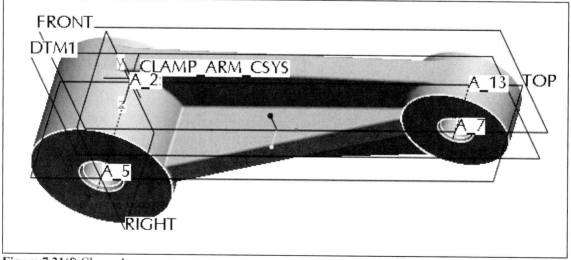

Figure 7.21(f) Clamp Arm

Create a new datum through the bottom of the parts' circular surfaces. Rotate the model ⇒ ⟦⟧ **Datum Plane Tool** ⇒ pick on the cut face of the part [Fig. 7.22(a)] ⇒ click on Surf:F15(EXTRUDE_4) in the Datum Plane dialog box ⇒ [Surf:F15(EXTRUDE_4) Offset / Through / Offset / Parallel / Normal] ⇒ **Through** [Fig. 7.22(a)] ⇒ **OK** ⇒ **Ctrl+D** [Fig. 7.22(b)]

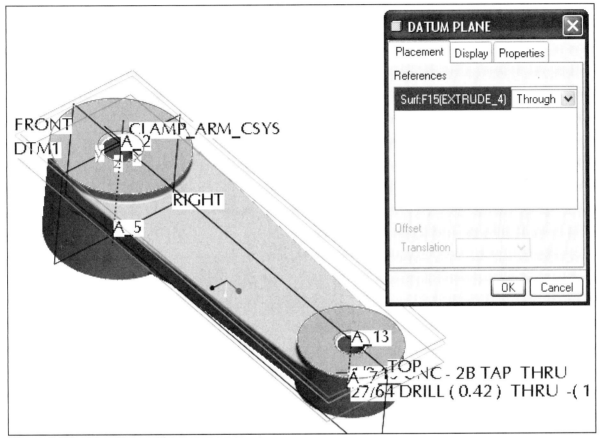

Figure 7.22(a) New Datum Plane

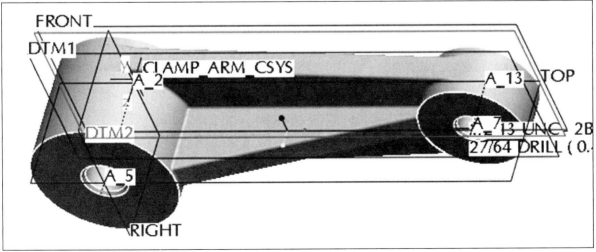

Figure 7.22(b) DTM2

Set the datums for geometric tolerancing. Click: **Edit ⇒ Setup ⇒ Geom Tol ⇒ Set Datum ⇒** pick **TOP** from the model [Fig. 7.23(a)] ⇒ Name- type **B ⇒ OK ⇒** pick **RIGHT ⇒** Name- type **C ⇒ OK ⇒** pick **DTM1 ⇒** Name- type **A ⇒ OK ⇒ Done/Return ⇒ Done** [Fig. 7.23(b)] ⇒ 🔍

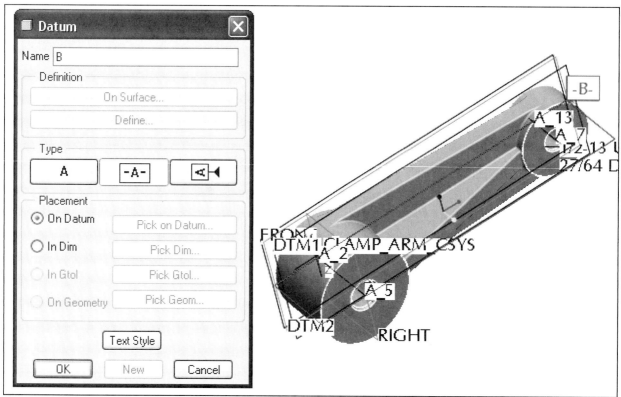

Figure 7.23(a) FRONT Datum Plane set as Datum B

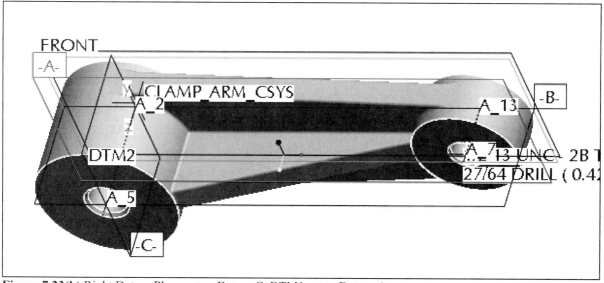

Figure 7.23(b) Right Datum Plane set as Datum C, DTM1 set as Datum A

View ⇒ Shade ⇒ Info (from the menu bar) ⇒ **Feature** ⇒ click on Datum A in the Model Tree [Fig. 7.23(c)] ⇒ **LMB** to deselect ⇒ close the Browser ⇒ **Ctrl+R** ⇒ **Ctrl+D** ⇒ **Ctrl+S** ⇒ **MMB** ⇒ **File** ⇒ **Delete** ⇒ **Old Versions** ⇒ **MMB** ⇒ **Window** ⇒ **Close**

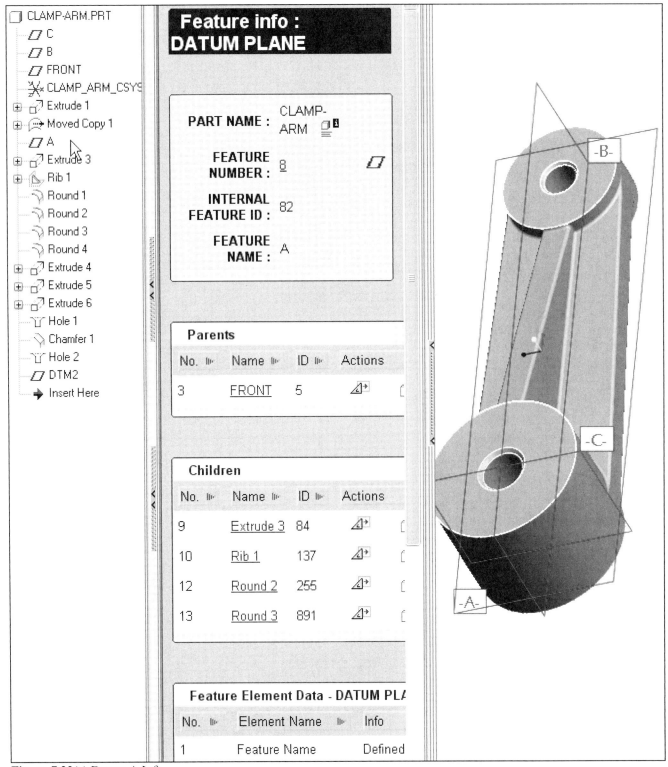

Figure 7.23(c) Datum A Info

Pro/MANUFACTURING

You now have two separate models, a casting (workpiece) and a machined part (design part). During the manufacturing process, the workpiece is merged (assembled) into the design part thereby creating a manufacturing model [Fig. 7.24(a)]. The difference between the two objects is the difference between the volume of the casting and the volume of the machined part. The manufacturing model is used to machine the part [Fig. 7.24(b)]. If you have the Pro/MANUFACTURING module, Pro/NC, and or Expert Machinist; you could now machine the part.

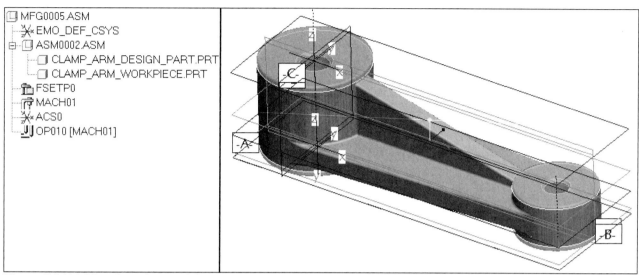

Figure 7.24(a) Manufacturing Model

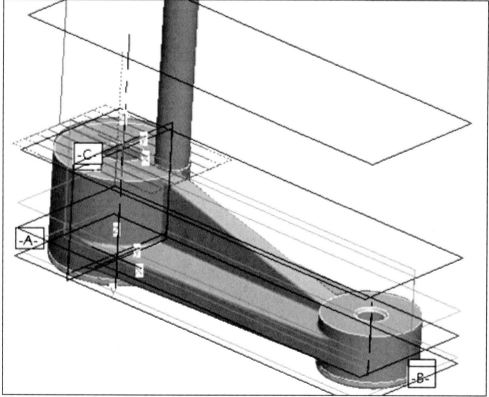

Figure 7.24(b) Facing

This Lesson is now complete. If you wish to model a project without instructions, a complete set of projects and illustrations are available at ***www.cad-resources.com*** ⇒ ***Downloads.***

Lesson 8 Assemblies

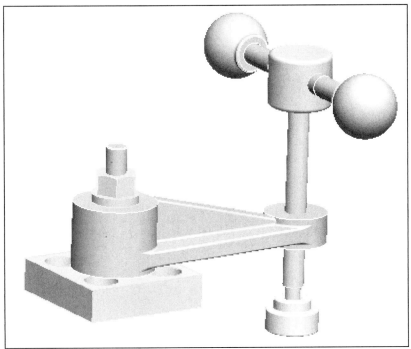

Figure 8.1(a) Swing Clamp Assembly

OBJECTIVES

- **Assemble** components to create an assembly
- Create a **subassembly**
- Understand and use a variety of **Assembly Constraints**
- **Modify** a component constraint
- **Edit** a constraint value
- Check for **clearance** and **interference**

ASSEMBLY CONSTRAINTS

Assembly mode allows you to place together components and subassemblies to create an assembly [Fig. 8.1(a)]. Assemblies [Fig. 8.1(b)] can be modified, reoriented, documented, or analyzed. An assembly can be assembled into another assembly, thereby becoming a subassembly.

Figure 8.1(b) Swing Clamp (**CARRLANE** at www.carrlane.com)

Placing Components

To assemble components, use: 🗔 **Add component to the assembly or Insert ⇒ Component ⇒ Assemble**. After selecting a component from the Open dialog box, the dashboard [Fig. 8.2(a)] opens and the component appears in the assembly window. Alternatively, you can select a component from a browser window and drag it into the Pro/E window. If there is an assembly in the window, Pro/E will begin to assemble the component into the current assembly.

Using icons in the dashboard, you can specify the screen window in which the component is displayed while you position it. You can change window options at any time using: 🗔 **Show a component in separate window while specifying constraints** or 🗔 **Show component in the assembly window while specifying constraints**.

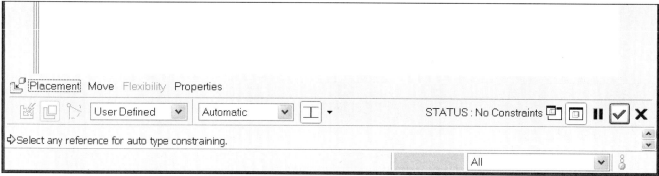

Figure 8.2(a) Component Placement Dashboard

The Automatic placement constraint [Automatic ▼] is selected by default when a new component is introduced into an assembly for placement. After you select a pair of valid references from the component and the assembly, Pro/E automatically selects a constraint type appropriate to the specified references.

Before selecting references, you may also change the type of constraint. Clicking on the current constraint in the Constraint Type box [Fig. 8.2(b)] shows the list of constraints.

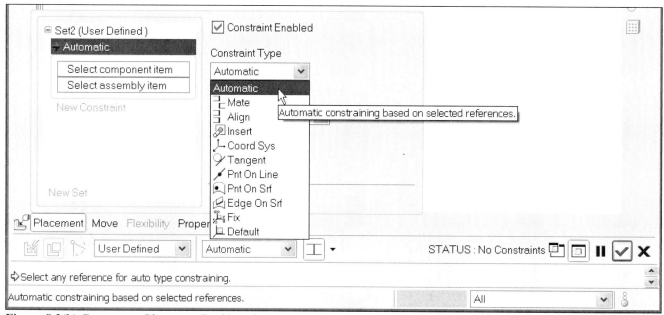

Figure 8.2(b) Component Placement Dashboard, Constraint Type List

Lesson 8 STEPS

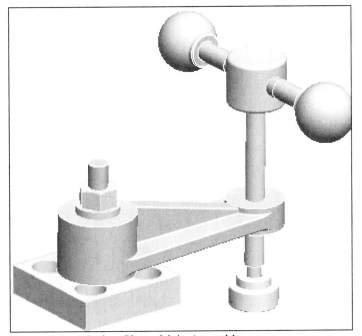

Figure 8.3(a) Swing Clamp Main Assembly

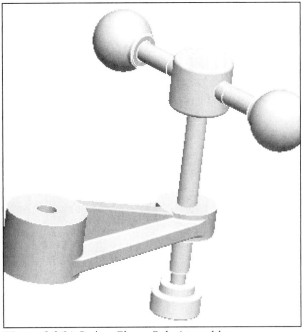

Figure 8.3(b) Swing Clamp Sub-Assembly

Swing Clamp Assembly

The parts required in this lesson are from this text. *If you have not modeled these parts previously, please do so before you start the following systematic instructions.* The other components required for the assembly are standard *off-the-shelf* hardware items that you can get from Pro/E by accessing the Pro/Library. If your system does not have a Pro/LIBRARY license for the Basic and Manufacturing libraries, model the parts using the detail drawings provided in this text or download them from the online PTC Catalog *(the Student Edition and the Tryout Edition do not allow access to the catalog parts)*. The **Flange Nut**, the **3.50 Double-ended Stud**, and the **5.00 Double-ended Stud** are standard items. The **Clamp Plate** component is the first component of the main assembly and will be modeled later when completing the main assembly [Fig. 8.3(a)] using the *top-down design* approach.

Because you will be creating the sub-assembly [Fig. 8.3(b)] using the *bottom-up design* approach, all the components must be available before any assembling starts. *Bottom-up design* means that existing parts are assembled, one by one, until the assembly is complete. The assembly starts with a set of default datum planes and a coordinate system. The parts are constrained to the datum features of the assembly. The sequence of assembly will determine the parent-child relationships between components.

Top-down design is the design of an assembly where one or more component parts are created in Assembly mode as the design unfolds. Some existing parts are available, such as standard components and a few modeled parts. The remaining design evolves during the assembly process. The main assembly will involve creating one part using the *top-down design* approach.

Regardless of the design method, the assembly default datum planes and coordinate system should be on their own separate *assembly layer*. Each part should also be placed on separate assembly layers; the part's datum features should already be on *part layers*.

Before starting the assembly, you will be modeling each part or retrieving *standard parts* from the library and saving them under unique names into *your* working directory. *Unless instructed to do so, **do not use the library parts directly in the assembly***. Start this process by retrieving the standard parts from Pro/Library [Figs. 8.3(c-e)] or the online PTC Catalog.

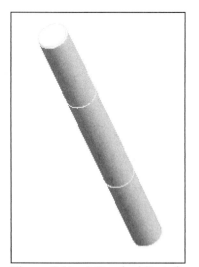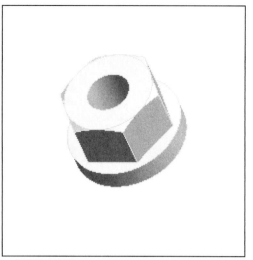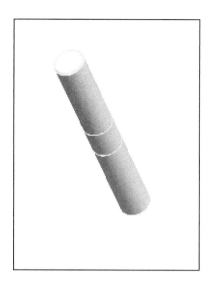

Figure 8.3(c-e) Standard Parts from Pro/LIBRARY

File ⇒ Set Working Directory ⇒ select your working directory **⇒ OK ⇒ File ⇒ Open ⇒** navigate to **Pro Library** [PRO LIBRARY] **⇒ mfglib** [MFG LIBRARY] **⇒ Open ⇒ fixture_lib** [MFG FIXTURES] **⇒ Open ⇒ nuts_bolts_screws** [NT BT WSH&SCR] **⇒ Open ⇒ st.prt ⇒ Open ⇒ By Parameter** tab (from Select Instance dialog box) **⇒ d0,thread_dia ⇒ .500 ⇒ d8,stud_length ⇒ 3.500** [Figs. 8.4(a-b)] **⇒ Open ⇒ File ⇒ Save a Copy ⇒ CLAMP_STUD35 ⇒ OK ⇒ File ⇒ Erase ⇒ Current ⇒ Yes**

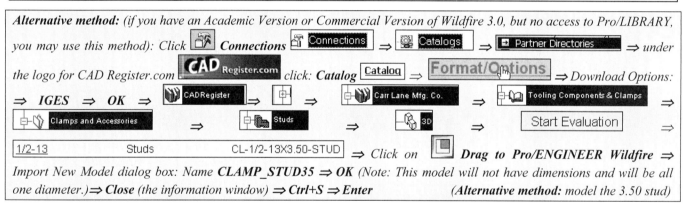

Alternative method: (if you have an Academic Version or Commercial Version of Wildfire 3.0, but no access to Pro/LIBRARY, you may use this method): Click **Connections** [Connections] **⇒** [Catalogs] **⇒** [Partner Directories] **⇒** *under the logo for CAD Register.com* [CAD Register.com] *click:* **Catalog** [Catalog] **⇒** [Format/Options] **⇒** *Download Options:* **⇒ IGES ⇒ OK ⇒** [CADRegister] **⇒** [+] **⇒** [Carr Lane Mfg. Co.] **⇒** [Tooling Components & Clamps] **⇒** [Clamps and Accessories] **⇒** [Studs] **⇒** [3D] **⇒** [Start Evaluation] **⇒** [1/2-13   Studs   CL-1/2-13X3.50-STUD] **⇒** *Click on* [▢] **Drag to Pro/ENGINEER Wildfire ⇒** *Import New Model dialog box: Name* **CLAMP_STUD35 ⇒ OK** *(Note: This model will not have dimensions and will be all one diameter.)* **⇒ Close** *(the information window)* **⇒ Ctrl+S ⇒ Enter**   *(Alternative method: model the 3.50 stud)*

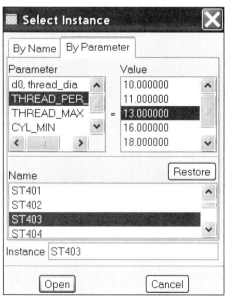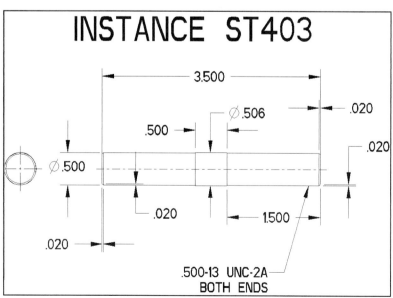

Figure 8.4(a-b) Select Instance Dialog Box and Double-ended Stud ⌀**.500** by **3.50** length: INSTANCE = ST403

268

File ⇒ Open ⇒ navigate to **Pro Library** ⇒ **mfglib** ⇒ Open ⇒ **fixture_lib** ⇒ Open ⇒ **nuts_bolts_screws** ⇒ Open ⇒ **st.prt** ⇒ Open ⇒ **By Parameter** (from Select Instance dialog box) ⇒ **d0,thread_dia** ⇒ **.500** ⇒ **d8,stud_length** ⇒ **5.00** [Figs. 8.4(c-d)] ⇒ Open ⇒ File ⇒ Save a Copy ⇒ **CLAMP_STUD5** ⇒ OK ⇒ File ⇒ Erase ⇒ Current ⇒ Yes
Alternative method: download the part from the CARR Lane Catalog   *Alternative method: model the 5.00 stud*

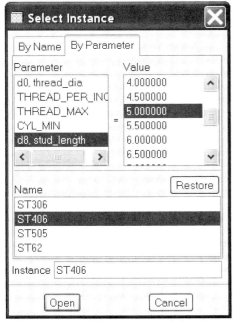
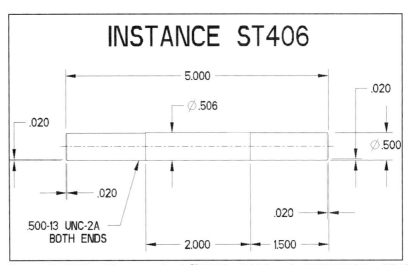

Figure 8.4(c) Select Instance Dialog Box   **Figure 8.4(d)** Double-ended Stud ∅**.500** by **5.00** length: INSTANCE = ST406

File ⇒ Open ⇒ navigate to **Pro Library** ⇒ **mfglib** ⇒ Open ⇒ **fixture_lib** ⇒ Open ⇒ **nuts_bolts_screws** ⇒ Open ⇒ **fn.prt** ⇒ Open ⇒ **By Parameter** (from Select Instance dialog box) ⇒ **d4,thread_dia** ⇒ **.500** ⇒ Open ⇒ File ⇒ Save a Copy ⇒ **CLAMP_FLANGE_NUT** [Figs. 8.4(e-f)] ⇒ OK ⇒ File ⇒ Erase ⇒ Current ⇒ Yes
Alternative method: download the part from the CARR Lane Catalog   *Alternative method: model the flange nut*

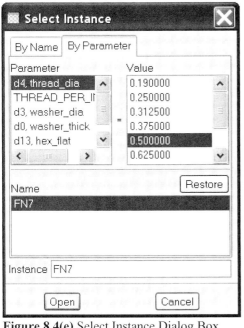
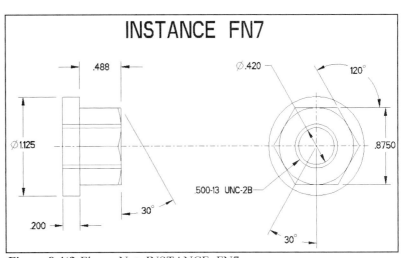

Figure 8.4(e) Select Instance Dialog Box   **Figure 8.4(f)** Flange Nut: INSTANCE=FN7

Open the Clamp_Arm [Fig. 8.5(a)], the Clamp_Swivel [Fig. 8.5(b)], the Clamp_Ball [Fig. 8.5(c)] and the Clamp_Foot [Fig. 8.5(d)] ⇒ review the components and standard parts for correct color, layering, coordinate system naming, and datum planes. Close all windows. Components remain "in session".

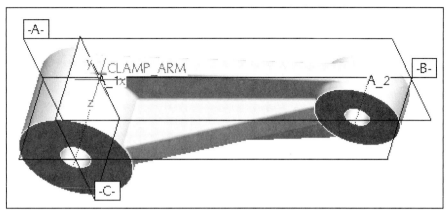

Figure 8.5(a) Clamp_Arm

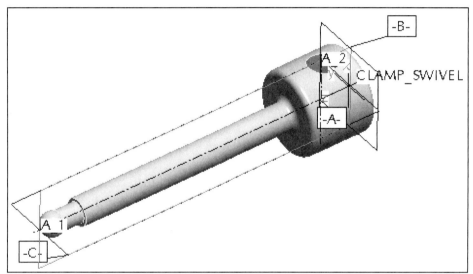

Figure 8.5(b) Clamp_Swivel

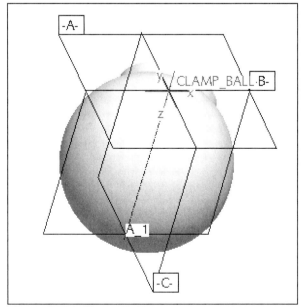

Figure 8.5(c) Clamp_Ball

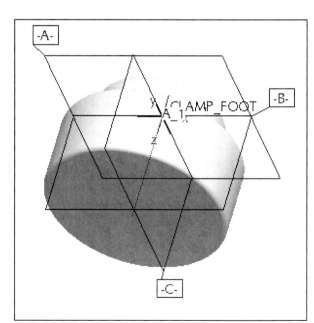

Figure 8.5(d) Clamp_Foot

You now have eight components (two identical Clamp_Ball components are used) required for the assembly. The ninth component- the Clamp_Plate (Fig. 8.6) will be created using *top-down design* procedures when you start the main assembly. *All parts must be in the same working directory used for the assembly.*

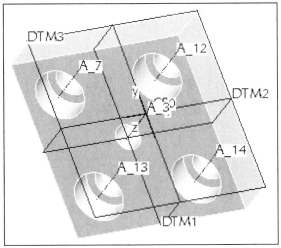

Figure 8.6 Clamp_Plate

A subassembly will be created first. The main assembly is created second. The subassembly will be added to the main assembly to complete the project. Note: the assembly and the components can have different units. Therefore, you must check and correctly set the assembly units before creating or assembling components or sub-assemblies.

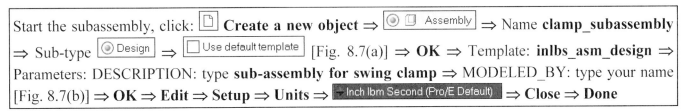

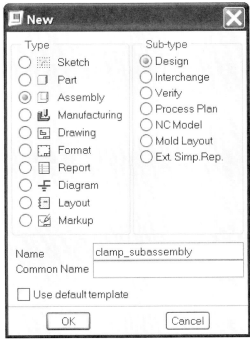

Figure 8.7(a) New Dialog Box

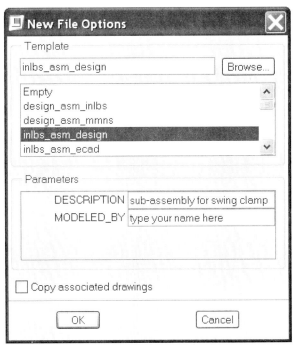

Figure 8.7(b) New File Options Dialog Box

Click: **Settings ▼** from the Navigator ⇒ **Tree Filters...** ⇒ check all Display options on [Fig. 8.7(c)] ⇒ **Apply ⇒ OK ⇒ 💾 ⇒ OK**

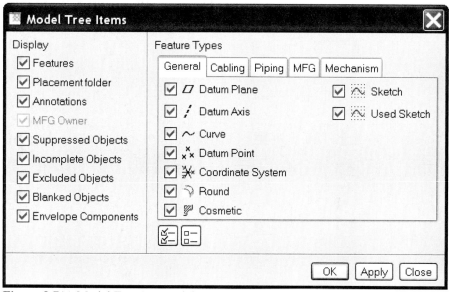

Figure 8.7(c) Model Tree Items Dialog Box

Datum planes and the coordinate system are created per the template provided by Pro/E. The datum planes will have the default names, **ASM_RIGHT**, **ASM_TOP**, and **ASM_FRONT**.

Change the coordinate system name: slowly double-click on **ASM_DEF_CSYS** in the Model Tree ⇒ type new name **SUB_ASM_CSYS** **SUB_ASM_CSYS** ⇒ pick in the graphics window [Fig. 8.7(d)]

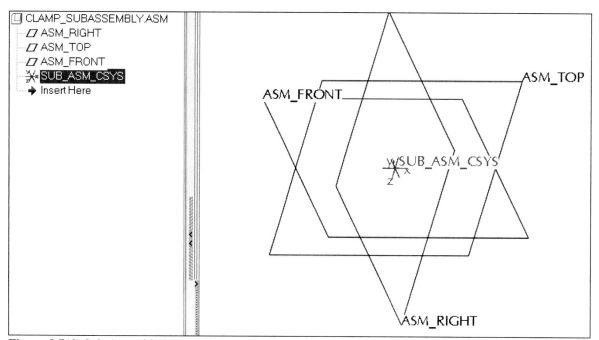

Figure 8.7(d) Sub-Assembly Datum Planes and Coordinate System

Regardless of the design methodology, the assembly datum planes and coordinate system should be on their own separate *assembly layer*. Each part should also be placed on separate assembly layers; the part's datum features should already be on *part layers*. Look over the default template for assembly layering.

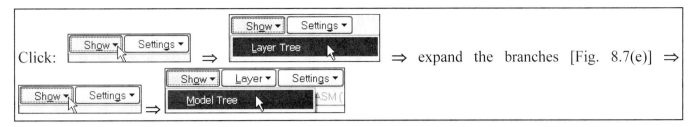

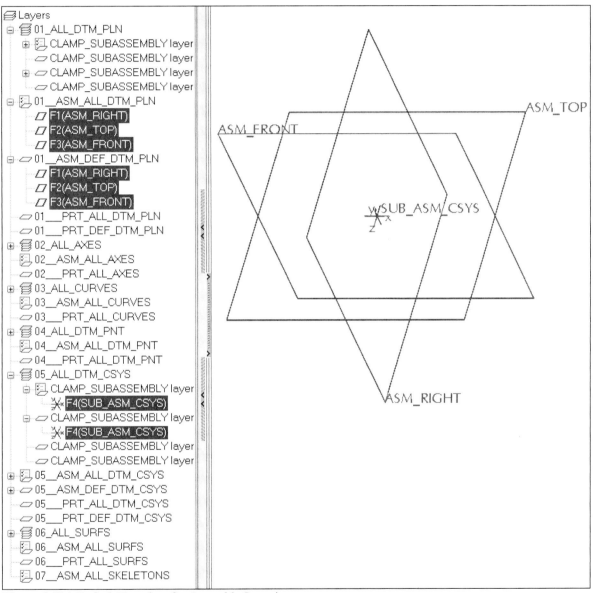

Figure 8.7(e) Default Template for Assembly Layering

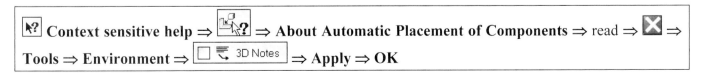

The first component to be assembled to the subassembly is the Clamp_Arm. The simplest and quickest method of adding a component to an assembly is to match the coordinate systems. The first component assembled is usually where this *constraint* is used, because after the first component is established, few if any of the remaining components are assembled to the assembly coordinate system (with the exception of *top-down design*) or, for that matter, other parts' coordinate systems. Make sure all your models are in the same working directory before you start the assembly process.

Click: **Add component to the assembly** ⇒ pick the **clamp_arm.prt** ⇒ **Preview** >>> [Fig. 8.8(a)] ⇒ **Open** ⇒ **CLAMP_ARM_DESIGN** [Fig. 8.8(b)] ⇒ **Open** ⇒ **Show component in a separate window while specifying constraints** [Fig. 8.8(c)] *(you may need to resize and reposition your windows)*

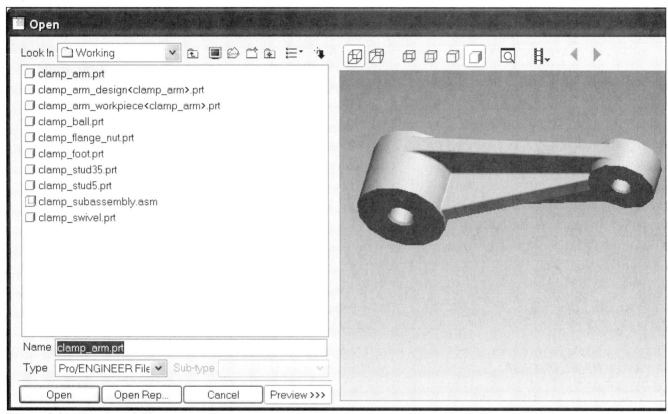

Figure 8.8(a) Previewed Clamp_Arm Component

Figure 8.8(b) Clamp_Arm_Design Instance

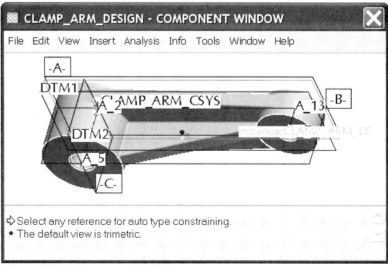

Figure 8.8(c) Component Window

When an assembly is complicated, or it is difficult to select constraining geometry, a separate window aids in the selection process. For simple assemblies (as is the Clamp Assembly and Clamp Subassembly) working in the assembly window is more convenient. Also, since this is the first component being added to the assembly, a separate window is not needed.

Click: [icon] to toggle off the component window ⇒ **RMB** ⇒ **Default Constraint** [Fig. 8.8(d)] (this option puts the component in the default position on the assembly model, which is the same as using the Coordinate System constraint) ⇒ **MMB** ⇒ in the Model Tree, click: [CLAMP_ARM_DESIGN.PRT] ⇒ [Placement] ⇒ [Set3] (your set # may be different) ⇒ [Default] ⇒ **Tools** ⇒ **Environment** ⇒ Standard Orient: **Isometric** ⇒ **Apply** ⇒ **OK** [Fig. 8.8(e)]

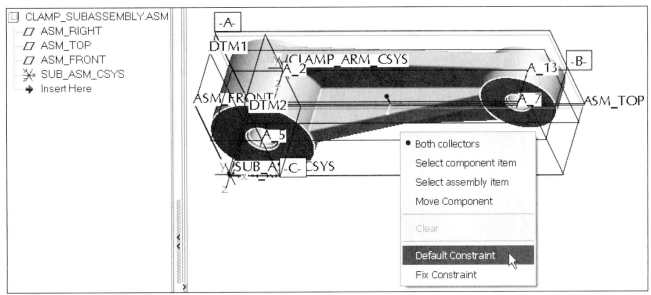

Figure 8.8(d) Default Constraint

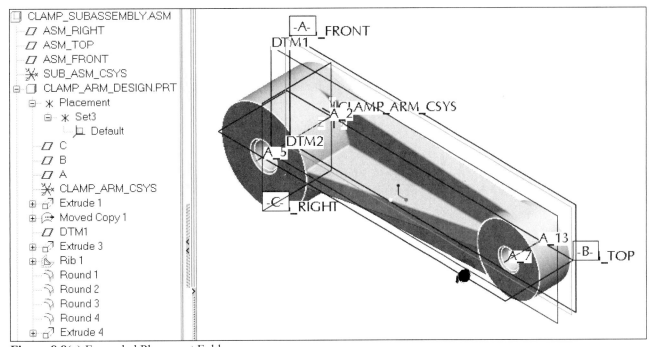

Figure 8.8(e) Expanded Placement Folder

The Clamp_Arm has been assembled to the default location. This means to align the default Pro/E-created coordinate system of the component to the default Pro/E-created coordinate system of the assembly. Pro/E places the component at the assembly origin. By using the constraint, Coord Sys and selecting the assembly and then the component's coordinate systems would have accomplished the same thing, but with more picks.

The next component to be assembled is the Clamp_Swivel. Two constraints will be used with this component: Insert and Mate (Offset). *Placement constraints* are used to specify the relative position of a *pair of surfaces/references* between two components. The Mate, Align, Insert, commands are placement constraints. The two surfaces/references must be of the same type. When using a datum plane as a placement constraint, specify Mate or Align. When using Mate (Offset) or Align (Offset), enter the *offset distance*. The *offset direction* is displayed with an arrow in the graphics window. If you need an offset in the opposite direction, enter a negative value.

Click: **Insert** from menu bar ⇒ **Component** ⇒ **Assemble** ⇒ pick the **clamp_swivel.prt** from the Open dialog box ⇒ **Preview** ⇒ **Open** ⇒ toggle off the datum planes, datum point and coordinate systems (leave datum axis on) ⇒ **Move** tab [Fig. 8.9(a)] ⇒ **Translate** ⇒ **Rotate** ⇒ **LMB** on the "ball" end of the Clamp_Swivel shaft (hint: release **LMB**) and move the mouse to rotate the Clamp_Swivel ⇒ **LMB** to position the component ⇒ **Rotate** ⇒ **Translate** ⇒ **LMB** on the component again and move (translate) until oriented similar to that shown in Figure 8.9(b) ⇒ **LMB** to position the component ⇒ **Move** tab to close ⇒ **RMB** ⇒ **Both collectors**

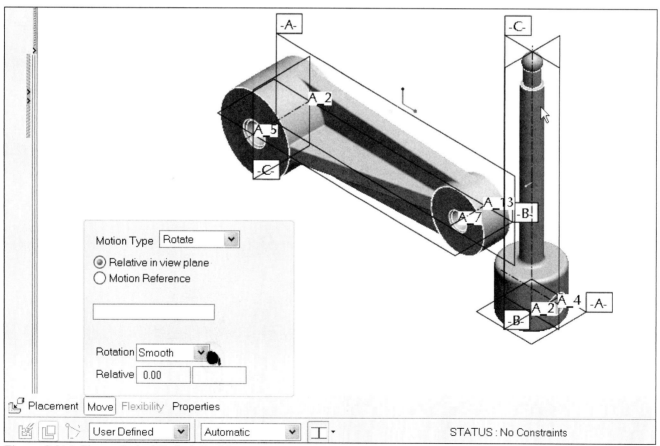

Figure 8.9(a) Clamp_Swivel Default Position

Pick the cylindrical surface of the Clamp_Swivel [Fig. 8.9(b)] ⇒ pick the hole surface of the Clamp_Arm [Fig. 8.9(c)] *constraint type becomes Insert* [Fig. 8.9(d)]

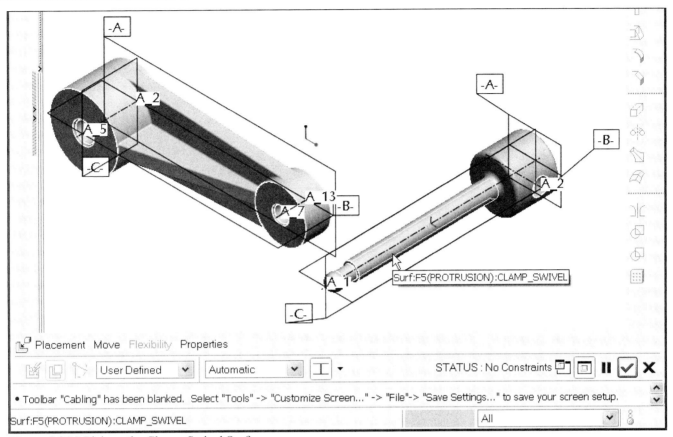

Figure 8.9(b) Pick on the Clamp_Swivel Surface

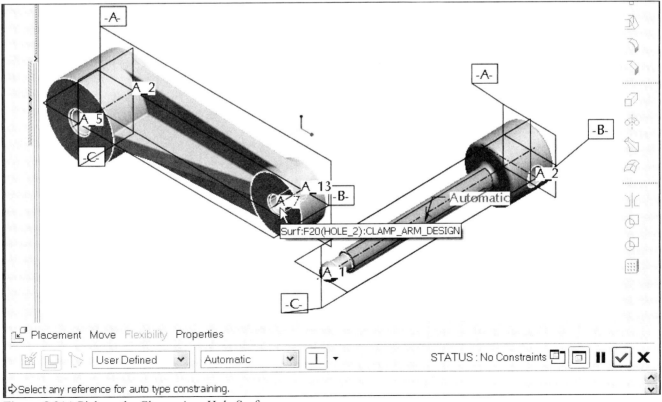

Figure 8.9(c) Pick on the Clamp_Arm Hole Surface

277

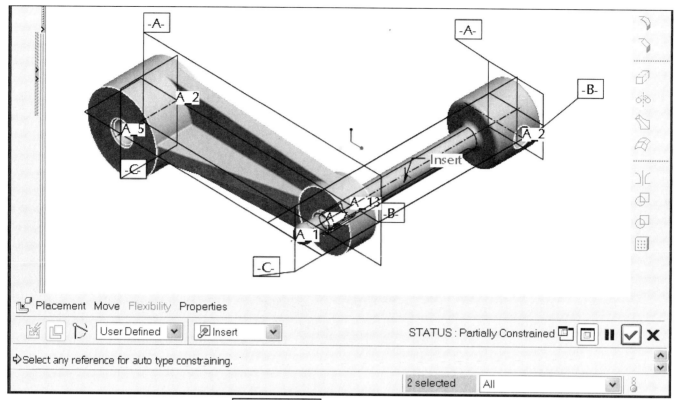

Figure 8.9(d) Insert Constraint Completed

Click: **RMB ⇒ New Constraint** [Fig. 8.9(e)]

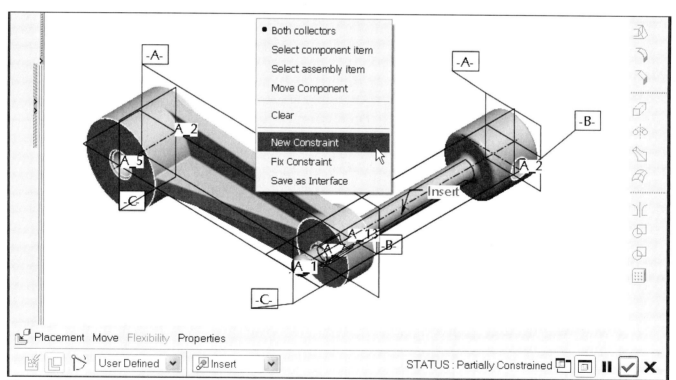

Figure 8.9(e) RMB ⇒ New Constraint

Pick the surface of the Clamp_Swivel [Fig. 8.9(f)] ⇒ pick the surface of the Clamp_Arm [Fig. 8.9(g)]

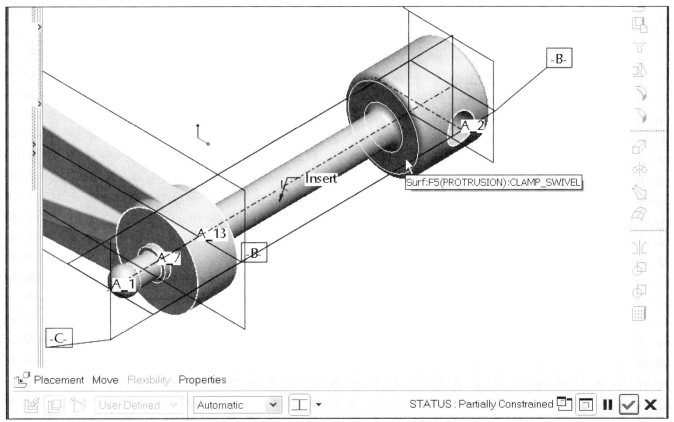

Figure 8.9(f) Select Underside Surface of the Clamp_Swivel

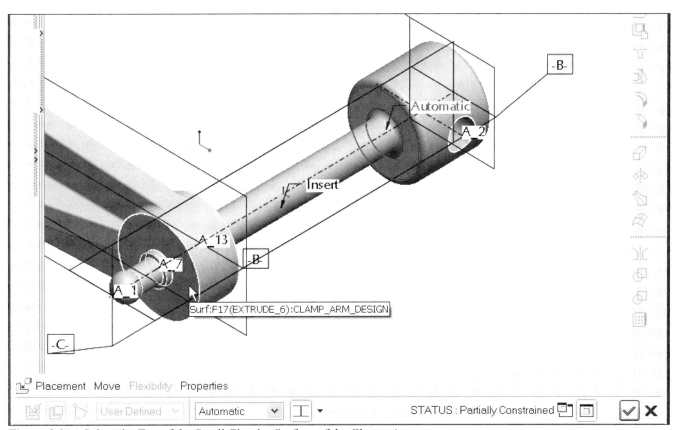

Figure 8.9(g) Select the Top of the Small Circular Surface of the Clamp_Arm

Click: **RMB** ⇒ **Flip** [Fig. 8.9(h)] *the Clamp_Swivel reverses* ⇒ **LMB** drag the Clamp_Swivel until it is **1.50** offset from the Clamp_Arm surface [Fig. 8.9(i)]

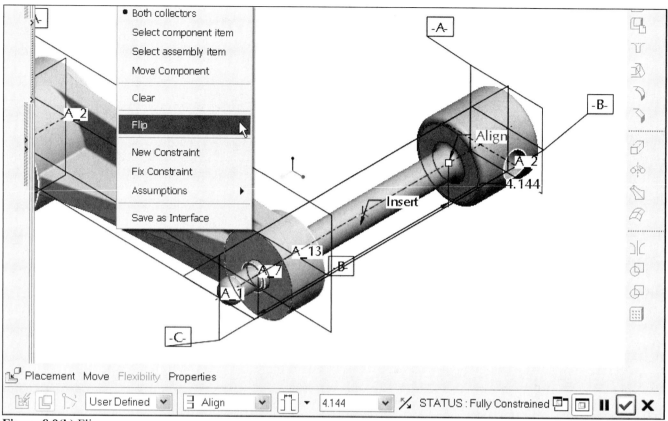

Figure 8.9(h) Flip

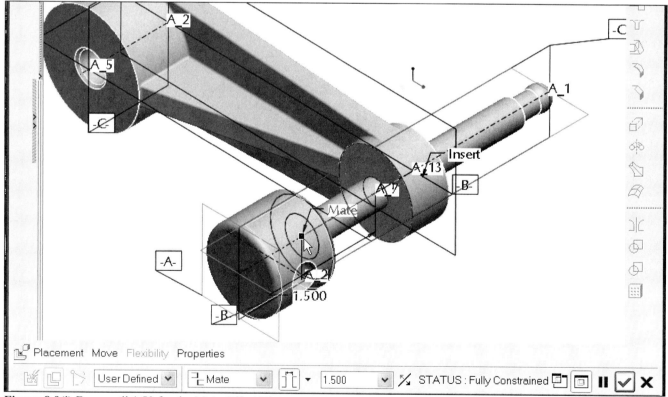

Figure 8.9(i) Drag until **1.50** for the Mate Offset Distance

Click: **Placement** tab [Fig. 8.9(j)] the Placement Status shows the component is `STATUS : Fully Constrained` since `☑ Allow Assumptions` is checked by default ⇒ ☑ ⇒ **Edit** ⇒ **Regenerate** ⇒ **View** ⇒ **Repaint** ⇒ **View** ⇒ **Orientation** ⇒ **Refit** ⇒ **Ctrl+S** ⇒ **MMB** ⇒ **File** ⇒ **Delete** ⇒ **Old Versions** ⇒ **MMB**

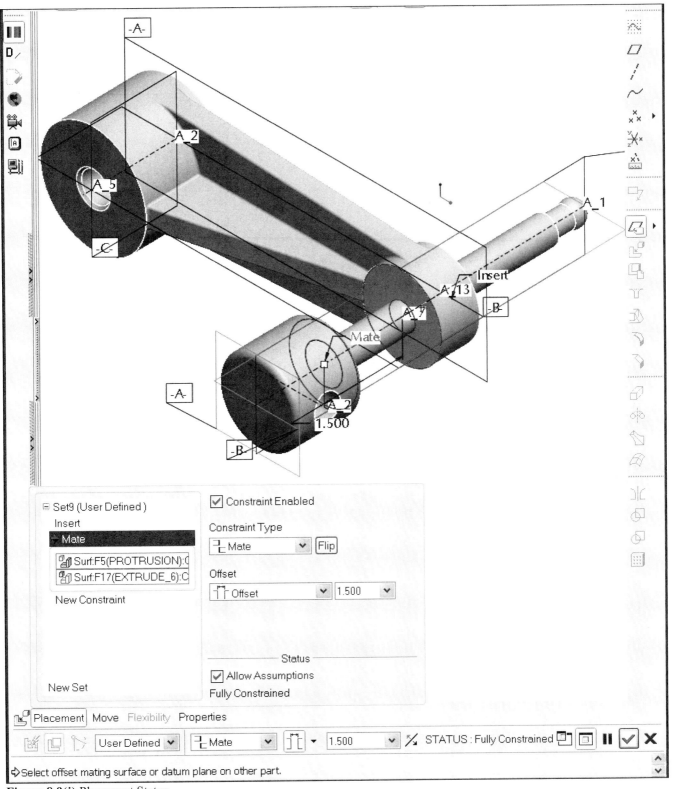

Figure 8.9(j) Placement Status

Regenerating Models

You can use Regenerate to find bad geometry, broken parent-child relationships, or any other problem with a part feature or assembly component. In general, it is a good idea to regenerate the model every time you make a change, so that you can see the effects of each change in the graphics window as you build the model. By regenerating often, it helps you stay on course with your original design intent by helping you to resolve failures as they happen.

When Pro/E regenerates a model, it recreates the model feature by feature, in the order in which each feature was created, and according to the hierarchy of the parent-child relationship between features.

In an assembly, component features are regenerated in the order in which they were created, and then in the order in which each component was added to the assembly. Pro/E regenerates a model automatically in many cases, including when you open, save, or close a part or assembly or one of its instances, or when you open an instance from within a Family Table. You can also use the Regenerate command to manually regenerate the model.

The Regenerate command, located on the Edit menu or using the icon **Regenerates Model**, lets you recalculate the model geometry, incorporating any changes made since the last time the model was saved. If no changes have been made, Pro/E informs you that the model has not changed since the last regeneration. The Custom Regenerate command (Assembly Mode), located on the Edit menu or using the icon, **Specify the list of modified features or components to regenerate**, opens the Regeneration Manager dialog box *if there are features or components that have been changed that require regeneration.* A column next to the Regeneration list indicates which entries you have selected to skip regeneration or to be regenerated.

In the dialog box you can:

- Select all features/components for regeneration, select Regenerate and then Select All.
- Omit all features/components from regeneration, select Skip Regen and then Select All.
- Determine the reason an object requires regeneration, select Highlight and select an entry in the Regeneration list.

Double-click on the Clamp_Swivel [Fig. 8.10(a)] ⇒ double-click on the **1.50** dimension and modify it to **1.75** [Fig. 8.10(b)] ⇒ **Enter** ⇒ **Specify the list of modified features or components to regenerate** ⇒ Regeneration Manager dialog box opens [Fig. 8.10(c)] ⇒ ⓞ Regenerate ⇒ **Select All** ⇒ **OK**

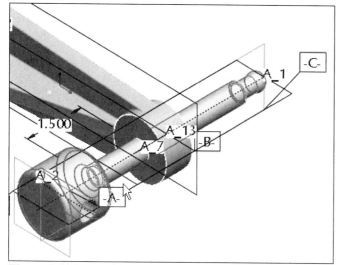

Figure 8.10(a) Double-click on the Clamp_Swivel

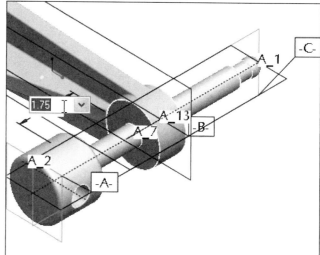

Figure 8.10(b) Modify the Offset Value to **1.75**

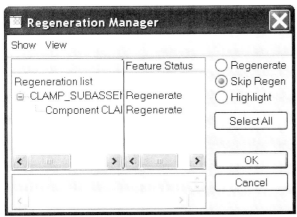

Figure 8.10(c) Regeneration Manager Dialog Box

Double-click on the Clamp_Swivel again ⇒ double-click on the **1.75** dimension and change it back to **1.50** ⇒ **Enter** ⇒ 🗗 **Regenerates Model** ⇒ 🔍 ⇒ 💾 ⇒ **Enter** (Fig. 8.11)

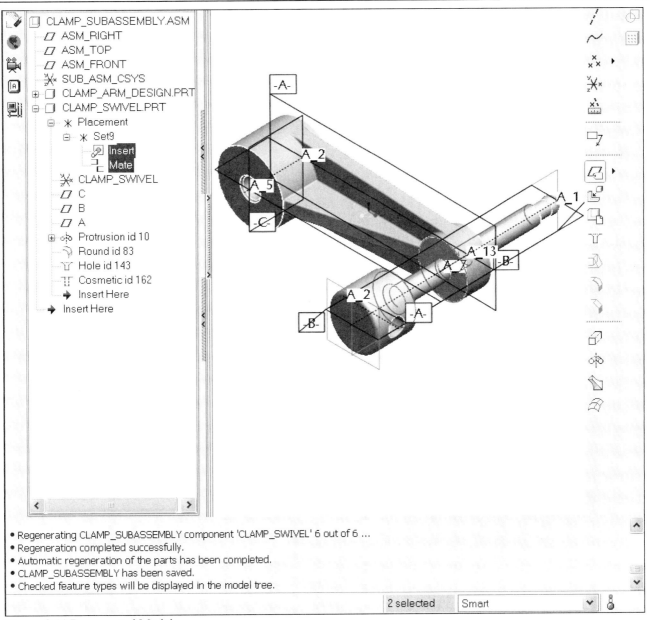

Figure 8.11 Regenerated Model

The next component to be assembled is the Clamp_Foot. Click: **Add component to the assembly** ⇒ pick the **clamp_foot.prt** from the Open dialog box ⇒ **Preview**>>> [Fig. 8.12(a)] ⇒ **MMB** rotate the model ⇒ **Open** ⇒ **Show component in a separate window while specifying constraints** ⇒ spin and zoom in on your model and resize your windows as desired ⇒ Constraints Automatic ⇒ pick the internal cylindrical surface of the Clamp_Foot [Figs. 8.12(b-c)]

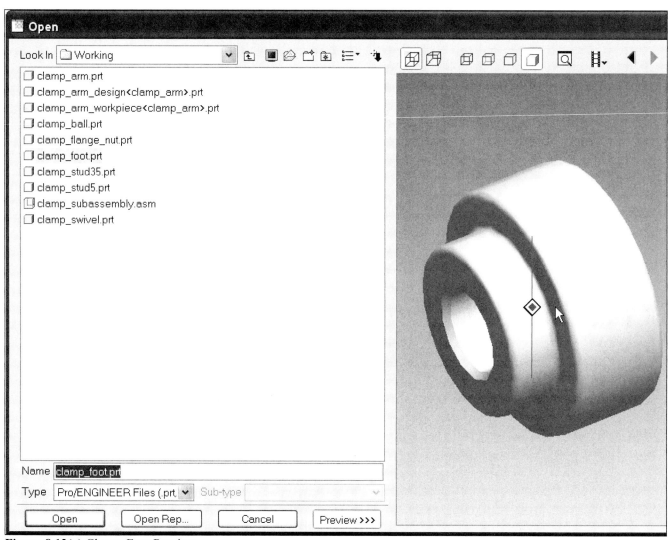

Figure 8.12(a) Clamp_Foot Preview

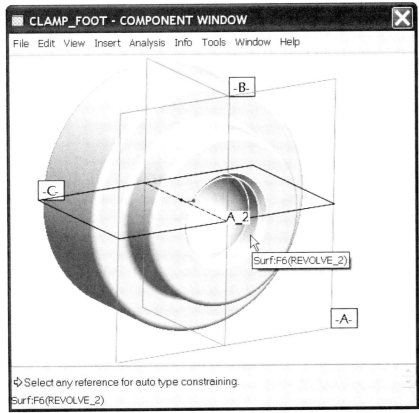

Figure 8.12(b) Pick on the Internal Cylindrical Surface

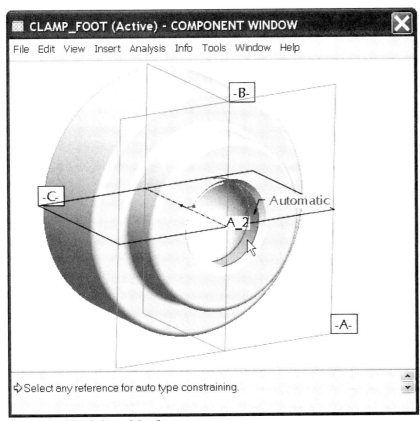

Figure 8.12(c) Selected Surface

Pick the external cylindrical surface of the Clamp_Swivel [Fig. 8.12(d)] *constraint becomes Insert* [Insert] [Fig. 8.12(e)] ⇒ **Placement** tab [Placement]

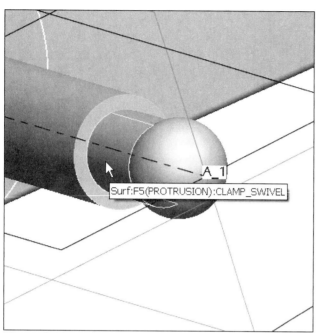

Figure 8.12(d) Pick on the External Cylindrical Surface

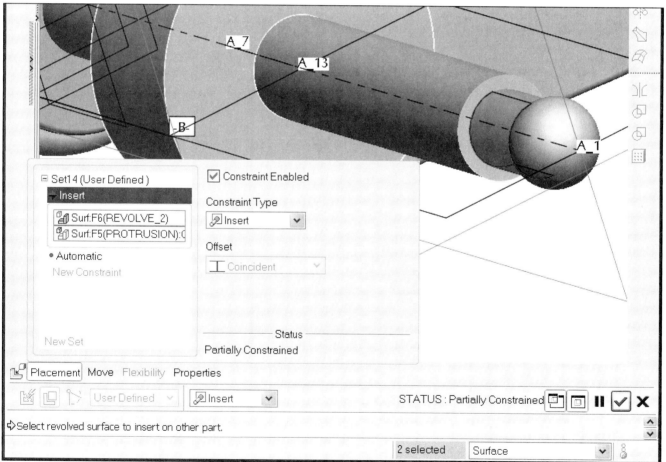

Figure 8.12(e) Constraint Type Insert

Click: **RMB** ⇒ **New Constraint** ⇒ pick the spherical hole of the Clamp_Foot [Figs. 8.12(f-g)] ⇒ pick the spherical end of the Clamp_Swivel [Fig. 8.12(h)] *constraint becomes Mate* [Mate] [Fig. 8.12(i)]

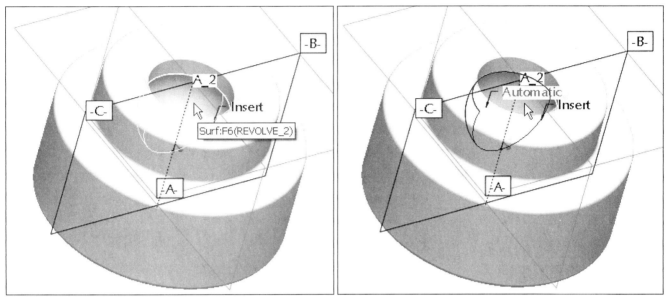

Figure 8.12(f) Pick on the Internal Spherical Hole Surface   **Figure 8.12(g)** Selected Surface

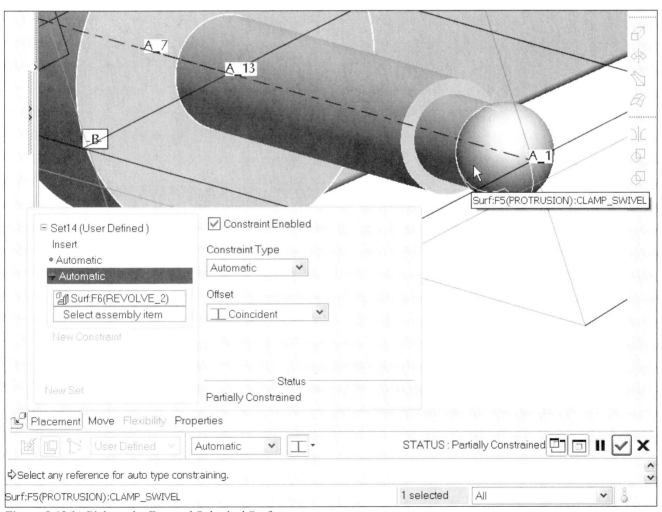

Figure 8.12(h) Pick on the External Spherical Surface

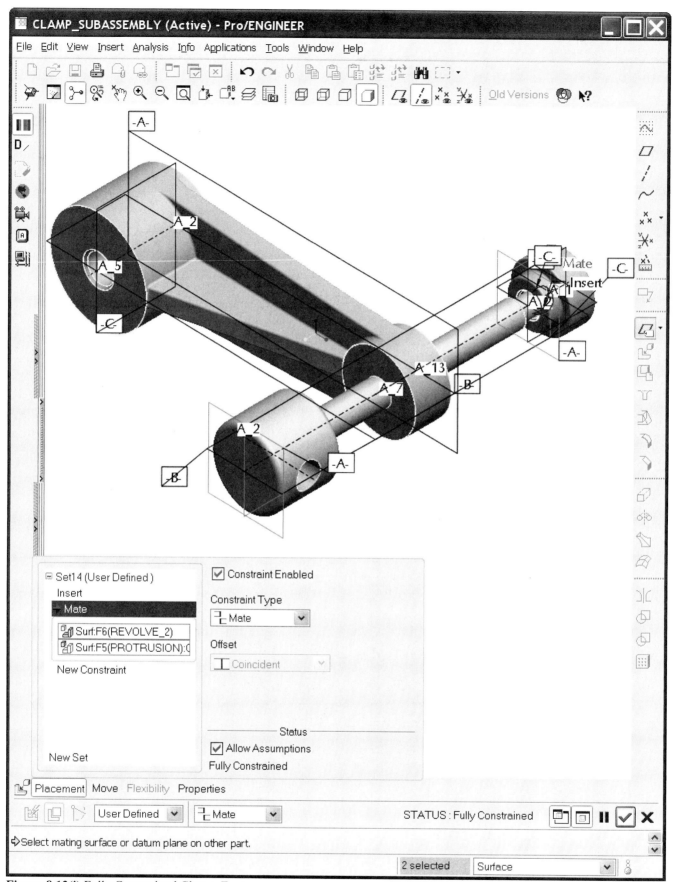

Figure 8.12(i) Fully Constrained Clamp_Foot

Click: **MMB** to complete the placement ⇒ **Ctrl+R** ⇒ **Ctrl+D** ⇒ **Ctrl+S** ⇒ **Enter** ⇒ Rotate and zoom in on the model. The Clamp_Foot may be facing the wrong direction [Fig. 8.12(j)]. ⇒ pick on the **Clamp_Foot** ⇒ **RMB** ⇒ **Edit Definition** [Fig. 8.12(k)]

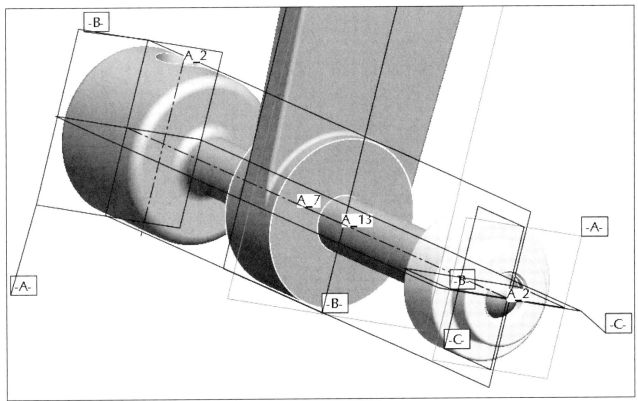

Figure 8.12(j) Clamp _Foot is facing the Wrong Direction

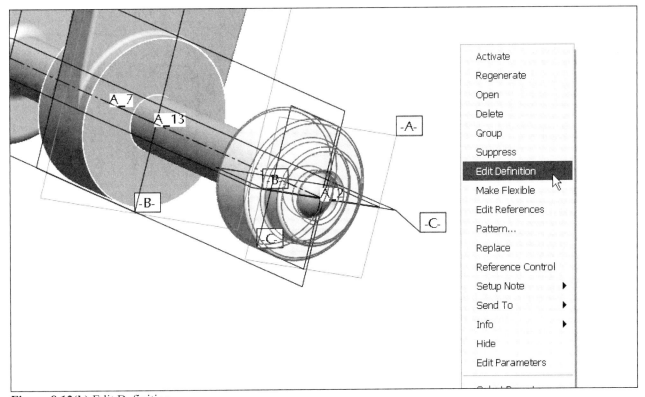

Figure 8.12(k) Edit Definition

Click: **RMB** ⇒ **New Constraint** ⇒ **Placement** tab ⇒ pick on the face of the Clamp_Foot [Fig. 8.12(l)] ⇒ pick on the face of the Clamp_Arm [Fig. 8.12(m)]

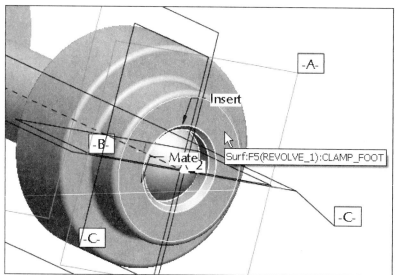

Figure 8.12(l) Pick on the Surface of the Clamp_Foot

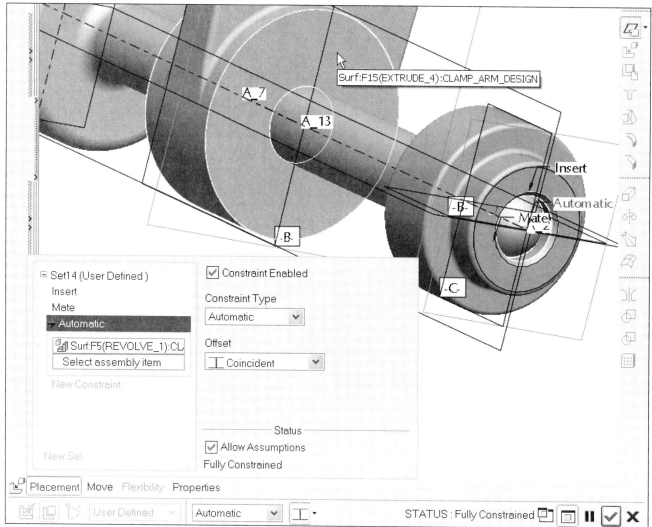

Figure 8.12(m) Pick on the Surface of the Clamp_Arm

Click: **Oriented** [Fig. 8.12(n)] ⇒ if needed, **Flip** [Fig. 8.12(o)] ⇒ **MMB** ⇒ **Ctrl+D** ⇒ **Ctrl+S** ⇒ **MMB**

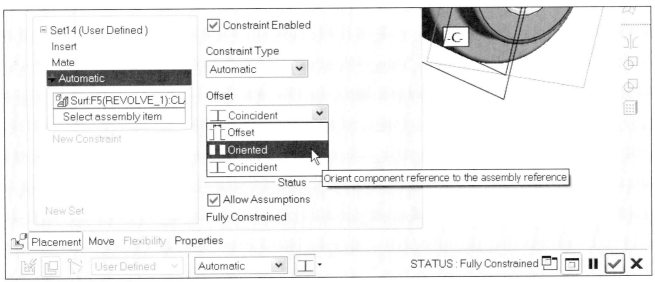

Figure 8.12(n) Oriented

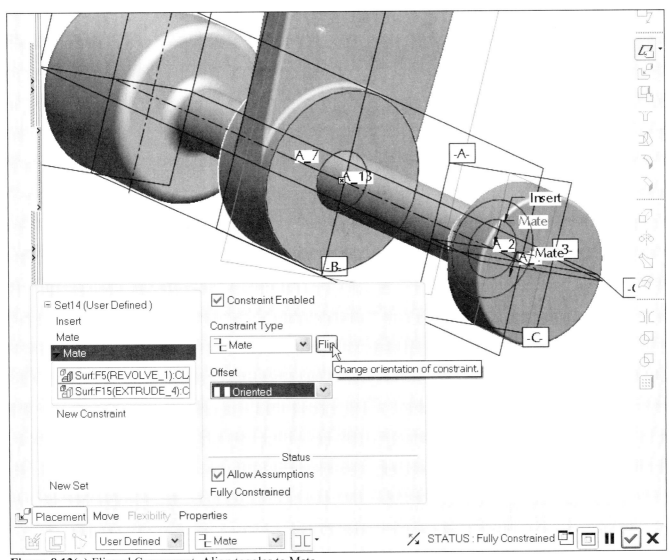

Figure 8.12(o) Flipped Component. Align toggles to Mate

Assemble the **5.00** double-ended stud. Click: **Add component to the assembly** ⇒ pick **clamp_stud5.prt** ⇒ **Open** ⇒ **Show component in a separate window** off ⇒ pick the external cylindrical surface of the Clamp_Stud5 [Fig. 8.13(a)] ⇒ pick the internal cylindrical surface of the hole of the Clamp_Swivel [Fig. 8.13(b)] *constraint becomes Insert* Insert

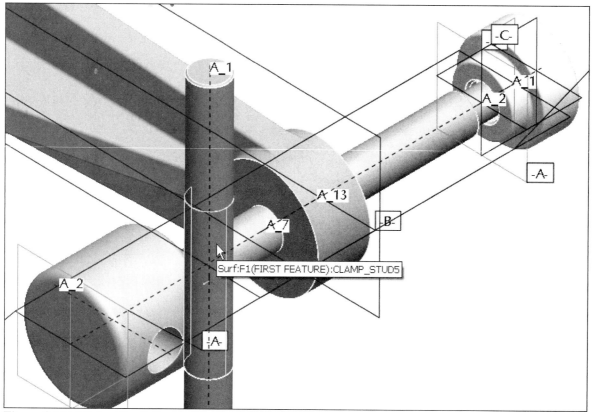

Figure 8.13(a) Select the Cylindrical Surface of the Clamp_Stud5

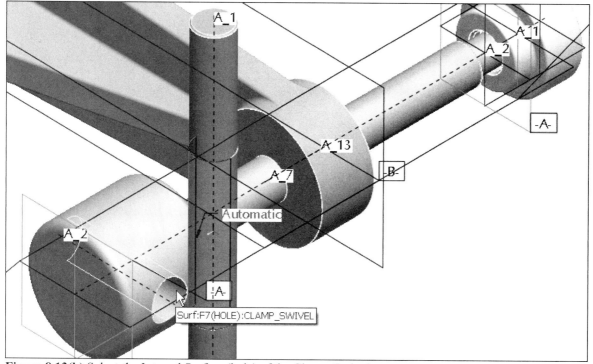

Figure 8.13(b) Select the Internal Surface (hole) of the Clamp_Swivel

Click: **RMB** ⇒ **Move Component** ⇒ pick the Clamp_Stud5 (hint: release LMB) and slide the component [Fig. 8.13(c)] ⇒ **LMB** to position the component ⇒ **RMB** ⇒ **New Constraint** [Fig. 8.13(d)]

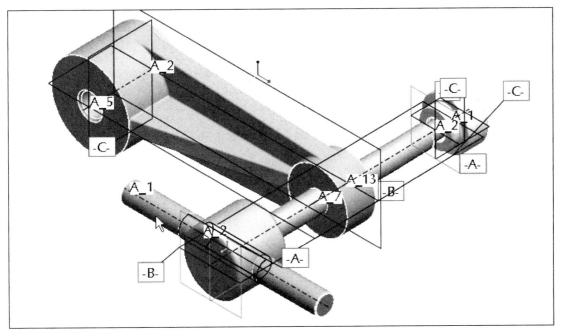

Figure 8.13(c) Move the Component

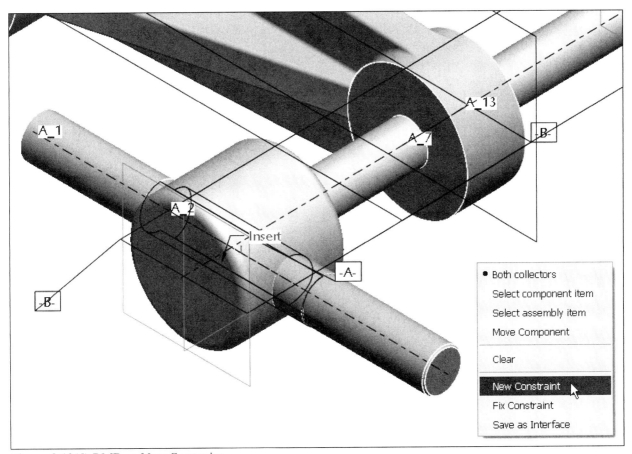

Figure 8.13(d) RMB ⇒ New Constraint

Pick the end surface of the Clamp_Stud5 [Fig. 8.14(a)] ⇒ pick datum **C** of the Clamp_Swivel [Fig. 8.14(b)] ⇒ [icon] ⇒ [icon] ⇒ change offset to **2.50** [icon] [Fig. 8.14(c)] ⇒ **Enter** ⇒ **MMB**

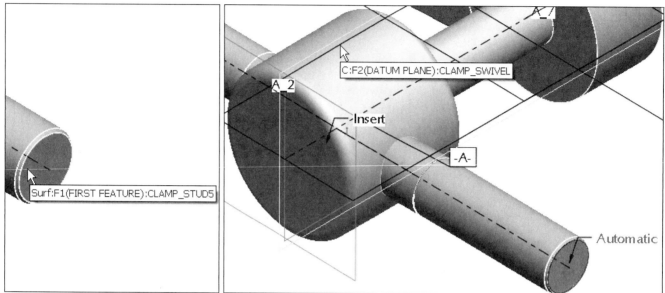

Figure 8.14(a) Select the End Surface **Figure 8.14(b)** Select Datum C from the Clamp_Swivel

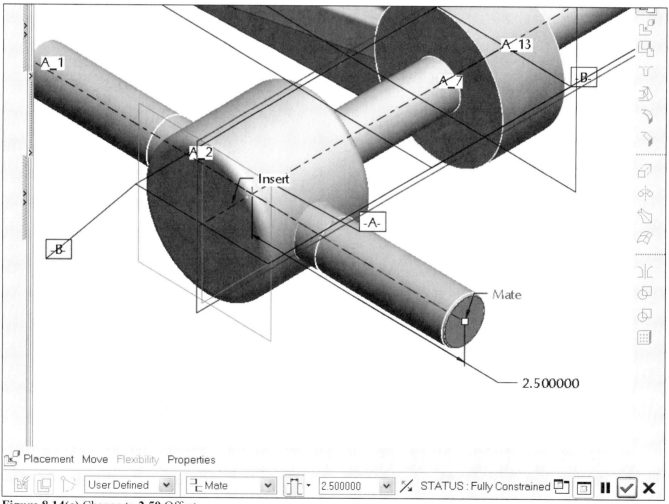

Figure 8.14(c) Change to **2.50** Offset

Click: **Settings** from the Navigator ⇒ **Tree Columns** ⇒ use [>>] to add the columns names to be Displayed [Fig. 8.15(a)] ⇒ **Apply** ⇒ **OK** ⇒ resize the Model Tree column widths [Fig. 8.15(b)]

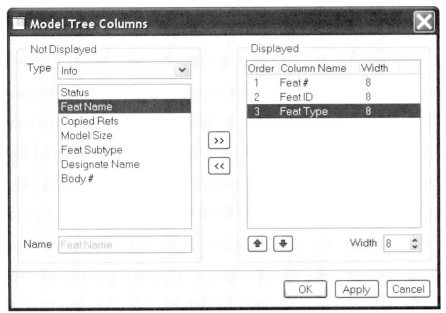

Figure 8.15(a) Model Tree Columns Dialog Box

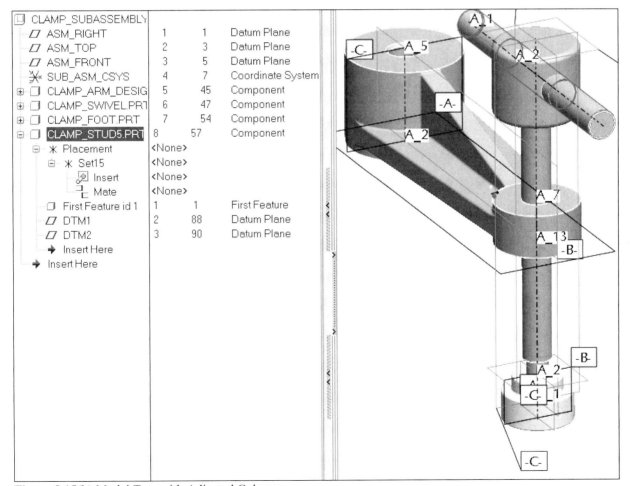

Figure 8.15(b) Model Tree with Adjusted Columns

295

The Clamp_Ball handles are the last components of the Clamp_Subassembly.

Click: **Add component to the assembly** ⇒ pick the **clamp_ball.prt** ⇒ **Preview>>>** ⇒ **MMB** spin the model to see the hole ⇒ **Open** [Fig. 8.16(a)] ⇒ **Ctrl+D** ⇒ close the Navigator shade [Fig. 8.16(b)]

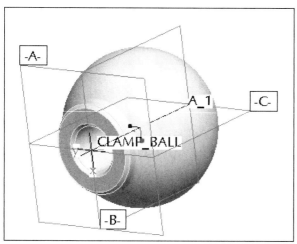

Figure 8.16(a) Clamp_Ball

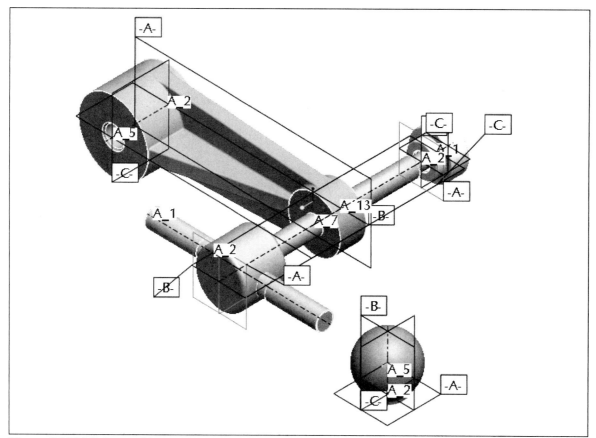

Figure 8.16(b) Subassembly

Press and hold **Ctrl+Alt** ⇒ **MMB** on the Clamp_Ball and rotate ⇒ zoom in on the Clamp_Ball and Clamp_Stud5 ⇒ pick the internal cylindrical surface of the hole of the Clamp_Ball [Fig. 8.16(c)] ⇒ pick the external cylindrical surface of the Clamp_Stud5 [Fig. 8.16(d)] *(constraint becomes Insert)*

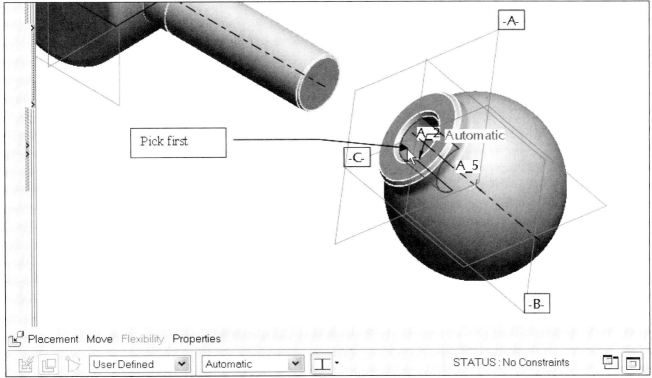

Figure 8.16(c) Select the Hole's Surface

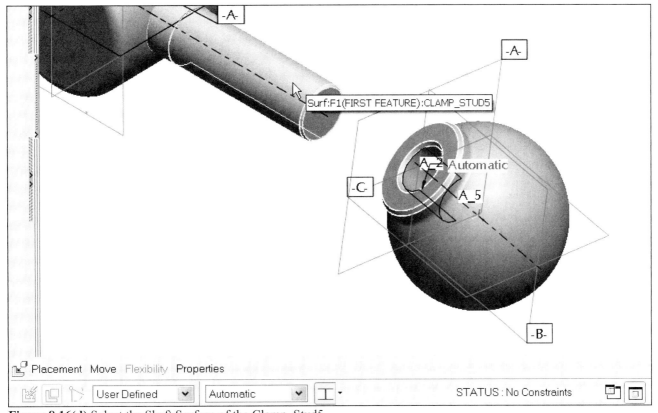

Figure 8.16(d) Select the Shaft Surface of the Clamp_Stud5

Click: **Placement** tab ⇒ **New Constraint** ⇒ pick the end surface of the Clamp_Stud5 [Fig. 8.16(e)] ⇒ **MMB** rotate the assembly to see the surface of the Clamp_Ball *or filter through the possible selections with your RMB* ⇒ pick the flat end of the Clamp_Ball [Fig. 8.16(f)]

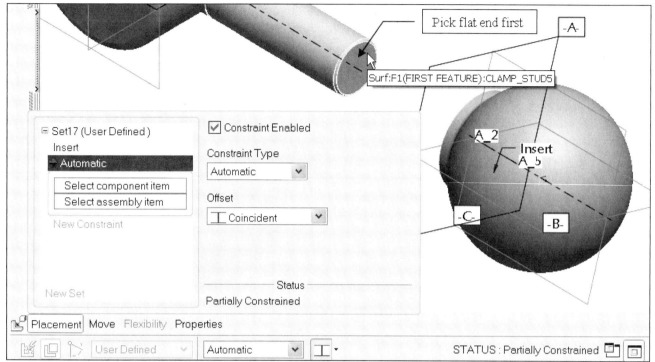

Figure 8.16(e) Pick the End Surface of the Clamp_Stud5

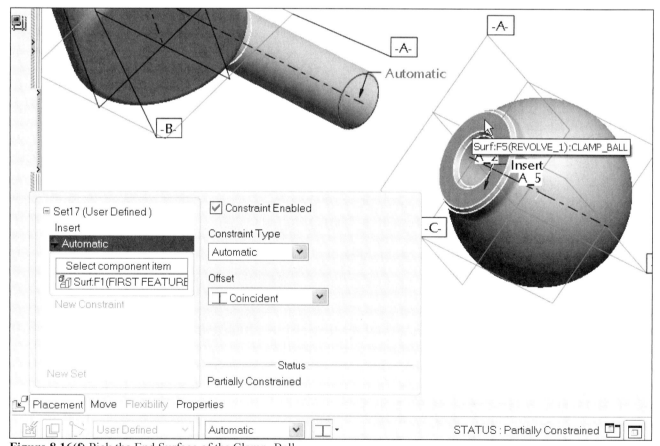

Figure 8.16(f) Pick the End Surface of the Clamp_Ball

Click: **Offset:** [Offset] [1.351] ⇒ type *negative* **-.500** for the Offset distance ⇒ **Enter** [Fig. 8.16(g)] ⇒ **Ctrl+D** ⇒ **View** ⇒ **Shade** [Fig. 8.16(h)] ⇒ **Ctrl+R** ⇒ **MMB** ⇒ [💾] ⇒ **MMB**

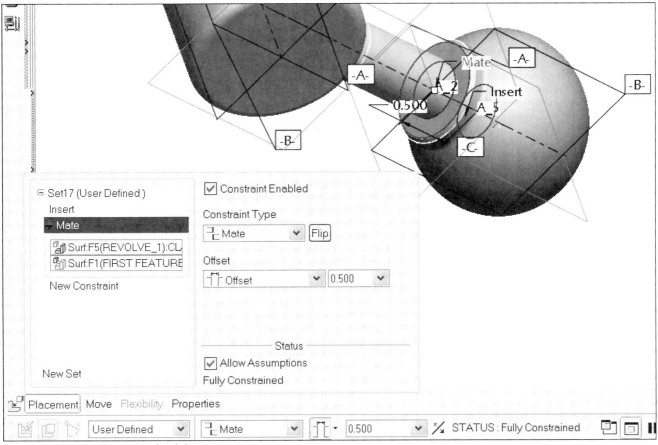

Figure 8.16(g) Fully Constrained Component

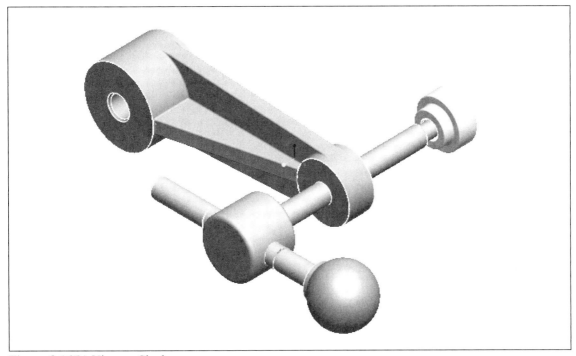

Figure 8.16(h) View ⇒ Shade

Pick on the **Clamp_Ball** component to select it *(highlights in red)* [Fig. 8.16(i)] ⇒ 🗗 **Hidden Line** ⇒ **Ctrl+C** ⇒ **Ctrl+V** ⇒ **Placement** tab ⇒ pick the opposite end cylindrical surface of the Clamp_Stud5 as the new reference [Fig. 8.16(j)]

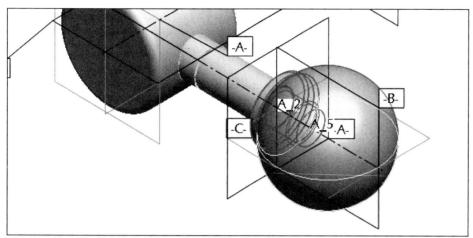

Figure 8.16(i) First Clamp_Ball Selected

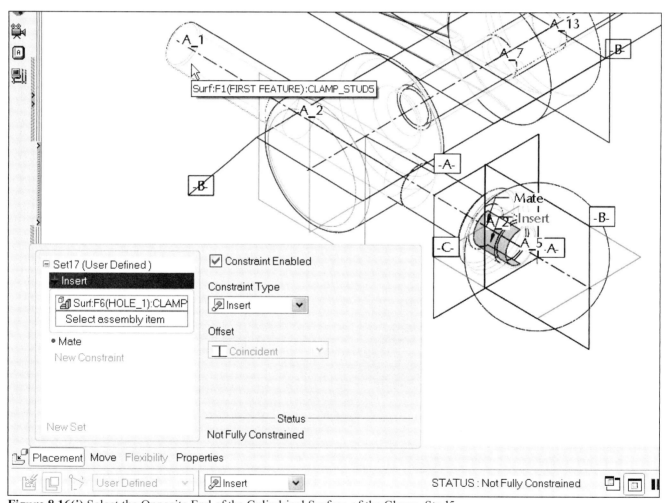

Figure 8.16(j) Select the Opposite End of the Cylindrical Surface of the Clamp_Stud5

Pick the opposite end (flat) surface of the Clamp_Stud5 [Fig. 8.16(k)] ⇒ **Shade** ⇒ in the graphics window, double-click on **.500** dimension ⇒ type **.4375** to modify the distance from the end of the shaft so that it does not bottom-out [Fig. 8.16(l)] ⇒ **Enter** ⇒ **MMB** ⇒ **LMB** to deselect

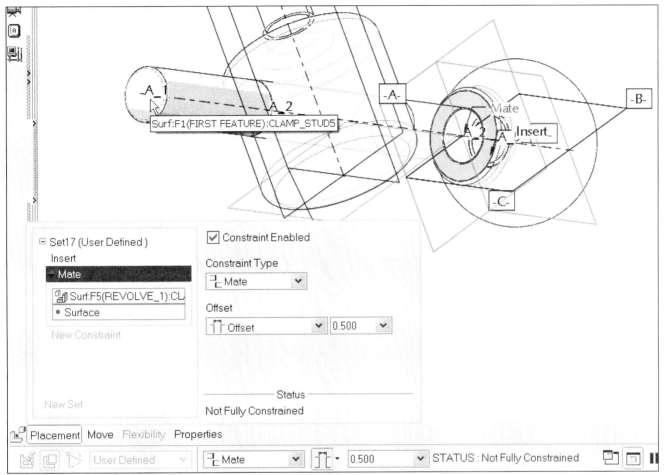

Figure 8.16(k) Pick the Opposite End (Flat) Surface of the Clamp_Stud5

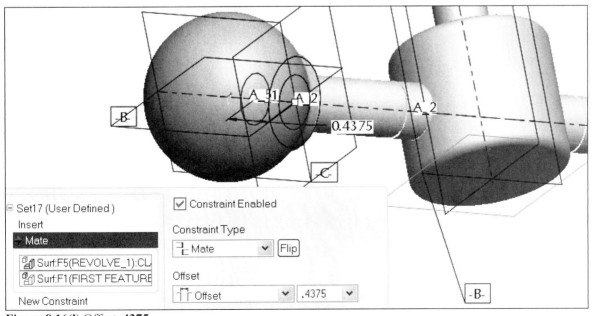

Figure 8.16(l) Offset **.4375**

Click: **Ctrl+D** ⇒ **View** ⇒ **Shade** [Fig. 8.17(a)] ⇒ **Info** from menu bar ⇒ **Bill of Materials** ⇒ ⦿ Top Level [Fig. 8.17(b)] ⇒ **OK** [Fig. 8.17(c)] ⇒ close the Browser shade ⇒ 🗔

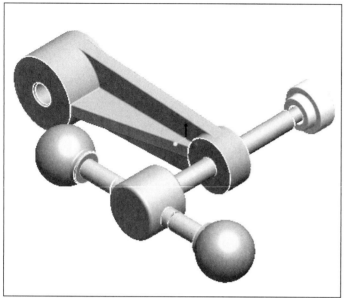

Figure 8.17(a) Completed Clamp_Subassembly

Figure 8.17(b) BOM Dialog Box

Bom Report : CLAMP_SUBASSEMBLY

Assembly CLAMP_SUBASSEMBLY contains:

Quantity	Type	Name	Actions
1	Part	CLAMP_ARM_DESIGN	
1	Part	CLAMP_SWIVEL	
1	Part	CLAMP_FOOT	
1	Part	CLAMP_STUD5	
2	Part	CLAMP_BALL	

Summary of parts for assembly CLAMP_SUBASSEMBLY:

Quantity	Type	Name	Actions
1	Part	CLAMP_ARM_DESIGN	
1	Part	CLAMP_SWIVEL	
1	Part	CLAMP_FOOT	
1	Part	CLAMP_STUD5	
2	Part	CLAMP_BALL	

Figure 8.17(c) Bill of Materials (BOM)

Click: ⌂ **Datum planes** on ⇒ 🞋 **Start the view manager** ⇒ **Xsec** tab [Fig. 8.18(a)] ⇒ **New** [Fig. 8.18(b)] ⇒ type name **A** [Fig. 8.18(c)] ⇒ **Enter** ⇒ **Model** ⇒ **Planar** ⇒ **Single** [Fig. 8.18(d)] ⇒ **Done** ⇒ **Plane** ⇒ pick datum **ASM_TOP** [Fig. 8.18(e)]

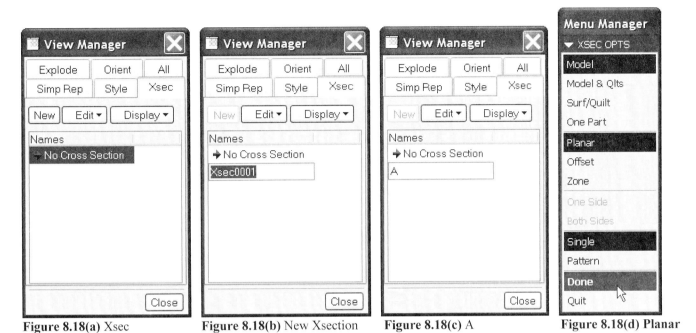

Figure 8.18(a) Xsec   **Figure 8.18(b)** New Xsection   **Figure 8.18(c)** A   **Figure 8.18(d)** Planar

Figure 8.18(e) Select Assembly Datum Top

Click: **Datum planes** off ⇒ **Display** ⇒ **Set Active** [Fig. 8.18(f)] ⇒ **Display** ⇒ **Visibility** ⇒ **Edit** ⇒ **Redefine** ⇒ **Hatching** ⇒ **Spacing** ⇒ **Half** ⇒ **MMB** ⇒ **MMB** ⇒ **No Cross Section** No Cross Section ⇒ **Display** ⇒ **Set Active** [Fig. 8.18(g)] ⇒ **A** ⇒ **RMB** ⇒ **Visibility** (toggles off) ⇒ **Close** ⇒ **Ctrl+D** ⇒ **Ctrl+S** ⇒ **OK** ⇒ **Window** ⇒ **Close** (close the subassembly)

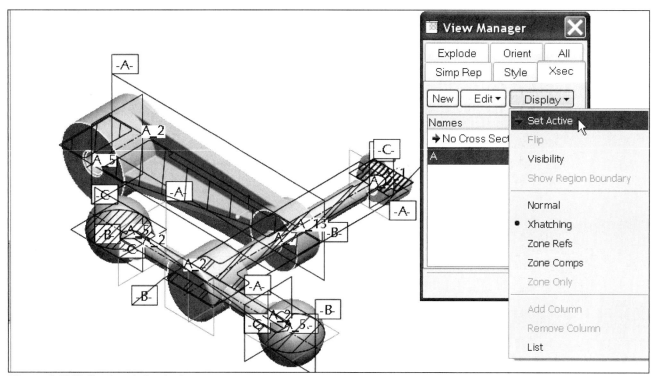

Figure 8.18(f) Section A Set Active

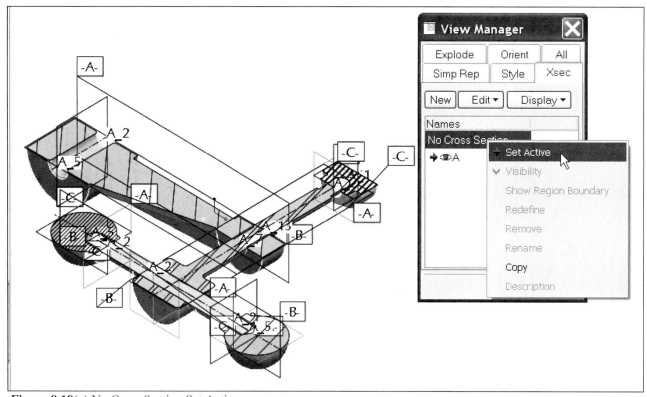

Figure 8.18(g) No Cross Section Set Active

Swing Clamp Assembly

The first features for the main assembly will be the default datum planes and coordinate system. The first part assembled on the main assembly will be the Clamp_Plate, which will be created using *top-down design*; where the assembly is active, and the component is created within the assembly mode. The subassembly is still *"in session-in memory"* even though it does not show on the screen after its window is closed. For the two standard parts of the main assembly, you will use specific constraints instead of using Automatic, which allows Pro/E to default to an appropriate constraint.

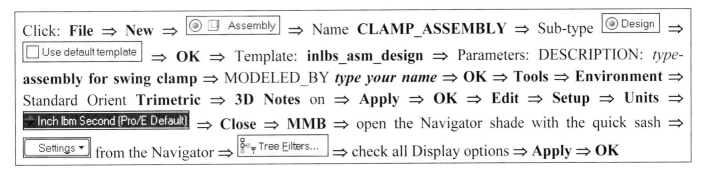

Change the coordinate system name: slowly double-click on [ASM_DEF_CSYS] in the Model Tree ⇒ type new name **ASM_CSYS** [ASM_CSYS] ⇒ pick anywhere in the graphics window ⇒ slowly double-click on each of the datum identifiers in the Model Tree and add **CL_** as a prefix for each (i.e. **CL_ASM_TOP**) ⇒ [icons] on (Fig. 8.19) ⇒ [save] ⇒ **OK**

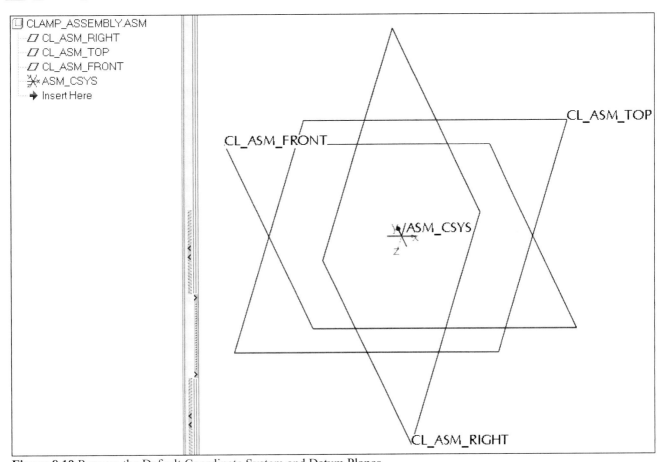

Figure 8.19 Rename the Default Coordinate System and Datum Planes

Creating Components in the Assembly Mode

The Clamp_Subassembly is now complete. The subassembly will be added to the assembly after the Clamp_Plate is created and assembled. Using the Component Create dialog box (Fig. 8.20), you can create different types of components: parts, subassemblies, skeleton models, and bulk items. You cannot reroute components created in the Assembly mode.

The following methods allow component creation in the context of an assembly without requiring external dependencies on the assembly geometry:

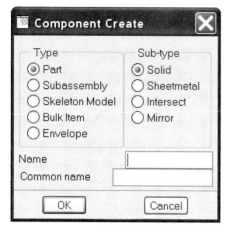

- Create a component by copying another component or existing start part or start assembly
- Create a component with default datums
- Create an empty component
- Create the first feature of a new part; this initial feature is dependent on the assembly
- Create a part from an intersection of existing components
- Create a mirror copy of an existing part or subassembly
- Create Solid or Sheetmetal components
- Mirror Components

Figure 8.20 Component Create Dialog Box

Main Assembly, Top-Down Design

The Clamp_Plate component is the first component of the main assembly. The Clamp_Plate is a new part. You will be modeling the plate *"inside"* the assembly using *top-down design*. The drawing in Figure 8.21 provides the dimensions necessary to model the Clamp_Plate.

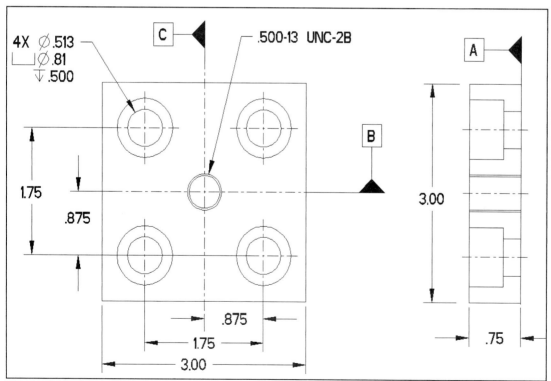

Figure 8.21 Clamp_Plate Detail Drawing

Click: ⬚ **Create a component in assembly mode** Component Create dialog box displays ⇒ Type ⦿Part ⇒ Sub-type ⦿Solid ⇒ Name **CLAMP_PLATE** [Fig. 8.22(a)] ⇒ **OK** Creation Options dialog box displays [Fig. 8.22(b)] ⇒ ⦿Locate Default Datums ⇒ ⦿Align Csys To Csys ⇒ **OK** ⇒ pick **ASM_CSYS** from the graphics window, ⬚CLAMP_PLATE.PRT displays in the Model Tree [Fig. 8.23(a)], *the small green symbol/icon* ⬚ *indicates that this component is now active* ⇒ expand and highlight the items in the Model Tree for the CLAMP_PLATE.PRT

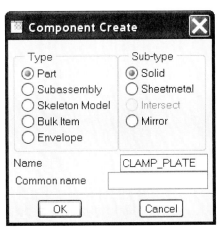

Figure 8.22(a) Component Create Dialog Box

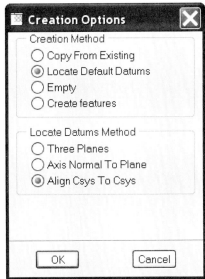

Figure 8.22(b) Creation Options Dialog Box

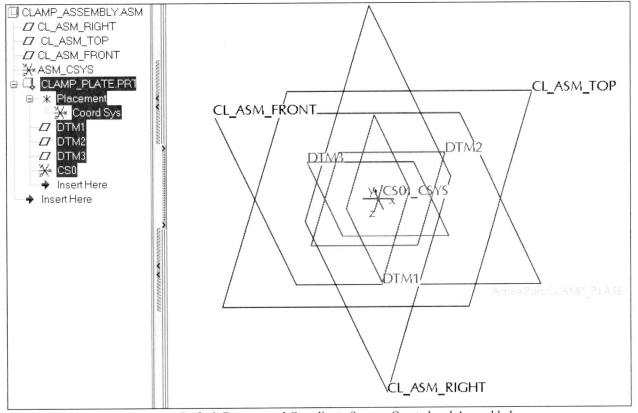

Figure 8.23(a) New Component Default Datums and Coordinate System Created and Assembled

Assembly Tools are now *unavailable* in the Right Toolchest. The current component is the Clamp_Plate. *You are now effectively in Part mode*, except that you can see the assembly features and components. Be sure to reference only part features as you model (part datum planes and part coordinate system), otherwise you will create unwanted external references. There are many situations where external references are needed and desired, but in this case, you are simply modeling a new part.

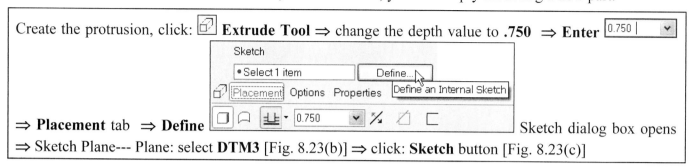

Create the protrusion, click: **Extrude Tool** ⇒ change the depth value to **.750** ⇒ **Enter** ⇒ **Placement** tab ⇒ **Define** Sketch dialog box opens ⇒ Sketch Plane--- Plane: select **DTM3** [Fig. 8.23(b)] ⇒ click: **Sketch** button [Fig. 8.23(c)]

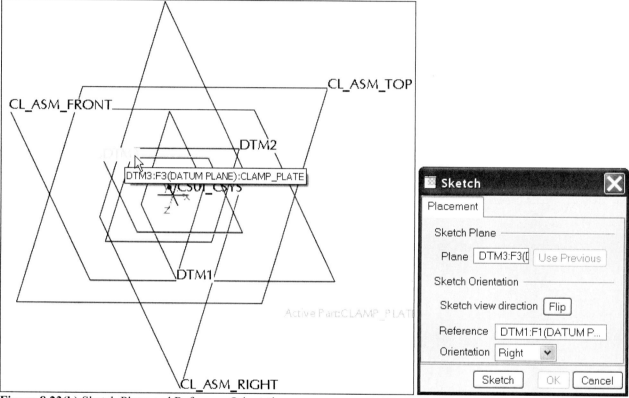

Figure 8.23(b) Sketch Plane and Reference Orientation   **Figure 8.23(c)** Sketch Dialog Box

Click: **RMB** ⇒ **Centerline** ⇒ sketch a vertical centerline ⇒ sketch a horizontal centerline ⇒ **Create rectangle** ⇒ pick the two corners of the rectangle *(if you select the corners carefully you will get a square with only one dimension, if not, then use the = constraint to make the two sides equal)* [Fig. 8.23(d)] ⇒ pick the dimension ⇒ **RMB** ⇒ **Modify** modify the value to **3.00** [Fig. 8.23(e)] ⇒ **Enter** ⇒ ✓ ⇒ **Ctrl+D** ⇒ ✓ [Fig. 8.23(f)] ⇒ **MMB** [Fig. 8.23(g)] ⇒ click on the CLAMP_ASSEMBLY.ASM in the Model Tree ⇒ **RMB** ⇒ **Activate** ⇒ ⇒ **Enter** ⇒ click on the CLAMP_PLATE.PRT in the Model Tree ⇒ **RMB** ⇒ **Activate** (the Clamp_Plate is now active) ⇒ **Ctrl+S** ⇒ **Enter**

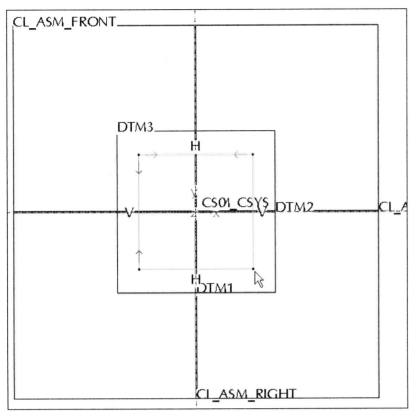

Figure 8.23(d) Sketch a Square Section

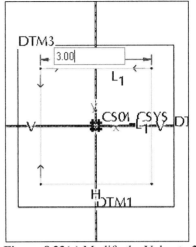

Figure 8.23(e) Modify the Value to **3.00**

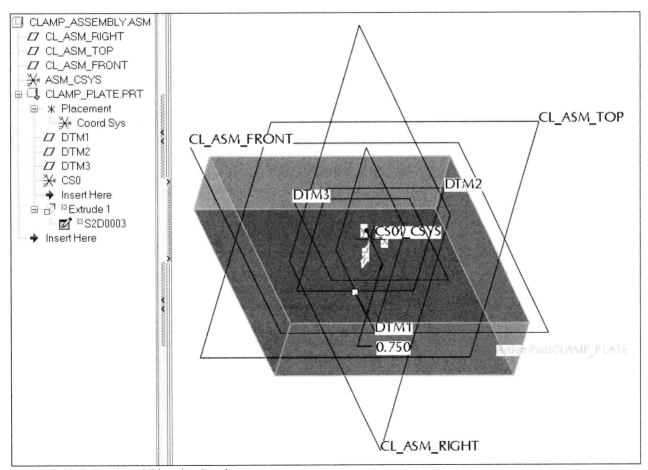

Figure 8.23(f) Depth and Direction Preview

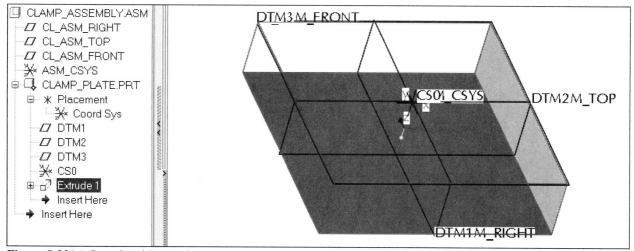

Figure 8.23(g) Completed Protrusion

Throughout the lesson, follow the steps as given, even if it seems that your picks create something different than what is described. Drag the handles to the appropriate references and let Pro/E select the references automatically *even if you think the selections are incorrect*. Later, we will address these inconsistencies, but if you correct them on your own, steps described later will not work (or be needed).

Create the tapped hole in the center of the part, click: ⛉ **Hole Tool** ⇒ 🔩 ⇒ 𝖸 toggle off ⇒ ⧉ toggle off ⇒ ≡ ⇒ ⊕ on ⇒ select the tap and drill size **1/2-13** `1/2-13` ⇒ **Shape** tab ⇒ ☑ Include thread surface ⇒ ⦿ Thru Thread ⇒ **Properties** tab [Fig 8.23 (h)] ⇒ **Placement** tab [Fig 8.23 (i)] ⇒ ▣ **Hidden Line** ⇒ Pick on the top surface of the protrusion as the placement plane. The preview hole displays with drag handles for hole position, diameter adjustment, depth adjustment, and two reference handles for establishing the dimensioning scheme [Fig 8.23(j)].

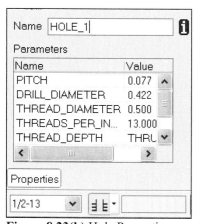

Figure 8.23(h) Hole Properties

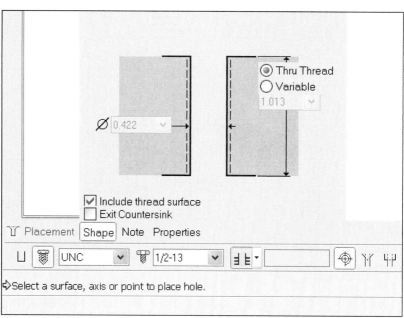

Figure 8.23(i) Hole Shape

Move one reference drag handle to datum **DTM1** and the other drag handle to datum **DTM2** respectively ⇒ change *both* linear dimensions to **Align** [Fig 8.23(k)] ⇒ ☑ ⇒ ◩ ⇒ **LMB** in the graphics window to deselect [Fig. 8.23(l)]

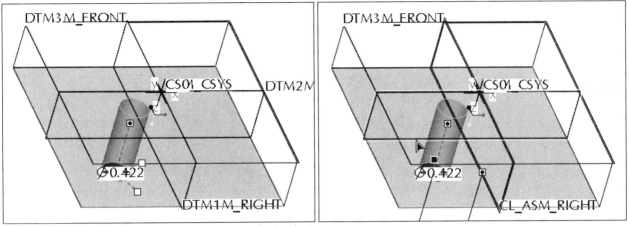

Figure 8.23(j) Initial Hole Placement (at point of selection)

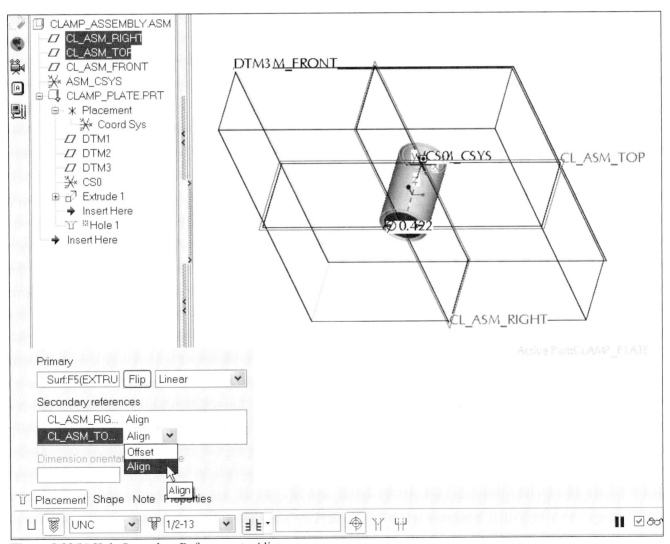

Figure 8.23(k) Hole Secondary References to Align

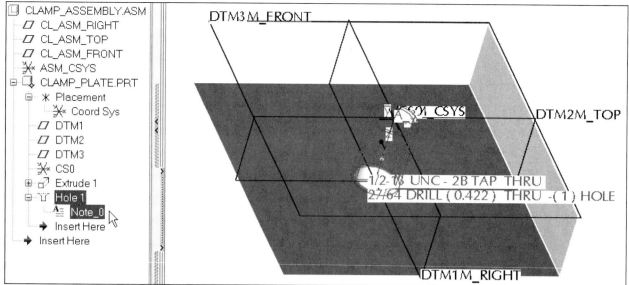

Figure 8.23(l) Completed Tapped Hole

Create a counterbore hole, click: **Hole Tool** ⇒ ⇒ 1/2-13 ⇒ toggle off ⇒ **Adds counterbore** on ⇒ **Drill to intersect with all surfaces** ⇒ toggle off ⇒ **Shape** tab ⇒ leave the defaults for the counterbores sizes [Fig. 8.23(m)] *the detail drawing callouts are slightly different* [Fig. 8.23(n)] ⇒ pick on the front surface to place the counterbore hole *in the first quadrant*

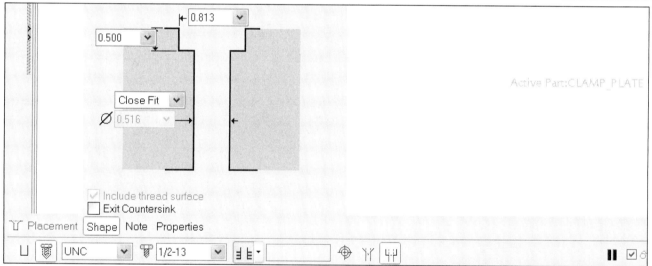

Figure 8.23(m) Counterbore Specifications

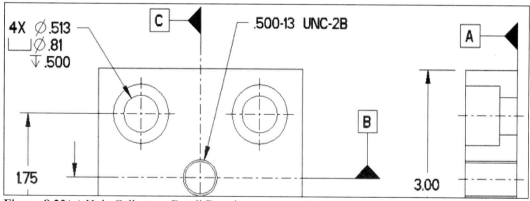

Figure 8.23(n) Hole Callout on Detail Drawing

Click: 🗔 **Hidden Line** ⇒ **Placement** tab ⇒ **RMB** ⇒ **Secondary References Collector** ⇒ *do the following quickly without regard to whether or not the correct vertical or horizontal plane is selected--* pick on the vertical **DTM1** ⇒ press **Ctrl** and pick on the horizontal **DTM2** [Fig 8.23(o)] ⇒ change linear dimensions to **-.875** *(negative .875 if you picked in the first quadrant, if not then .875)* ⇒ **Enter** and **.875** (0.875) ⇒ **Enter** ⇒ ✓ ⇒ 🗔 [Fig. 8.23(p)]

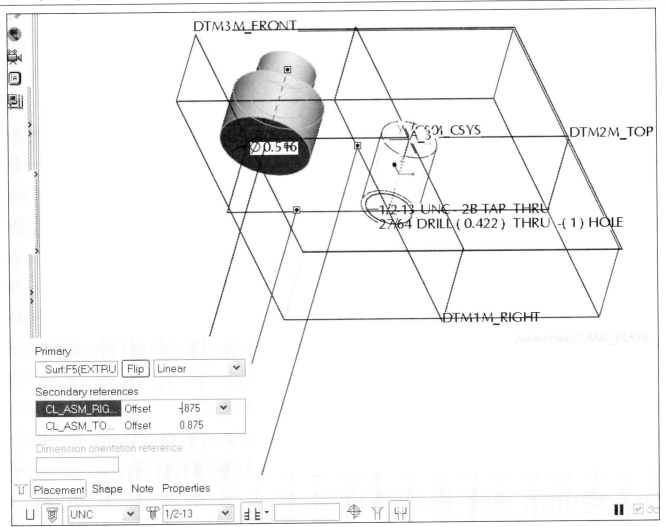

Figure 8.23(o) Secondary References

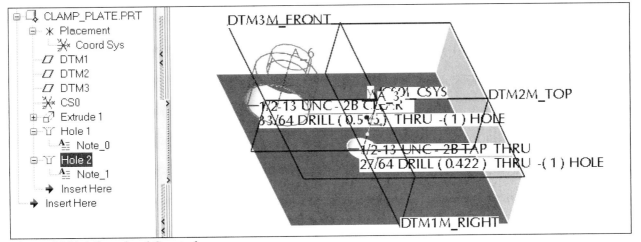

Figure 8.23(p) Completed Counterbore

The counterbore "seems" correct, but still check for external references. Click: **Info** ⇒ **Global Reference Viewer** [Fig 8.23(q)] ⇒ **Show Filters** tab ⇒ activate all Display Objects [Fig. 8.23(r)] ⇒ **Apply** ⇒ **OK** ⇒ expand as necessary [Fig. 8.23(s)] ⇒ click on **Hole 2** [Hole 2] ⇒ **Actions** ⇒ **Set Current** ⇒ expand as necessary

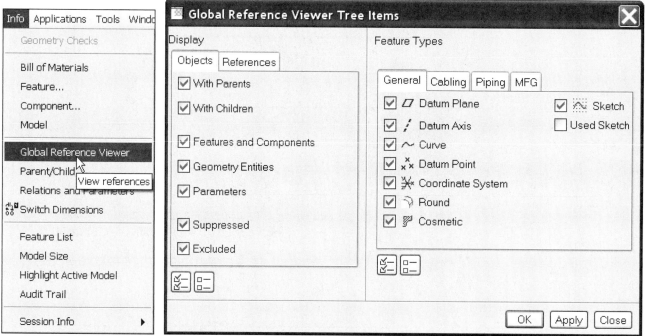

Figure 8.23(q) Global

Figure 8.23(r) Global Reference Viewer Tree Items Dialog Box

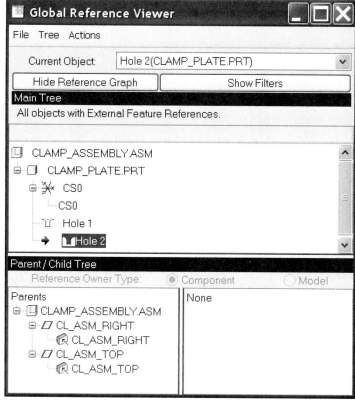

Figure 8.23(s) Global Reference Viewer Dialog Box

If you look back at Figure 8.23(k) and Figure 8.23(o) you will see that the assembly datums were used to place the holes instead of DTM1 and DTM2. The following commands will show you how to edit the references of one of the holes. Redo the references of the other hole using similar steps.

From the Global Reference Viewer Dialog box, click: **File** ⇒ **Close** ⇒ **RMB** on the hole in the Model Tree ⇒ **Edit Definition** [Fig. 8.23(t)]

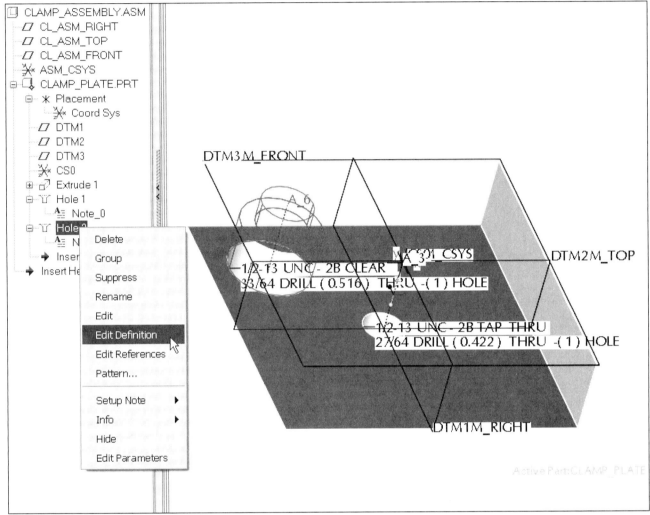

Figure 8.23(t) Edit Definition

Click: **Hidden Line** ⇒ **Placement** tab ⇒ click in the Secondary references field ⇒ **RMB** ⇒ **Remove All** [Fig. 8.23(u)] ⇒ place your cursor over datum CL_ASM_TOP to highlight ⇒ **RMB** ⇒ DTM2 highlights [Fig. 8.23(v)] ⇒ pick on **DTM2**

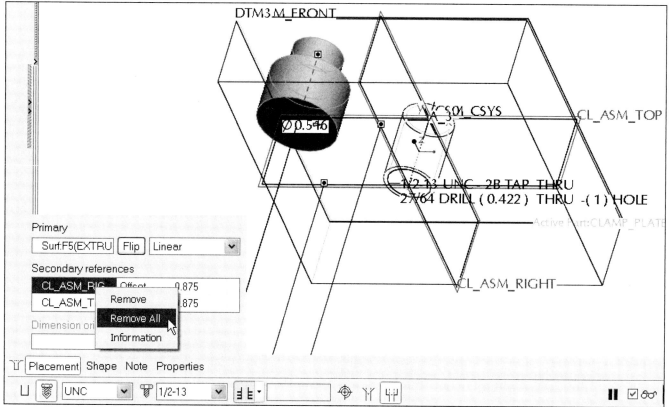

Figure 8.23(u) Remove Secondary References

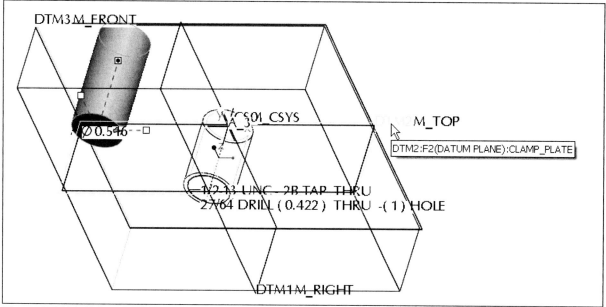

Figure 8.23(v) Pick on DTM2

Press and hold **Ctrl** key and place your cursor over CL_ASM_RIGHT (highlights) ⇒ **RMB** DTM1 highlights [Fig. 8.23(w)] ⇒ pick on **DTM1** [Fig. 8.23(x)] ⇒ **MMB** ⇒ **Ctrl+S** ⇒ **MMB**

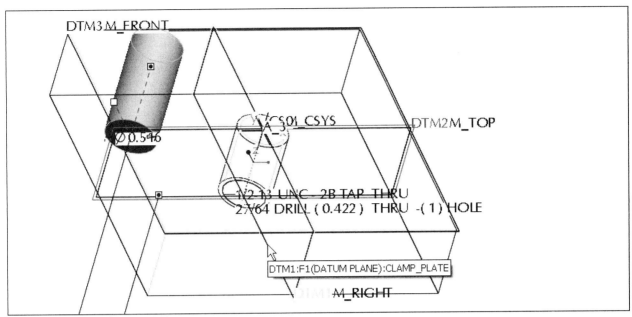

Figure 8.23(w) Pick on DTM1

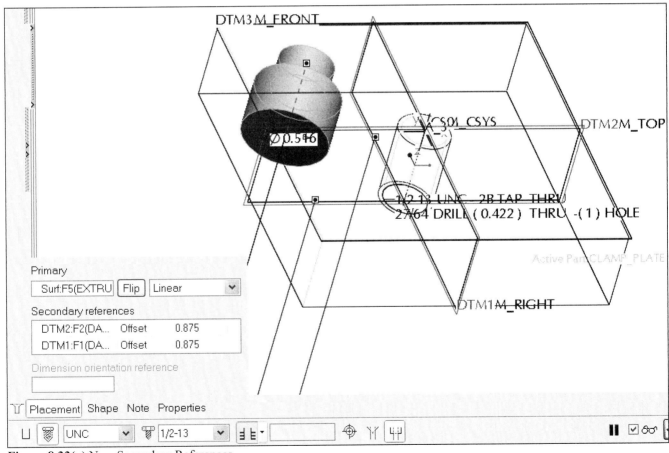

Figure 8.33(x) New Secondary References

If you have difficulty with this method, simply click on the three assembly datum planes in the Model Tree ⇒ RMB ⇒ Hide. Then go and redo the references. Unhide the assembly datums after you are finished changing the references.

Click: **Info** ⇒ **Global Reference Viewer** [Fig. 8.33(y)] *only one hole shows as having an external reference* ⇒ click on **Hole 1** ⇒ **RMB** ⇒ **Set Current** ⇒ **File** ⇒ **Close** ⇒ repeat the previous process to select the correct references (DTM1 and DTM2) for the tapped hole in the center of the part [Fig. 8.23(z)] ⇒ **Ctrl+S** ⇒ **MMB**

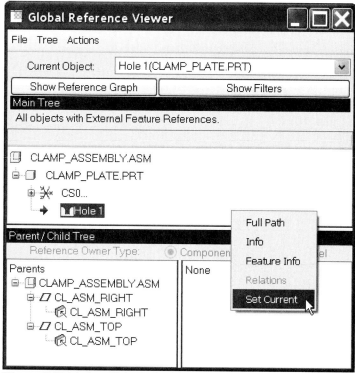

Figure 8.23(y) One Hole Left with External References

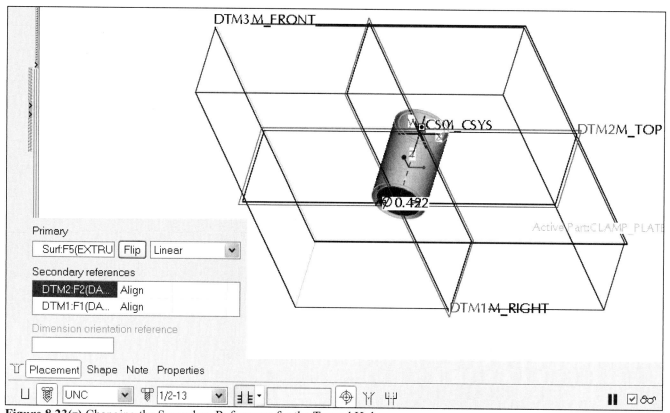

Figure 8.23(z) Changing the Secondary References for the Tapped Hole

Click: **View** ⇒ **Orientation** ⇒ **Standard Orientation** ⇒ ⬚ ⇒ **Tools** ⇒ **Environment** ⇒ ☐ 3D Notes ⇒ **Apply** *removes the 3D Notes from the screen* (Fig. 8.24) ⇒ **OK** ⇒ 💾 ⇒ **OK**

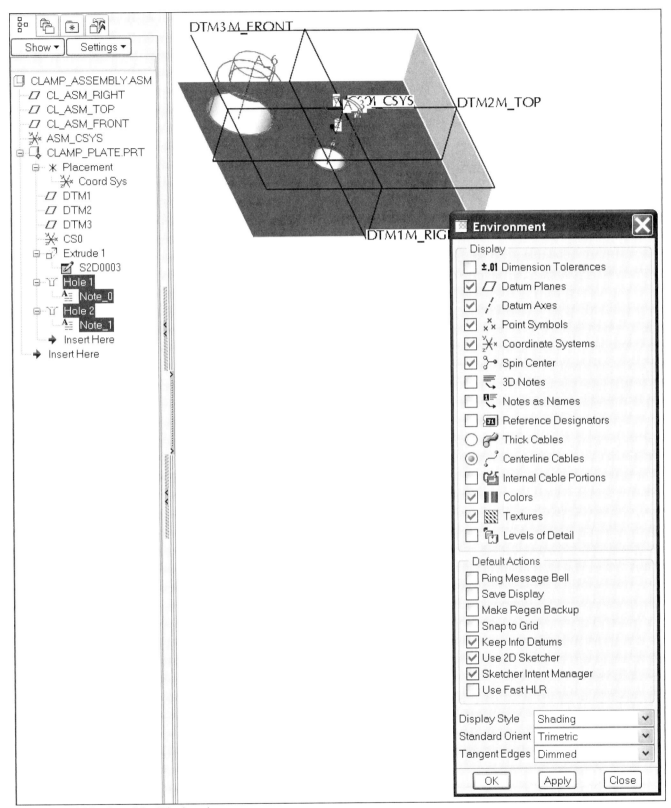

Figure 8.24 3D Notes Not Displayed

Pattern the counterbore holes. Click on the last hole in the Model Tree [Hole id] ⇒ **RMB** ⇒ **Pattern** [Fig. 8.25(a)] ⇒ **Dimensions** tab ⇒ pick on the horizontal **.875** dimension ⇒ highlight value ⇒ type **–1.75** [Fig. 8.25(b)] ⇒ **Enter**

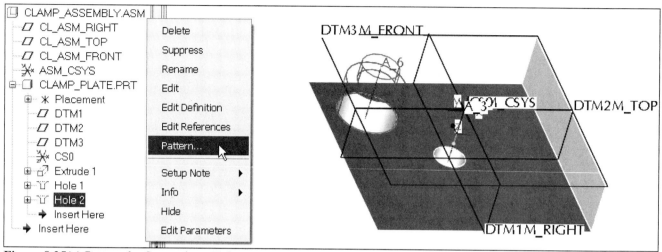

Figure 8.25(a) Pattern the Counterbore Hole

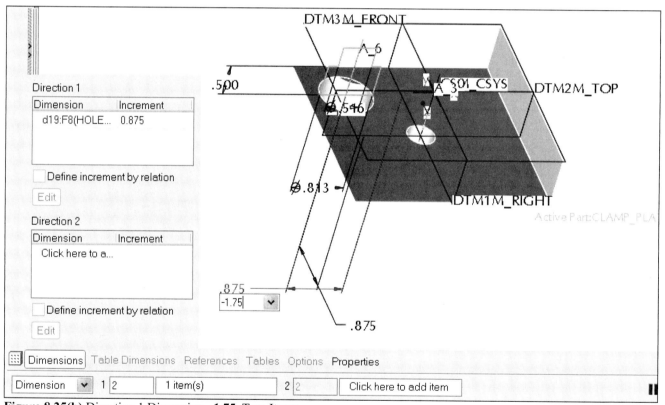

Figure 8.25(b) Direction 1 Dimension **–1.75**, Two Items

In Direction 2 collector box, pick [Click here to a...] ⇒ pick on the vertical **.875** dimension [Fig. 8.26(a)] ⇒ highlight value and type **–1.75** ⇒ **Enter** ⇒ **MMB** ⇒ **Tools** ⇒ **Environment** ⇒ [Standard Orient | Isometric] ⇒ **Apply** ⇒ **OK** [Fig. 8.26(b)] ⇒ **Edit** ⇒ **Regenerate** ⇒ **Ctrl+S** ⇒ **MMB**

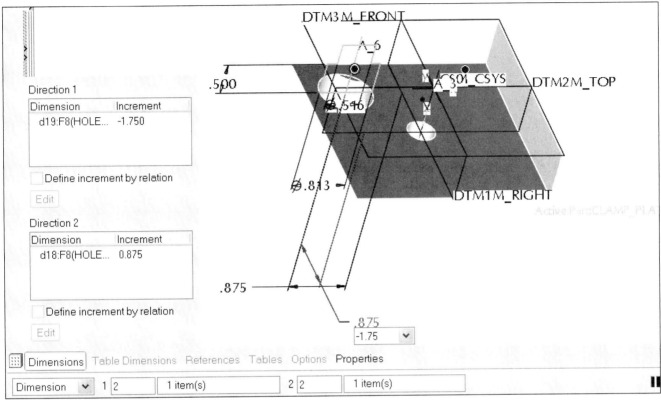

Figure 8.26(a) Direction 2 Dimension **–1.75**, Two Items

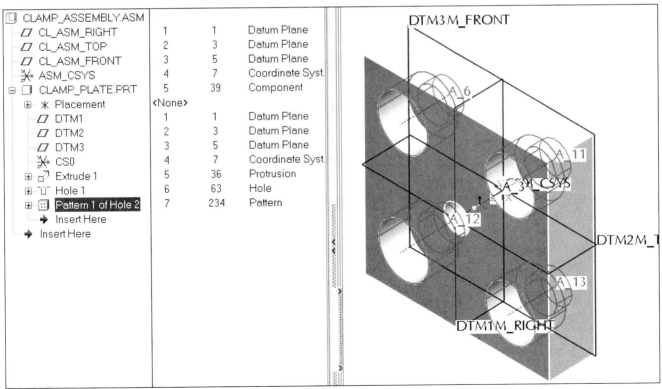

Figure 8.26(b) Patterned Counterbore Holes

Click [CLAMP_ASSEMBLY.ASM] in the Model Tree ⇒ **RMB** ⇒ **Activate** ⇒ pick datum tag [CL_ASM_FRONT] in the graphics window [Fig. 8.27(a)] ⇒ **RMB** ⇒ **Move Datum Tag** ⇒ pick a new position [Fig. 8.27(b)] ⇒ repeat this process and move the other assembly datum tags ⇒ [⌕] ⇒ [🗎] **Add component to the assembly** ⇒ pick the **clamp_subamssembly.asm** ⇒ **Preview>>>** [Fig. 8.28(a)] ⇒ **Open**

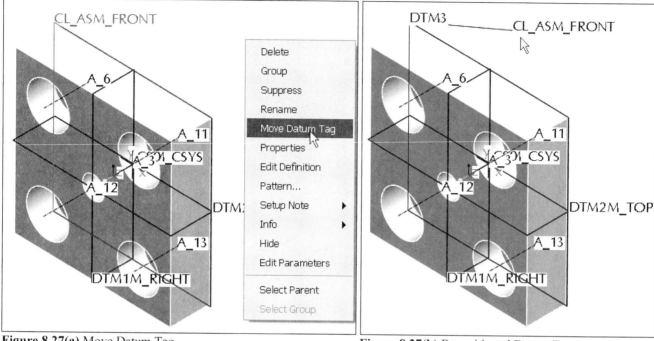

Figure 8.27(a) Move Datum Tag   **Figure 8.27(b)** Repositioned Datum Tag

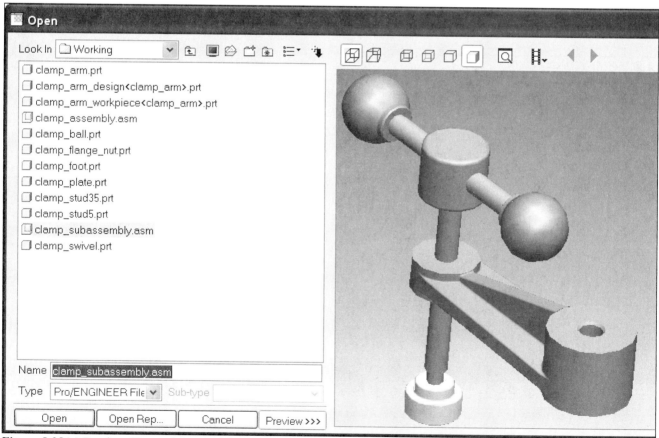

Figure 8.28(a) Preview Clamp_Subassembly

Click: **Datum planes** off ⇒ **Coordinate system** off ⇒ toggle off component window ⇒ Automatic ⇒ **Mate** [Fig. 8.28(b)] (Mate) ⇒ spin the model ⇒ **View** ⇒ **Shade** ⇒ pick the bottom surface of the Clamp_Arm (of the Clamp_Subassembly) [Fig. 8.28(c)]

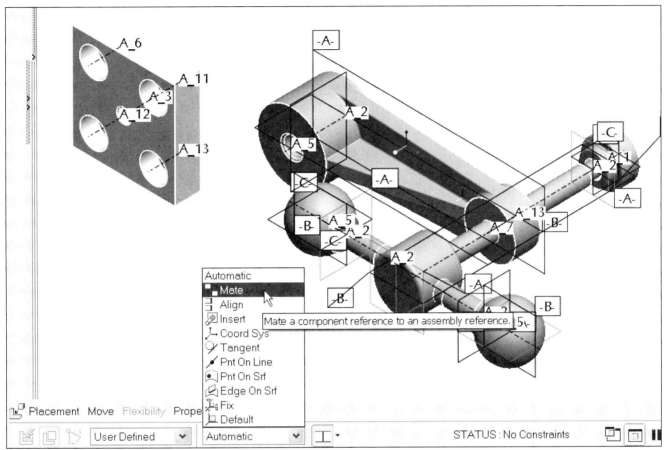

Figure 8.28(b) Assembling the Clamp_Subassembly

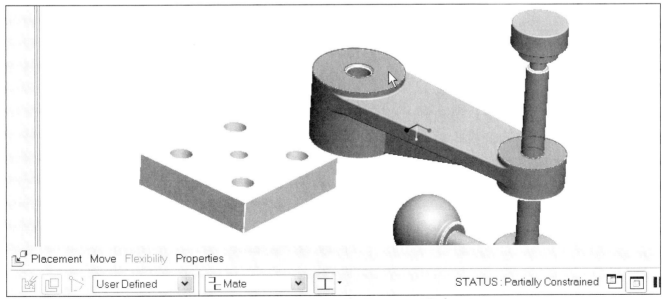

Figure 8.28(c) Bottom Surface of the Clamp_Arm is selected as the Component Reference

Spin the model ⇒ **View** ⇒ **Shade** ⇒ **Placement** tab ⇒ pick the non-counterbore surface of the Clamp_Plate [Fig. 8.28(d)] ⇒ **New constraint** ⇒ [Automatic] ⇒ **Align** ([Align]) ⇒ [icon] ⇒ pick the hole axis of the Clamp_Arm [Fig. 8.28(e)] ⇒ pick the center tapped hole axis of the Clamp_Plate [Fig. 8.28(f)] ⇒ [Fig. 8.28(g)] *(note that you could have picked the two respective hole surfaces to achieve the same result using an Insert Constraint)*

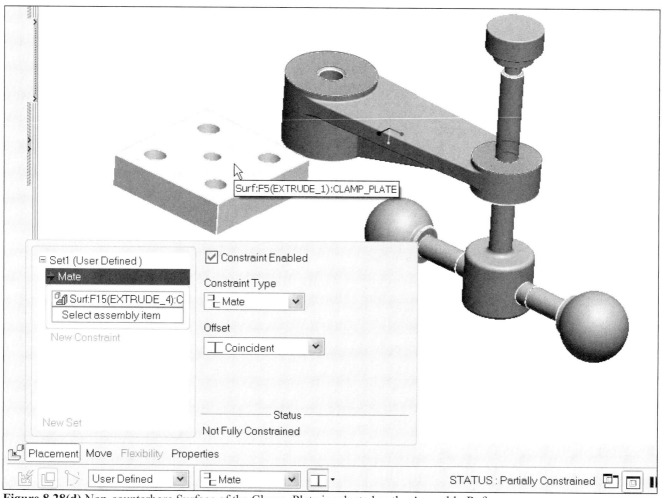

Figure 8.28(d) Non-counterbore Surface of the Clamp_Plate is selected as the Assembly Reference

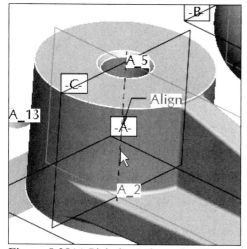

Figure 8.28(e) Pick the Axis on the Clamp_Arm

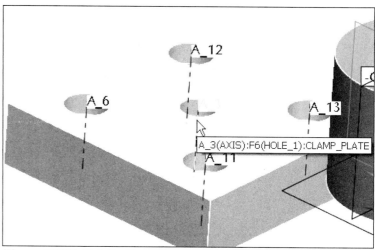

Figure 8.28(f) Pick the Tapped Hole Axis on the Clamp_Plate

Click: ☑ ⇒ 🗒 ⇒ **FRONT** ⇒ **View** ⇒ **Shade** ⇒ **View** ⇒ **Orientation** ⇒ **Standard Orientation** ⇒ 💾 ⇒ **MMB**

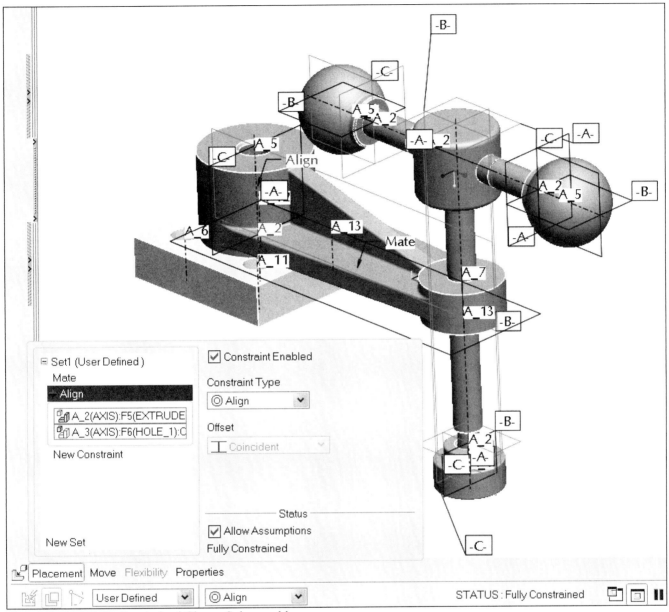

Figure 8.28(g) Fully Constrained Clamp_Subassembly

Click: **Add component to the assembly** ⇒ pick the **clamp_stud35.prt** ⇒ **Preview>>>** [Fig. 8.29(a)]
⇒ **Open** ⇒ **Placement** tab [Fig. 8.29(b)]

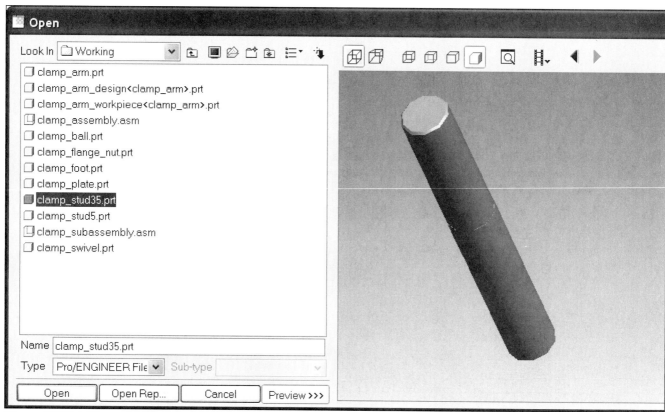

Figure 8.29(a) Clamp_Stud35

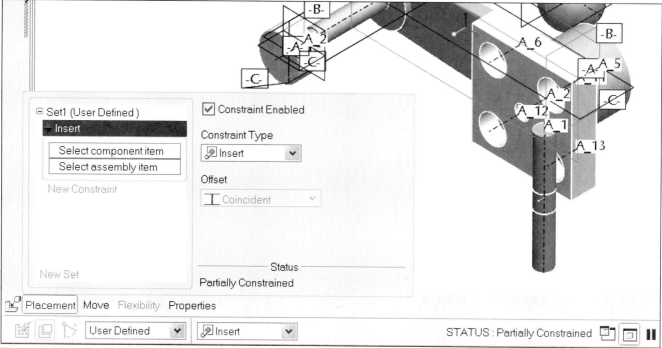

Figure 8.29(b) Clamp_Stud35

326

Constraints `Automatic` ⇒ `Insert` ⇒ **MMB** rotate the model ⇒ pick the cylindrical surface of the Clamp_Stud35 [Fig. 8.29(c)] ⇒ pick the hole surface of the Clamp_Arm [Fig. 8.29(d)]

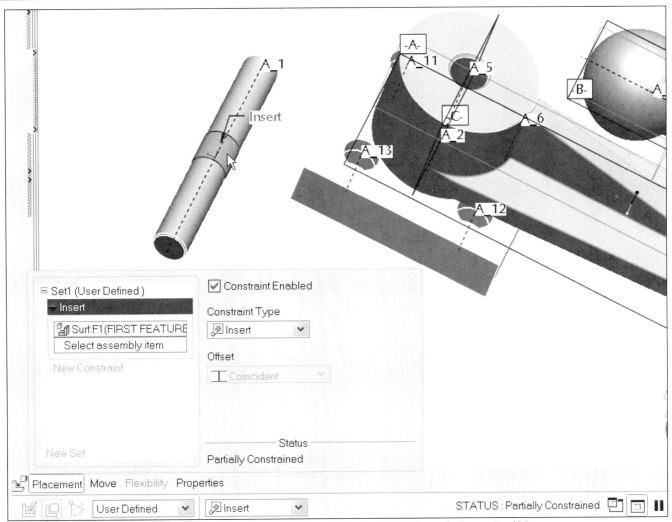

Figure 8.29(c) Component Placement Dialog Box, Placement Tab, Select Surface of Clamp_Stud35

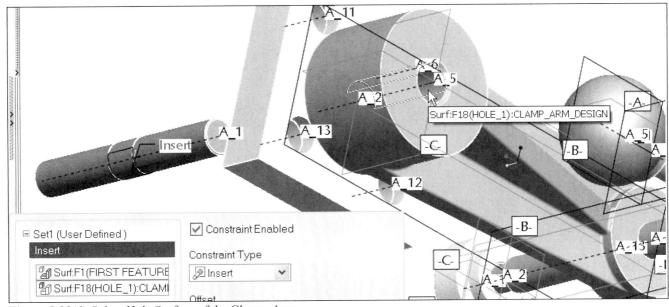

Figure 8.29(d) Select Hole Surface of the Clamp_Arm

Click: **Ctrl+D** ⇒ **Move** tab ⇒ Motion Type [Translate ▼] ⇒ pick the Clamp_Stud35 (hint: release LMB) and slide it deeper into the hole ⇒ **LMB** [Fig. 8.39(e)] ⇒ [⌘] ⇒ **TOP** ⇒ [◻] **Hidden Line** ⇒ pick Clamp_Stud35 and translate (slide) it until it is inside the Clamp_Plate [Fig. 8.39(f)] ⇒ **LMB**

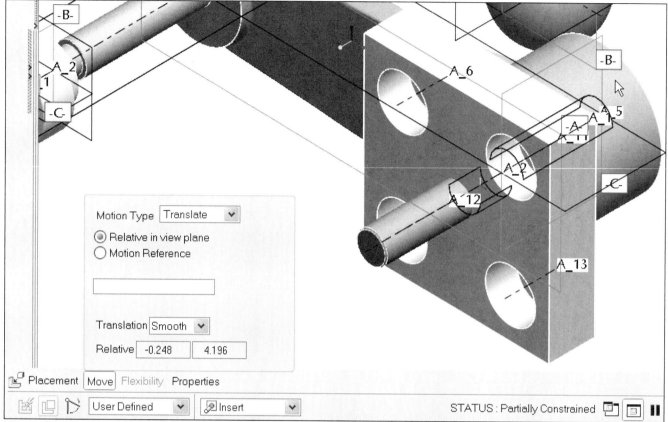

Figure 8.29(e) Move Tab Motion Type Translate

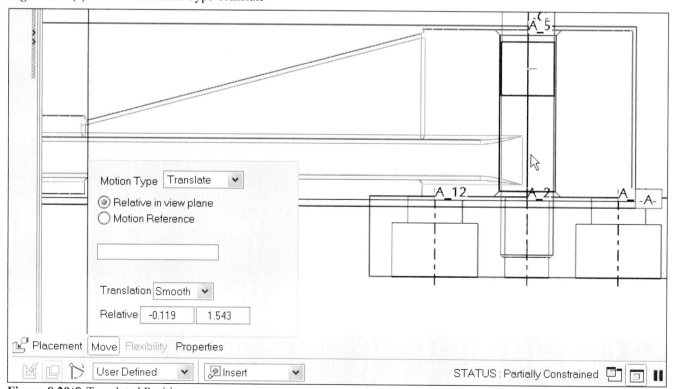

Figure 8.29(f) Translated Position

Click: **Placement** tab ⇒ **New Constraint** ⇒ Constraint Type [Fix] **Fix component to current position** ⇒ View ⇒ Shade ⇒ MMB ⇒ [icon] ⇒ **Standard Orientation** [Fig. 8.29(g)] ⇒ **Ctrl+S** ⇒ MMB ⇒ [icon] **Shading**

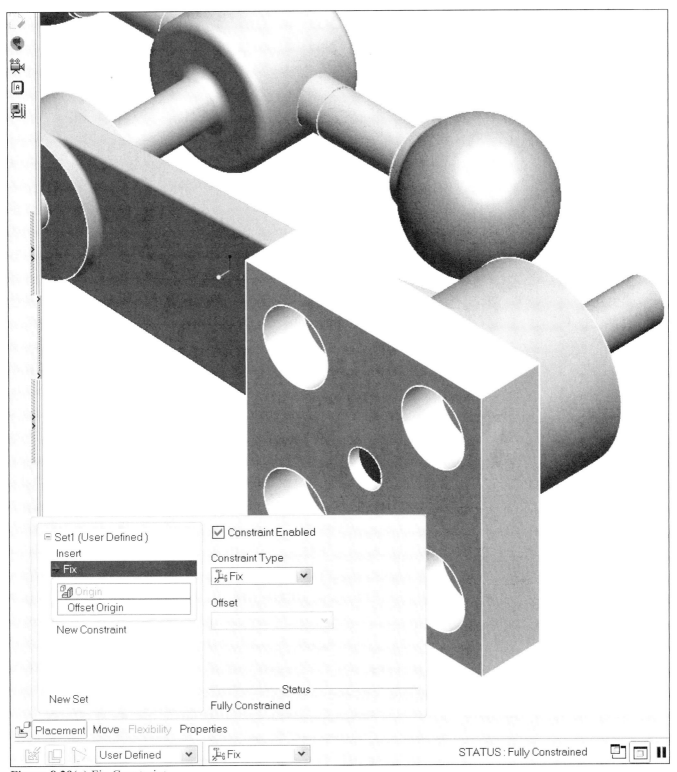

Figure 8.29(g) Fix Constraint

Click: **Add component to the assembly** ⇒ **clamp_flange_nut.prt** [Fig. 8.30(a)] ⇒ **MMB** ⇒ **Ctrl+Alt** ⇒ **RMB** on the Flange_Nut and move closer to the assembly [Fig. 8.30(b)]

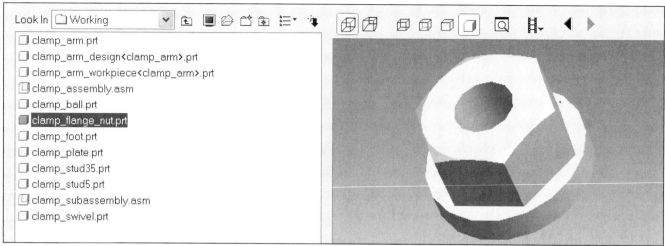

Figure 8.30(a) Clamp_Flange_Nut

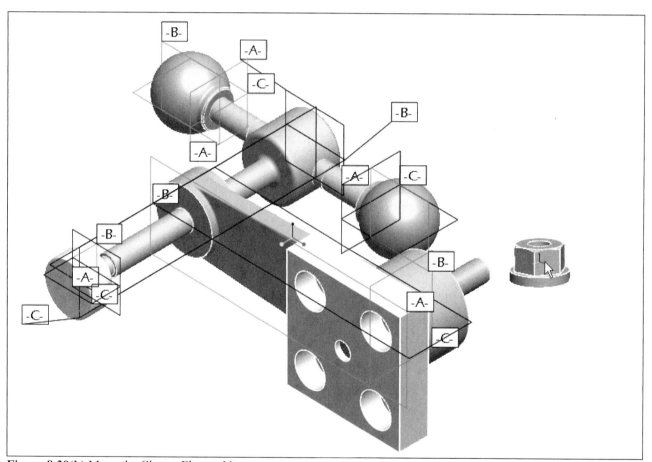

Figure 8.30(b) Move the Clamp_Flange_Nut

Ctrl+Alt ⇒ **MMB** on the Flange_Nut and rotate [Fig. 8.30(c)] ⇒ **RMB** ⇒ **Select component item** [Fig. 8.30(d)]

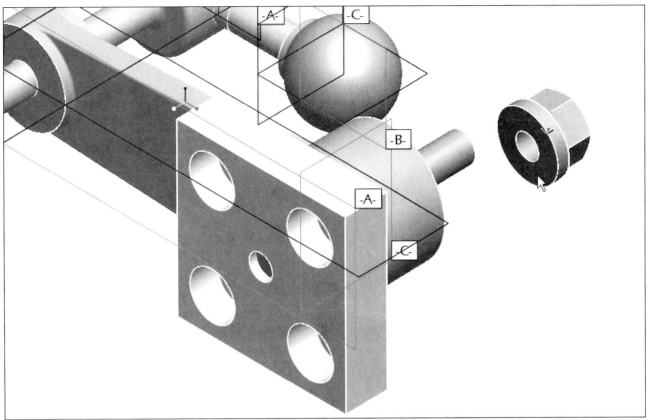

Figure 8.30(c) Rotate the Clamp_Flange_Nut

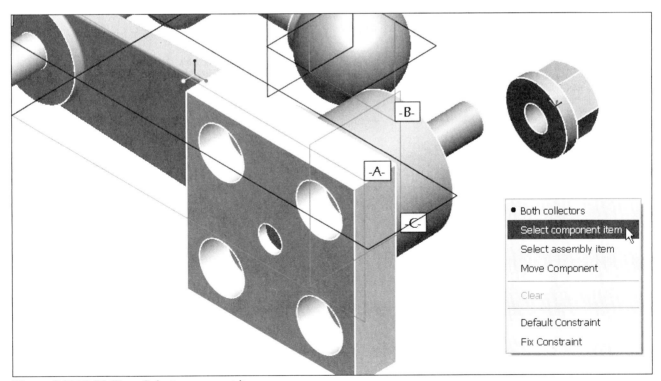

Figure 8.30(d) RMB ⇒ Select component item

Pick the hole surface of the Clamp_Flange_Nut ⇒ pick the shaft surface of the Clamp_Stud35 ⇒ **Placement** tab [Fig. 8.30(e)]

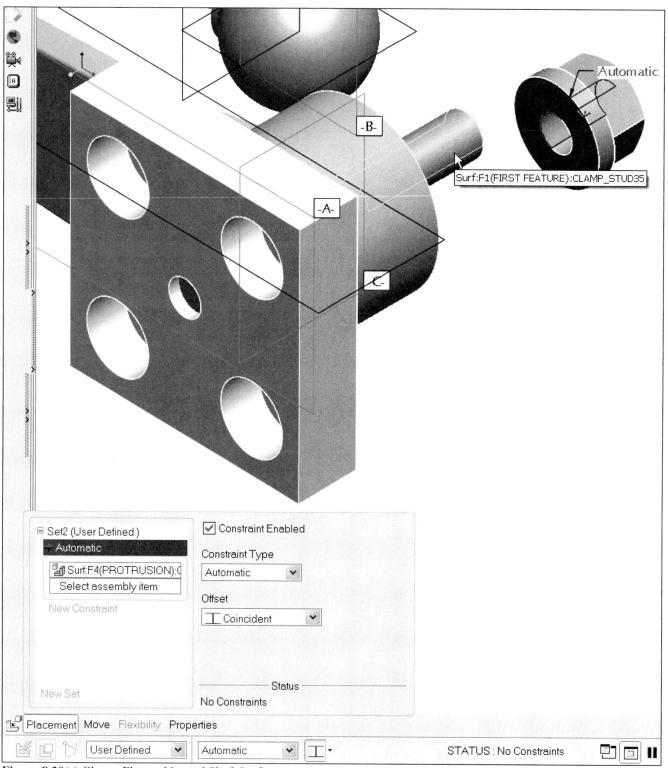

Figure 8.30(e) Clamp_Flange_Nut and Shaft Surface

Ctrl+Alt ⇒ **RMB** on the Flange_Nut and slide it [Fig. 8.30(f)] ⇒ **New Constraint** ⇒ rotate the model as required ⇒ pick on the Clamp_Arm surface [Fig. 8.30(g)] ⇒ pick on the Clamp_Flange_Nut bearing surface [Fig. 8.30(h)]

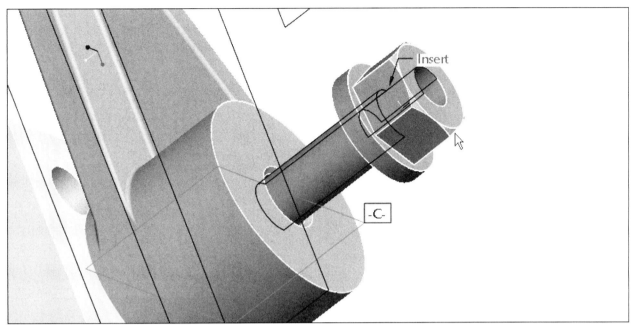

Figure 8.30(f) Moving the Clamp_Flange_Nut

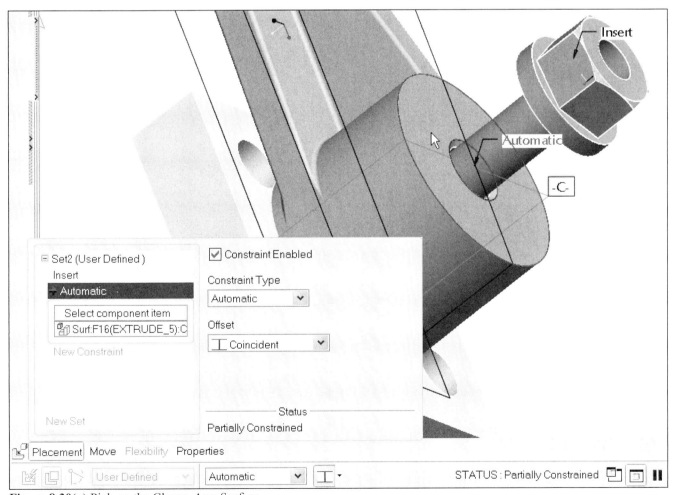

Figure 8.30(g) Pick on the Clamp_Arm Surface

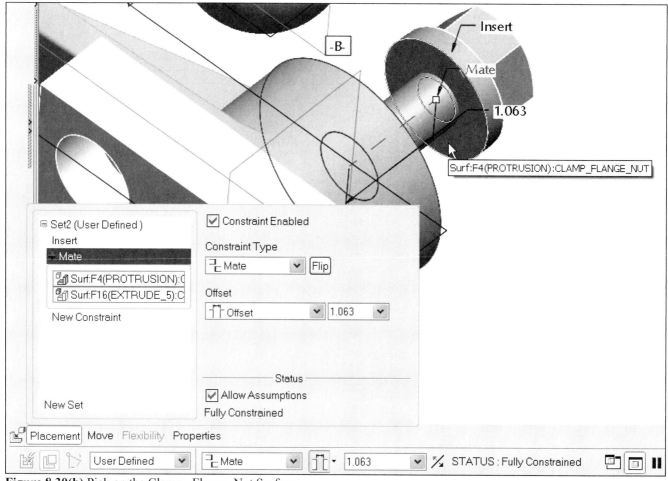

Figure 8.30(h) Pick on the Clamp_Flange_Nut Surface

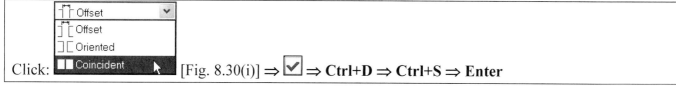

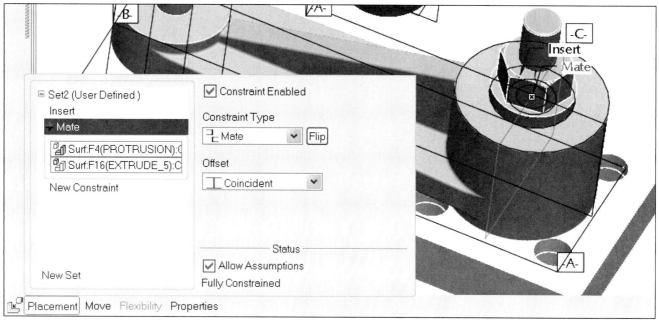

Figure 8.30(i) Offset: Coincident

Perform a check on the assembly using the **Analysis** command. As an example, if you look at the long (**5.00**) stud's detail drawing, you will see that the shaft diameter is greater than **.500** at the center (**.506**) of the stud and that each end has a **.500-13 UNC** thread. The hole in the swivel is **.500** in diameter. This means that there should be a slight interference (press fit) between the two components.

(Note: if you used the CARR Lane stud from the PTC catalog, the iges file part has a constant .500 diameter).

Click: **Analysis** ⇒ **Model** ⇒ **Global Interference** ⇒ [👓] **Compute current analysis for preview** (Fig. 8.31) ⇒ `2: CLAMP_S... CLAMP_S... 0.00690137 INCH^3` *(Note: the interference will be different or zero if you used a PTC catalog CARR Lane stud)* ⇒ [✓]

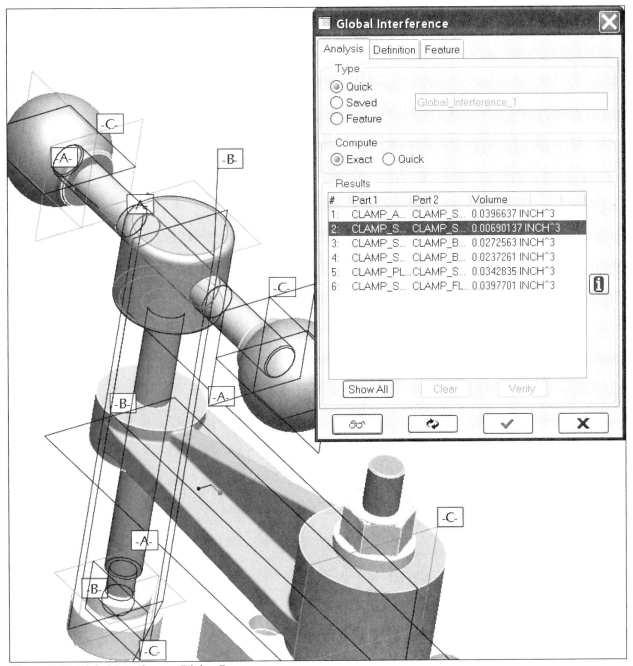

Figure 8.31 Global Interference Dialog Box

Rotating Components

To rotate an existing component or set of components of an assembly (or subassembly), you select a coordinate system to use as a reference, pick one or more components, and give the rotation angle about a chosen axis of the reference coordinate system. The Clamp_Subassembly has a Clamp_Swivel, Clamp_Foot, Clamp_Stud5, and two Clamp_Ball components that can be rotated about the Clamp_Arm during normal operation of the assembly. You will rotate these components so that the Clamp_Stud5 and two Clamp_Ball components are perpendicular to the Clamp_Arm. This position looks better when you are displaying the assembly as exploded (and in its unexploded state). The components will be rotated in the subassembly, and the change will be propagated to the assembly.

Click: ▭CLAMP_SUBASSEMBLY.ASM in the Model Tree [Fig. 8.32(a)] ⇒ **RMB** ⇒ **Open** ⇒ **Window** in the Clamp_Subassembly menu bar ⇒ **Activate** to make sure you are working on the correct object ⇒ ▭ **Datum planes** off ⇒ ▭ **Coordinate systems** on ⇒ ▭ **Redraw the current view** [Fig. 8.32(b)]

Note: Pay attention- you are working on the subassembly NOT the assembly. [Fig. 8.32(c)]

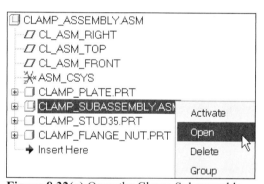

Figure 8.32(a) Open the Clamp Subassembly

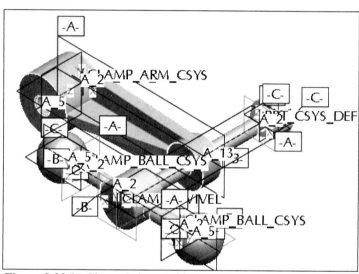

Figure 8.32(b) Clamp Subassembly

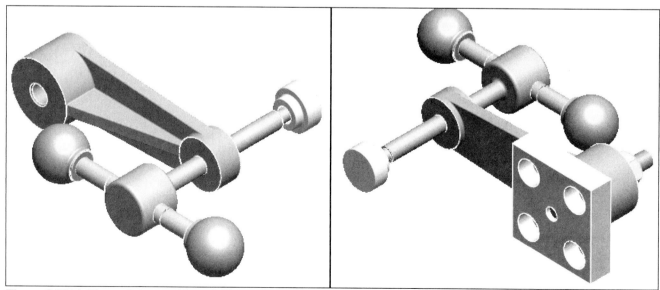

Figure 8.32(c) Clamp_Subassembly and Clamp_Assembly are Open in Different Windows

Click: **Edit** ⇒ **Component Operations** ⇒ **Transform** ⇒ Select coordinate system: pick coordinate system **CLAMP_SWIVEL_CSYS** [Fig. 8.32(d)] *or choose the coordinate system from the Model Tree* ⇒ Select components to move: pick the **CLAMP_SWIVEL** component [Fig. 8.32(e)] *you may also choose the component from the Model Tree* ⇒ **OK** ⇒ **Rotate** ⇒ *select the appropriate axis (points down the length of the shaft) on "your" component* ⇒ type **90** as the angle of rotation ⇒ **Enter** ⇒ **Done Move** ⇒ **Done/Return** ⇒ **Regenerates Model**

Only the Clamp_Swivel was selected, but the Clamp_Stud5, both Clamp_Ball components, and the Clamp_Foot were rotated **90** degrees. These other components were children of the Clamp_Swivel and therefore rotated through the same **90** degree angle.

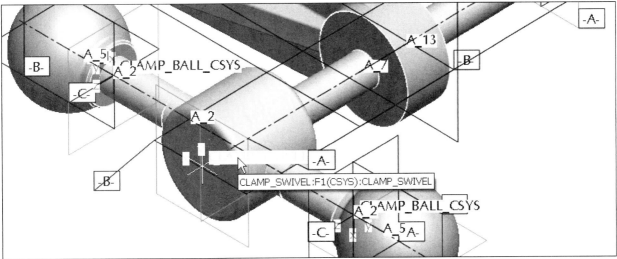

Figure 8.32(d) Select the Coordinate System CLAMP_SWIVEL _CSYS

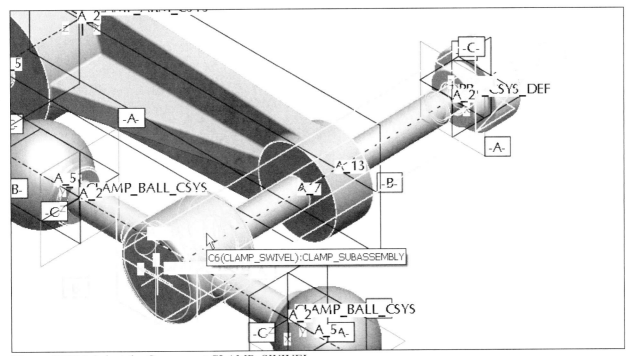

Figure 8.32(e) Select the Component CLAMP_SWIVEL

Click: **Window** ⇒ 1 CLAMP_ASSEMBLY.ASM ⇒ **Window** ⇒ Activate ⇒ 🔍 **Refit object** [Fig. 8.32(f)] ⇒ **MMB** rotate the model ⇒ double-click on the first Clamp_Ball [Fig. 8.32(g)] ⇒ double-click on the **.500** dimension ⇒ modify value to **.4375** [Fig. 8.32(h)] ⇒ **Enter** ⇒ **Ctrl+G** ⇒ **Ctrl+D** ⇒ **Ctrl+S** ⇒ **Enter**

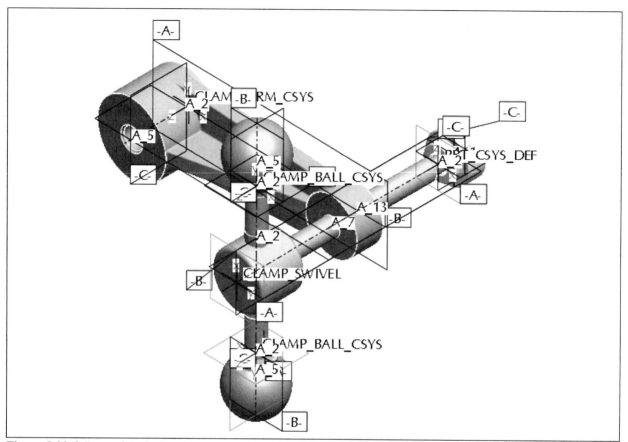

Figure 8.32(f) Rotated Swivel and Component Children

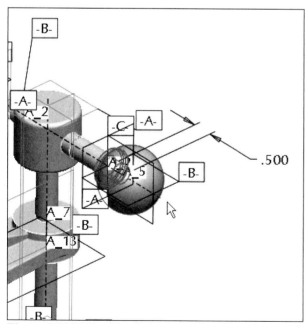

Figure 8.32(g) Double-click on the First Clamp_Ball

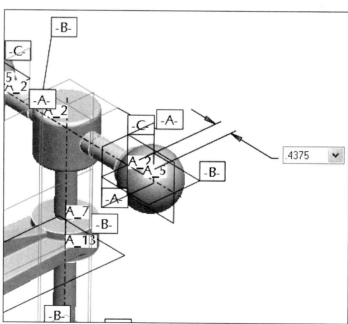

Figure 8.32(h) Modify the Offset to **.4375**

Bill of Materials

The Bill of Materials (BOM) lists all parts and part parameters in the current assembly or assembly drawing. It can be displayed in HTML or text format and is separated into two parts: breakdown and summary. The Breakdown section lists what is contained in the current assembly or part. The BOM HTML breakdown section lists quantity, type, name (hyperlink), and three actions (highlight, information and open) about each member or sub-member of your assembly:

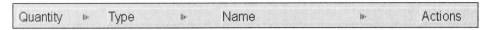

- **Quantity** Lists the number of components or drawings
- **Type** Lists the type of the assembly component (part or sub-assembly)
- **Name** Lists the assembly component and is hyper-linked to that item. Selecting this hyperlink highlights the component in the graphics window
- **Actions** is divided into three areas:
    - **Highlight** Highlights the selected component in the assembly graphics window
    - **Information** Provides model information on the relevant component
    - **Open** Opens the component in another Pro/ENGINEER window

---

A bill of materials (BOM) can be seen by clicking: **Info** ⇒ **Bill of Materials** [Fig. 8.33(a)] ⇒ **Top Level** ⇒ check all Include options [Fig. 8.33(b)] ⇒ **OK** [Fig. 8.33(c)] ⇒ click quick sash to close the Browser ⇒ **File** ⇒ **Close Window** ⇒ **File** ⇒ **Close Window**

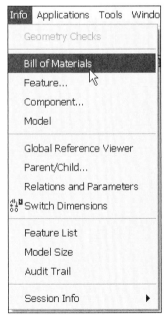

Figure 8.33(a) Info ⇒ Bill of Materials

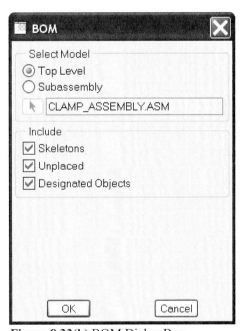

Figure 8.33(b) BOM Dialog Box

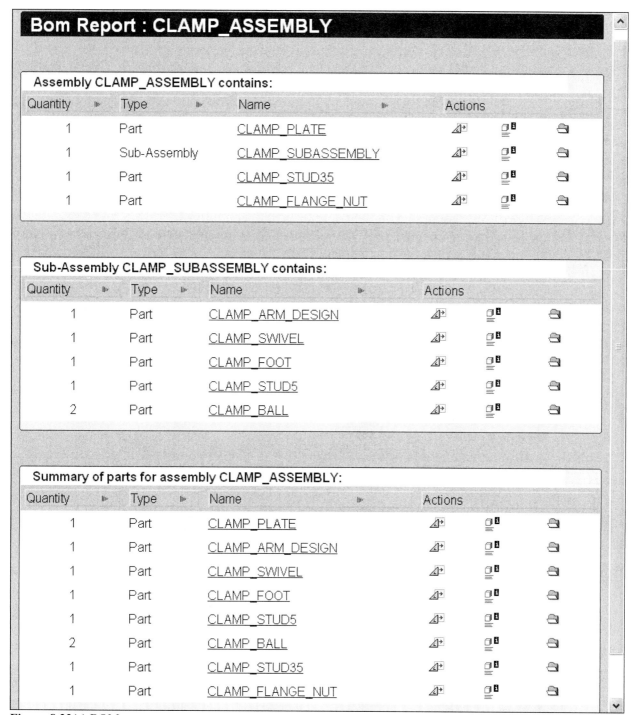

Figure 8.33(c) BOM

Upon closer inspection the assembly is not assembled correctly. The Clamp_subassembly is assembled on the wrong side of the Clamp_Plate. When using Pro/E, you can edit items, features, components, etc., at any time. The Edit Definition command is used to change the constraint reference of assembled components.

If necessary, Open the Clamp_Assembly, or click: **Window** ⇒ **Activate** (Fig. 8.34) to activate the top assembly model ⇒ click on the **CLAMP_SUBASSEMBLY** in the Model Tree ⇒ **RMB** ⇒ **Edit Definition** [Fig. 8.35(a)]

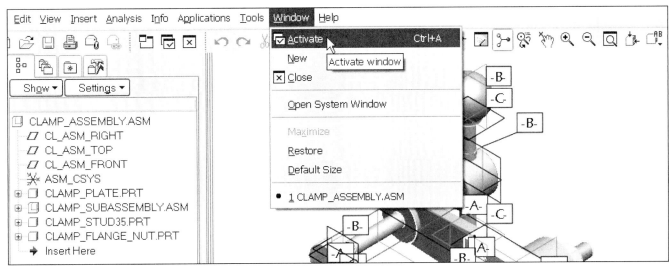

Figure 8.34 Activate the Assembly

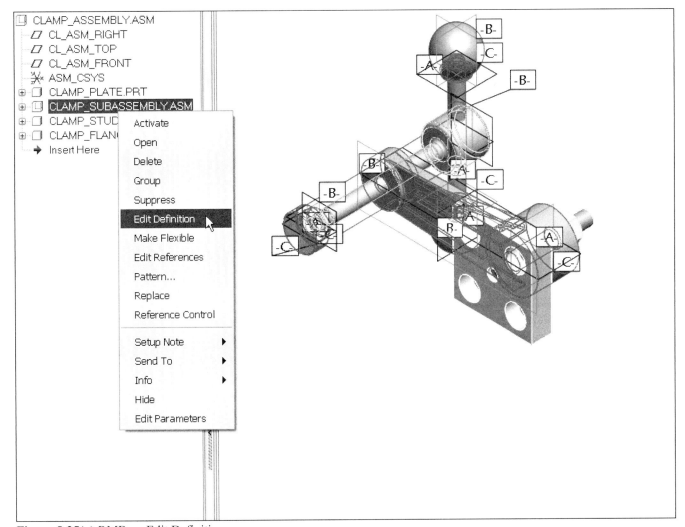

Figure 8.35(a) RMB ⇒ Edit Definition

Click: **Placement** tab, to see the existing Mate references ⇒ pick the second (assembly) reference `Surf:F5(EXTRUDE_1):CL` from the Placement dialog box [Fig. 8.35(b)] *(turns light yellow to indicate that it is active, check the Tool Tip `Surf:F5(EXTRUDE_1):CLAMP_PLATE` to make sure you selected the Clamp_Plate reference)*

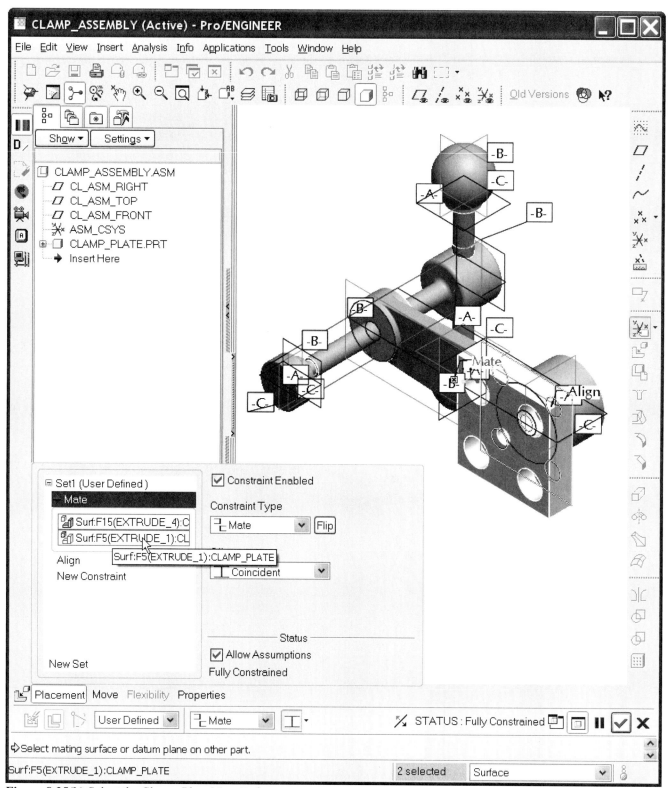

Figure 8.35(b) Select the Clamp_Plate Mate Reference

Pick the counterbore side of the CLAMP_PLATE as the new mate reference [Fig. 8.35(c)] and the subassembly reverses [Fig. 8.35(d)] ⇒ ✓ [Fig. 8.35(e)]

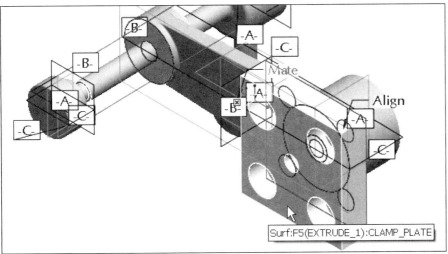

Figure 8.35(c) Pick the Counterbore Side of the Plate

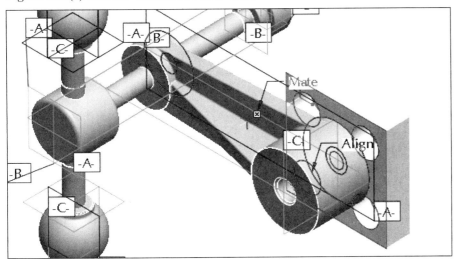

Figure 8.35(d) Clamp_Subassembly Reverses Position

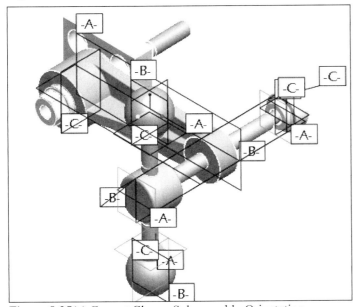

Figure 8.35(e) Correct Clamp_Subassembly Orientation

Click on the **Clamp_Stud35** in the Model Tree ⇒ **RMB** ⇒ **Edit** [Fig. 8.36(a)] ⇒ **MMB** to rotate the model so that the offset dimension can be seen clearly ⇒ double-click on your offset dimension [2.888] ⇒ type **-.125** [Fig. 8.36(b)] [-.125] ⇒ **Enter** ⇒ **Ctrl+G** ⇒ **Ctrl+D**

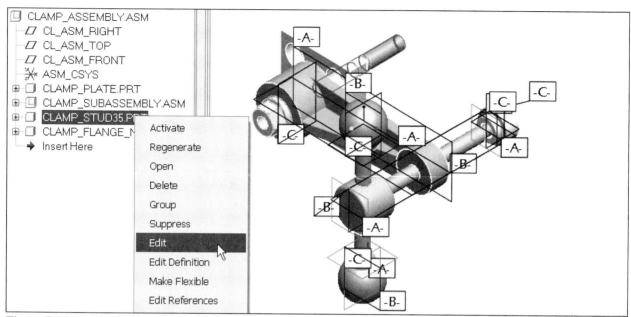

Figure 8.36(a) Click on the Clamp_Stud35 in the Model Tree ⇒ RMB ⇒ Edit

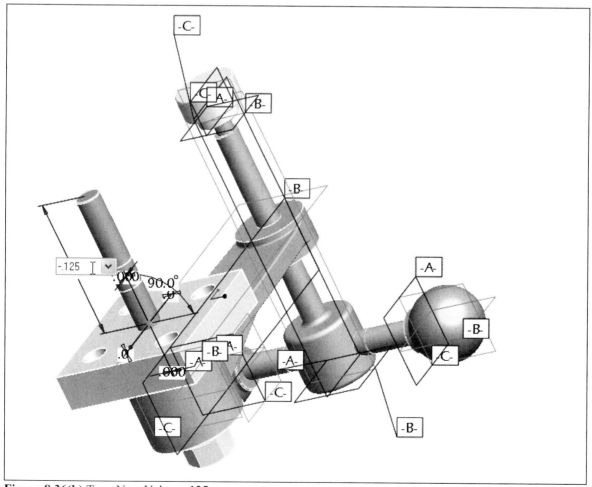

Figure 8.36(b) Type New Value: **-.125**

Click: **Ctrl+S** ⇒ **Enter** ⇒ **File** ⇒ **Delete** ⇒ **Old Versions** [Fig. 8.37(a)] ⇒ **Enter** ⇒ expand Model Tree Items ⇒ **Spin center** off ⇒ **View** ⇒ **Shade** [Fig. 8.37(b)]

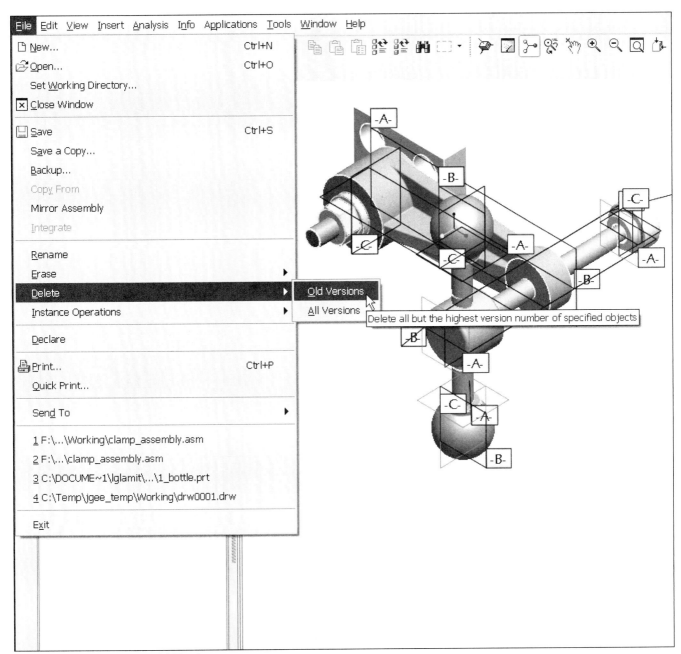

Figure 8.37(a) Delete ⇒ Old Versions

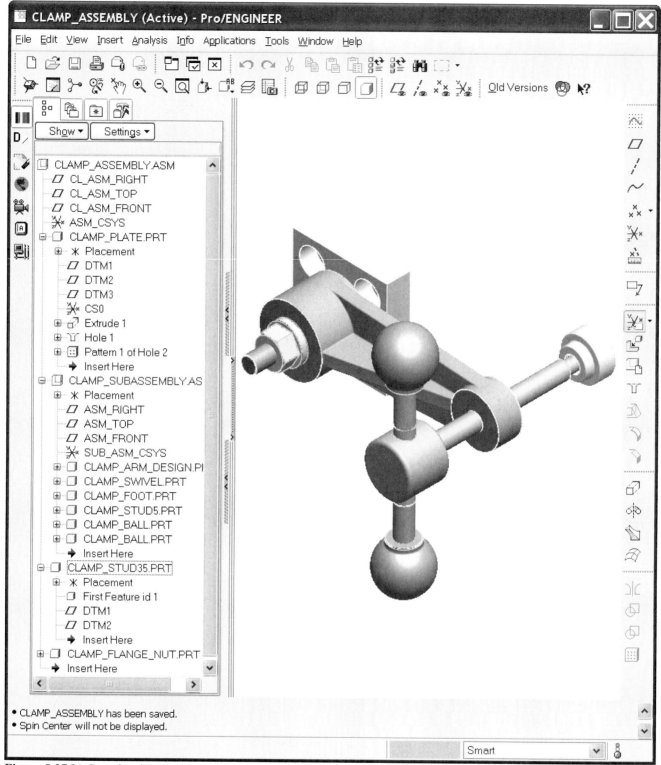

Figure 8.37(b) Completed Assembly

This lesson is now complete. You may notice that the Pro/E window shown in Figure 8.37(b) has Icons present in the Toolchests that are not the default for the interface. See Appendix to learn how to add icons and to customize your interface.

If you wish to model an assembly project (without step-by-step) instructions, a complete set of projects and illustrations are available at *www.cad-resources.com* ⇒ *Downloads*.

Lesson 9 Exploded Assemblies and View Manager

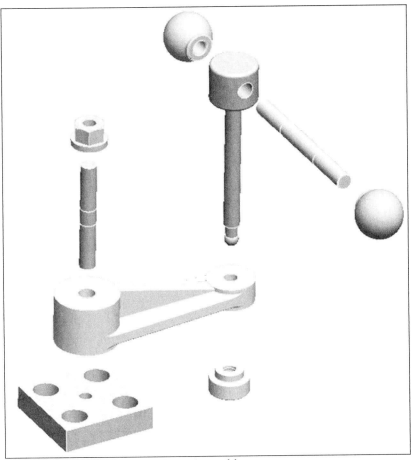

Figure 9.1 Exploded Swing Clamp Assembly

OBJECTIVES

- Create **Exploded Views**
- Create **Explode States**
- Utilize the **View Manager** to organize and control views states
- Create unique component **visibility settings (Style States)**
- **Move** and **Rotate** components in an assembly
- Add a **URL** to a **3D Note**
- Create **Perspective** views of the model
- View the **BOM** for an assembly

EXPLODED ASSEMBLIES

Pictorial illustrations, such as exploded views, are generated directly from the 3D model database (Fig. 9.1). The model can be displayed and oriented in any position. Each component in the assembly can have a different display type: wireframe, hidden line, no hidden, and shading. You can select and orient the component to provide the required view orientation to display the component from underneath or from any side or position. Perspective projections are made with selections from menus. The assembly can be spun around, reoriented, and even clipped to show the interior features. You have the choice of displaying all components and subassemblies or any combination of components in the design.

Creating Exploded Views

Using the **Explode State** option in the **View Manager**, you can automatically create an exploded view of an assembly (Fig. 9.2). Exploding an assembly affects only the display of the assembly; it does not alter true design distances between components. Explode states are created to define the exploded positions of all components. For each explode state, you can toggle the explode status of components, change the explode locations of components, and create and modify explode offset lines to show how explode components align when they are in their exploded positions. The Explode State Explode Position functionality is similar to the Package/Move functionality.

You can define multiple explode states for each assembly and then explode the assembly using any of these explode states at any time. You can also set an explode state for each drawing view of an assembly. Pro/E gives each component a default explode position determined by the placement constraints. By default, the reference component of the explode is the parent assembly (top-level assembly or subassembly).

To explode components, you use a drag-and-drop user interface similar to the Package/Move functionality. You select the motion reference and one or more components, and then drag the outlines to the desired positions. The component outlines drag along with the mouse cursor. You control the move options using a Preferences setting. Two types of explode instructions can be added to a set of components. The children components follow the parent component being exploded or they do not follow the parent component. Each explode instruction consists of a set of components, explode direction references, and dimensions that define the exploded position from the final (installed) position with respect to the explode direction references.

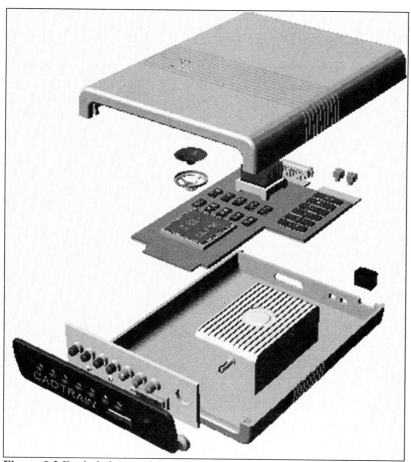

Figure 9.2 Exploded Assemblies (CADTRAIN, COAch for Pro/ENGINEER)

When using the explode functionality, keep in mind the following:

- You can select individual parts or entire subassemblies from the Model Tree or main window.
- If you explode a subassembly, in the context of a higher-level assembly, you can specify the explode state to use for each subassembly.
- You do not lose component explode information when you turn the status off. Pro/E retains the information so that the component has the same explode position if you turn the status back on.
- All assemblies have a default explode state called "Default Explode", which is the default explode state Pro/E creates from the component placement instructions.
- Multiple occurrences of the same subassembly can have different explode characteristics at a higher-level assembly.

Component Display

Style accessed through using: View ⇒ View Manager [Fig. 9.3(a)] ⇒ Style [Fig. 9.3(b)], manages the display styles of an assembly. (Orient and Explode [Fig. 9.3(c)] are also available on separate tabs from this dialog. Wireframe, hidden line, no hidden, shaded, or blanked display styles can be assigned to each component. The components will be displayed according to their assigned display styles in the current style state (that is, blanked, shaded, drawn in hidden line color, and so on). The current setting, in the Environment menu, controls the display of unassigned components.

Figure 9.3(a) View Manager

Figure 9.3(b) View Manager- Style

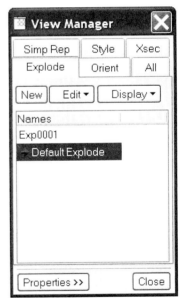
Figure 9.3(c) View Manager- Explode

Components appear in the currently assigned style state (Fig. 9.4). The current setting is indicated in the display style column in the Model Tree window. Component display or style states can be modified without using the View Manager. You can select desired models from the graphics window, Model Tree, or search tool, and then use the View ⇒ Display Style commands to assign a display style (wireframe, shaded, and so forth) to the selected models. The Style representation is temporarily changed. These temporary changes can then be stored to a new style state, or updated to an existing style state. You can also define default style states. If the default style state is updated to reflect changes different from that of the master style state, then that default style state will be reflected each time the model is retrieved.

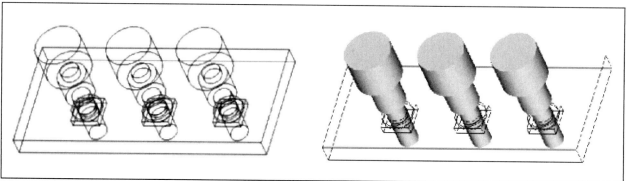

Figure 9.4 Style States

Types of Representations

The main types of simplified representations: **master**, **geometry**, **graphics**, and **symbolic**, designate which representation appears, using the commands in the View Manager dialog box.

Graphics and geometry representations speed up the retrieval process of large assemblies. All simplified representations provide access to components in the assembly and are based upon the Master Representation.

You cannot modify a feature in a graphics representation, but you can do so in a geometry representation.

Assembly features are displayed when you retrieve a model. Subtractive assembly features such as cuts and holes are represented in graphics and geometry representations, making it possible to use these simplified representations for performance improvement while still displaying on screen a completely accurate geometric model.

You can access model information for graphics and geometry representations of part models from the Information menu and from the Model Tree. Because part graphics and geometry representations do not contain feature history of the part model, information for individual features of the part is not accessible from these representations.

- The *Master Representation* always reflects the fully detailed assembly, including all of its members. The Model Tree lists all components in the Master Representation of the assembly, and indicates whether they are included, excluded, or substituted.
- The *Graphics Representation* contains information for display only and allows you to browse through a large assembly quickly. You cannot modify or reference graphics representations. The type of graphic display available depends on the setting of the *save_model_display* configuration option, the last time the assembly was saved:
    - *wireframe* (default) The wireframe of the components appear.
    - *shading_low*, *shading_med*, *shading_high* A shaded version of the components appears. The different levels indicate the density of the triangles used for shading.
    - *shading_lod* The level of detail depends on the setting in the View Performance dialog box. To access the View Performance dialog box, click View ⇒ Display Settings ⇒ Performance.

While in a simplified representation, Pro/E applies changes to an assembly, such as creation or assembly of new components, to the Master Representation. It reflects them in all of the simplified representations (Pro/PROGRAM processing also affects the Master Representation). It applies all suppressing and resuming of components to the Master Representation. However, it applies the actions of a simplified representation only to currently resumed members, that is, to members that are present in the BOM of the Master Representation.

Lesson 9 STEPS

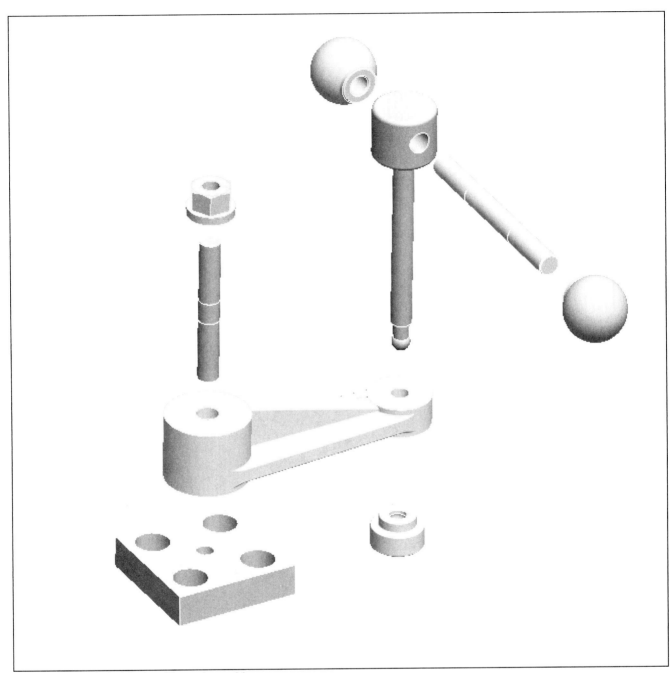

Figure 9.5 Exploded Swing Clamp Assembly

Exploded Swing Clamp Assembly

In this lesson, you will use the previously created subassembly and assembly (Fig. 9.5) to establish and save new views, exploded views, and views with component style states that differ from one another. The View Manager Dialog box will be employed to control and organize a variety of different states.

You will also be required to move and rotate components of the assembly before cosmetically displaying the assembly in an exploded state. The creation and assembly of new components will not be required. A bill of materials will also be displayed using the Info command. A 3D Note with a URL will be added to the model as the last information feature.

URLs and Model Notes

Since the assembly (Fig. 9.6) you created is a standard clamp from CARR LANE Manufacturing Co.; you could go to CARR LANE's Website and see the assembly, order it (about $60.00 U.S.) or download a 3D IGES model. Create and attach a 3D Note to the assembly, identifying the manufacturer and Website.

Click: **File ⇒ Open ⇒ clamp_assembly.asm ⇒ Open ⇒ Edit ⇒ Setup ⇒ Notes ⇒ New ⇒** Name **CARR_LANE ⇒** Text: type text lines (Fig. 9.7) ⇒ ☑ Place Note Flat to Screen ⇒ URL **Hyperlink ⇒** type **www.carrlane.com** (Fig. 9.8) ⇒ **ScreenTip ⇒** Screen tip text: **CARR LANE Manufacturing** (Fig. 9.9) ⇒ **OK ⇒ OK ⇒ Place ⇒ No Leader ⇒ Standard ⇒ Done ⇒** ⇨Select LOCATION for note. ⇒ 📄 pick near the assembly in the graphics window ⇒ **OK ⇒ Settings ⇒ Tree Filters ⇒** ☑ Annotations (Fig. 9.10) ⇒ **OK ⇒ Tools ⇒ Environment ⇒** ☑ 3D Notes ⇒ **OK** (Fig. 9.11) ⇒ **Done/Return ⇒ Done ⇒ Ctrl+R**

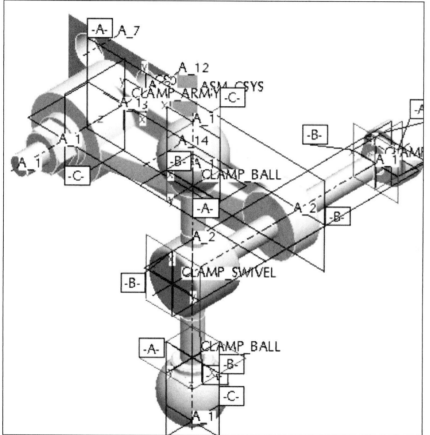

Figure 9.6 Model

Figure 9.7 Note Dialog Box

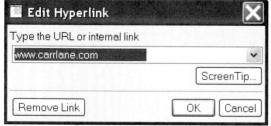

Figure 9.8 Edit Hyperlink Dialog Box

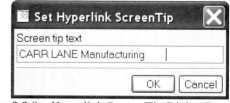

Figure 9.9 Set Hyperlink Screen Tip Dialog Box

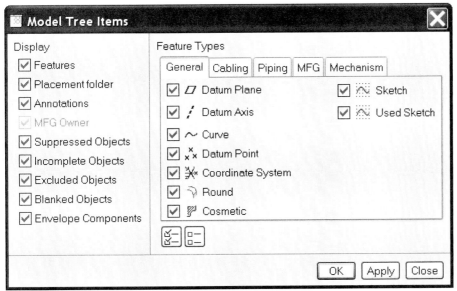

Figure 9.10 Model Tree Items Checked to Display

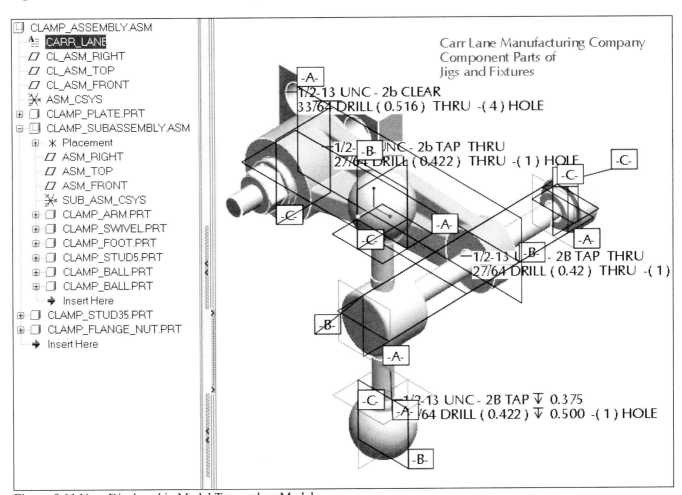

Figure 9.11 Note Displayed in Model Tree and on Model

To open a Hyperlink defined in a model note and to launch the embedded Web browser and go to a World Wide Web URL (Universal Resource Locator) associated with a model note, use one of three methods.

1. Click on the note in the graphics window or the Model Tree ⇒ **RMB** [Fig. 9.12(a)] the shortcut menu displays ⇒ **Open URL** [Fig. 9.12(b)] the associated URL opens in the Pro/E embedded browser ⇒ navigate the site to locate the Swing Clamp [Figs. 9.12(c-f)] ⇒ use the scroll bars to the see the drawing [Figs. 9.12(g-h)]) ⇒ close the Browser by clicking on the quick sash ⇒ 🖫 **Save** ⇒ **MMB**
(Note that to pick the 3D Note in the graphics window; you may have to change the Selection Filter to Annotation)
or
2. *Move the cursor over the hyperlinked note* ⇒ *press* **Ctrl** *the cursor changes to a hand icon* ⇒ *click* **LMB** *the associated URL opens in the Pro/E embedded browser* ⇒ *navigate to locate the Swing Clamp*
or
3. *Click:* **Edit** ⇒ **Setup** ⇒ **Notes** ⇒ **Modify** ⇒ *pick on the note in the Model Tree or from the screen to see the note before opening the URL* ⇒ **OK** *to close the Note dialog box* ⇒ **Open URL** ⇒ *pick the 3D Note from the screen or the Model Tree* ⇒ *navigate to locate the Swing Clamp*

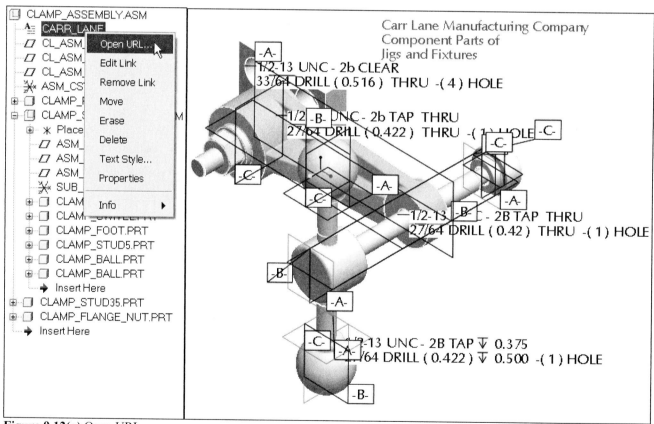

Figure 9.12(a) Open URL

Figure 9.12(b) Carr Lane Site [Address: http://www.carrlane.com/]

Figures 9.12(c-f) Online Catalog ⇒ Clamps and Accessories ⇒ Swing Clamps ⇒ Swing Clamp Assemblies Flange Mounted

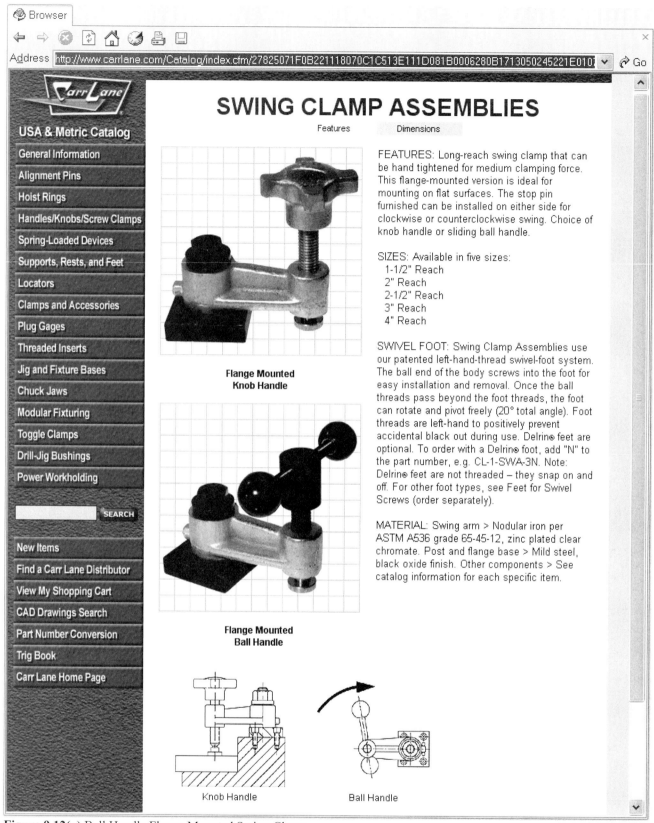

Figure 9.12(g) Ball Handle Flange Mounted Swing Clamp

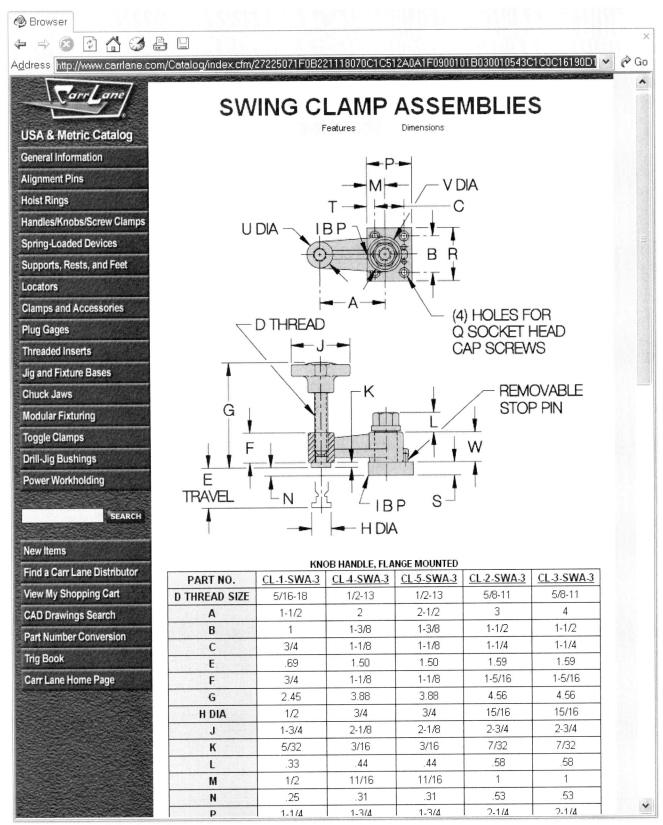

Figure 9.12(h) Click Dimensions

Click: **Tools** ⇒ **Environment** ⇒ [3D Notes] ⇒ **Apply** ⇒ **OK** ⇒ [🔍] ⇒ [▢]

Views: Perspective, Saved, and Exploded

You will now create a variety of cosmetically altered view states. Cosmetic changes to the assembly do not affect the model itself, only the way it is displayed on the screen. One type of view that can be created is the perspective view.

Perspective creates a single-vanishing-point perspective view of a shaded or wireframe model. These views allow you to observe an object as the view location follows a curve, axis, cable, or edge through or around an object. To add perspective to a model view, you select a viewing path and then control the viewing position along the path in either direction. You can also rotate the perspective view in any direction, zoom the view in or out, and change the view angle at any point along the path.

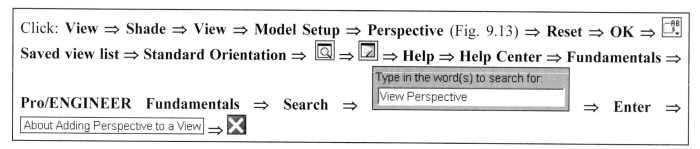

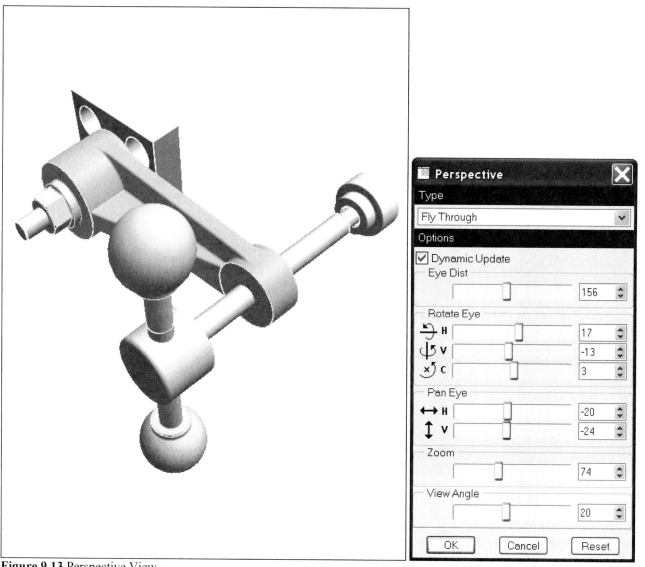

Figure 9.13 Perspective View

Saved Views

You need to create a saved view and explode state to use later on the assembly drawing. When using the View Manager to set display and explode states, it is a good idea to create one or more saved views to be used later for exploding. The default trimetric and isometric views do not adequately represent the assembly in its functional position.

Using the **MMB**, rotate the model from its default position [Fig. 9.14(a)] to one where the Clamp_Swivel is vertical [Fig. 9.14(b)] ⇒ **Reorient view** [Fig. 9.14(c)] ⇒ **Saved Views** [Fig. 9.14(d)] ⇒ Name: type **EXPLODE1** ⇒ **Save** [Fig. 9.14(e)] ⇒ **OK**

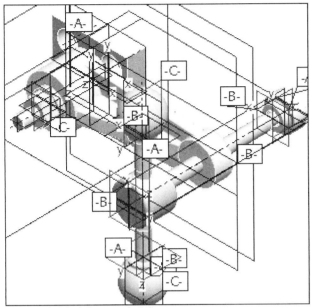

Figure 9.14(a) Isometric View

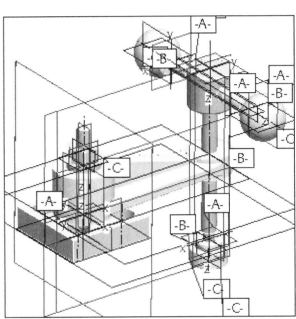

Figure 9.14 (b) Reoriented View

Figure 9.14 (c) Orientation Dialog Box

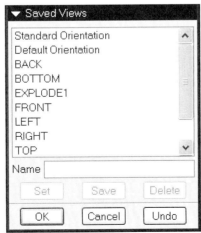

Figure 9.14(d) Saved Views

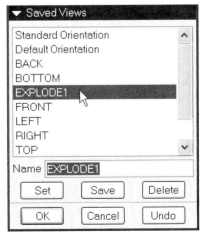

Figure 9.14(e) EXPLODE1 View

Default Exploded Views

When you create an *exploded view*, Pro/E moves apart the components of an assembly to a set default distance. The default position is seldom the most desirable.

Click: ⊞ ⊞ ⊞ off ⇒ **View** ⇒ **Explode** [Fig. 9.15(a)] ⇒ **Explode View** [Fig. 9.15(b)] ⇒ **View** ⇒ **Shade** [Fig. 9.15(c)] ⇒ **View** ⇒ **Explode** ⇒ **Unexplode View** ⇒ ⊞ ⇒ ⊞ ⇒ **MMB**

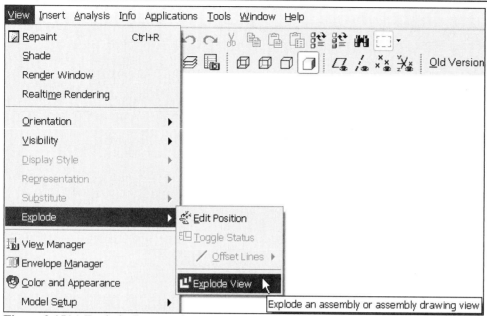

Figure 9.15(a) Explode Command

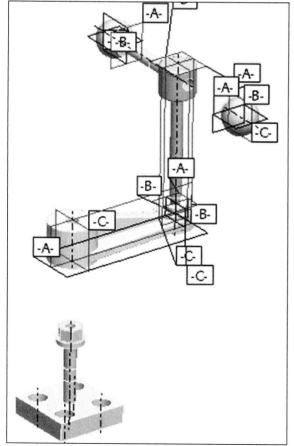

Figure 9.15(b) Default Exploded View

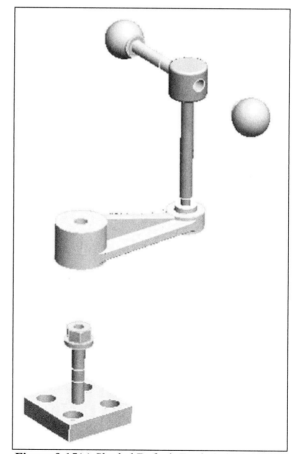

Figure 9.15(c) Shaded Default Exploded View

View Manager

The View Manager allows you to see and control the models representation [Fig. 9.16(a)], orientation [Fig. 9.16(b)], explode [Fig. 9.16(c)], style [Fig. 9.16(d)], all (combined) states [Fig. 9.16(e)], and Xsec [Fig. 9.16(f)].

Figure 9.16(a) Representation

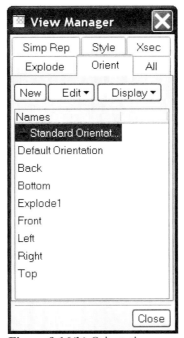

Figure 9.16(b) Orientation

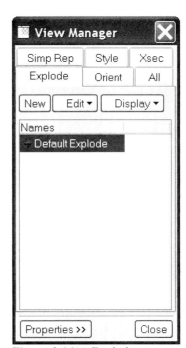

Figure 9.16(c) Explode

Figure 9.16(d) Style

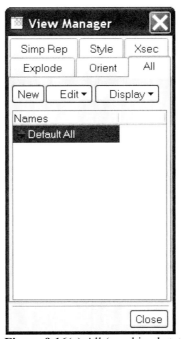

Figure 9.16(e) All (combined state)

Figure 9.16(f) Xsec

Click: **Start the view manager** ⇒ **Simp Rep** tab ⇒ **Master Rep** ⇒ **Display** ⇒ **Set Active** ⇒ **Orient** tab ⇒ **Explode1** ⇒ **Display** ⇒ **Set Active** [Fig. 9.17(a)] ⇒ **Explode** tab ⇒ **New** ⇒ **Enter** to accept the default name: Exp0001 ⇒ **RMB** ⇒ **Set Active** [Fig. 9.17(b)] ⇒ **Properties>>** ⇒ **Edit position** [Figs. 9.17(c-d)]

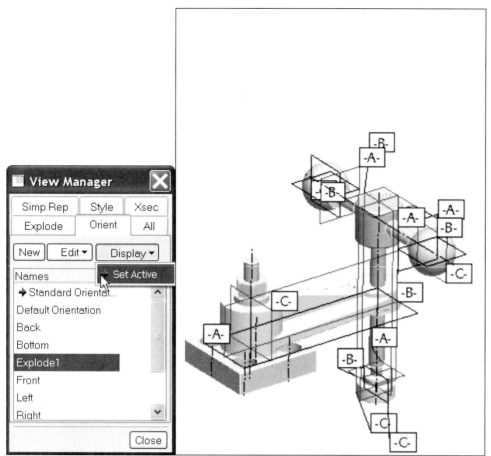

Figure 9.17(a) View Manager Orient

Figure 9.17(b) Set Active

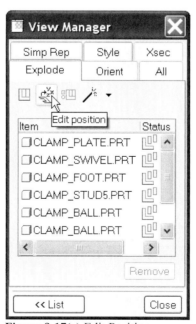

Figure 9.17(c) Edit Position

Figure 9.17(d) Edit Position Dialog Box

Click: Motion Reference ▣ [Figs. 9.18(a-b)] ⇒ select the vertical axis of the CLAMP_STUD35 [Figs. 9.18(c-d)] *or any vertical edge of the CLAMP_PLATE* ⇒ **Preferences** ⇒ **Move Many** ⇒ **Close** ⇒ select the **CLAMP_STUD35** ⇒ press and hold the **Ctrl** key and select **CLAMP_FLANGE_NUT** [Fig. 9.18(e)] ⇒ **MMB** ⇒ click **LMB** in the graphics window and start dragging [Fig. 9.18(f)] ⇒ click **LMB** to place

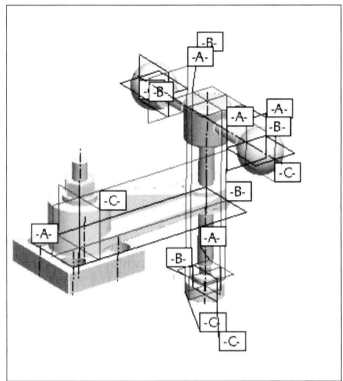

Figure 9.18(a) Assembly

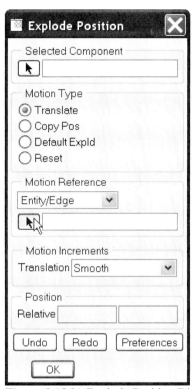

Figure 9.18(b) Explode Position Dialog Box

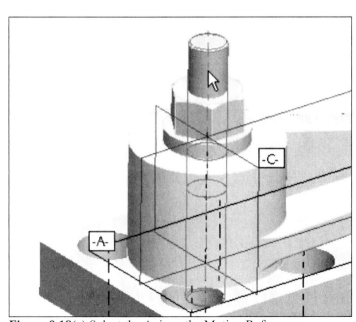

Figure 9.18(c) Select the Axis as the Motion Reference

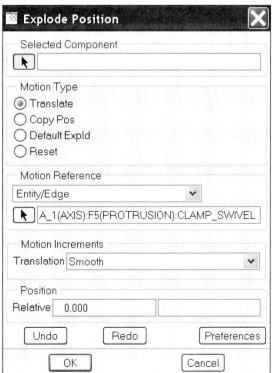

Figure 9.18(d) Motion Reference Selected

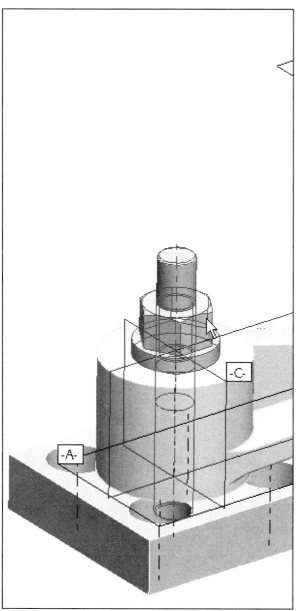

Figure 9.18(e) Select Components

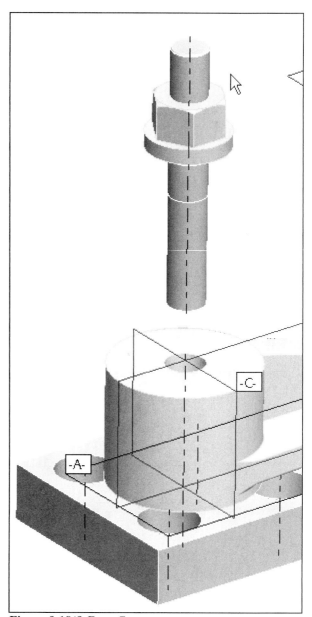

Figure 9.18(f) Drag Components

Select **CLAMP_FLANGE_NUT** [Fig. 9.18(g)] ⇒ **MMB** ⇒ click **LMB** in the graphics window and start dragging [Fig. 9.18(h)] ⇒ click **LMB** to place

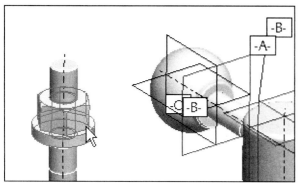

Figure 9.18(g) Select CLAMP_FLANGE_NUT

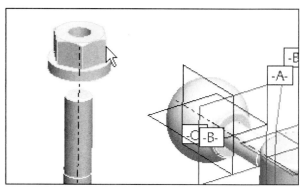

Figure 9.18(h) Drag CLAMP_FLANGE_NUT

Select **CLAMP_PLATE** [Fig. 9.18(i)] ⇒ **MMB** ⇒ click **LMB** in the graphics window and start dragging [Fig. 9.18(j)] ⇒ click **LMB** to place ⇒ **Preferences** ⇒ **Move With Children** ⇒ **Close** ⇒ select **CLAMP_SWIVEL** [Fig. 9.18(k)] ⇒ start dragging [Fig. 9.18(l)] ⇒ click **LMB** to place ⇒ **Preferences** ⇒ **Move One** ⇒ **Close** ⇒ select **CLAMP_FOOT** [Fig. 9.18(m)] ⇒ start dragging [Fig. 9.18(n)] ⇒ **LMB** to place

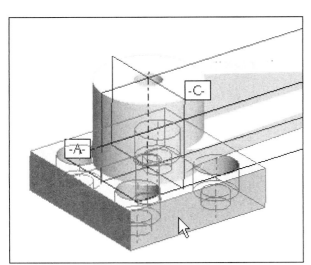

Figure 9.18(i) Select CLAMP_PLATE

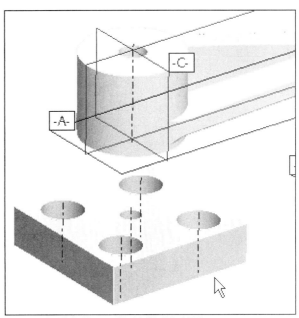

Figure 9.18(j) Drag CLAMP_PLATE

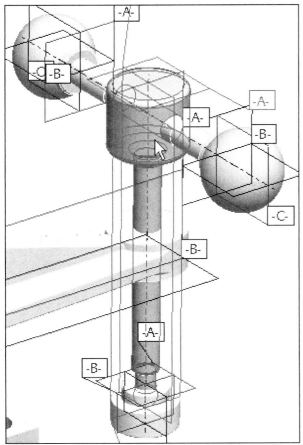

Figure 9.18(k) Select CLAMP_SWIVEL

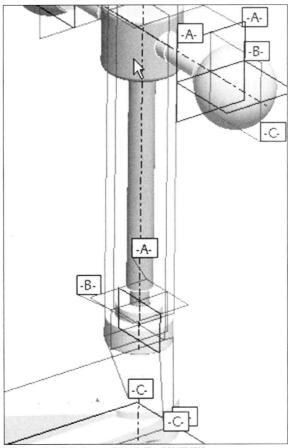

Figure 9.18(l) Drag CLAMP_SWIVEL

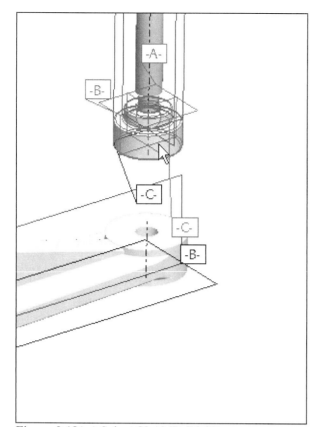

Figure 9.18(m) Select CLAMP_FOOT

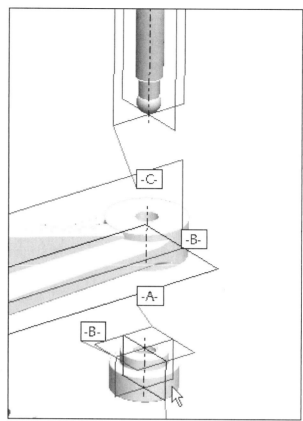

Figure 9.18(n) Drag CLAMP_FOOT

Motion Reference 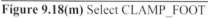 ⇒ select the horizontal edge of the CLAMP_PLATE [Fig. 9.18(o)] ⇒ select **CLAMP_BALL** ⇒ start dragging ⇒ click **LMB** to place [Fig. 9.18(p)] ⇒ select the second **CLAMP_BALL** ⇒ start dragging ⇒ click **LMB** to place [Fig. 9.18(q)] ⇒ select the **CLAMP_STUD5** ⇒ start dragging ⇒ click **LMB** to place [Fig. 9.18(r)] ⇒ **OK** [Fig. 9.18 (s)] ⇒ 🔍 ⇒ <<**List** ⇒ **Close** ⇒ **Ctrl+S** ⇒ **Enter**

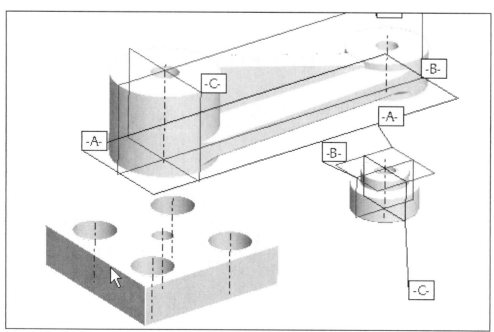

Figure 9.18(o) Select the Horizontal Edge of the CLAMP_PLATE

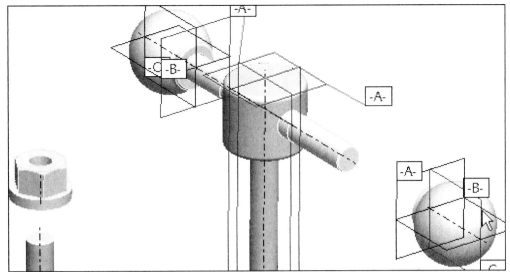

Figure 9.18(p) Select and Move the CLAMP_BALL

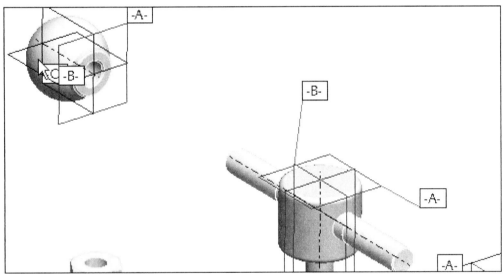

Figure 9.18(q) Select and Move the Second CLAMP_BALL

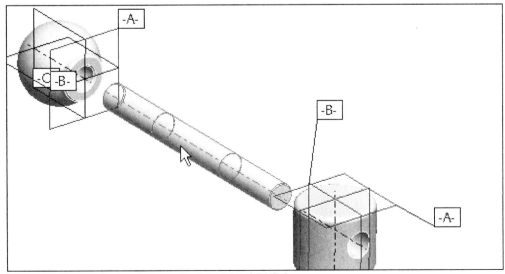

Figure 9.18(r) Select and Move the CLAMP_STUD5

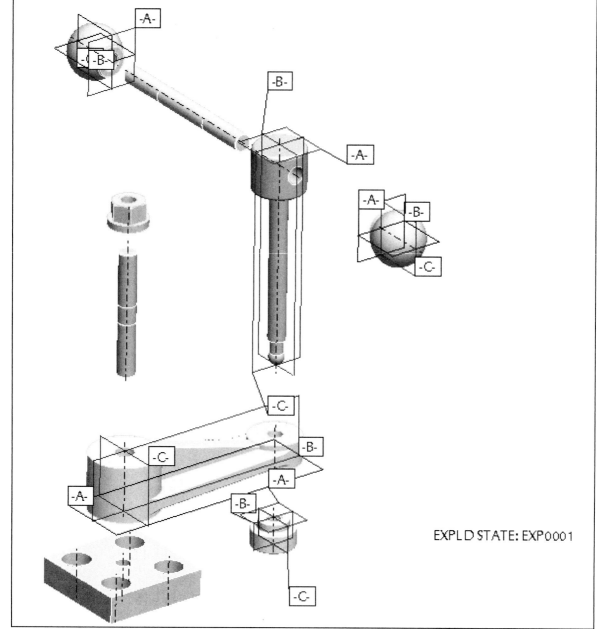

Figure 9.18(s) Completed Exploded State EXP0001

View Style

The components of an assembly (whether exploded or not) can be displayed individually with **Wireframe**, **Hidden Line**, **No Hidden**, **Shading**, or **Blank**. Style is used to manage the display styles of an assembly's components. The components will be displayed according to their assigned display style in the current style state. The setting, in the Environment menu, controls the display style of unassigned components and they appear according to the current mode of the environment.

Click: **Start the view manager** from Top Toolchest ⇒ **Style** tab ⇒ **New** ⇒ **Enter** [Fig. 9.19(a)] to accept the default name: Style0001 ⇒ **Show** tab [Fig. 9.19(b)] ⇒ Wireframe ⇒ select **CLAMP_SWIVEL** and **CLAMP_STUD35** ⇒ ✓ ✗ 👓 **Update**

368

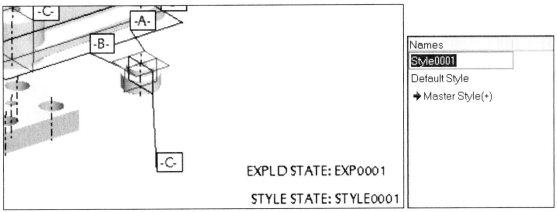

Figure 9.19(a) Style State STYLE0001

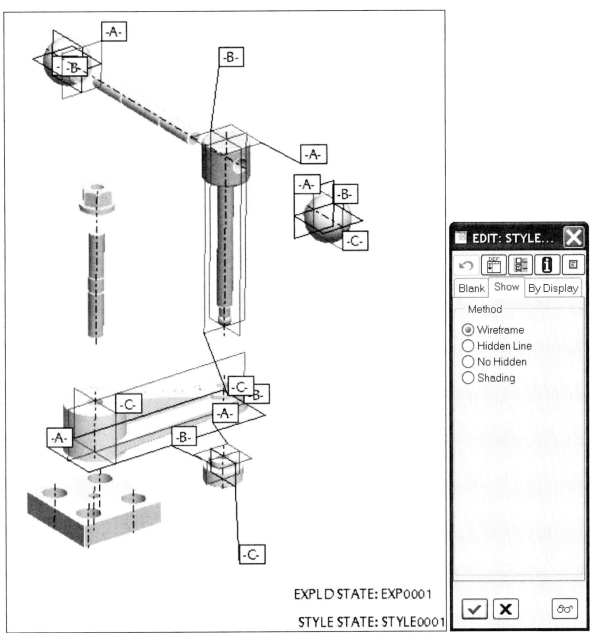

Figure 9.19(b) Edit Dialog Box

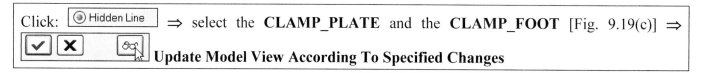

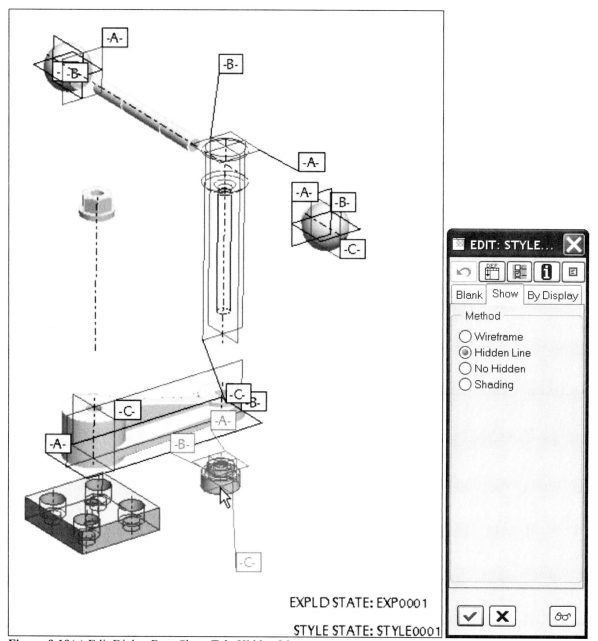

Figure 9.19(c) Edit Dialog Box, Show Tab, Hidden Line

Click: [No Hidden] ⇒ select the two **CLAMP_BALL** components ⇒ [👓] Update ⇒ [Shading] ⇒ select **CLAMP_FLANGE_NUT**, **CLAMP_ARM**, and **CLAMP_STUD5** ⇒ [👓] Update Model View According To Specified Changes [Fig. 9.19(d)] ⇒ [✓][✗] [👓]

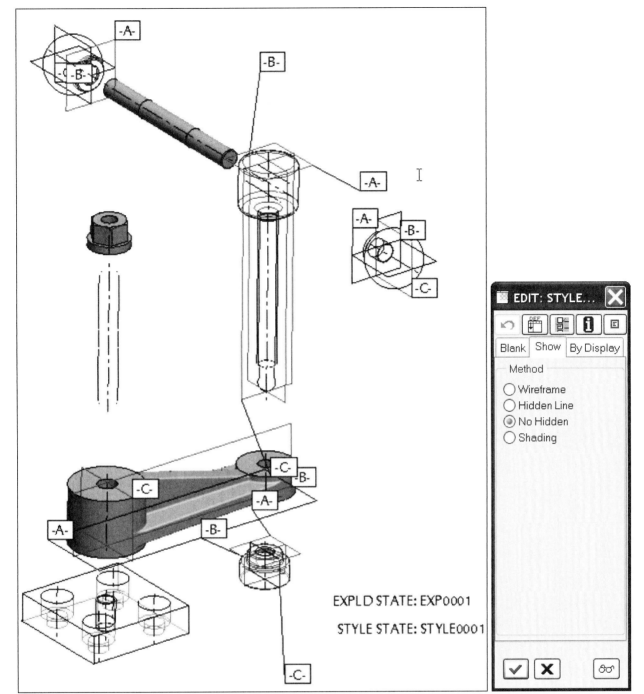

Figure 9.19(d) Completed Style State

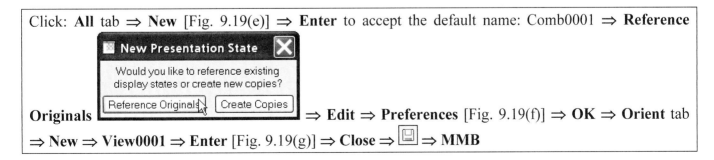

Click: **All** tab ⇒ **New** [Fig. 9.19(e)] ⇒ **Enter** to accept the default name: Comb0001 ⇒ **Reference Originals** ⇒ **Edit** ⇒ **Preferences** [Fig. 9.19(f)] ⇒ **OK** ⇒ **Orient** tab ⇒ **New** ⇒ **View0001** ⇒ **Enter** [Fig. 9.19(g)] ⇒ **Close** ⇒ 💾 ⇒ **MMB**

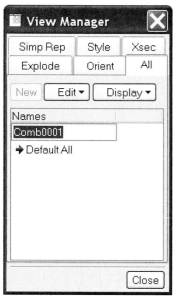

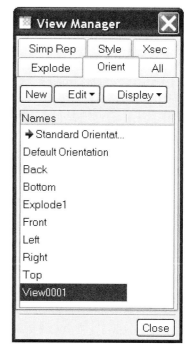

Figure 9.19(e) All Tab **Figure 9.19(f)** All Edit Preferences **Figure 9.19(g)** Orient Tab

Model Tree

You can display component style states, and explode states in the Model Tree.

Click: **Settings** from Model Tree ⇒ **Tree Columns** ⇒ Type: click ⬇ [Fig. 9.20(a)] ⇒ **Display Styles** ⇒ **STYLE0001** ⇒ >> [Fig. 9.20(b)] ⇒ **Apply** ⇒ **OK** ⇒ **View** ⇒ **Shade** ⇒ adjust your column format [Fig. 9.20(c)]

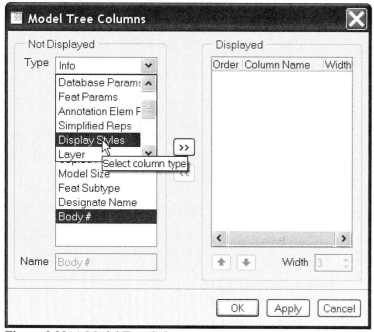

Figure 9.20(a) Model Tree Columns

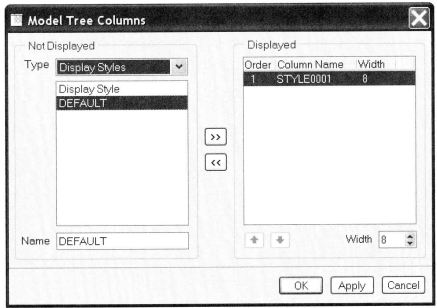

Figure 9.20(b) Adding Display Style STYLE0001 to the Model Tree Columns

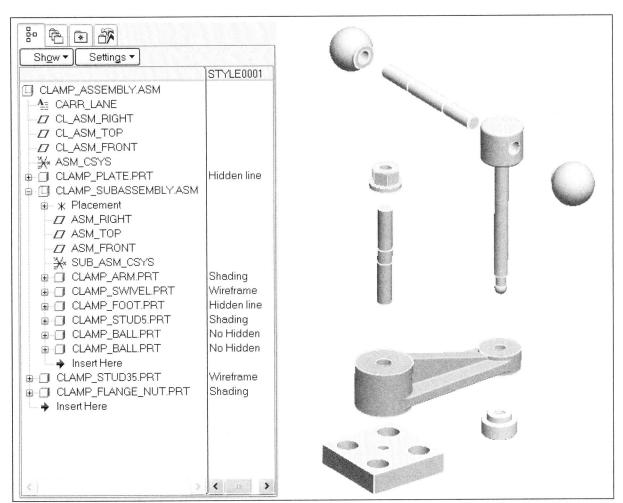

Figure 9.20(c) Adjusted Model Tree

Download projects from *www.cad-resources.com* ⇒ *Downloads*.

Notes:

Lesson 10 Introduction to Drawings

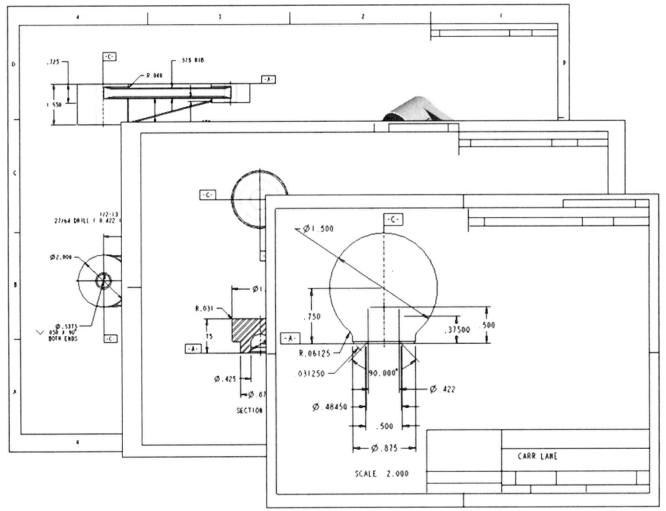

Figure 10.1 Clamp Arm Drawing

OBJECTIVES

- Create part drawings
- Display **standard views** using the template
- **Add, Move, Erase,** and **Delete** views
- Retrieve a standard **format**
- **Display** gtol datums, centerlines, and dimensions
- **Cleanup** dimension positions
- Specify and retrieve standard **Format** paper size and units
- Change the **Scale** of a view

Introduction to Drawings

This lesson will introduce you to the basic concepts and procedure for creating detail drawings (Fig. 10.1). Using parts modeled for the Swing Clamp Assembly, you will quickly create drawings, change formats, delete and add views, and display dimensions and centerlines. Since this lesson is meant to introduce drawing mode concepts and procedures, a minimum of cleanup, reformatting and repositioning of dimensions and centerlines will be required.

FORMATS, TITLE BLOCKS, AND VIEWS

Formats are user-defined drawing sheet layouts. A drawing can be created with an empty format and have a standard size format added as needed. Formats can be added to any number of drawings. The format can also be modified or replaced in a Pro/E drawing at any time.

Title Blocks are standard or sketched line entities that can contain parameters (object name, tolerances, scale, and so on) that will show when the format is added to the drawing.

Views created by Pro/E are identical to views constructed manually by a designer on paper. The same rules of projection apply; the only difference is that you choose commands in Pro/E to create the views as needed.

Formats

Formats consist of draft entities, not model entities. There are two types of formats: standard and sketched. You can select from a list of **Standard Formats** (**A-F** and **A0-A4**) or create a new size by entering values for length and width.

Sketched Formats created in Sketcher mode (Fig. 10.2) may be parametrically modified, enabling you to create nonstandard-size formats or families of formats.

Formats can be altered to include note text, symbols, tables, and drafting geometry, including drafting cross sections and filled areas.

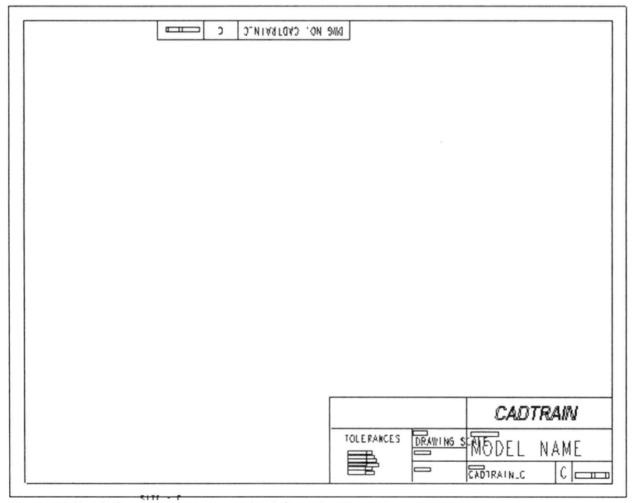

Figure 10.2 Sketched Format (CADTRAIN, COAch for Pro/ENGINEER)

With Pro/E, you can do the following in Format mode:

- Create draft geometry and notes (Fig. 10.3)
- Move, mirror, copy, group, translate, and intersect geometry
- Use and modify the draft grid
- Enter user attributes
- Create drawing tables
- Use interface tools to create plot, DXF, SET, and IGES files
- Import IGES, DXF, and SET files into the format
- Create user-defined line styles
- Create, use, and modify symbols
- Include drafting cross sections in a format

Whether you use a standard format or a sketched format, the format is added to a drawing that is created for a set of specified views of a parametric 3D model.

> When you place a format on your drawing, the system automatically writes the appropriate notes based on information in the model you use.

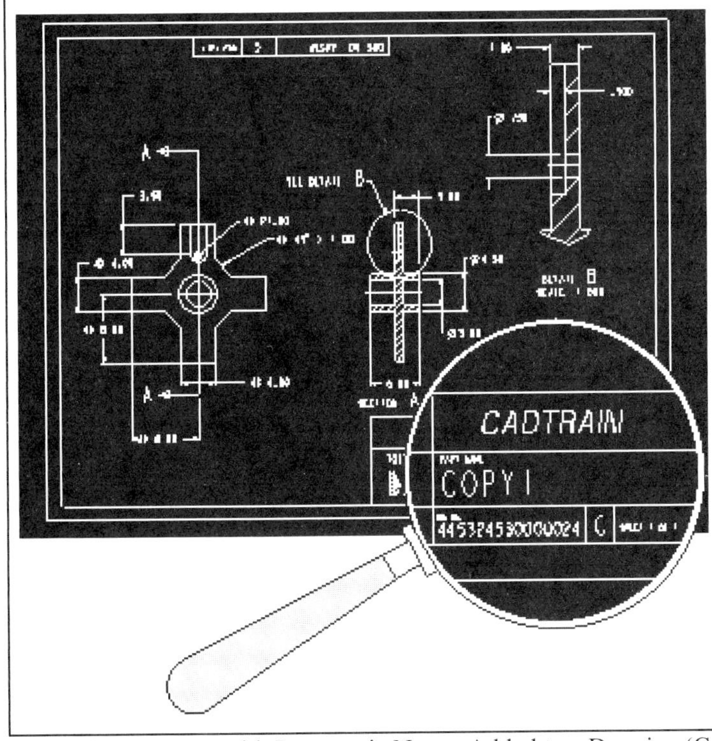

Figure 10.3 Format with Parametric Notes, Added to a Drawing (CADTRAIN, COAch for Pro/ENGINEER)

Specifying the Format Size when Creating a New Drawing

If you want to use an existing template, select a listed template [Fig. 10.4(a)], or select a template from the appropriate directory. If you want to use the existing format [Fig. 10.4(b)], using Browse will open Pro/E's System Formats folder [Fig. 10.4(c)]. Select from the list of standard formats or navigate to a directory with user or company created formats.

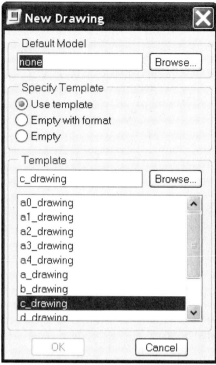

Figure 10.4(a) New Drawing Dialog Box (Use Template)

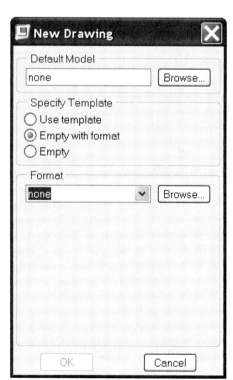

Figure 10.4(b) Empty with format

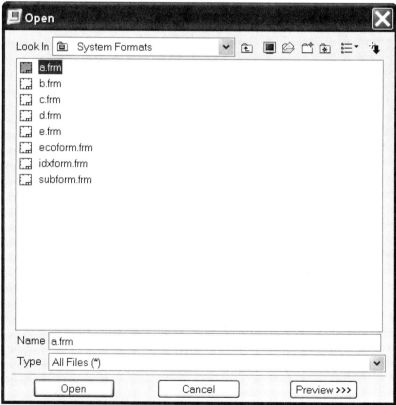

Figure 10.4(c) System Formats

If you want to create your own variable size format [Figs. 10.4(d-e)], enter values for width and length. The main grid spacing and format text units depend on the units selected for a variable size format. The New Drawing dialog box also provides options for the orientation of the format sheet:

- **Portrait** Uses the larger of the dimensions of the sheet size for the format's height; uses the smaller for the format's width
- **Landscape** Uses the larger of the dimensions of the sheet size for the format's width; uses the smaller for the format's height [Fig. 10.4(d)]
- **Variable** Select the unit type, Inches or Millimeters, and then enter specific values for the Width and Height of the format [Fig. 10.4(e)]

A0	841 X 1189	mm	A	8.5 X 11	in.
A1	594 X 841	mm	B	11 X 17	in.
A2	420 X 594	mm	C	17 X 22	in.
A3	297 X 420	mm	D	22 X 34	in.
A4	210 X 297	mm	E	34 X 44	in.
			F	28 X 40	in.

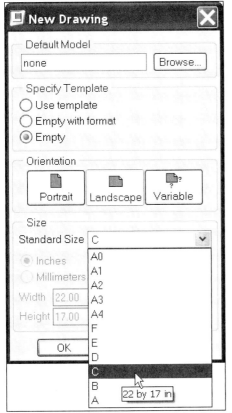

Figure 10.4(d) New Drawing Dialog Box (Empty- Landscape)

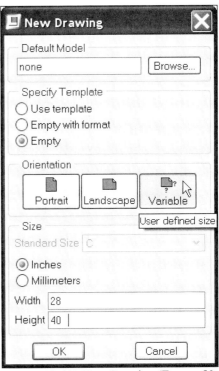

Figure 10.4(e) New Drawing (Empty- Variable)

Drawing Templates

Drawing templates may be referenced when creating a new drawing. Templates can automatically create the views, set the desired view display, create snap lines, and show model dimensions. Drawing templates contain three basic types of information for creating new drawings. The first type is basic information that makes up a drawing but is not dependent on the drawing model, such as notes, symbols, and so forth. This information is copied from the template into the new drawing. The second type is instructions used to configure drawing views and the actions that are performed on their views. The instructions are used to build a new drawing object. The third type is a parametric note. Parametric notes are notes that update to new drawing model parameters and dimension values. The notes are re-parsed or updated when the template is instantiated. Use the templates to:

- Define the layout of views
- Set view display
- Place notes
- Place symbols
- Define tables
- Create snap lines
- Show dimensions

Template View

You can also create customized drawing templates for the different types of drawings that you create. Creating a template allows you to create portions of drawings automatically, using the customizable template. The Template View Instructions dialog box (Fig. 10.5) is accessed through Applications ⇒ Template ⇒ Insert ⇒ Template View, when in the Drawing mode.

Figure 10.5 Template View Instructions Dialog Box

You can use the following options in the Template View Instructions dialog box to customize your drawing templates:

- **View Name** Set the name of the drawing view that will be used as the view symbol label
- **View Orientation** Create a General view or a Projection view
- **Model "Saved View" Name** Orient the view based on a named view in the model
- **Place View** Places the view after you have set the appropriate options and values
- **Edit View Symbol** Allows you to edit the view symbol using the Symbol Instance dialog box
- **Replace View Symbol** Allows you to replace the view symbol using the Symbol Instance dialog box

View Options	View Values
View States—Specify the type of view. By default, the **View States** check box is selected and the view state is displayed when the **Template View Instructions** dialog box opens. The value of **Orientation** is FRONT, by default. If you select **Combination State**, the **Orientation**, **Simplified Rep**, **Explode** and **Cross Section** boxes display the text Defined by Combination State. The **Arrow Placement View** and **Show X-Hatching** check boxes also become available for selection after you specify the Combination State.	Combination State, Orientation, Simplified Rep, Explode, Cross Section, Arrow Placement View
Scale—Type a new value for the scale or use the default value.	View Scale
Process Step—Set the process step for the view. You can specify the step number and set the view of the tool by checking the **Tool View** check box under **Process Step**.	Step Number, Tool View
Model Display—Set the view display for the drawing view.	Follow Environment, Wireframe, Hidden Line, No Hidden, Shading
Tan Edge Display—Set the tangent edge display.	Tan Solid, No Disp Tan, Tan Ctrln, Tan Phantom, Tan Dimmed, Tan Default
Snap Lines—Set the number, spacing, and offset of the snap lines.	Number, Incremental Spacing, Initial Offset
Dimensions—Show dimensions on the view.	Create Snap Lines, Incremental Spacing, Initial Offset
Balloons—Show balloons on the view.	

Views

A wide variety of views (Fig. 10.6) can be derived from the parametric model. Among the most common are projection views. Pro/E creates projection views by looking to the left of, to the right of, above, and below the picked view location to determine the orientation of a projection view. When conflicting view orientations are found, you are prompted to select the view that will be the parent view. A view will then be constructed from the selected view.

At the time when they are created, projection, auxiliary, detailed, and revolved views have the same representation and explosion offsets, if any, as their parent views. From that time onward, each view can be simplified, be restored, and have its explosion distance modified without affecting the parent view. The only exception to this is detailed views, which will always be displayed with the same explosion distances and geometry as their parent views.

Once a model has been added to the drawing, you can place views of the model on a sheet. When a view is placed, you can determine how much of the model to show in a view, whether the view is of a single surface or shows cross sections, and how the view is scaled. You can then show the associative dimensions passed from the 3D model, or add reference dimensions as necessary.

Basic view types used by Pro/Engineer include general, projection, auxiliary, and detailed:

- **General** Creates a view with no particular orientation or relationship to other views in the drawing. The model must first be oriented to the desired view orientation established by you.
- **Projection** Creates a view that is developed from another view by projecting the geometry along a horizontal or vertical direction of viewing (orthographic projection). The projection type is specified by you in the drawing setup file and can be based on third-angle (default) or first-angle rules.
- **Auxiliary** Creates a view that is developed from another view by projecting the geometry at right angles to a selected surface or along an axis. The surface selected from the parent view must be perpendicular to the plane of the screen.
- **Detailed** Details a portion of the model appearing in another view. Its orientation is the same as that of the view it is created from, but its scale may be different so that the portion of the model being detailed can be better visualized.

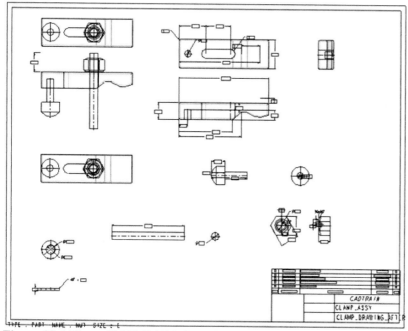

Figure 10.6 Views (CADTRAIN, COAch for Pro/ENGINEER)

The view options that determine how much of the model is visible in the view are:

- **Full View** Shows the model in its entirety.
- **Half View** Removes a portion of the model from the view on one side of a cutting plane.
- **Broken View** Removes a portion of the model from between two selected points and closes the remaining two portions together within a specified distance.
- **Partial View** Displays a portion of the model in a view within a closed boundary. The geometry appearing within the boundary is displayed; the geometry outside of it is removed.

The options that determine whether the view is of a single surface or has a cross section are:

- **Section** Displays an existing cross section of the view if the view orientation is such that the cross-sectional plane is parallel to the screen (Fig. 10.7).
- **No Xsec** Indicates that no cross section is to be displayed.
- **Of Surface** Displays a selected surface of a model in the view. The single-surface view can be of any view type except detailed.

The options that determine whether the view is scaled are:

- **Scale** Allows you to create a view with an individual scale shown under the view. When a view is being created, Pro/E will prompt you for the scale value. This value can be modified later. General and detailed views can be scaled.
- **No Scale** A view will be scaled automatically using a pre-defined scale value.
- **Perspective** Creates a perspective general view.

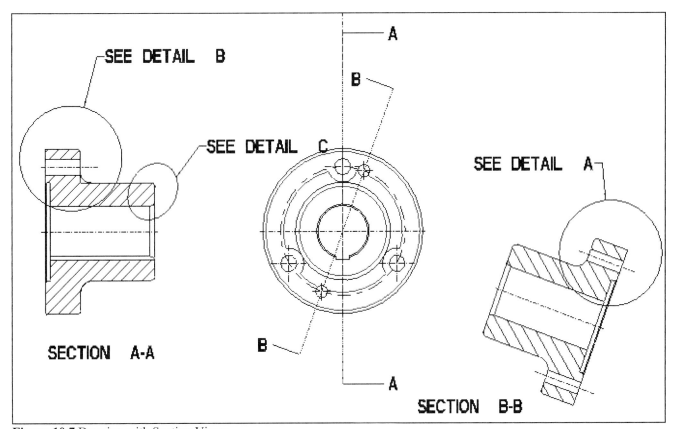

Figure 10.7 Drawing with Section Views

Lesson 10 STEPS

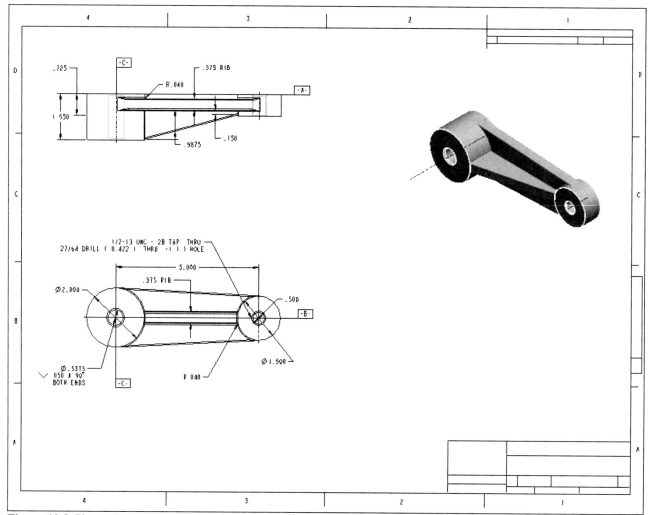

Figure 10.8 Clamp Arm

Creating a Drawing of the Clamp Arm

Before starting the drawing you will need to set your working directory to the folder that contains the Swing Clamp Assembly (and therefore all the components of the assembly). The standard three views of the Clamp_Arm (Fig. 10.8) will display automatically on the drawing, since you will use the default drawing template for this project.

Click: **File ⇒ Set Working Directory** ⇒ select the directory where the **clamp_assembly.asm** was saved ⇒ **OK** ⇒ **Create a new object** ⇒ Drawing ⇒ Name **clamp_arm** ⇒ Use default template [Fig. 10.9(a)] ⇒ **OK** [Fig. 10.9(b)] ⇒ Default Model **Browse** ⇒ pick **clamp_arm.prt** ⇒ **Open** ⇒ Template: **c_drawing** [Fig. 10.9(c)] ⇒ **OK** ⇒ CLAMP_ARM_DESIGN [Fig. 10.9(d)] ⇒ **Open** [Fig. 10.9(e)]

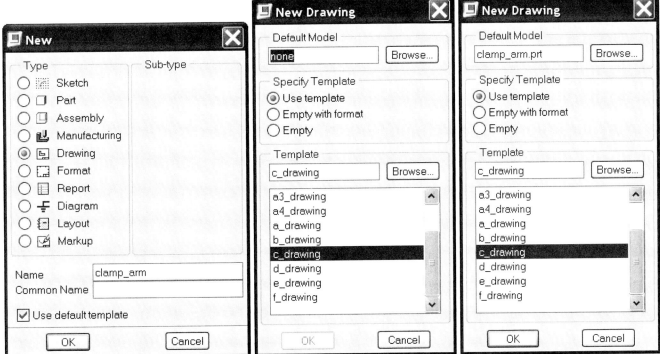

Figure 10.9(a) Type: Drawing   **Figure 10.9(b)** New Drawing Dialog   **Figure 10.9(c)** Default Model

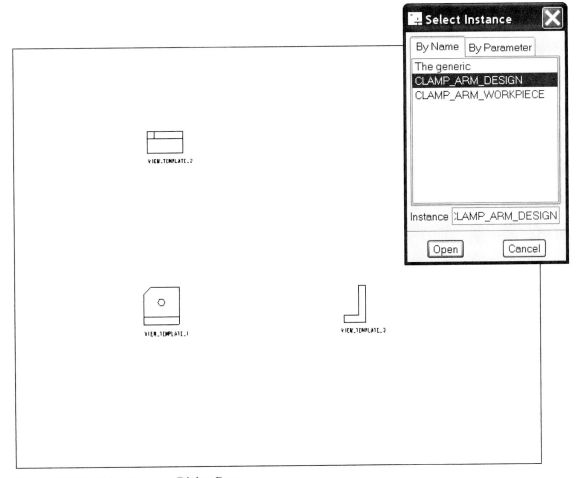

Figure 10.9(d) Select Instance Dialog Box

Click: 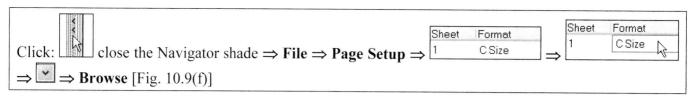 close the Navigator shade ⇒ **File** ⇒ **Page Setup** ⇒ ⇒ ⇒ **Browse** [Fig. 10.9(f)]

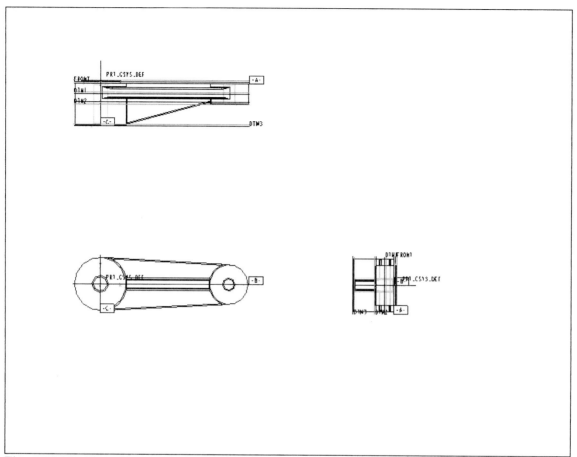

Figure 10.9(e) Three Standard Views Displayed

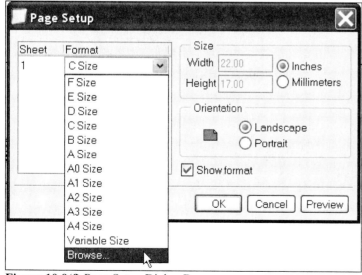

Figure 10.9(f) Page Setup Dialog Box

System Formats Open dialog box opens [Fig. 10.9(g)] ⇒ pick **c.frm** ⇒ **Open** [Fig. 10.9(h)] ⇒ **OK** [Fig. 10.9(i)] ⇒ 🖫 ⇒ **OK**

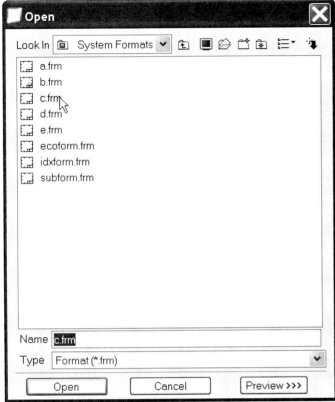

Figure 10.9(g) Open c.frm

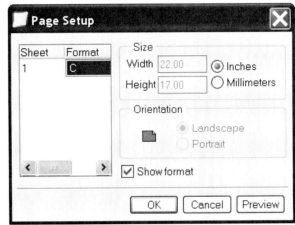

Figure 10.9(h) Standard C Size Format

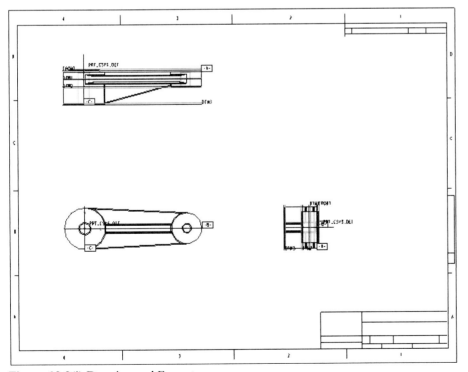

Figure 10.9(i) Drawing and Format

Click: 🔍 ⇒ 🗐 ⇒ **RMB** ⇒ **Insert General View** [Fig. 10.9(j)] ⇒ pick a position for the new view [Fig. 10.9(k)]

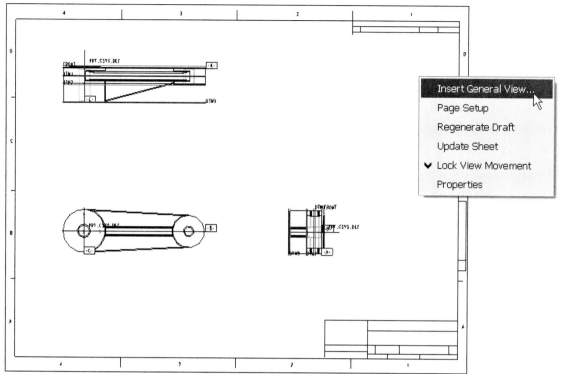

Figure 10.9(j) Insert General View

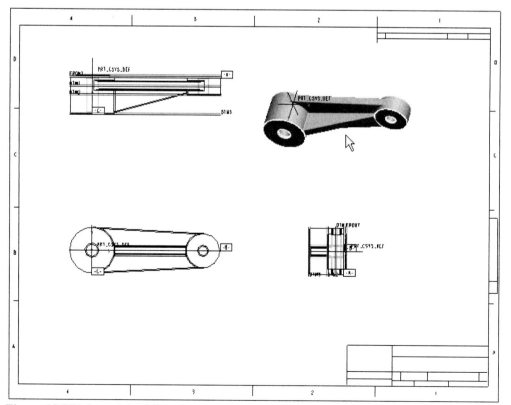

Figure 10.9(k) Insert New View

Default orientation: **Isometric** [Figs. 10.9(l-m)] ⇒ **OK** ⇒ 🖫 ⇒ **OK** ⇒ **LMB** to deselect

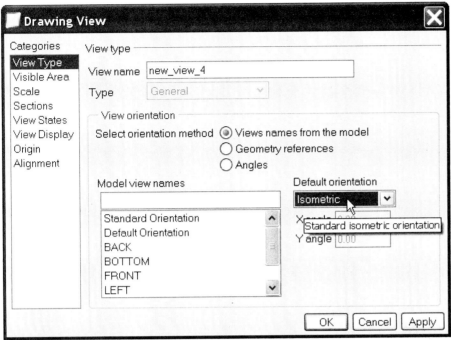

Figure 10.9(l) Default orientation: Isometric

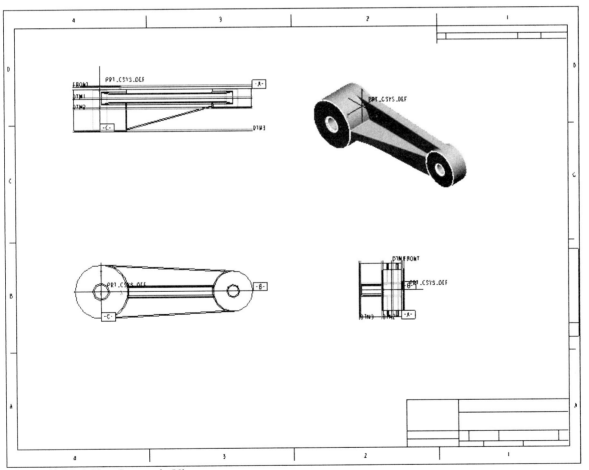

Figure 10.9(m) New Isometric View

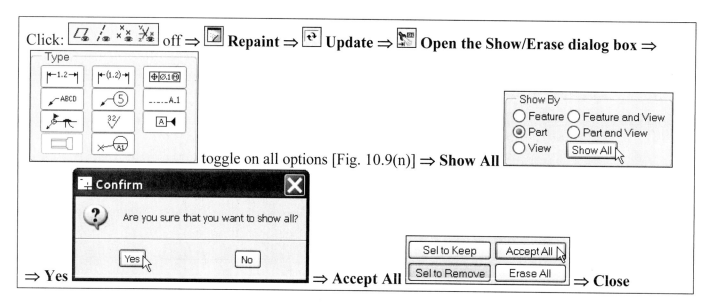

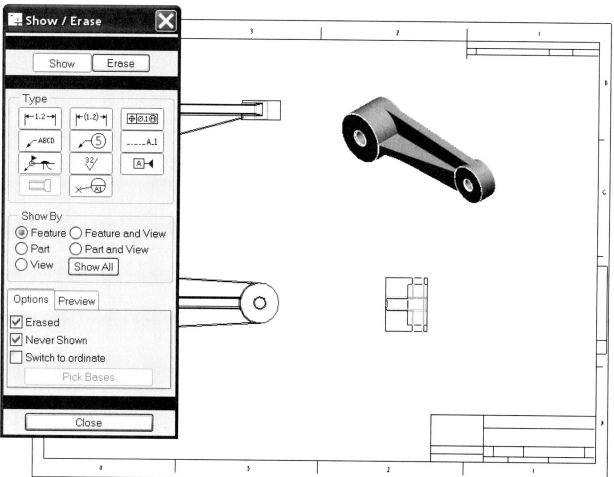

Figure 10.9(n) Show/Erase Dialog Box

Click: **RMB** ⇒ **Cleanup Dimensions** [Fig. 10.9(o)] ⇒ ☐ Create Snap Lines [Fig. 10.9(p)] ⇒ **Apply** ⇒ **Close**

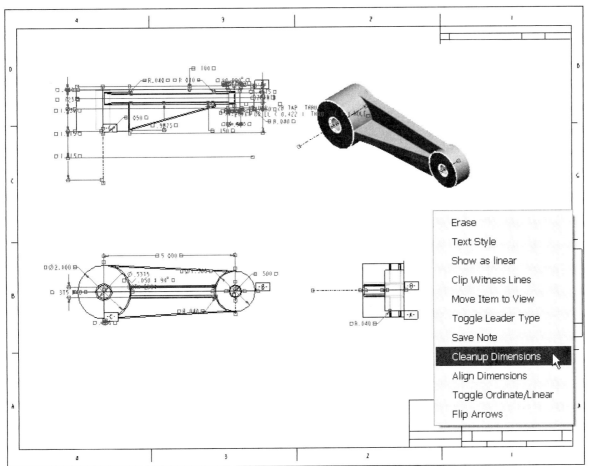

Figure 10.9(o) Cleanup Dimensions

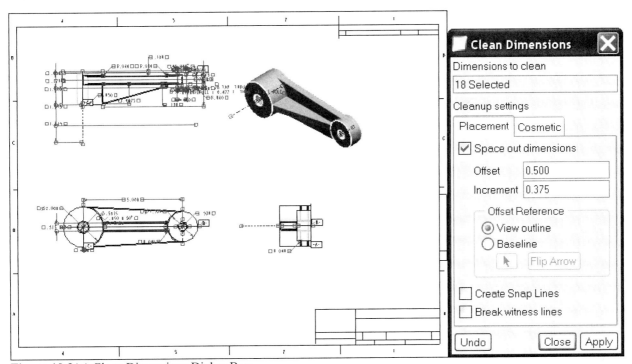

Figure 10.9(p) Clean Dimensions Dialog Box

Click: **LMB** ⇒ **RMB** ⇒ [Lock View Movement] ⇒ [Lock View Movement] ⇒ pick on the isometric view ⇒ [✥] [Fig. 10.9(q)] ⇒ hold down **LMB** and move view away from the top view [Fig. 10.9(r)] ⇒ **LMB** outside of the view to set the position ⇒ **Ctrl+S** ⇒ **OK**

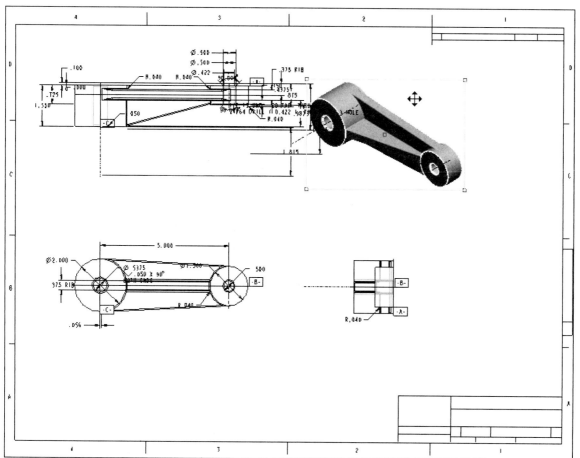

Figure 10.9(q) Move View

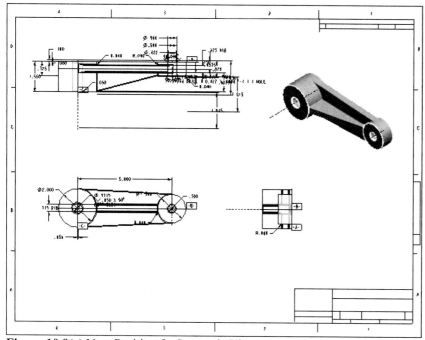

Figure 10.9(r) New Position for Isometric View

Pick on the right side view ⇒ ⊕ ⇒ **RMB** ⇒ **Delete** [Fig. 10.9(s)] ⇒ **Ctrl+S** ⇒ **OK** ⇒ 🔍 **Zoom In** on the front view [Fig. 10.9(t)] ⇒ **MMB** to end zoom box

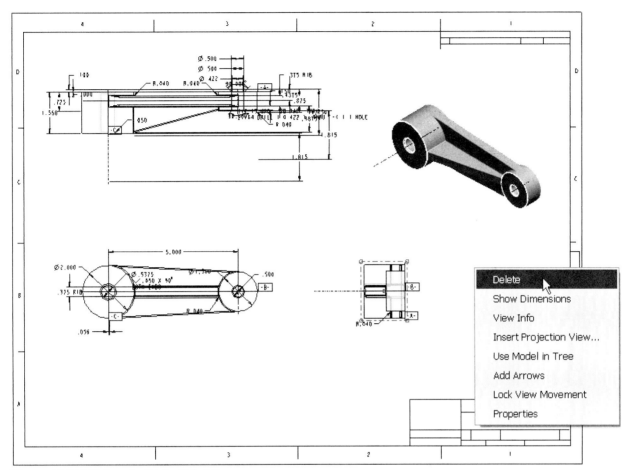

Figure 10.9(s) Delete the Right Side View

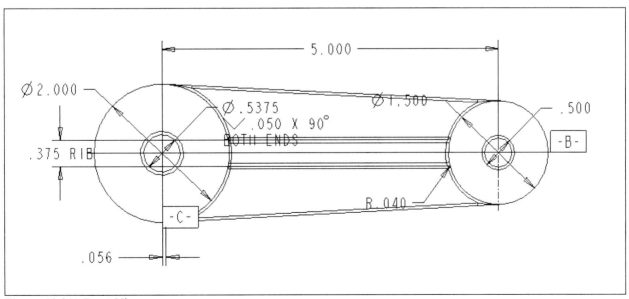

Figure 10.9(t) Front View

Pick on the hole note and move to a new location [Fig. 10.9(u)] ⇒ reposition dimensions as needed [Fig. 10.9(v)] ⇒ reposition each datum tag by picking on the item and then moving it to a new location ⇒ to erase an item, click on it ⇒ **RMB** ⇒ **Erase** ⇒ **LMB** to accept [Fig. 10.9(v)]

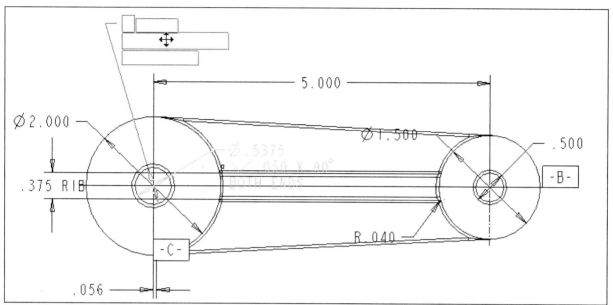

Figure 10.9(u) Move the Note

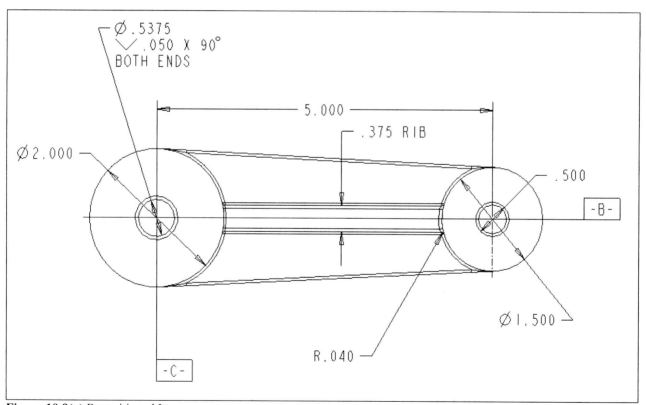

Figure 10.9(v) Repositioned Items

Erase unneeded dimensions and reposition dimensions, notes, and datum tags as needed [Fig. 10.9(w)] ⇒ pick on the note [Fig. 10.9(w)] ⇒ **RMB** ⇒ **Move Item to View** [Fig. 10.9(x)] ⇒ pick on the front view

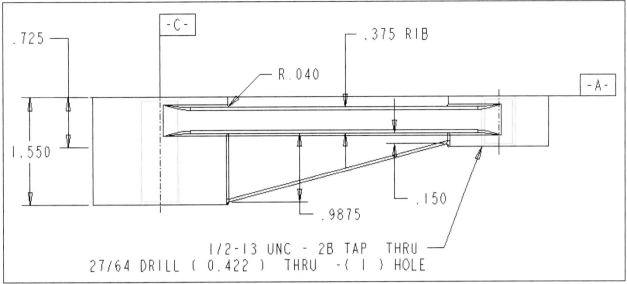

Figure 10.9(w) Top View Cleaned Up

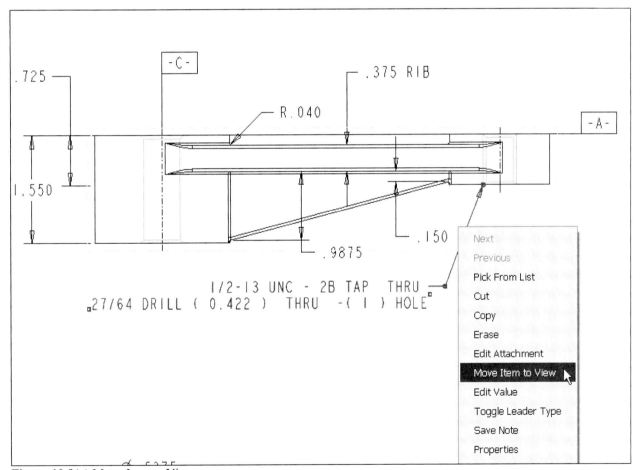

Figure 10.9(x) Move Item to View

Reposition the notes as needed [Fig. 10.9(y)] ⇒ 🔍 **Refit ⇒ Ctrl+S ⇒ Enter ⇒ Window ⇒ Close**

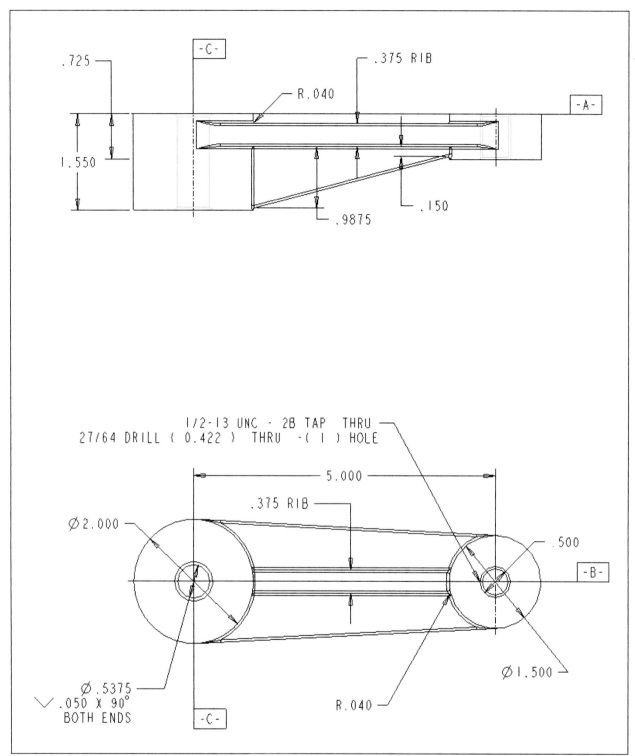

Figure 10.9(y) Move Item

Though not complete, or correct to AMSE 14.5 standards, we will leave this drawing as is. You may finalize the drawing after completing the next lesson.

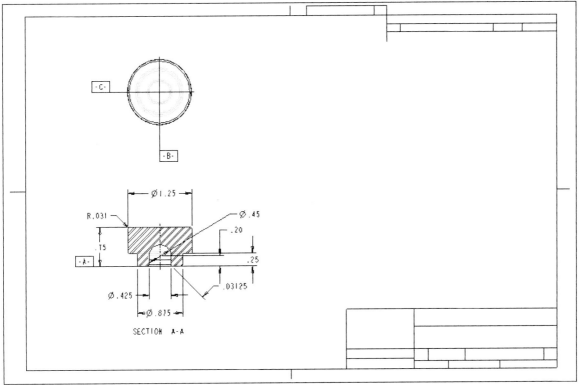

Figure 10.10 Clamp Foot Drawing

Creating a Drawing of the Clamp Foot

Before starting the drawing you will need to set your directory to the folder that contains the Swing Clamp Assembly (and therefore all the components of the assembly), and this time you will bring the Clamp_Foot (Fig. 10.10) in session by opening it. The standard three views of the model will display automatically on the drawing, since you will use the default drawing template for this project.

Click: **File** ⇒ **Set Working Directory** ⇒ select the directory where the **clamp_assembly.asm** was saved ⇒ **OK** ⇒ **File** ⇒ **Open** ⇒ **clamp_foot.prt** ⇒ **Open** (Fig. 10.11) ⇒ Create a new object ⇒ Drawing ⇒ Name **clamp_foot** ⇒ Use default template ⇒ **OK** ⇒ Template: **b_drawing** [Fig. 10.12(a)] ⇒ **OK** [Fig. 10.12(b)]

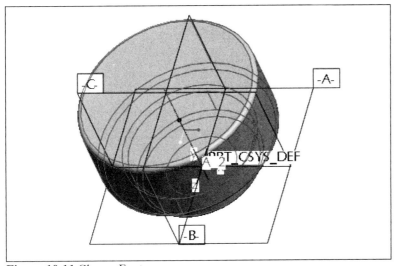

Figure 10.11 Clamp_Foot

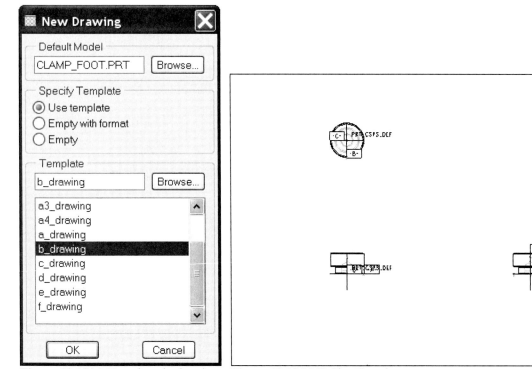

Figure 10.12(a) Template: b_drawing   **Figure 10.12(b)** B-Size Drawing

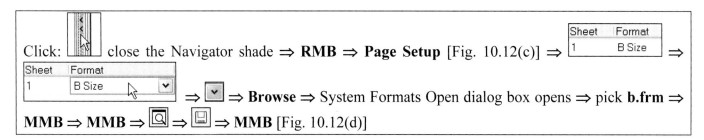

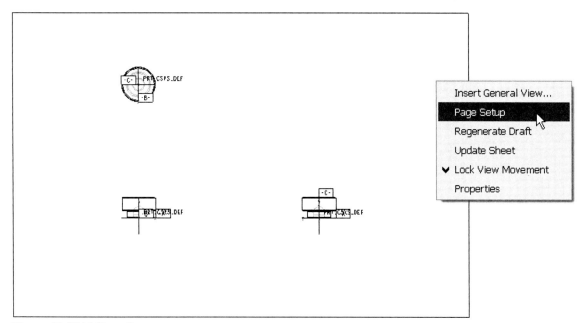

Figure 10.12(c) Page Setup

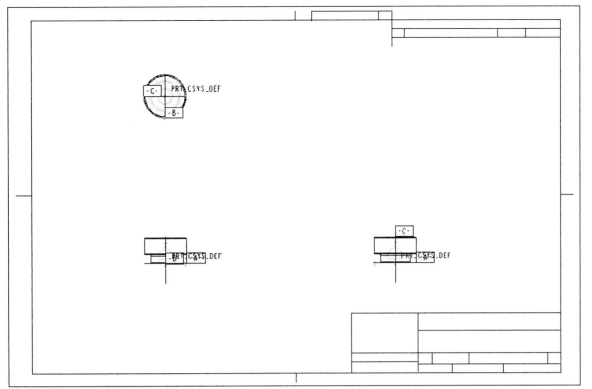

Figure 10.12(d) Clamp Foot Drawing

Double-click on **SCALE: 1.000** [Fig. 10.12(e)] ⇒ type **1.50** ⇒ **Enter** [Fig. 10.12(f)]

Figure 10.12(e) Scale

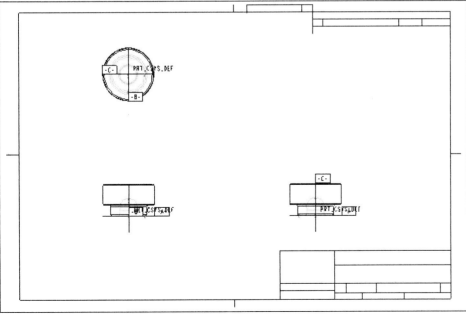

Figure 10.12(f) Resized Views

Click: [icons] off ⇒ Update ⇒ Repaint ⇒ Open the Show/Erase dialog box ⇒ [Type dialog] toggle on all options ⇒ **Show All** ⇒ **Yes** ⇒ **Accept All** ⇒ **Close** ⇒ **RMB** ⇒ **Cleanup Dimensions** ⇒ ☐ Create Snap Lines ⇒ **Apply** ⇒ **Close** [Fig. 10.12(g)] ⇒ pick on the right side view ⇒ **RMB** [Fig. 10.12(h)] ⇒ **Delete**

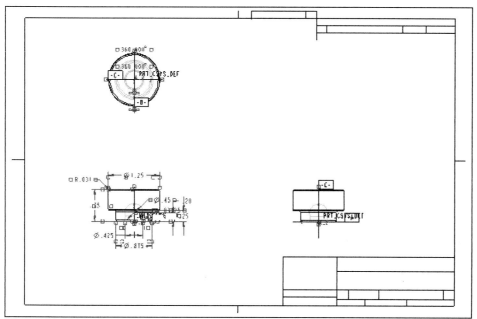

Figure 10.12(g) Cleaned Dimensions

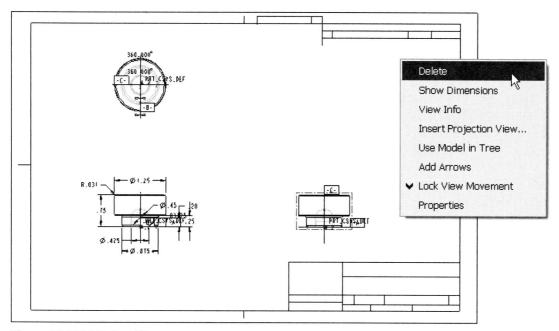

Figure 10.12(h) Delete View

400

Click: **Window** [Fig. 10.12(i)] ⇒ **CLAMP_FOOT.PRT** [Fig. 10.12(j)] ⇒ **Start the view manager** ⇒ **Xsec** ⇒ **New** ⇒ **A** [Fig. 10.12(k)] ⇒ **Enter** ⇒ **Done** ⇒ pick on datum **C** [Fig. 10.12(l)] ⇒ **Close** ⇒ **Window** ⇒ **CLAMP_FOOT.DRW** [Fig. 10.12(m)] ⇒ **Ctrl+S** ⇒ **MMB** [Fig. 10.12(n)]

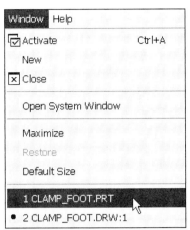
Figure 10.12(i) Switch to the CLAMP_FOOT Window

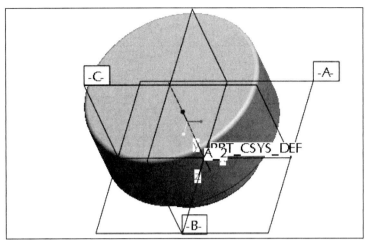
Figure 10.12(j) Part Window Activated

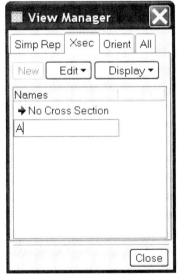
Figure 10.12(k) View Manager Dialog Box

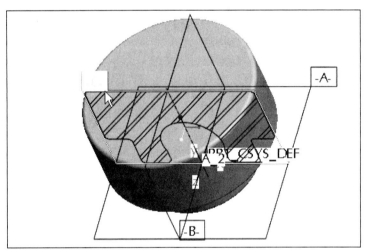
Figure 10.12(l) Part Xsection A

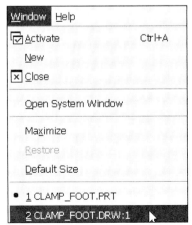
Figure 10.12(m) Drawing Active

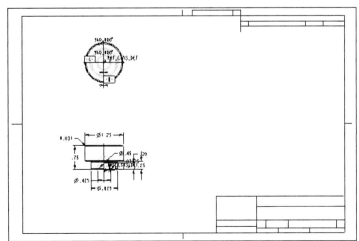
Figure 10.12(n) Two Views

Pick on the front view ⇒ **RMB** ⇒ **Properties** [Fig. 10.12(o)] ⇒ **Sections** ⇒ [2D cross-section] ⇒ [+] ⇒ [A] ⇒ **Apply** [Fig. 10.12(p)] ⇒ **Close**

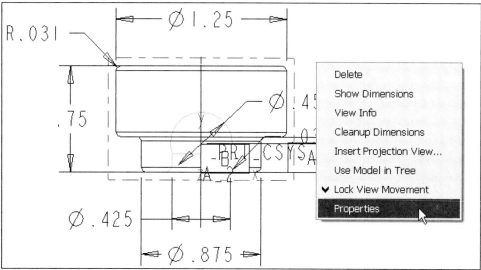

Figure 10.12(o) Properties

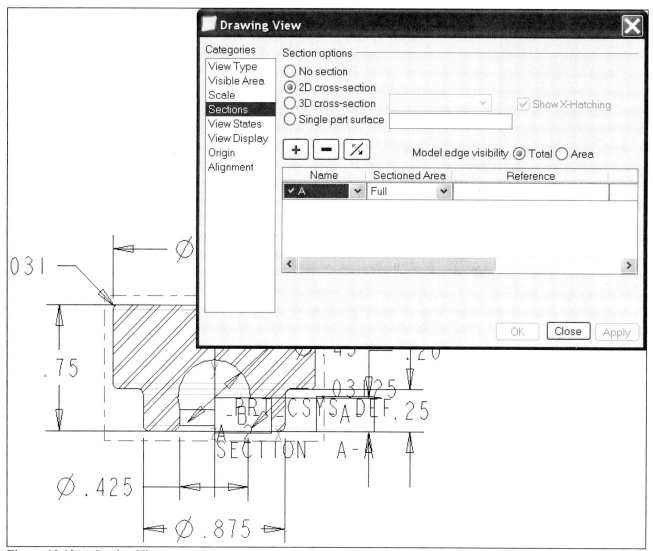

Figure 10.12(p) Section View

Click: [icons] off ⇒ Cleanup the drawing views by erasing unneeded dimensions, moving datums, etc. [Fig. 10.12(q)] ⇒ **Update** ⇒ [icon] ⇒ **Ctrl+S** ⇒ **MMB** ⇒ **Window** ⇒ **Close** ⇒ **Window** ⇒ **Close**

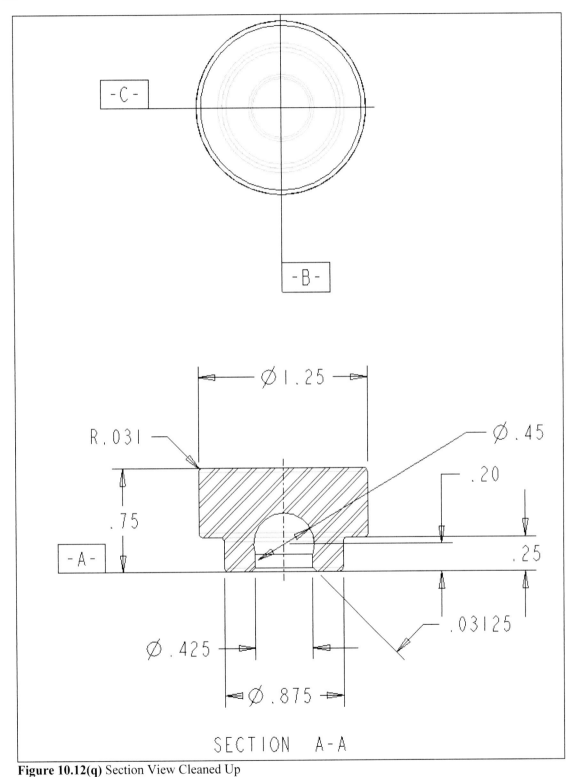

Figure 10.12(q) Section View Cleaned Up

Though not complete, or correct to AMSE 14.5 standards, we will leave this drawing as is. You may finalize the drawing after completing the detail drawing in the next lesson.

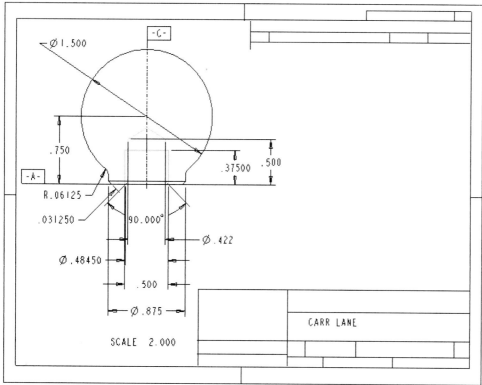

Figure 10.13 Clamp Ball Drawing

Creating a Drawing of the Clamp Ball

Before starting the Clamp Ball drawing (Fig. 10.13), you will create a template that can be used on any drawing.

Click: **File** ⇒ **Set Working Directory** (Fig. 10.15) ⇒ select the directory where the **clamp_assembly.asm** was saved ⇒ **OK** ⇒ ▢ **Create a new object** ⇒ ⦿ ▭ Drawing ⇒ Name: **detail_template** ⇒ ☐ Use default template ⇒ **OK** ⇒ Default Model **Browse** ⇒ pick **clamp_ball.prt** ⇒ **Open** ⇒ ⦿ Empty with format ⇒ Format: **Browse** ⇒ **a.frm** ⇒ **Open** [Fig. 10.15(a)] ⇒ **OK**

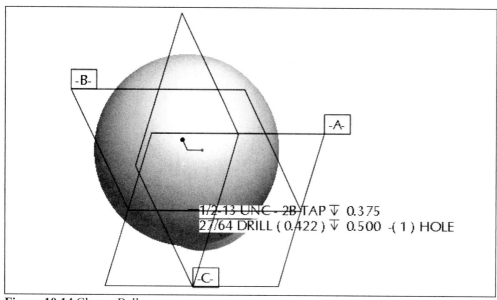

Figure 10.14 Clamp_Ball

Click: **Applications** from menu bar ⇒ **Template** [Fig. 10.15(b)] ⇒ **Insert** from menu bar ⇒ **Template View** [Fig. 10.15(c)]

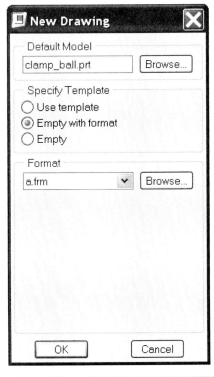

Figure 10.15(a) New Drawing Dialog Box

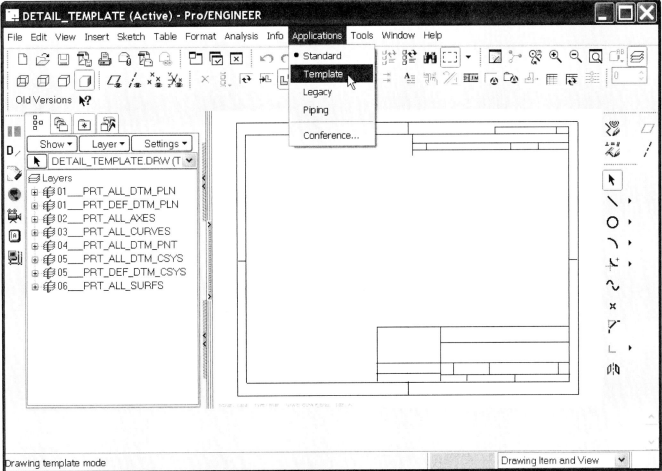

Figure 10.15(b) Applications ⇒ Template

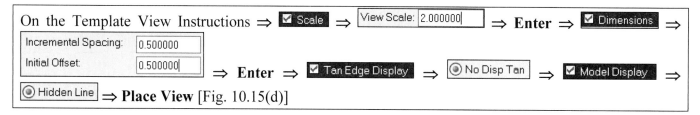

Figure 10.15(c) Template View Instructions Dialog Box

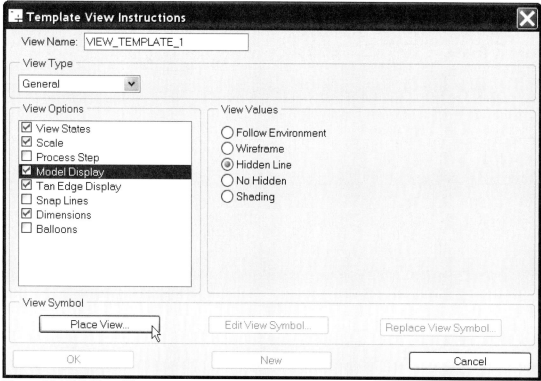

Figure 10.15(d) Place View

Pick a position for the front view [Fig. 10.15(e)] ⇒ **OK** ⇒ **Ctrl+S** ⇒ **Enter** ⇒ **File** ⇒ **Close Window** ⇒ **File** ⇒ **Erase** ⇒ **Not Displayed** 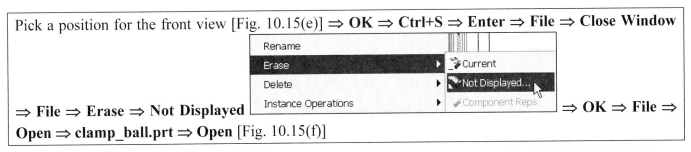 ⇒ **OK** ⇒ **File** ⇒ **Open** ⇒ **clamp_ball.prt** ⇒ **Open** [Fig. 10.15(f)]

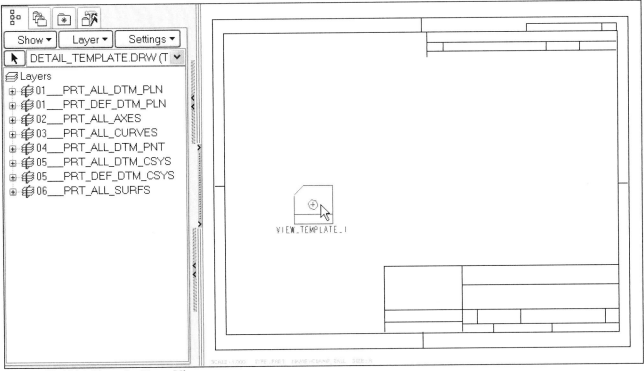

Figure 10.15(e) Place the Front View

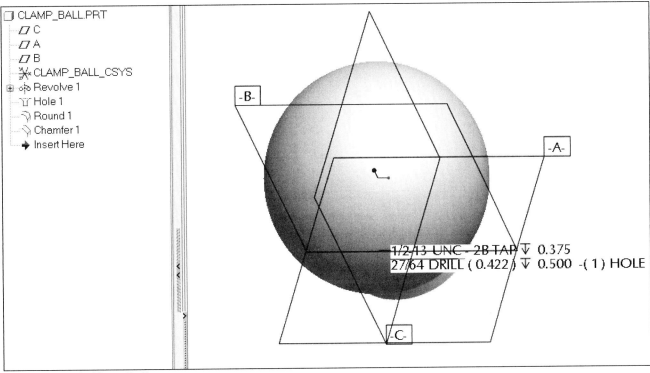

Figure 10.15(f) Clamp Ball

Click: ⬛ **Create a new object** ⇒ ⊙ 🗎 Drawing ⇒ Name **clamp_ball** ⇒ ☐ Use default template ⇒ **OK** ⇒ ⊙ Use template ⇒ **Browse** ⇒ 🗎 detail_template.drw ⇒ **Open** [Fig. 10.15(g)] ⇒ **OK** [Fig. 10.15(h)] ⇒ **RMB** ⇒ **Lock View Movement** uncheck ⇒ pick on the view and move it to a better position [Fig. 10.15(i)] ⇒ **Ctrl+S** ⇒ **MMB** ⇒ **LMB** to deselect

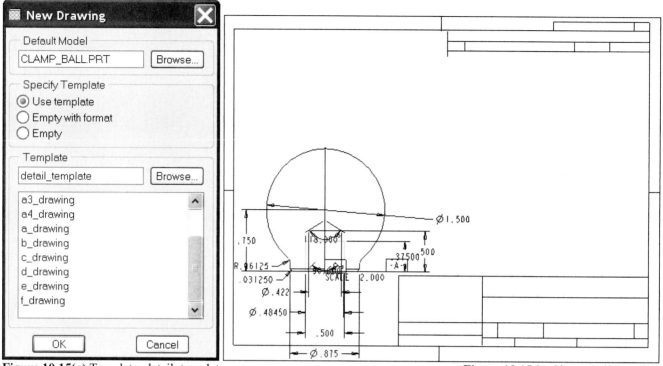

Figure 10.15(g) Template: detail_template

Figure 10.15(h) Clamp Ball Drawing

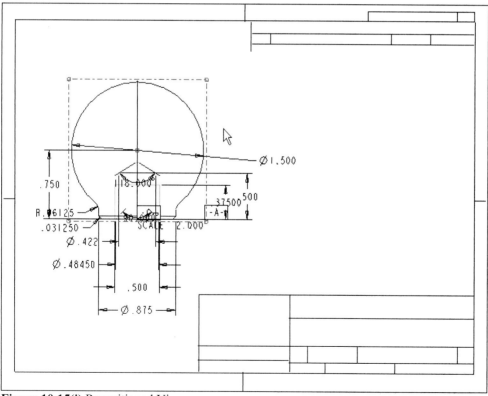

Figure 10.15(i) Repositioned View

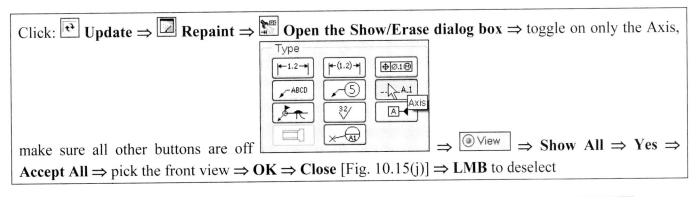

Click: **Update** ⇒ **Repaint** ⇒ **Open the Show/Erase dialog box** ⇒ toggle on only the Axis, make sure all other buttons are off ⇒ **View** ⇒ **Show All** ⇒ **Yes** ⇒ **Accept All** ⇒ pick the front view ⇒ **OK** ⇒ **Close** [Fig. 10.15(j)] ⇒ **LMB** to deselect

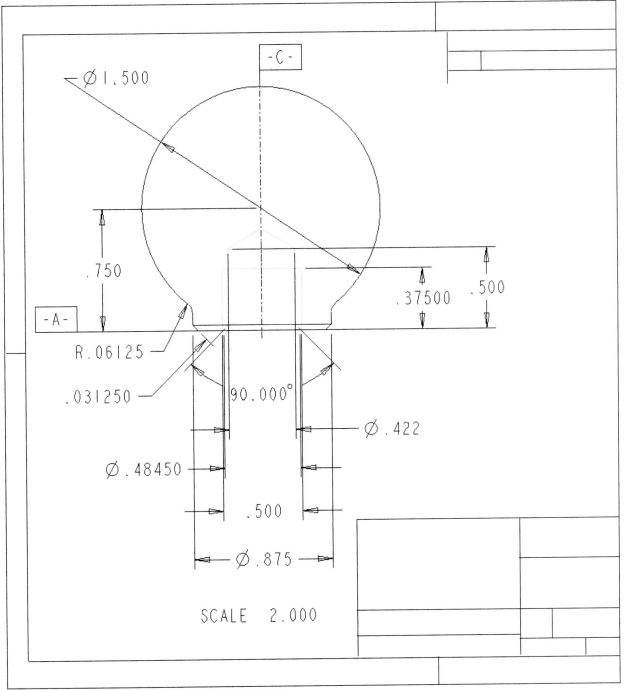

Figure 10.15(j) Axis Added to View

Click: **Create a note** ⇒ **Make Note** ⇒ Select LOCATION for note [Fig. 10.15(k)] ⇒ type Enter NOTE: CARR LANE ⇒ ☑ ⇒ ☑ ⇒ **Done/Return** [Fig. 10.15(l)] ⇒ **LMB** to deselect ⇒ 🔍 ⇒ **Ctrl+S** ⇒ **MMB** [Fig. 10.15(m)] ⇒ **File** ⇒ **Delete** ⇒ **Old Versions** ⇒ **MMB** ⇒ **File** ⇒ **Close Window** ⇒ **File** ⇒ **Close Window**

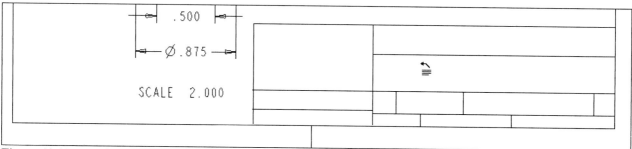

Figure 10.15(k) Pick a Position for the Note

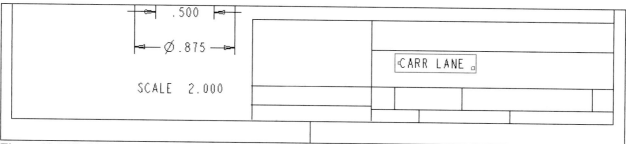

Figure 10.15(l) CARR LANE Note

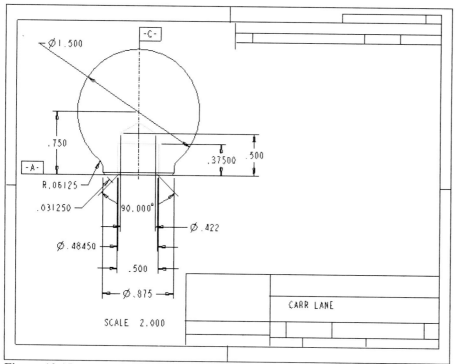

Figure 10.15(m) Completed Drawing

You may finalize the drawing after completing the next lesson. You can create drawings of the remaining assembly parts. And, a variety of part, assemblies, and drawing projects can be downloaded from the website **www.cad-resources.com**.

Lesson 11 Part Drawings

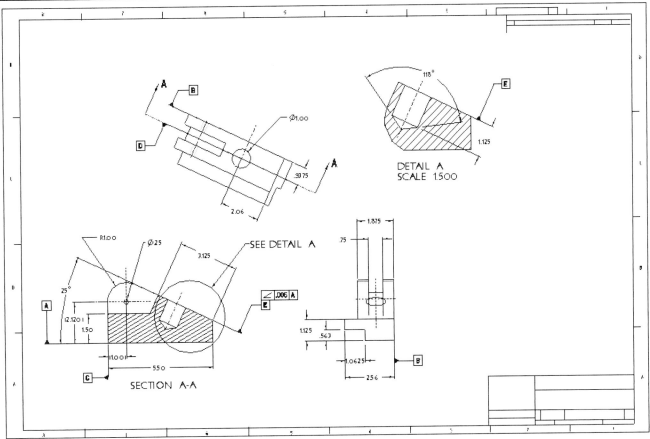

Figure 11.1 Anchor Drawing

OBJECTIVES

- Establish a **Drawing Options** file to use when detailing
- Identify the need for **views** to clarify interior features of a part
- Create **Cross Sections** using datum planes
- Produce **Auxiliary Views**
- Create **Detail Views**
- Use **multiple drawing sheets**
- Apply **standard drafting conventions** and linetypes to illustrate interior features

PART DRAWINGS

Designers and drafters use drawings to convey design and manufacturing information. Drawings consist of a **Format** and views of a part (or assembly) (Fig 11.1). Standard views, sectional views, detail views, and auxiliary views are utilized to describe the objects' features and sizes. **Sectional views**, also called **sections**, are employed to clarify and dimension the internal construction of an object. Sections are needed for interior features that cannot be clearly described by hidden lines in conventional views. **Auxiliary views** are used to show the *true shape/size* of a feature or the relationship of features that are not parallel to any of the principal planes of projection. Many objects have inclined surfaces and features that cannot be adequately displayed and described by using principal views alone. To provide a clearer description of these features, it is necessary to draw a view that will show the *true shape/size*.

Lesson 11 STEPS

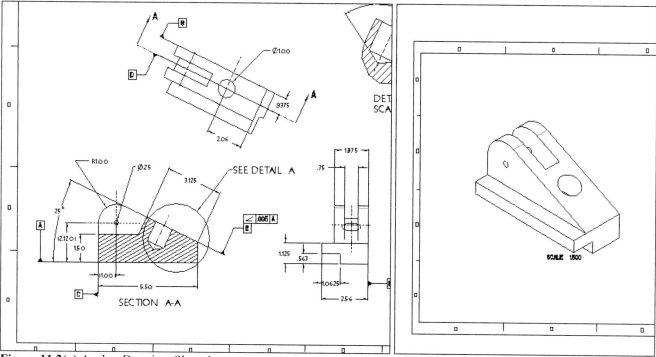

Figure 11.2(a) Anchor Drawing, Sheet 1

Figure 11.2(b) Anchor Drawing, Sheet 2

Anchor Drawing

You will be creating a multiple sheet detail drawing of the Anchor [Figs. 11.2(a-b)]. The front view will be a full section. A right side view and an auxiliary view are required to detail the part. Views will be displayed according to visibility requirements per ANSI standards, such as no hidden lines in sections. The part is to be dimensioned according to ASME Y14.5M. You will add a standard format. Detailed views of other parts will be introduced to show the wide variety of view capabilities.

Click: **File** ⇒ **Set Working Directory** ⇒ select the directory where the **anchor.prt** was saved ⇒ **OK** ⇒ ▯ ⇒ ◉ Drawing ⇒ Name **ANCHOR** ⇒ ☐ Use default template *(note: if you keep the "Use default template" checked; the Front, Top, and Right views will be automatically created for you)* ⇒ **OK** ⇒ Default Model **Browse** ⇒ pick **anchor.prt** ⇒ **Open** ⇒ Standard Size **D** [Fig. 11.2(c)] ⇒ **OK** ⇒ **File** ⇒ **Page Setup** ⇒ D Size [Fig. 11.2(d)] ⇒ ▾ ⇒ **Browse** ⇒ System Formats Open dialog box opens [Fig. 11.2(e)] ⇒ pick **d.frm** ⇒ **Open** ⇒ **OK** ⇒ ▯ ⇒ **OK**

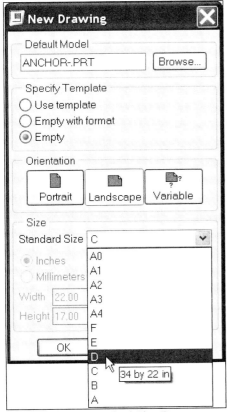
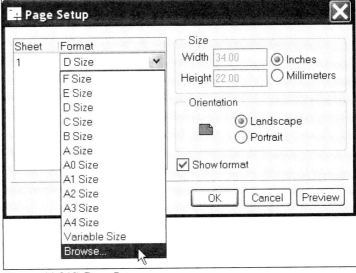

Figure 11.2(c) Standard Size D   **Figure 11.2(d)** Page Setup

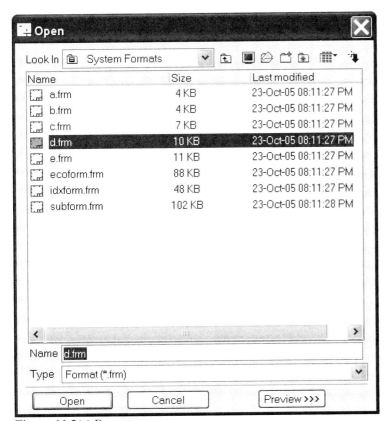

Figure 11.2(e) Formats

Click: **RMB** ⇒ **Properties** [Fig. 11.2(f)] ⇒ **Drawing Options** [Fig. 11.2(g)] the Options dialog box opens ⇒ create a new **.dtl** file, *(or* 📂 ⇒ *pick a previously saved **.dtl** from your directory list* ⇒ *Open)* Change the following options to the values listed below:

Option: *drawing_text_height* Value: *.25* ⇒ **Add/Change**
Option: *default_font* Value: *filled* ⇒ **Add/Change**
Option: *draw_arrow_style* Value: *filled* ⇒ **Add/Change**
Option: *allow_3d_dimensions* Value: *yes* ⇒ **Add/Change**

Click: **Apply** ⇒ Sort: **Alphabetical** [Fig. 11.2(h)] ⇒ 💾 **Save a copy of the currently displayed configuration file** ⇒ type a unique name for your file (*draw_options*) ⇒ **Ok** ⇒ **Close** ⇒ **Done/Return** ⇒ **Ctrl+S** ⇒ **MMB**

Figure 11.2(f) "D" Size Format

Figure 11.2(g) Drawing Options

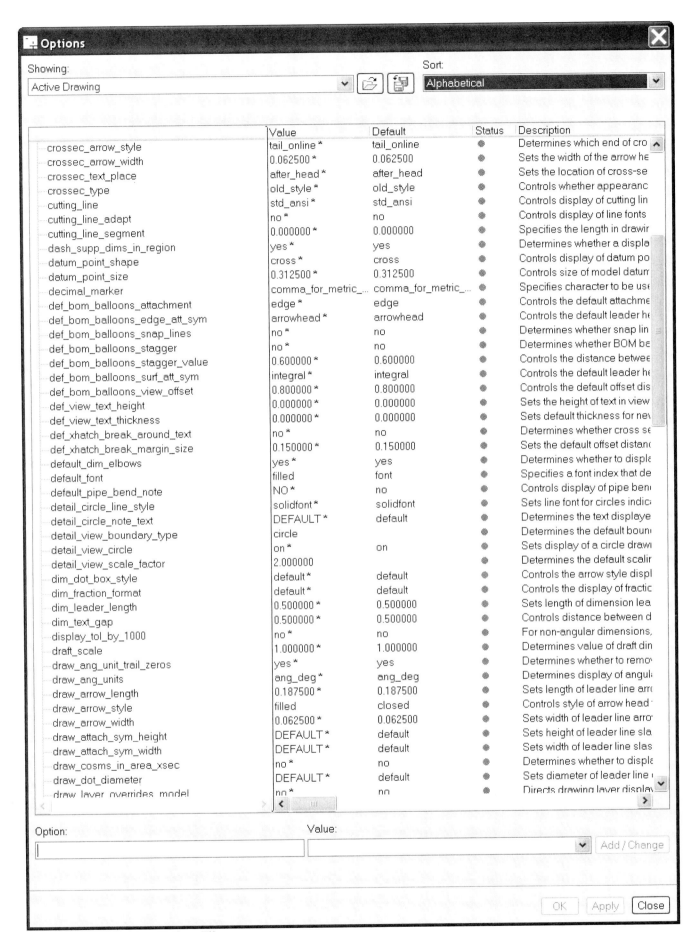

Figure 11.2(h) Drawing Options File

Click the quick sash to close your Navigator (normally close the Navigator so you can see a larger drawing) ⇒ **Hidden Line** ⇒ **Create a general view** ⇒ pick a position for the view [Fig. 11.3(a)] ⇒ **Geometry references** [Fig. 11.3(b)] ⇒ **Redraw the current view**

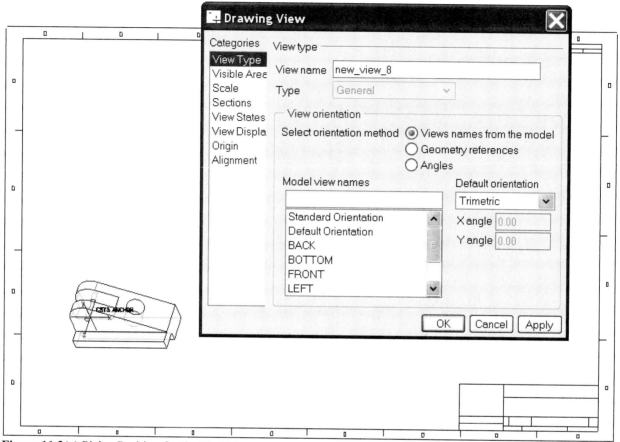

Figure 11.3(a) Pick a Position for the First View

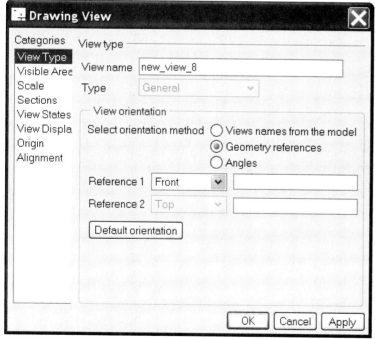

Figure 11.3(b) Drawing View Dialog Box, Geometry references Selected

Reference 1: **Front** ⇒ pick datum **B** [Fig. 11.3(c)] ⇒ Reference 2: **Top** [Fig. 11.3(d)] ⇒ pick datum **A** ⇒ **OK** [Fig. 11.3(e)] ⇒ 🔍 ⇒ 💾 ⇒ **MMB** ⇒ **LMB** to deselect ⇒ 🔲 **Datum planes** off ⇒ 📐 **Datum axes** off ⇒ ⁂ **Datum points** off ⇒ ⁂ **Coordinate systems** off ⇒ 🔲

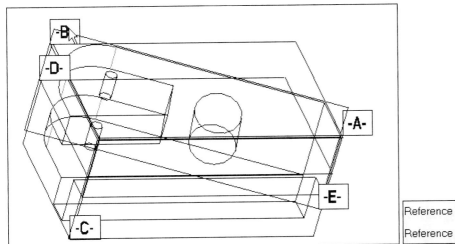

Figure 11.3(c) Reference 1 Front, Datum B

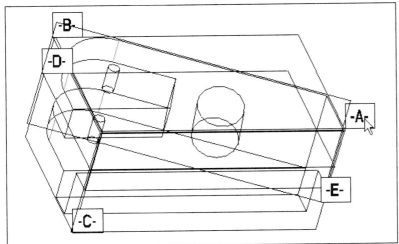

Figure 11.3(d) Reference 2 Top, Datum A

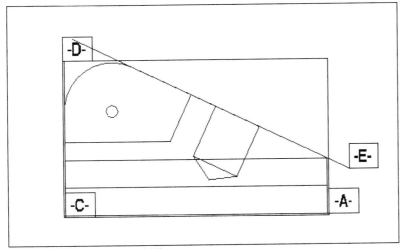

Figure 11.3(e) Reoriented View

Add two more views, click: **Insert** ⇒ **Drawing View** ⇒ **Projection** ⇒ [Select CENTER POINT for drawing view.] pick a position for the right side view [Fig. 11.3(f)] ⇒ **LMB** to deselect ⇒ pick on the Front view ⇒ **RMB** ⇒ **Insert Projection View** ⇒ [Select CENTER POINT for drawing view.] pick a position for the top view [Fig. 11.3(g)] ⇒ **LMB** ⇒ 🔒 **Disallow the movement of drawing views with the mouse** (unlock) ⇒ pick on a view, hold down the **LMB**, and reposition as needed ⇒ **LMB** to deselect ⇒ 🔄 **Update** ⇒ 🔍 ⇒ 💾 ⇒ **MMB** [Fig. 11.3(h)]

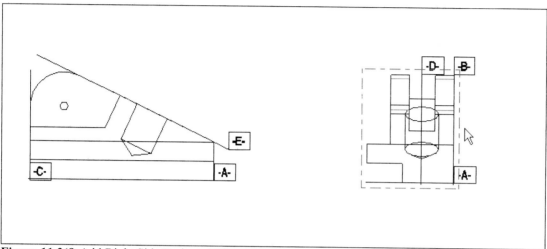

Figure 11.3(f) Add Right Side Projected View

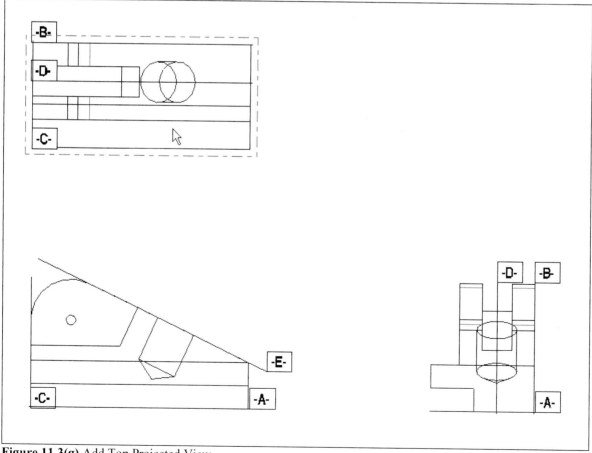

Figure 11.3(g) Add Top Projected View

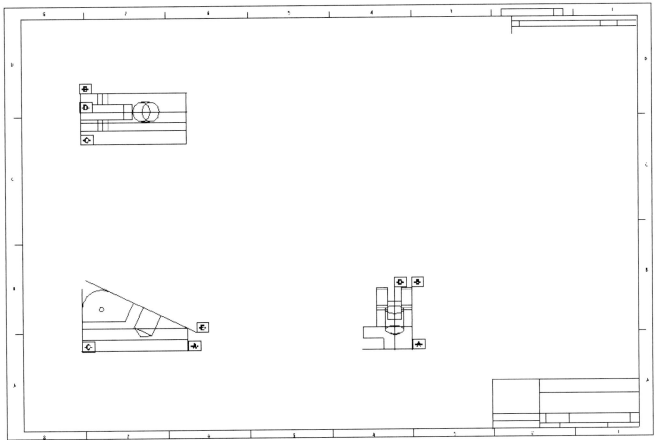

Figure 11.3(h) Repositioned Views

The top view does not help in the description of the part's geometry. Delete the top view before adding an auxiliary view that will display the true shape of the angled surface.

Pick on the Top view ⇒ **RMB** [Fig. 11.3(i)] ⇒ **Delete** [Fig. 11.3(j)] ⇒ 🔄 **Update** ⇒ 🔍

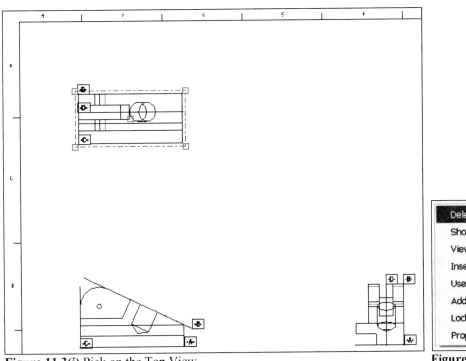

Figure 11.3(i) Pick on the Top View   **Figure 11.3(j)** Delete

Click: **Insert** ⇒ **Drawing View** ⇒ **Auxiliary** ⇒ ⇨Select edge of or axis through, or datum plane as, front surface on main view. pick on datum **E** [Fig. 11.3(k)] ⇒ ⇨Select CENTER POINT for drawing view. [Fig. 11.3(l)] ⇒ **LMB** to place the view [Fig. 11.3(m)] ⇒ **LMB** to deselect ⇒ **RMB** ⇒ **Update Sheet** ⇒ 🖫 ⇒ **MMB**

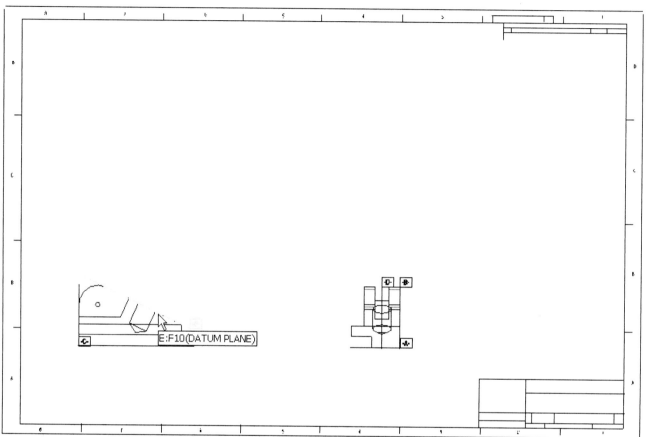

Figure 11.3(k) Select Datum E of the Angled Surface

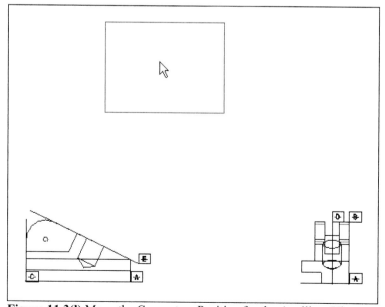

Figure 11.3(l) Move the Cursor to a Position for the Auxiliary View

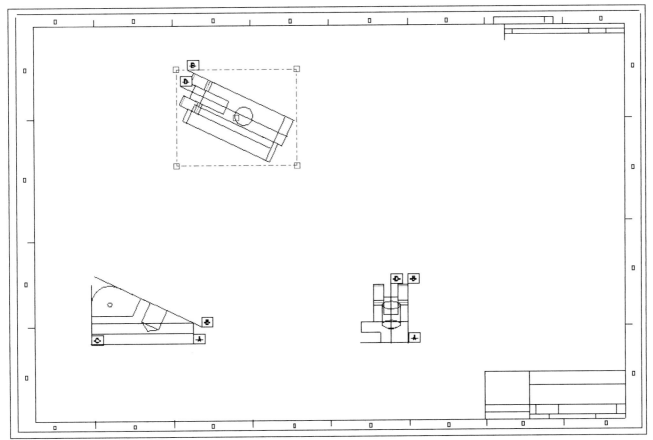

Figure 11.3(m) Auxiliary View

The correct standard was not selected when you made changes to the Drawing Options. ASME symbols for datum planes (gtol_datums) are the correct style standard used on drawings. The ANSI style [Fig. 11.3(n)] was discontinued in 1994, though retained by some companies as an "in house" standard. For outside vendors and for manufacturing internationally, the ISO-ASME standards should be applied to all manufacturing drawings.

Click: **RMB** ⇒ **Properties** ⇒ **Drawing Options** ⇒ [gtol_datums | std_asme] ⇒ **Add/Change** ⇒ **Apply** ⇒ **Close** ⇒ **Done/Return** ⇒ [Update] ⇒ [📷] ⇒ [📷] [Fig. 11.3(o)] ⇒ [💾] ⇒ **MMB** ⇒ **Sketch** ⇒ **Sketcher Preferences** ⇒ [Grid intersection] off ⇒ **Close**

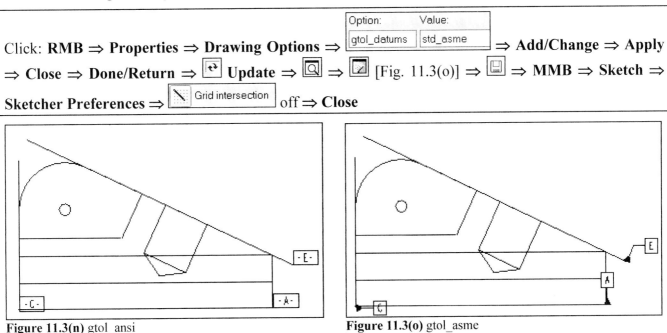

Figure 11.3(n) gtol_ansi   **Figure 11.3(o)** gtol_asme

Next, change the front view into a sectional view. Section **A** was created in the Part mode.

Pick on the front view as the view to be modified [Fig. 11.4(a)] ⇒ **RMB** ⇒ **Properties** [Fig. 11.4(b)] ⇒ Categories **Sections** ⇒ [2D cross-section] ⇒ [+] **Add cross-section to view** ⇒ pick section **A** from the Name list [Fig. 11.4(c)] ⇒ **Apply** [Fig. 11.4(d)] ⇒ **Close** ⇒ **RMB** ⇒ **Add Arrows** [Fig. 11.4(e)] ⇒ pick the auxiliary view [Fig. 11.4(f)] ⇒ **LMB** ⇒ [↻] ⇒ [🔍] ⇒ [▨] ⇒ [💾] ⇒ **MMB**

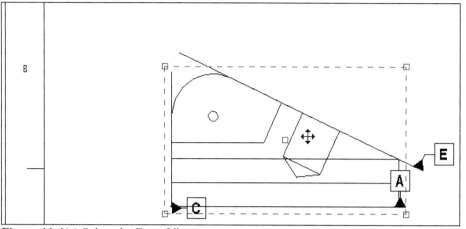

Figure 11.4(a) Select the Front View

Figure 11.4(b) RMB Properties

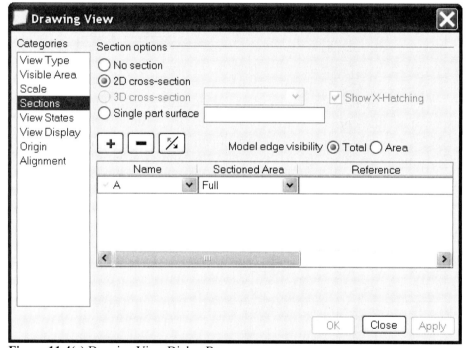

Figure 11.4(c) Drawing View Dialog Box

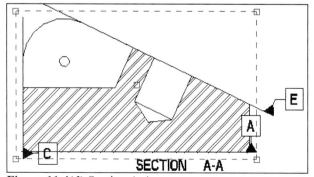

Figure 11.4(d) Section A-A

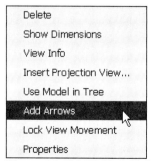

Figure 11.4(e) RMB Add Arrows

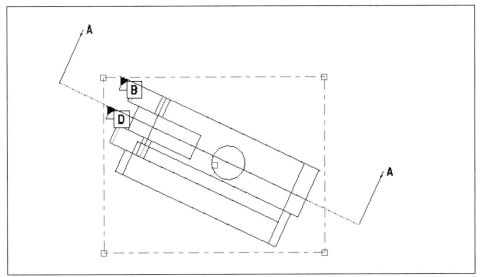

Figure 11.4(f) Section A-A Arrows

Modify the visibility of the views to remove all hidden lines. While pressing and holding down the **Ctrl** key, pick on all three views [Fig. 11.5(a)] ⇒ click **RMB** with the cursor outside of any view outlines ⇒ **Properties** [Fig. 11.5(b)] ⇒ Display style **No Hidden** [Fig. 11.5(c)] ⇒ Tangent edges display style **Dimmed** [Fig. 11.5(d)] ⇒ **Apply** ⇒ **Close** ⇒ **LMB** to deselect [Fig. 11.5(e)]

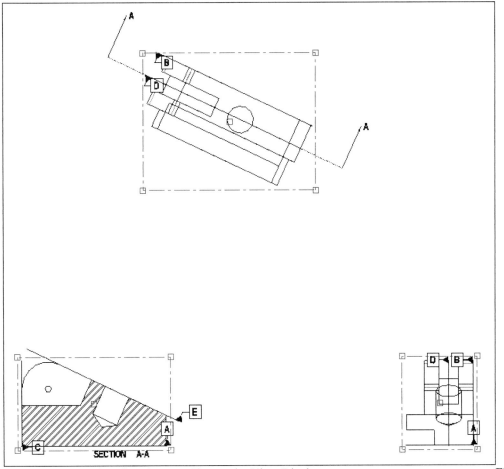

Figure 11.5(a) Select the Views to Change the View Display   **Figure 11.5(b)** RMB Properties

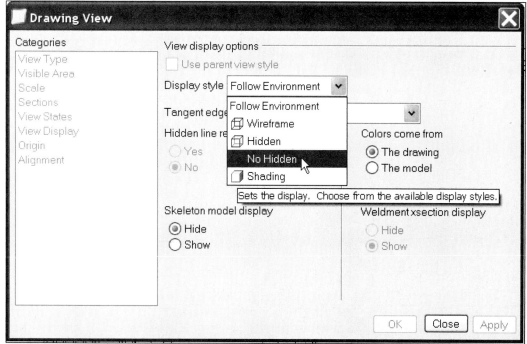

Figure 11.5(c) Drawing View Dialog Box, No Hidden

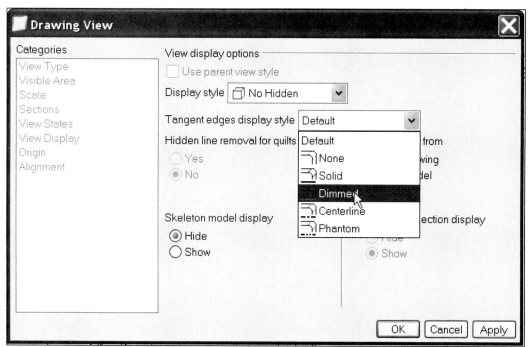

Figure 11.5(d) Drawing View Dialog Box, Dimmed

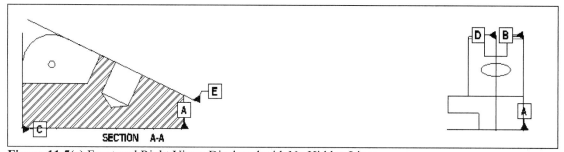

Figure 11.5(e) Front and Right Views Displayed with No Hidden Lines

Show all dimensions and axes (centerlines), click: **Open the Show/Erase dialog box** ⇒ Show ⇒ Dimension Axis Note Datum Plane [Fig. 11.6(a)] ⇒ **Show All** ⇒ **Yes** ⇒ **Accept All** ⇒ **Close** ⇒ **LMB** [Fig. 11.6(b)]

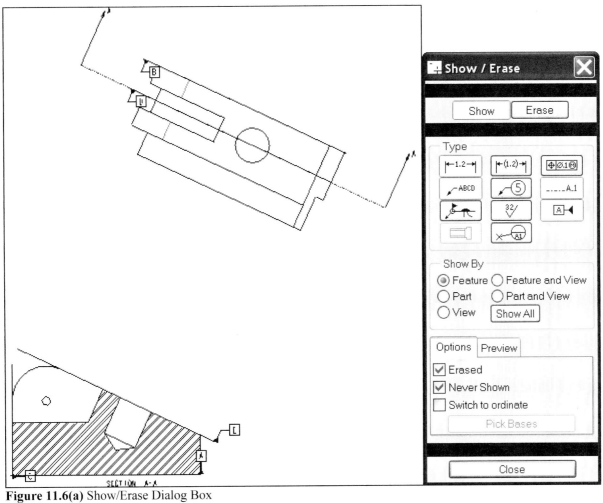

Figure 11.6(a) Show/Erase Dialog Box

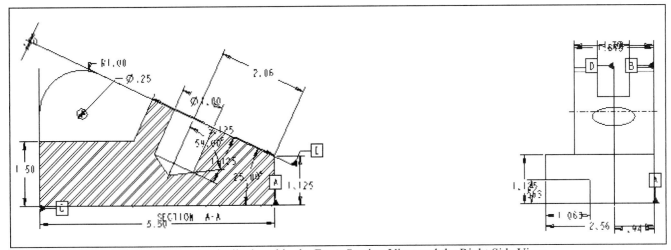

Figure 11.6(b) All but One Dimension is displayed in the Front Section View and the Right Side View

Click: ⊞ **Clean up position of dimensions around a view** *(or select a view ⇒ RMB ⇒ Cleanup Dimensions)* ⇒ select all dimensions by enclosing them in a selection box or picking them ⇒ **MMB** ⇒ ☐ Create Snap Lines ⇒ Increment .500 [Fig. 11.6(c)] ⇒ **Enter** ⇒ **Apply** [Fig. 11.6(d)] ⇒ **Close** ⇒ use **Ctrl+MMB** to zoom as needed ⇒ **MMB** to pan as needed

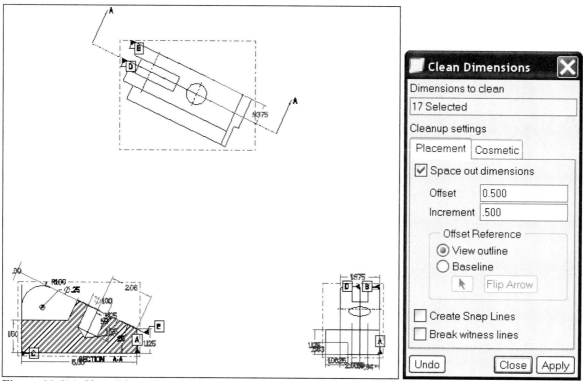

Figure 11.6(c) Clean Dimensions Dialog Box

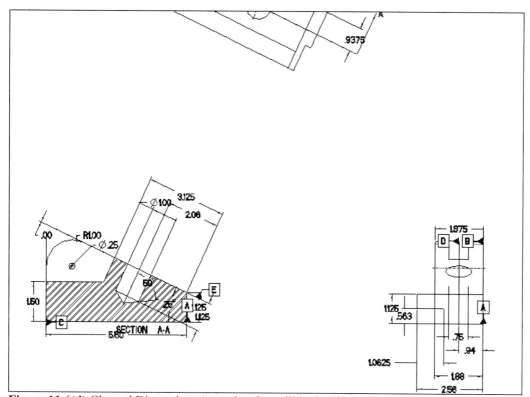

Figure 11.6(d) Cleaned Dimensions (your drawing will look different)

Pick on the **1.00** diameter hole dimension [Fig. 11.7(a)] ⇒ **RMB** ⇒ **Move Item to View** [Fig. 11.7(b)] ⇒ pick on the auxiliary view [Fig. 11.7(c)] ⇒ pick on and reposition the **1.00** dimension [Fig. 11.7(d)] ⇒ using the dimension handles to reposition, move dimension text, clip extension lines, or flip arrows where appropriate ⇒ move, erase, or clip axes and position datums as necessary ⇒ add a reference dimension to the small hole (2.120) [Fig. 11.8(a)] ⇒ **Insert** ⇒ **Reference Dimension** ⇒ **New References** ⇒ pick datum **A** ⇒ pick the horizontal axis of the small hole ⇒ **MMB** to place the reference dimension ⇒ **MMB** ⇒ **LMB**

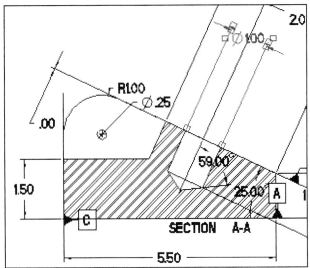 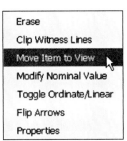

Figure 11.7(a) Pick on the **1.00** Diameter Dimension   **Figure 11.7(b)** Move Item to View

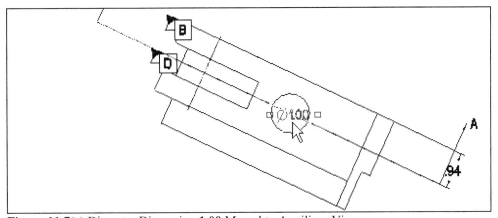

Figure 11.7(c) Diameter Dimension **1.00** Moved to Auxiliary View

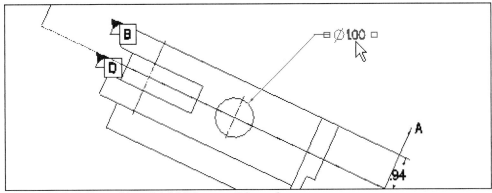

Figure 11.7(d) Repositioned **1.00** Diameter Dimension

Change the arrow length and width to be in proportion to the text height. Click: **RMB** ⇒ **Properties** ⇒ **Drawing Options** opens an Options dialog box ⇒ *draw_arrow_length .25*, *draw_arrow_width .10* ⇒ **Add/Change** ⇒ **Apply** ⇒ **Close** ⇒ **MMB** ⇒ 🔄 Update the display of all views in the active sheet [Figs. 11.8(a-c)]

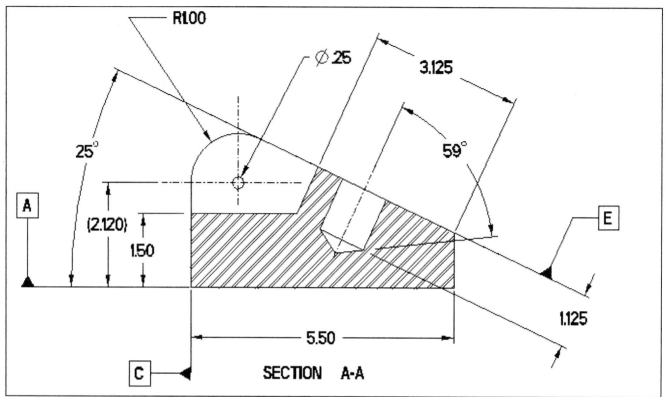

Figure 11.8(a) Edited Front Section View

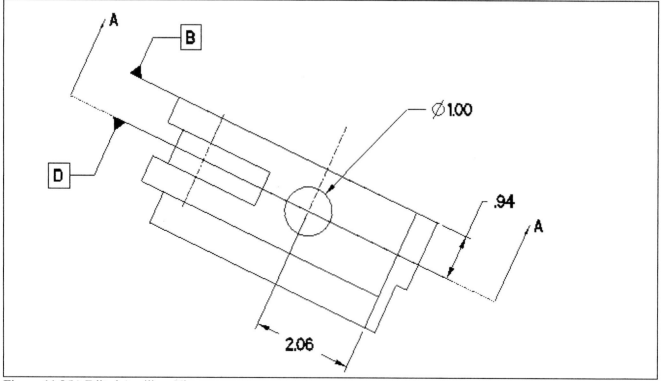

Figure 11.8(b) Edited Auxiliary View

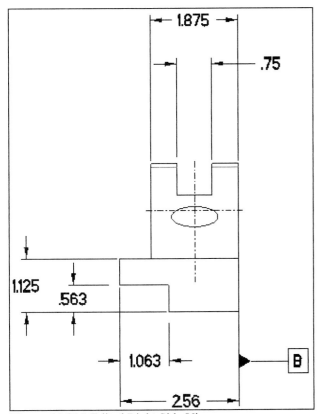

Figure 11.8(c) Edited Right Side View

Add the edges of the small through hole back into the auxiliary view. The edges will display in the graphics window as light gray, but print as dashed on the drawing plot.

Click: **View** from menu bar ⇒ **Drawing Display** ⇒ **Edge Display** ⇒ **Hidden Line** ⇒ press and hold the **Ctrl** key and pick near where the small through hole would show as hidden in the auxiliary view [Fig. 11.9(a)] ⇒ pick the opposite edge [Fig. 11.9(b)] ⇒ **MMB** ⇒ **Done** ⇒ 🔄 ⇒ 🔍 ⇒ 📝 ⇒ 💾 ⇒ **MMB** [Figs. 11.9(c-d)]

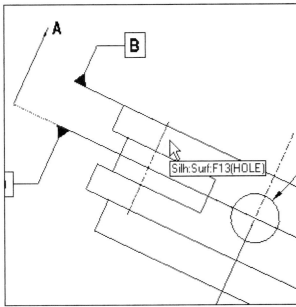

Figure 11.9(a) Pick near the Hole Edge

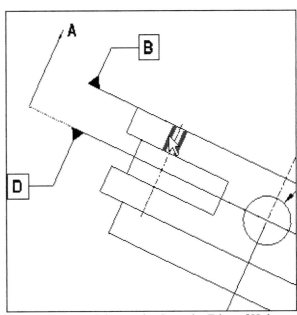

Figure 11.9(b) Pick near the Opposite Edge of Hole

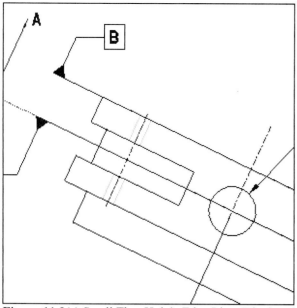
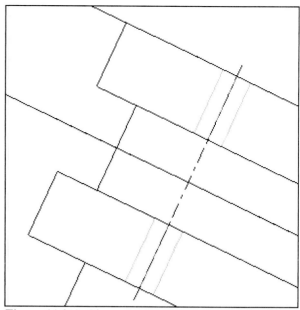

Figure 11.9(c) Small Thru Hole's Edge Lines Displayed

Figure 11.9(d) Close-up of the Thru Hole's Edge Lines

To increase the clarity of this drawing, you will need to master a number of capabilities. Partial views, detail views, using multiple sheets, and modifying section lining (crosshatch lines) are just a few of the many options available in Drawing mode. Create a Detail View.

Click: **Insert** ⇒ **Drawing View** ⇒ **Detailed** ⇒ `Select center point for detail on an existing view.` pick the top edge of the hole in the front section view [Fig. 11.10(a)] ⇒ `Sketch a spline, without intersecting other splines, to define an outline.` *each **LMB** pick adds a point to the spline* ⇒ **MMB** to end the spline [Fig. 11.10(b)] *(you do not have to "close" the spline sketch)* ⇒ `Select CENTER POINT for drawing view.` pick in the upper right of the drawing sheet to place the view

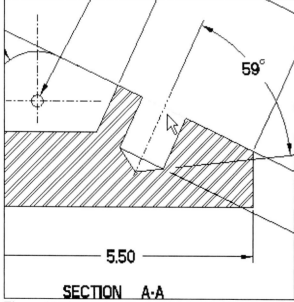
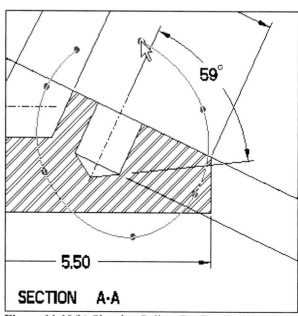

Figure 11.10(a) Select Center Point for Detail View

Figure 11.10(b) Sketch a Spline (Pro/E will close it)

RMB on detailed view ⇒ **Properties** [Fig. 11.10(c)] ⇒ **Scale** ⇒ ⦿ Custom scale 1.500 **1.500** [Fig. 11.10(d)] ⇒ **Enter** ⇒ **Apply** ⇒ **Close** ⇒ **LMB** to deselect

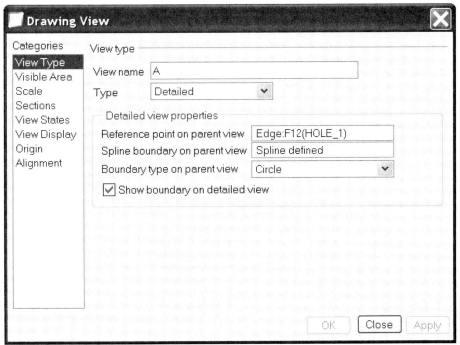

Figure 11.10(c) Drawing View Dialog Box

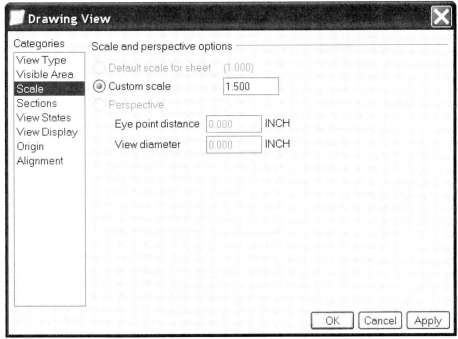

Figure 11.10(d) Custom Scale

Add an axis to the detail of the hole; click **View** from menu bar ⇒ **Show and Erase** ⇒ Show ⇒ only A.1 **Axis** active ⇒ Feature and View [Fig. 11.10(e)] ⇒ pick on the hole in the detail view ⇒ **OK** ⇒ **Show All** ⇒ **Yes** ⇒ **Accept All** [Fig. 11.10(f)] ⇒ **Close** [Fig. 11.10(g)] ⇒ **LMB**

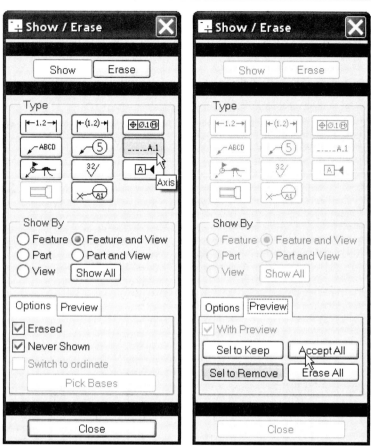

Figure 11.10(e) Show Axes   **Figure 11.10(f)** Accept All

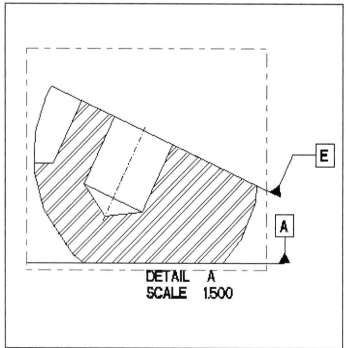

Figure 11.10(g) Axis Displayed

In DETAIL **A**, erase datum **A** and clip datum **E**. Also, clip the axis. ⇒ with the **Ctrl** key pressed, pick on the text items: **SECTION A-A**, **SEE DETAIL A** and **DETAIL A SCALE 1.500** [Fig. 11.11(a)] ⇒ **RMB** ⇒ **Text Style** ⇒ ☐ Default ⇒ Height **.375** [Fig. 11.11(b)] ⇒ **Apply** ⇒ **OK** [Fig. 11.11(c)] ⇒ **LMB**

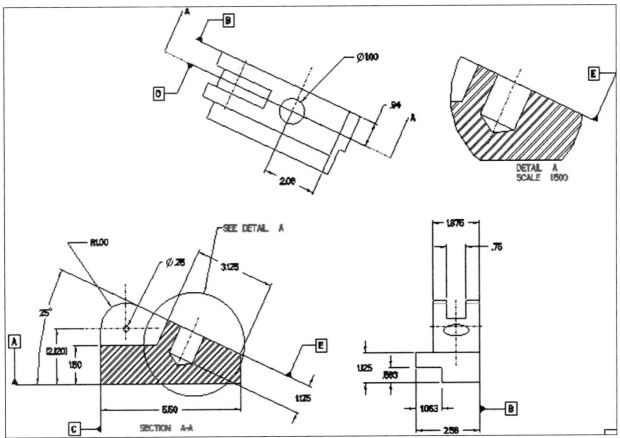

Figure 11.11(a) Select the Text Items and Change Their Height

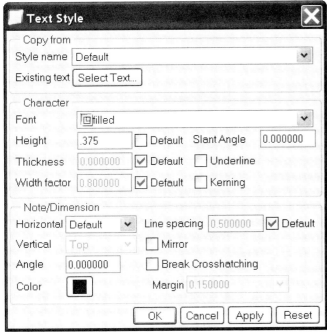

Figure 11.11(b) Text Style Dialog Box

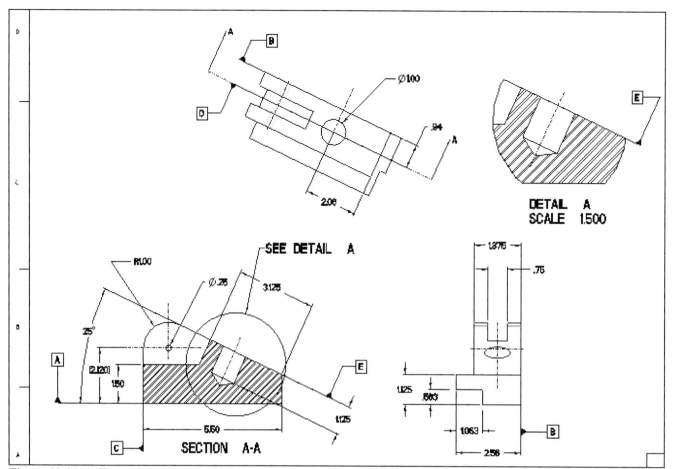

Figure 11.11(c) Changed Text Style (Height)

Change the height of the section identification lettering to **.375** [Fig. 11.11(d)] ⇒ pick on "A" ⇒ **RMB** ⇒ **Text Style** ⇒ ☐ Default ⇒ Height **.375** ⇒ **Apply** [Fig. 11.11(e)] ⇒ **OK** ⇒ **LMB**

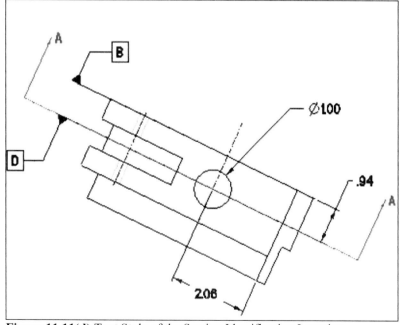

Figure 11.11(d) Text Style of the Section Identification Lettering

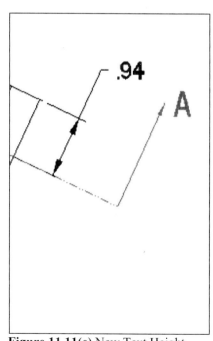

Figure 11.11(e) New Text Height

The Arrows for the cutting plane are now too small. These are controlled by the Drawing Options.

Click: **File** ⇒ **Properties** ⇒ **Drawing Options** ⇒ *crossec_arrow_length* *.375*, *crossec_arrow_width* *.125* [Fig. 11.12(a)] ⇒ **Enter** ⇒ **Apply** ⇒ **Close** ⇒ **MMB** ⇒ [Fig. 11.12(b)]

Figure 11.12(a) Cross Section Drawing Options

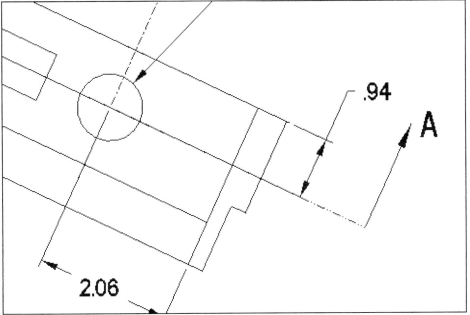

Figure 11.12(b) Section Arrows Length and Width Changed

Pick on the **1.125** dimension in the front section view [Fig. 11.13(a)] ⇒ **RMB** ⇒ **Move Item to View** ⇒ pick on view **DETAIL A** [Fig. 11.13(b)] ⇒ reposition and clip as needed [Fig. 11.13(c)] ⇒ **LMB**

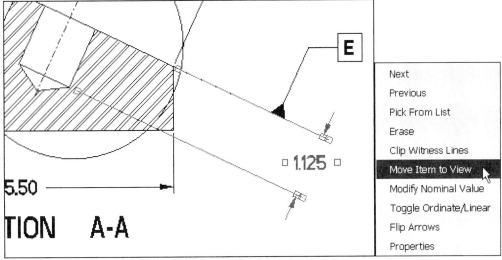

Figure 11.13(a) Pick on the **1.125** Hole Depth Dimension

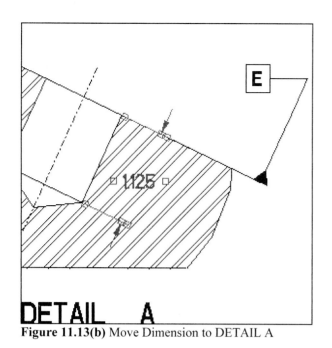
Figure 11.13(b) Move Dimension to DETAIL A

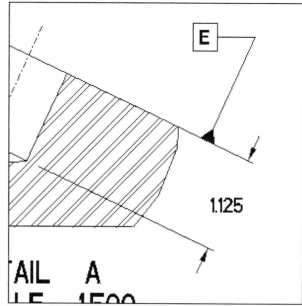
Figure 11.13(c) Reposition and Clip

Add another reference (horizontal) dimension to the small hole in the front view ⇒ keeping in mind ASME standards, cleanup the drawing as needed ⇒ **MMB**⇒ [icon] ⇒ [icon] ⇒ [icon] ⇒ [icon] ⇒ **MMB** ⇒ **File** ⇒ **Delete** ⇒ **Old Versions** ⇒ **MMB** (Fig. 11.14) ⇒ **LMB** to deselect

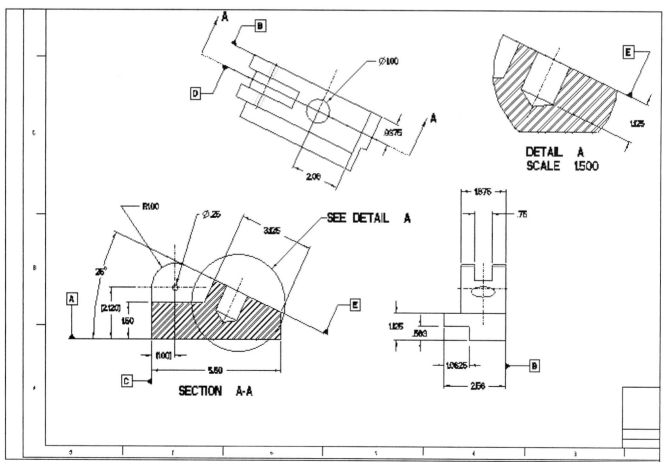
Figure 11.14 Anchor Drawing

Next, you will change the boundary of DETAIL A.

Pick on view **DETAIL A** ⇒ **RMB** (if you RMB inside the view outline, you get a pop-up list of options [Fig. 11.15(a)], whereas if you RMB outside the view outline, there are fewer options) [Fig. 11.15(b)] ⇒ **Properties** ⇒ click inside Spline area 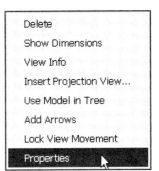 [Fig. 11.15(c)] ⇒ sketch the spline again in the front (SECTION A-A) view [Fig. 11.15(d)] ⇒ **MMB** to end spline ⇒ **Apply** [Fig. 11.15(e)] ⇒ **Close** ⇒ **LMB** [Fig. 11.15(f)] ⇒ 🔁 ⇒ 🔍 ⇒ 🗔 ⇒ 💾 ⇒ **MMB**

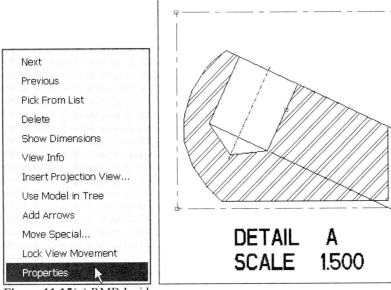

Figure 11.15(a) RMB Inside **Figure 11.15(b)** RMB Outside

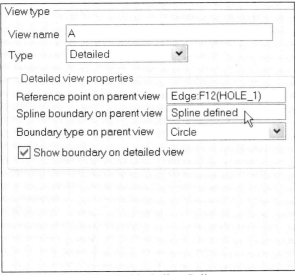

Figure 11.15(c) Pick inside Spline Collector

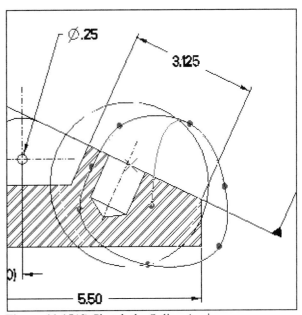

Figure 11.15(d) Sketch the Spline Again

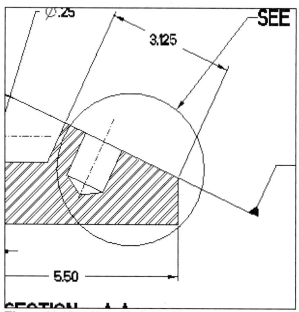

Figure 11.15(e) New Circle Position

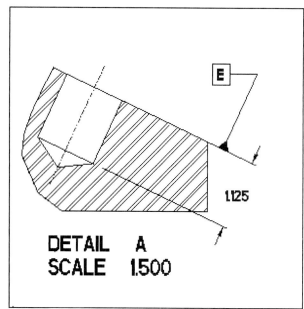

Figure 11.15(f) Updated DETAIL A

Add a dimension for the full angle of the hole's drill tip in **DETAIL A** ⇒ **Insert** ⇒ **Dimension** ⇒ **New References** ⇒ pick both edge lines of the drill tip [Fig. 11.16(a)] ⇒ **MMB** to place the dimension [Fig. 11.16(b)] ⇒ **MMB** ⇒ reposition and clip the dimension ⇒ pick on the **118.0** dimension [Fig. 11.16(c)] ⇒ **RMB** ⇒ **Properties** ⇒ change decimal places to **0** ⇒ **Enter** ⇒ **OK** [Figs. 11.16(d-e)]

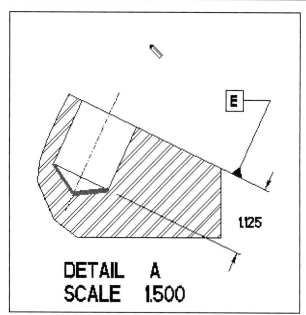

Figure 11.16(a) Drill Tip Edges

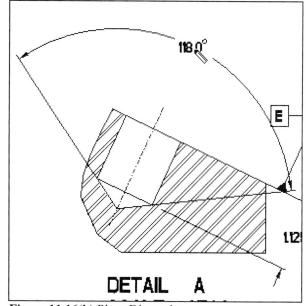

Figure 11.16(b) Place Dimension

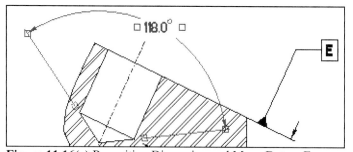

Figure 11.16(c) Reposition Dimension and Move Datum E

Figure 11.16(d) Change Dimension Properties of the **118** Degree Dimension

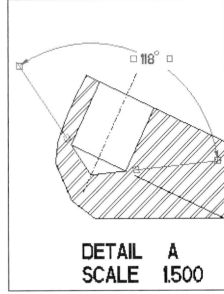

Figure 11.16(e) 118 Degrees

Pick on the **118.0** dimension (or **59** degree dimension) in the front view ⇒ **RMB** ⇒ **Erase** ⇒ **LMB** ⇒ Change the spacing and the angle of the section lining. Pick on the hatching in SECTION A-A ⇒ **RMB** ⇒ **Properties** [Fig. 11.17(a)] ⇒ **Delete Line** (if necessary) ⇒ **Angle** ⇒ **30** ⇒ **MMB** [Fig. 11.17(b)] ⇒ **LMB** ⇒ 🔄 ⇒ 🔍 ⇒ **View** ⇒ **Repaint** ⇒ 💾 ⇒ **MMB**

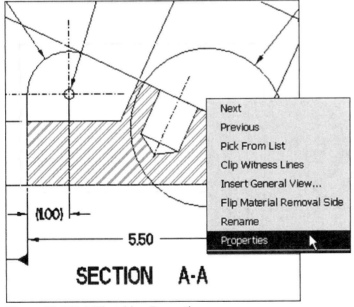

Figure 11.17(a) Xhatching Properties

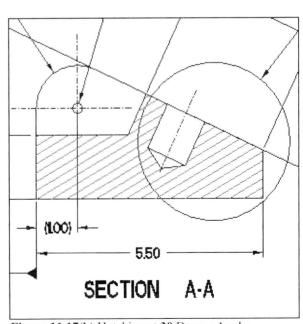

Figure 11.17(b) Hatching at **30** Degree Angle

Change the spacing and the angle of the section lining in **DETAIL A**. Pick on the hatching in **DETAIL A** ⇒ **RMB** ⇒ **Properties** [Fig. 11.18(a)] ⇒ **Det Indep** ⇒ **Hatch** ⇒ **Spacing** ⇒ **Double** ⇒ **Angle** ⇒ **45** ⇒ **MMB** [Fig. 11.18(b)] ⇒ **LMB** ⇒ **RMB** ⇒ **Update Sheet** ⇒ 🔍 ⇒ 🗔 ⇒ 💾 ⇒ **MMB**

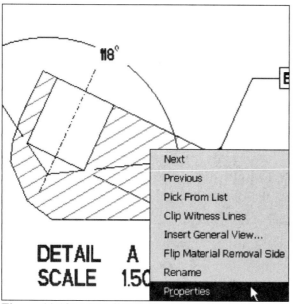

Figure 11.18(a) Detail Hatching Properties

Figure 11.18(b) Hatching at 45 Degree Angle

Change the text style used on the drawing. Click: ▸ ⇒ enclose all of the drawing text with a selection box [Fig. 11.19(a)] ⇒ **RMB** ⇒ **Text Style**

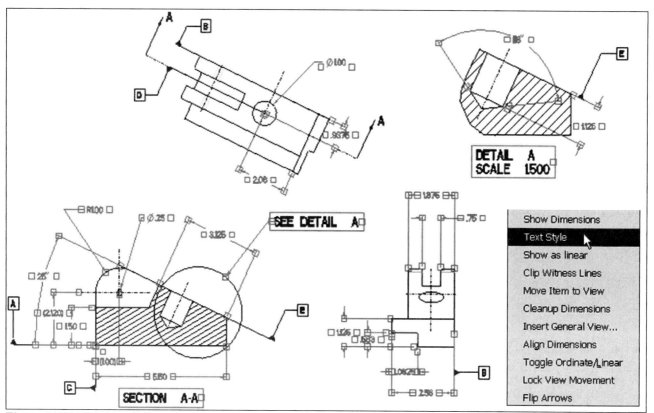

Figure 11.19(a) Select all Text

Font **Blueprint MT** [Figs. 11.19(b-c)] ⇒ **Apply** ⇒ **OK** ⇒ **LMB** [Fig. 11.19(d)] ⇒ 🖫 ⇒ **MMB**

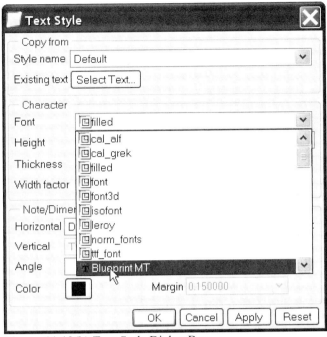

Figure 11.19(b) Text Style Dialog Box

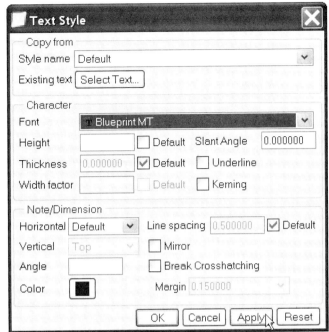

Figure 11.19(c) Character- Font- Blueprint MT

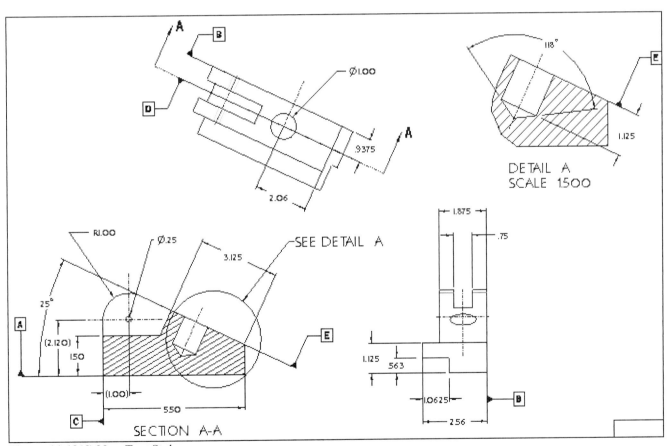

Figure 11.19(d) New Text Style

441

Add a geometric tolerance to the angled surface, click: **Create geometric tolerances** (*or Insert* ⇒ *Geometric Tolerance*) ⇒ Angularity ⇒ Type **Datum** [Fig. 11.20(a)] ⇒ Select Entity... ⇒ pick on datum **E** [Fig. 11.20(b)] ⇒ **Datum Refs** tab ⇒ **Primary** tab- Basic ⇒ pick on datum **A** in the front view as the Primary Reference Datum [Fig. 11.20(c)] ⇒ **Tol Value** tab ⇒ ☑ Overall Tolerance 0.005 [Fig. 11.20(d)] ⇒ **OK** ⇒ **LMB** ⇒ **RMB** ⇒ **Update Sheet** [Fig. 11.20(e)]

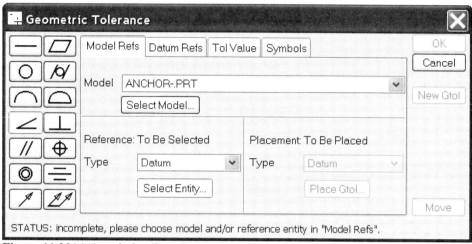

Figure 11.20(a) Angularity, Type Datum

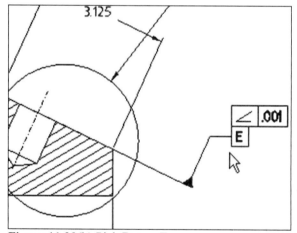

Figure 11.20(b) Pick Datum E

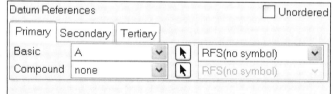

Figure 11.20(c) Primary Reference Datum A

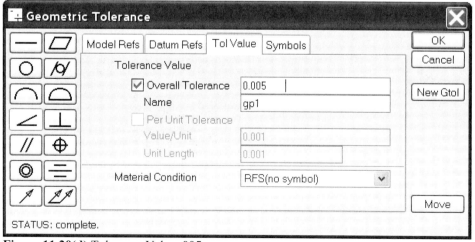

Figure 11.20(d) Tolerance Value **.005**

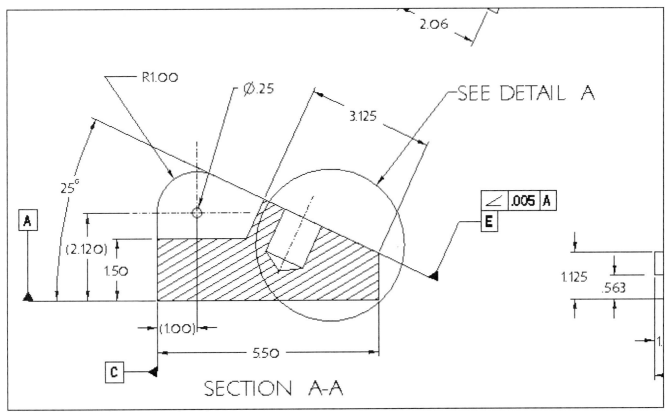

Figure 11.20(e) Added Geometric Tolerance

Create a *second sheet* with an isometric view of the Anchor by clicking: **Tools ⇒ Environment ⇒** `Standard Orient Isometric` **⇒ Apply ⇒ OK ⇒ Insert ⇒ Sheet ⇒ RMB ⇒ Page Setup ⇒** `D` **⇒** `▾` **⇒ Browse** [Fig. 11.21(a)] **⇒ System Formats c.frm** [Fig. 11.21(b)] **⇒ Open ⇒ OK ⇒** 🔍 [Fig. 11.21(c)]

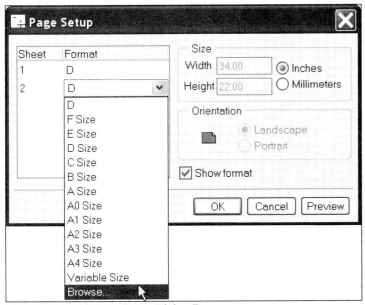

Figure 11.21(a) Page Setup Dialog Box

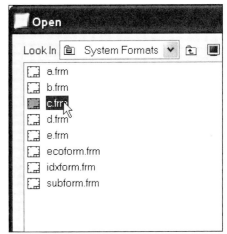

Figure 11.21(b) c.frm

Click: **Insert** ⇒ **Drawing View** ⇒ **General** ⇒ pick a center point for the view [Fig. 11.21(c)] ⇒ **Scale** ⇒ ◉ Custom scale 1.500 [Fig. 11.21(d)] ⇒ **Apply** ⇒ **Close** ⇒ **LMB** to deselect

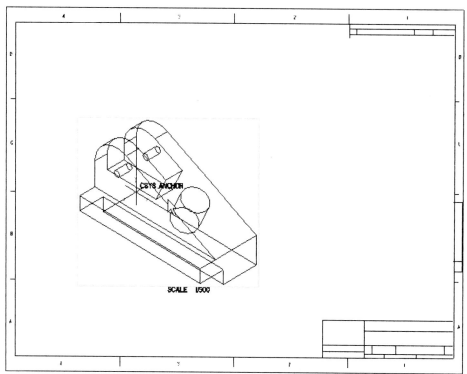

Figure 11.21(c) C Format and Isometric View

Figure 11.21(d) Drawing View Custom Scale

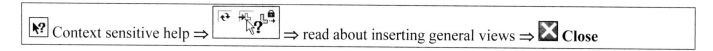

Pick on the pictorial view ⇒ **RMB** ⇒ **Properties** ⇒ **View Display** ⇒ Display style ⇒ **No Hidden** ⇒ Tangent edges display style ⇒ **Dimmed** ⇒ **Apply** ⇒ **Close** ⇒ **LMB** [Fig. 11.21(e)] ⇒ **File** ⇒ **Save** ⇒ **MMB**

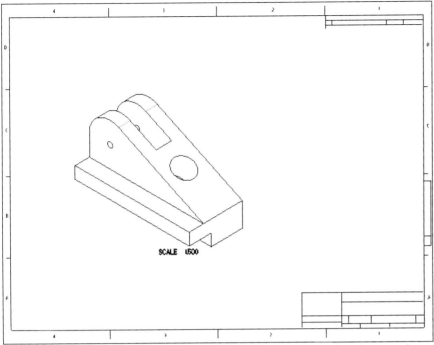

Figure 11.21(e) Sheet 2, Tangent Edges Dimmed and Hidden Lines Removed

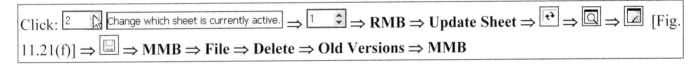

Click: 2 Change which sheet is currently active. ⇒ 1 ⇒ **RMB** ⇒ **Update Sheet** ⇒ [Fig. 11.21(f)] ⇒ ⇒ **MMB** ⇒ **File** ⇒ **Delete** ⇒ **Old Versions** ⇒ **MMB**

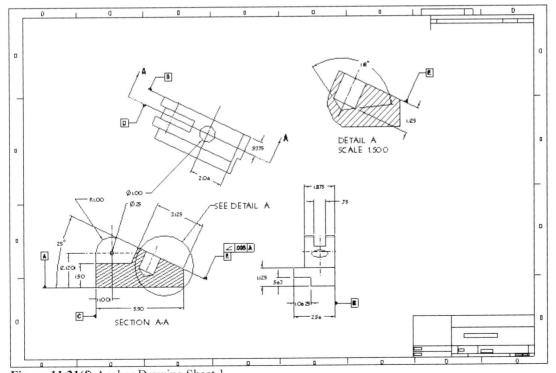

Figure 11.21(f) Anchor Drawing Sheet 1

Click: 🗁 **Open an existing object** ⇒ 🔽 ⇒ **System Formats** ⇒ **d.frm** ⇒ **Open** ⇒ **File** ⇒ **Save a Copy** ⇒ type a unique name for your format: New Name **DETAIL_FORMAT_D** ⇒ **OK** ⇒ **Window** ⇒ **Close** ⇒ 🗁 ⇒ pick **detail_format_d.frm** ⇒ **Open** ⇒ **RMB** in the graphics window ⇒ **Properties** Options dialog box opens ⇒ *drawing_text_height* **.25** ⇒ **Add/Change** ⇒ *default_font* **filled** ⇒ **Add/Change** ⇒ *draw_arrow_style* **filled** ⇒ **Add/Change** ⇒ *draw_arrow_length* **.25** ⇒ **Add/Change** ⇒ *draw_arrow_width* **.08** ⇒ **Add/Change** (● option in Status column shows Default column value active, ✳ Status column shows pending value) [Fig. 11.22(a)] ⇒ **Apply** ⇒ **Close** ⇒ **Ctrl+S** ⇒ **MMB**

The **format** will have a **.dtl** associated with it, and the **drawing** will have a different **.dtl** file associated with it. *They are separate .dtl files.* When you activate a drawing and then add a format, the **.dtl** for the format controls the font, etc. for the format only.

Active Drawing				
These options control text not subject to oth				
drawing_text_height	.25	0.156250	✳	Sets default text height for all text in the drawing usi
text_thickness	0.000000 *	0.000000	●	Sets default text thickness for new text after regene
text_width_factor	0.800000 *	0.800000	●	Sets default ratio between the text width and text he
draft_scale	1.000000 *	1.000000	●	Determines value of draft dimensions relative to actu
These options control text and line fonts				
default_font	filled	font	✳	Specifies a font index that determines default text fo
These options control leaders				
draw_arrow_length	.25	0.187500	✳	Sets length of leader line arrows.
draw_arrow_style	filled	closed	✳	Controls style of arrow head for all detail items invol
draw_arrow_width	.08	0.062500	✳	Sets width of leader line arrows. Drives these: "draw

Figure 11.22(a) Drawing Option **.dtl** File for Format

Zoom into the title block region, and create notes for the title text and parameter text, click: **Tools** ⇒ **Environment** ⇒ ☑ **Snap to Grid** ⇒ **Apply** ⇒ **OK** ⇒ **View** ⇒ **Draft Grid** ⇒ **Show Grid** ⇒ **Grid Params** ⇒ **X&Y Spacing** ⇒ type **.1** ⇒ **MMB** ⇒ **MMB** ⇒ **MMB** ⇒ **Create a note** [Fig. 11.22(b)] ⇒

Make Note ⇒ pick a point for the note in the largest area of the title block ⇒ type **TOOL ENGINEERING CO.** ⇒ **Enter** ⇒ **Enter** ⇒

Make Note ⇒ ⇒ type **DRAWN** ⇒ **Enter** ⇒ **Enter** ⇒

Make Note ⇒ ⇒ type **ISSUED** ⇒ **Enter** ⇒ **Enter** ⇒

Make Note ⇒ ⇒ type **&dwg_name** ⇒ **Enter** ⇒ **Enter** ⇒

Make Note ⇒ ⇒ type **&scale** ⇒ **Enter** ⇒ **Enter** ⇒

Make Note ⇒ ⇒ type **SHEET &current_sheet OF &total_sheets** ⇒ **Enter** ⇒ **Enter** ⇒ **Done/Return** ⇒ **LMB** ⇒ **Tools** ⇒ **Environment** ⇒ ☐ **Snap to Grid** ⇒ **Apply** ⇒ **OK** ⇒ modify some of the text height (**.10**) and the placement of the notes as needed [Fig. 11.22(b)] ⇒ **View** ⇒ **Update** ⇒ **Current Sheet** ⇒ **View** ⇒ **Orientation** ⇒ **Refit** ⇒ **Ctrl+S** ⇒ **MMB** ⇒ **File** ⇒ **Close Window**

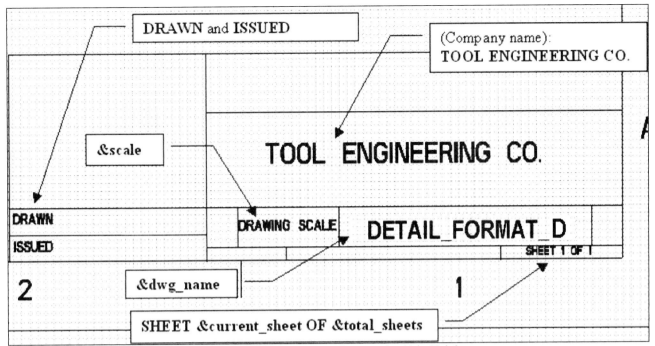

Figure 11.22(b) Parameters and Labels in the Title Block, Smaller Text is **.10** in Height

Click: **Window ⇒ Activate ⇒ RMB ⇒ Page Setup ⇒** [D ▼] ⇒ [▼] ⇒ **Browse ⇒** [📁] **Working Directory ⇒ detail_format_d.frm ⇒ Open** [DETAIL_FORMAT_D ▼] ⇒ **OK ⇒ Ctrl+S ⇒ MMB** [Fig. 11.22(c)]

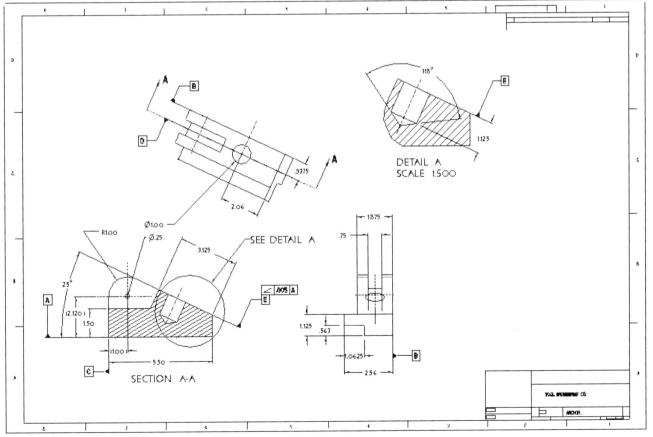

Figure 11.22(c) Completed Drawing

Click: **Info** ⇒ **Drawing View** ⇒ pick **Detail View A** from the screen (Fig. 11.23) ⇒ **Close** ⇒ **MMB** ⇒ **LMB** ⇒ **Ctrl+R** ⇒ **Window** ⇒ **Close**

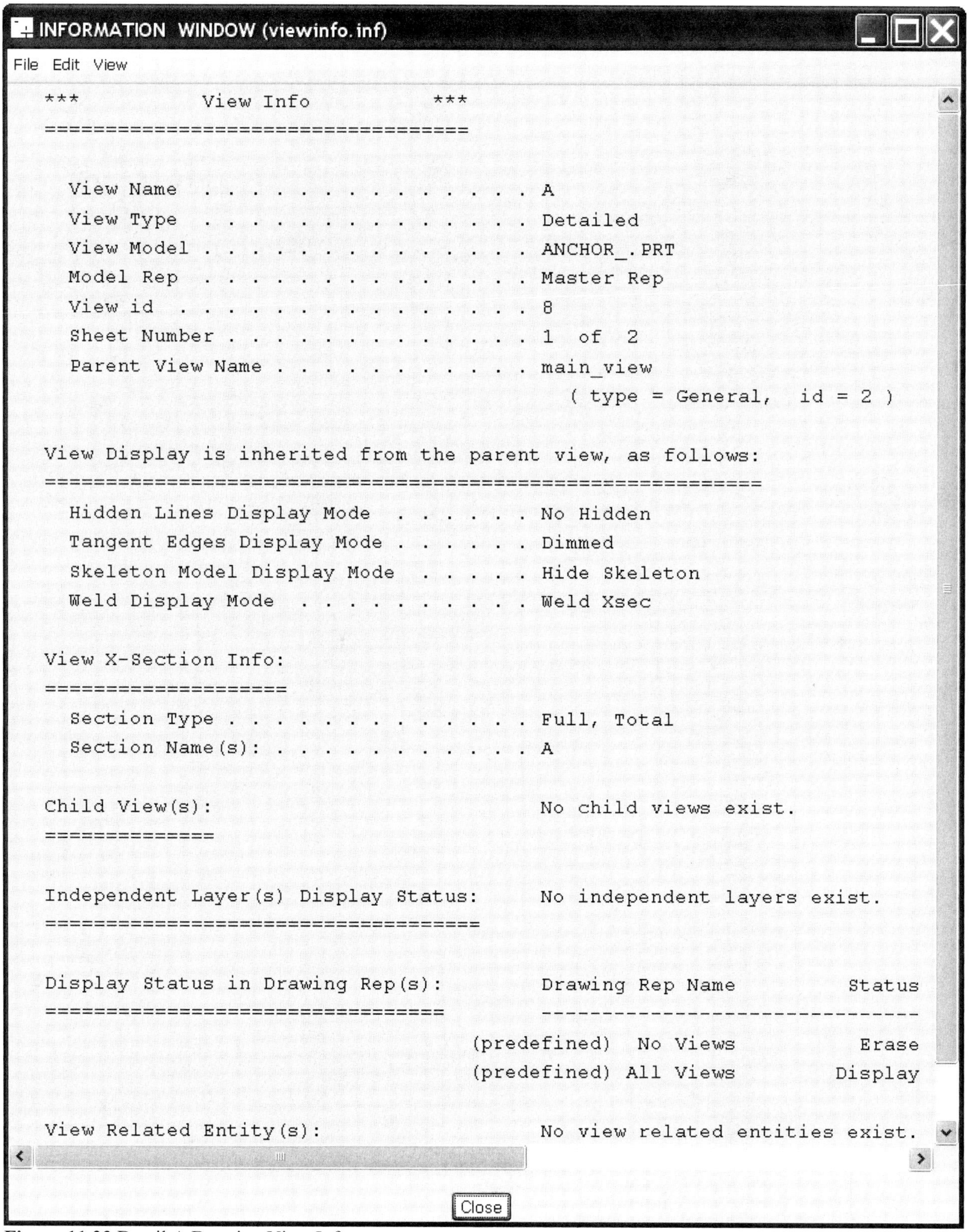

```
INFORMATION WINDOW (viewinfo.inf)
File  Edit  View

    ***          View Info           ***
    ==========================================

        View Name . . . . . . . . . . . . . . A
        View Type . . . . . . . . . . . . . . Detailed
        View Model . . . . . . . . . . . . .  ANCHOR_.PRT
        Model Rep . . . . . . . . . . . . . . Master Rep
        View id . . . . . . . . . . . . . . . 8
        Sheet Number . . . . . . . . . . . .  1 of 2
        Parent View Name . . . . . . . . . .  main_view
                                              ( type = General,  id = 2 )

    View Display is inherited from the parent view, as follows:
    ==============================================================
        Hidden Lines Display Mode . . . . . No Hidden
        Tangent Edges Display Mode . . . . .Dimmed
        Skeleton Model Display Mode . . . . Hide Skeleton
        Weld Display Mode . . . . . . . . . Weld Xsec

    View X-Section Info:
    ====================
        Section Type . . . . . . . . . . . . Full, Total
        Section Name(s): . . . . . . . . . . A

    Child View(s):                          No child views exist.
    ==============

    Independent Layer(s) Display Status:    No independent layers exist.
    ====================================

    Display Status in Drawing Rep(s):       Drawing Rep Name          Status
    =================================       -------------------------------
                                            (predefined)  No Views    Erase
                                            (predefined)  All Views   Display

    View Related Entity(s):                 No view related entities exist.

                              [ Close ]
```

Figure 11.23 Detail A Drawing View Info

The lesson is now complete. If you wish to detail a project without instructions, a complete set of projects and illustrations are available at ***www.cad-resources.com*** ⇒ ***Downloads.***

Lesson 12 Assembly Drawings

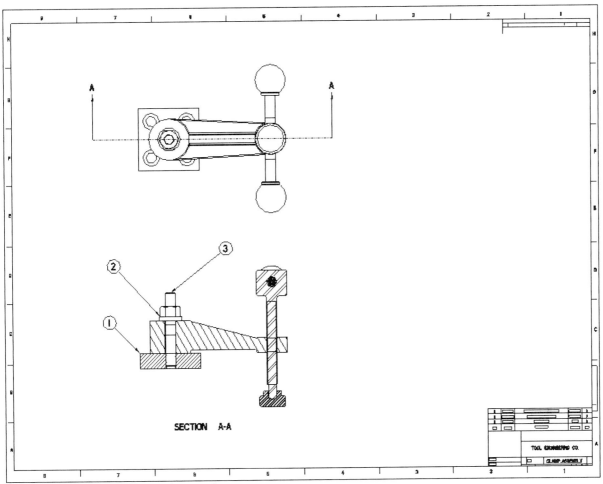

Figure 12.1(a) Clamp_Assembly Drawing

OBJECTIVES

- Create an **Assembly Drawing**
- Generate a **Parts List** from a **Bill of Materials (BOM)**
- **Balloon** an assembly drawing
- Create a **section assembly view** and change **component visibility**
- Add **Parameters** to parts
- Create a **Table** to generate a parts list automatically
- Use **Multiple Sheets**
- Make assembly **Drawing Sheets** with **multiple models**

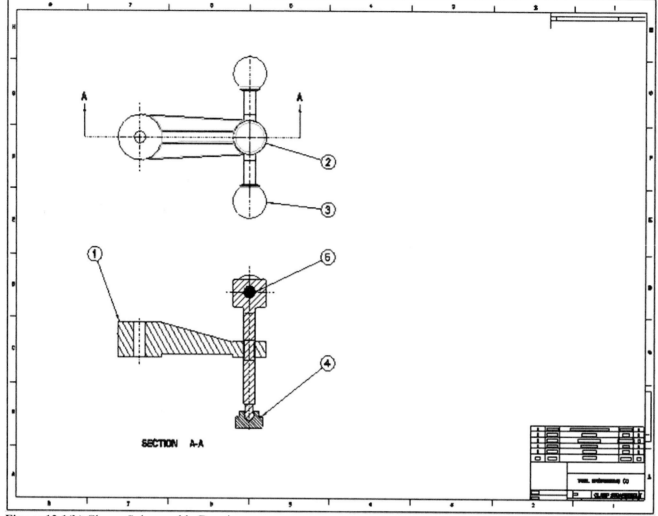

Figure 12.1(b) Clamp_Subassembly Drawing

ASSEMBLY DRAWINGS

Pro/E incorporates a great deal of functionality into drawings of assemblies [Figs. 12.1(a-b)]. You can assign parameters to parts in the assembly that can be displayed on a *parts list* in an assembly drawing. Pro/E can also generate the item balloons for each component on standard orthographic views or on an exploded view.

In addition, a variety of specialized capabilities allow you to alter the manner in which individual components are displayed in views and in sections. The format for an assembly is usually different from the format used for detail drawings. The most significant difference is the presence of a *Parts List*.

As part of this lesson, you will create a set of assembly formats and place your standard parts list in them. A parts list is actually a *Drawing Table object* that is formatted to represent a bill of materials in a drawing. By defining *parameters* in the parts in your assembly that agree with the specific format of the parts list, you make it possible for Pro/E to add pertinent data to the assembly drawing's parts list automatically as components are added to the assembly. After the parts list and parameters have been added, Pro/E can balloon the assembly drawing automatically.

| Lesson 12 STEPS |

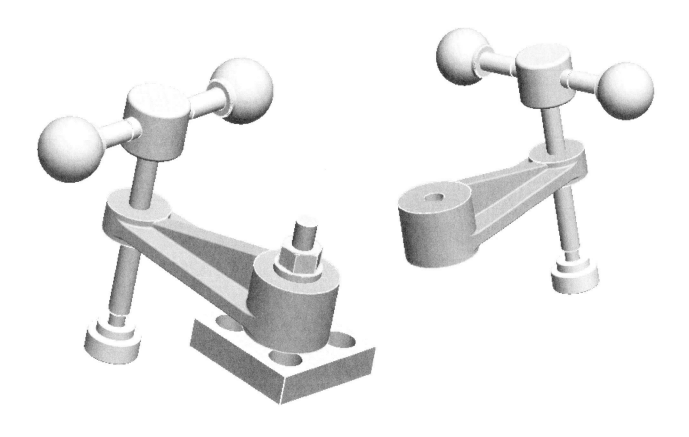

Figure 12.2 Swing Clamp Assembly and Subassembly

Swing Clamp Assembly Drawing

The format for an assembly drawing is usually different from the format used for detail drawings. The most significant difference is the presence of a parts list. You will create a standard "E" size format and place a standard parts list on it. You should create a set of assembly formats on "B", "C", "D", and "F" size sheets at your convenience.

A parts list is actually a *Drawing Table object* that is formatted to represent a bill of material (BOM) in a drawing. By defining parameters in the parts in your assembly that agree with the specific format of the parts list, you make it possible for Pro/E to add pertinent data to the assembly drawing parts list automatically, as components are added to the assembly.

After you create an "E" size format sheet with a parts list table, you will create two new drawings (each with two views) using your new assembly format. The Swing Clamp *subassembly* [Fig. 12.2(inset)] will be used in the first drawing. The second drawing will use the Swing Clamp *assembly* (Fig. 12.2). Both drawings use the "E" size format created in the first section of this lesson. The format will have a parameter-driven title block and an integral parts list.

Click: **File** ⇒ **Set Working Directory** ⇒ select working directory ⇒ **OK** ⇒ [icon] **Open an existing object** ⇒ [▼] ⇒ [System Formats] ⇒ [e.frm] ⇒ **Open** ⇒ **File** ⇒ **Save a Copy** ⇒ type a unique name for your format: New Name **ASM_FORMAT_E** ⇒ **OK** ⇒ **Window** ⇒ **Close** ⇒ [icon] ⇒ pick **asm_format_e.frm** ⇒ **Open**

The **format** will have a **.dtl** associated with it, and the **drawing** will have a different **.dtl** file associated with it. *They are separate .dtl files.* When you activate a drawing and then add a format, the **.dtl** for the format controls the font, etc. for the format only. The drawing **.dtl** file that controls items on the drawing needs to be established separately in the Drawing mode.

Click: **RMB** in the graphics window ⇒ **Properties** Options dialog box opens ⇒ *drawing_text_height .25* ⇒ Add/Change ⇒ *default_font filled* ⇒ Add/Change ⇒ *draw_arrow_style filled* ⇒ Add/Change ⇒ *draw_arrow_length .25* ⇒ Add/Change ⇒ *draw_arrow_width .08* ⇒ Add/Change (Fig. 12.3) ⇒ **Apply** ([●] option in Status column shows Default column value active, [✳] option in Status column shows pending value for this drawing) ⇒ **Close** ⇒ [icon] ⇒ [icon] ⇒ **MMB**

	Value	Default	Status	Description
Active Drawing				
These options control text not subject to oth				
drawing_text_height	.25	0.156250	✳	Sets default text height for all text in the drawing usin
text_thickness	0.000000 *	0.000000	●	Sets default text thickness for new text after regener
text_width_factor	0.800000 *	0.800000	●	Sets default ratio between the text width and text hei
draft_scale	1.000000 *	1.000000	●	Determines value of draft dimensions relative to actu
These options control text and line fonts				
default_font	filled	font	✳	Specifies a font index that determines default text for
These options control leaders				
draw_arrow_length	.25	0.187500	✳	Sets length of leader line arrows.
draw_arrow_style	filled	closed	✳	Controls style of arrow head for all detail items involvi
draw_arrow_width	.08	0.062500	✳	Sets width of leader line arrows. Drives these: "draw_
draw_attach_sym_height	DEFAULT *	default	●	Sets height of leader line slashes, integral signs, and
draw_attach_sym_width	DEFAULT *	default	●	Sets width of leader line slashes, integral signs, and b
draw_dot_diameter	DEFAULT *	default	●	Sets diameter of leader line dots. If set to "default," u
leader_elbow_length	0.250000 *	0.250000	●	Determines length of leader elbow (the horizontal leg
sort_method_in_region	delimited		●	Determines repeat regions sort mechanism. String_or
Miscellaneous options				
draft_scale	1.000000 *	1.000000	●	Determines value of draft dimensions relative to actu

Figure 12.3 New Drawing Options

Click: [?] **Context sensitive help** ⇒ [icon] click on Insert a Drawing Table icon ⇒ **To create a Drawing Table** ⇒ read ⇒ **About Drawing Tables** ⇒ read ⇒ [X] **Close** ⇒ [icon] close the Navigation area

Zoom into the title block region, and create notes for the title text and parameter text required to display the proper information.

Click: **Tools** ⇒ **Environment** ⇒ ☑ Snap to Grid ⇒ **Apply** ⇒ **OK** ⇒ **View** ⇒ **Draft Grid** ⇒ **Show Grid** ⇒ **Grid Params** ⇒ **X&Y Spacing** ⇒ type **.1** ⇒ **MMB** ⇒ **MMB** ⇒ **MMB** ⇒ Create a note ⇒ **Make Note** ⇒ pick point for the note in the largest area of the title block (Fig. 12.4) ⇒ type **TOOL ENGINEERING CO.** ⇒ **Enter** ⇒ **Enter** ⇒

Make Note ⇒ (Fig. 12.4) ⇒ type **DRAWN** ⇒ **Enter** ⇒ **Enter** ⇒

Make Note ⇒ (Fig. 12.4) ⇒ type **ISSUED** ⇒ **Enter** ⇒ **Enter** ⇒

Make Note ⇒ (Fig. 12.4) ⇒ type **&dwg_name** ⇒ **Enter** ⇒ **Enter** ⇒

Make Note ⇒ (Fig. 12.4) ⇒ type **&scale** ⇒ **Enter** ⇒ **Enter** ⇒

Make Note ⇒ (Fig. 12.4) ⇒ type **SHEET &current_sheet OF &total_sheets** ⇒ **Enter** ⇒ **Enter** ⇒ **Done/Return** ⇒ **LMB** ⇒ **Tools** ⇒ **Environment** ⇒ ☐ Snap to Grid ⇒ **Apply** ⇒ **OK** ⇒ modify the text height (**.10**) and the placement of the notes so that they are positioned correctly (Fig. 12.4) ⇒ ⇒ ⇒ ⇒ ⇒ **MMB** ⇒ **File** ⇒ **Delete** ⇒ **Old Versions** ⇒ **MMB**

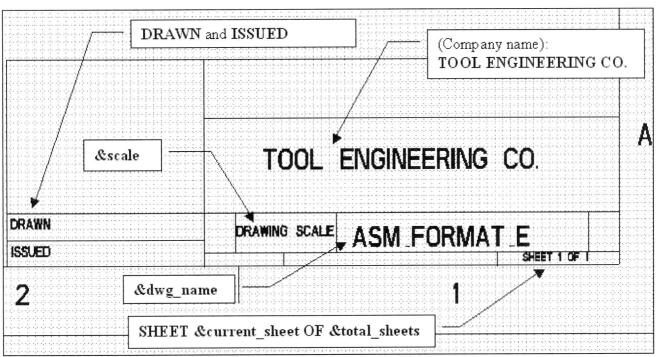

Figure 12.4 Parameters and Labels in the Title Block, Smaller Text is **.10** in Height

The parts list table can now be created and saved with this format. You can add and replace formats and still keep the table associated with the drawing. Start the parts list by creating a table.

Click: **View** ⇒ **Draft Grid** ⇒ **Hide Grid** ⇒ **MMB** ⇒ 🔲 **Insert a table by specifying the columns and rows sizes** ⇒ **Ascending** ⇒ **Rightward** ⇒ **By Length** ⇒ pick a point at the upper left-hand corner of the title block [Fig. 12.5(a)] ⇒

Enter the width of the first column in drawing units (INCH) [Quit] 1.00 ⇒ **MMB** ⇒

Enter the width of the next column in drawing units (INCH) [Done] 1.00 ⇒ **MMB** ⇒

Enter the width of the next column in drawing units (INCH) [Done] 4.00 ⇒ **MMB** ⇒

Enter the width of the next column in drawing units (INCH) [Done] 1.00 ⇒ **MMB** ⇒

Enter the width of the next column in drawing units (INCH) [Done] .75 ⇒ **MMB** ⇒ **MMB** ⇒

Enter the height of the first row in drawing units(INCH) [Quit] .50 ⇒ **MMB** ⇒

Enter the height of the next row in drawing units (INCH) [Done] .375 ⇒ **MMB** ⇒ **MMB** [Fig. 12.5(b)] ⇒ **LMB**

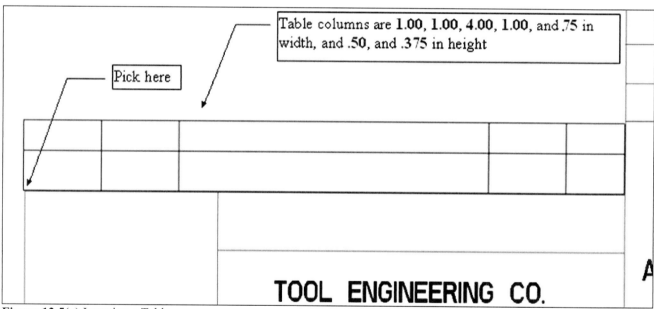

Figure 12.5(a) Inserting a Table

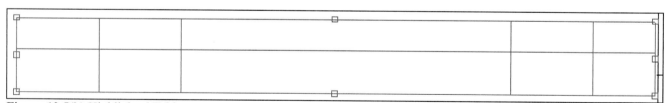

Figure 12.5(b) Highlighted Table

Click: **Table** from menu bar ⇒ **Repeat Region** ⇒ **Add** ⇒ pick in the left block [Fig. 12.5(c)] ⇒ pick in the right block [Fig. 12.5(c)] ⇒ **Attributes** ⇒ select the Repeat Region just created ⇒ **No Duplicates** ⇒ **Recursive** ⇒ **MMB** ⇒ **MMB** ⇒ **MMB** ⇒ 🗘 **Update the display of all views in the active sheet**

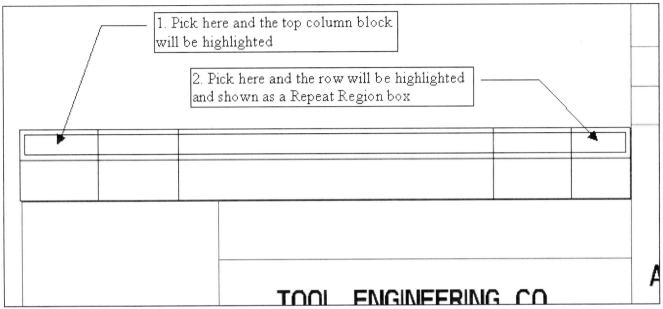

Figure 12.5(c) Repeat Region

Insert column headings using plain text: double-click on the first block [Fig. 12.6(a)] ⇒ type **ITEM** [Fig. 12.6(b)] ⇒ **OK**

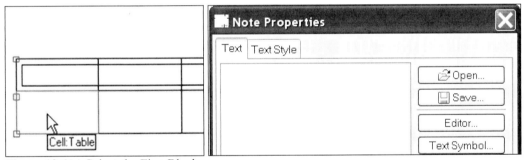

Figure 12.6(a) Select the First Block

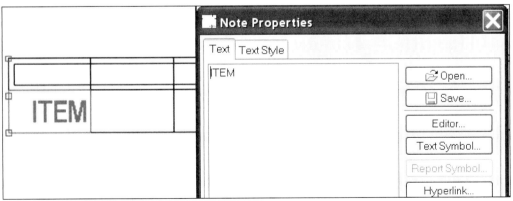

Figure 12.6(b) ITEM

The Repeat Region now needs to have some of its headings correspond to the parameters that will be created for each component model.

Double-click on the second block ⇒ type **PT NUM** [Fig. 12.6(c)] ⇒ **OK** ⇒ double-click on the third block ⇒ type **DESCRIPTION** ⇒ **OK** ⇒ double-click on the fourth block ⇒ type **MATERIAL** ⇒ **OK** ⇒ double-click on the fifth block ⇒ type **QTY** ⇒ **OK** [Fig. 12.6(d)]

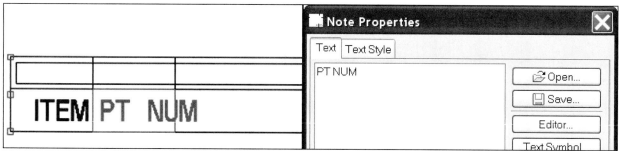

Figure 12.6(c) PT NUM

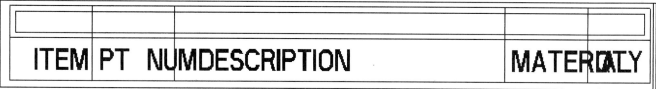

Figure 12.6(d) ITEM, PT MUM, DESCRIPTION, MATERIAL, QTY

Change the text height (**.125**), position (middle) and justification (centered) for all five titles.

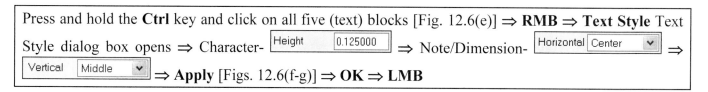

Figure 12.6(e) Select All Five Blocks

Figure 12.6(f) New Text Style

456

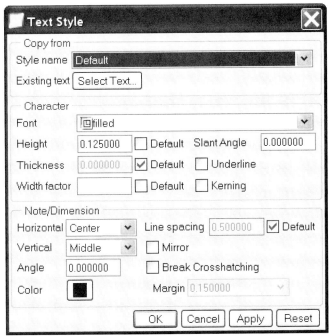

Figure 12.6(g) Text Style Dialog Box

Insert parametric text into each appropriate repeat region block.

Double-click on the first table cell of the Repeat Region [Fig. 12.7(a)] ⇒ click **rpt...** from the Report Symbol dialog box ⇒ click **index** [Figs. 12.7(b-c)]

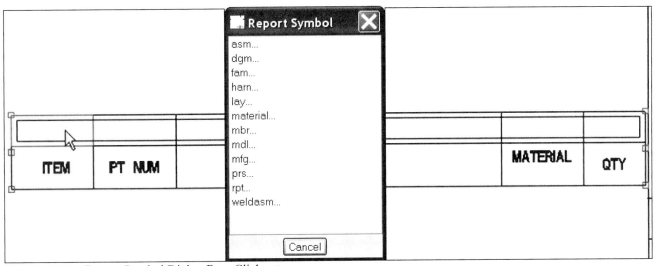

Figure 12.7(a) Report Symbol Dialog Box, Click rpt...

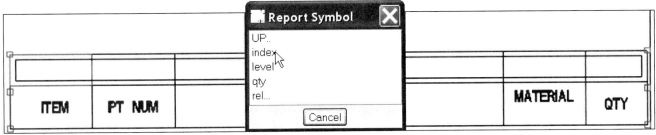

Figure 12.7(b) Click index

Figure 12.7(c) &rpt.index

Double-click on the fifth table cell of Repeat region ⇒ click **rpt...** [Fig. 12.7(d)] ⇒ **qty** [Figs. 12.7(e-f)]

Figure 12.7(d) Pick rpt...

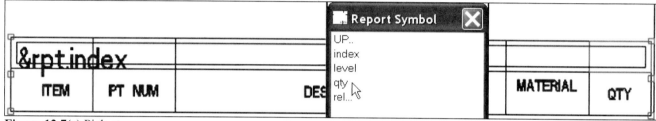

Figure 12.7(e) Pick qty

Figure 12.7(f) &rpt.qty

Double-click on the third (middle) table cell of the Repeat Region ⇒ **asm...** [Fig. 12.7(g)] ⇒ **mbr...** [Fig. 12.7(h)] ⇒ **User Defined** [Fig. 12.7(i)] ⇒ Enter symbol text: type **DSC** ⇒ **MMB** [Fig. 12.7(j)]

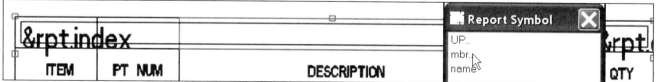

Figure 12.7(g) Pick asm...

Figure 12.7(h) Pick mbr...

Figure 12.7(i) Pick User Defined

Figure 12.7(j) &asm.mbr.DSC

Double-click on the fourth table cell of the Repeat Region ⇒ **asm…** ⇒ **mbr…** ⇒ **User Defined** ⇒ Enter symbol text: type **MAT** ⇒ **MMB** [Fig. 12.7(k)] ⇒ double-click on the second table cell of the Repeat Region ⇒ **asm…** ⇒ **mbr…** ⇒ **User Defined** ⇒ Enter symbol text: type **PRTNO** ⇒ **MMB** [Fig. 12.7(l)] ⇒ **LMB**

Figure 12.7(k) &asm.mbr.MAT

Figure 12.7(l) &asm.mbr.PRTNO

Change the text height of the Report Symbols in the table cells of the Repeat Region to **.125**.

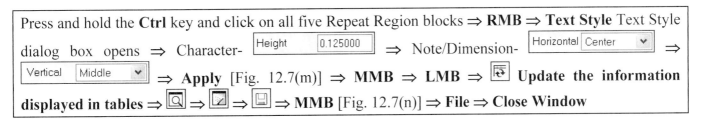

Press and hold the **Ctrl** key and click on all five Repeat Region blocks ⇒ **RMB** ⇒ **Text Style** Text Style dialog box opens ⇒ Character- Height 0.125000 ⇒ Note/Dimension- Horizontal Center ⇒ Vertical Middle ⇒ **Apply** [Fig. 12.7(m)] ⇒ **MMB** ⇒ **LMB** ⇒ Update the information displayed in tables ⇒ ⇒ ⇒ ⇒ **MMB** [Fig. 12.7(n)] ⇒ **File** ⇒ **Close Window**

Figure 12.7(m) Completed BOM

&rpt.index &asmmbr.PRTNO	&asmmbr.DSC	&asmmbr.MAT	&rpt.qty
ITEM PT NUM	DESCRIPTION	MATERIAL	QTY

```
                    TOOL ENGINEERING CO.

DRAWN
                DRAWING SCALE   ASM_FORMAT_E
ISSUED
                                           SHEET 1 OF 1
     2                              1
```

Figure 12.7(n) Title Block and BOM Table

Adding Parts List Data

When you save your standard assembly format, the Drawing Table that represents your standard parts list is now included. You must be aware of the titles of the parameters under which the data is stored, so that you can add them properly to your parts.

As you add components to an assembly, Pro/E reads the parameters from them and updates the parts list. You can also see the same effect by adding these parameters after the drawing has been created.

Pro/E also creates Item Balloons on the first view that was placed on the drawing. To improve their appearance, you can move these balloons to other views and alter the locations where they attach.

Retrieve the clamp arm, click: **File** ⇒ **Open** ⇒ select **clamp_arm.prt** ⇒ **Open** ⇒ **Open** [Fig. 12.8(a)]

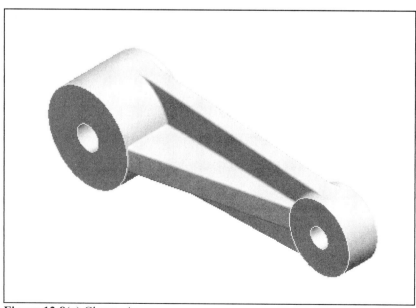

Figure 12.8(a) Clamp_Arm

Click: **Tools** from menu bar ⇒ **Parameters** [Fig. 12.8(b)] ⇒ **Parameters** [Fig. 12.8(c)] ⇒ **Add Parameter** from Parameters dialog box ⇒ click in Name field ⇒ type **PRTNO** [Figs. 12.8(d-e)]

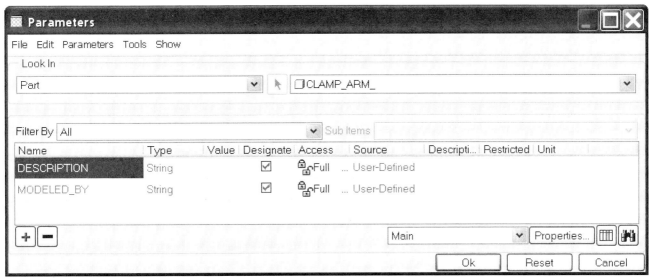

Figure 12.8(b) Parameters Dialog Box

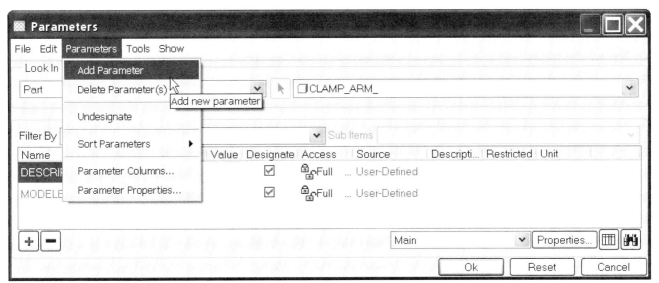

Figure 12.8(c) Add Parameter

Name	Type	Value	Designate	Access	Source	Descripti...	Restricted	Unit
DESCRIPTION	String		✓	Full	... User-Defined			
MODELED_BY	String		✓	Full	... User-Defined			
PARAMETER_1	Real Number	0.0	☐	Full	... User-Defined			

Figure 12.8(d) Adding Parameters

Name	Type	Value	Designate	Access	Source	Descripti...	Restricted	Unit
DESCRIPTION	String		✓	Full	... User-Defined			
MODELED_BY	String		✓	Full	... User-Defined			
PRTNO	Real Number	0.0	☐	Full	... User-Defined			

Figure 12.8(e) Name PRTNO

Click in Type field [Fig. 12.8(f)] ⇒ [Fig. 12.8(g)] ⇒ String ⇒ click in Value field ⇒ type **SW101-5AR** [Fig. 12.8(h)] ⇒ Designate ⇒ click in Description field ⇒ type **part number** [Fig. 12.8(i)]

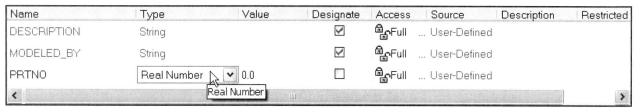

Figure 12.8(f) Click in Type Field- Real Number

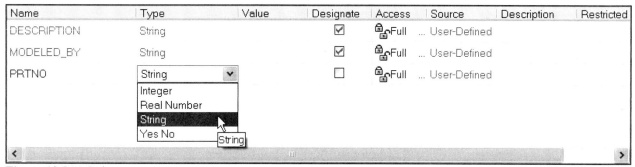

Figure 12.8(g) String

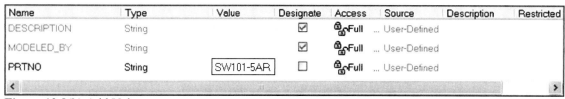

Figure 12.8(h) Add Value

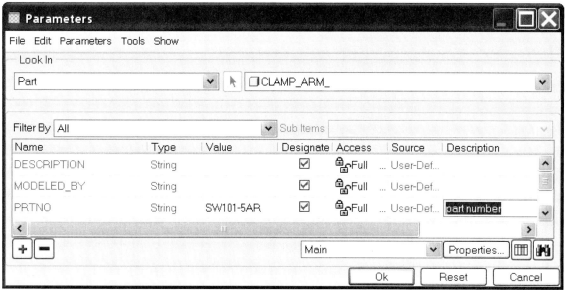

Figure 12.8(i) Add Description

Click: ⊕ **Add new Parameter** [Fig. 12.8(j)] ⇒ click in Name field ⇒ type **DSC** [Fig. 12.8(k)] ⇒ click in Type field ⇒ String ⇒ click in Value field ⇒ type **CLAMP ARM** ⇒ click in Description field ⇒ type **part name** [Fig. 12.8(l)] ⇒ click ⊕ **Add new Parameter** ⇒ click in Name field ⇒ type **MAT** ⇒ click in Type field String ⇒ click in Value field ⇒ type **STEEL** ⇒ click in Description field ⇒ type **material** [Fig. 12.8(m)] ⇒ Designate ☑ **DSC** and **MAT** ⇒ **Ok** ⇒ **File** ⇒ **Save** ⇒ **MMB**

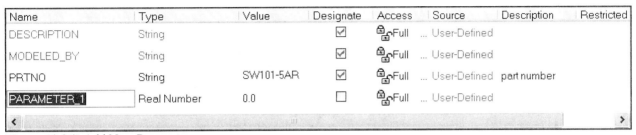

Figure 12.8(j) Add New Parameter

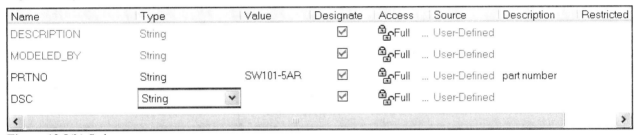

Figure 12.8(k) String

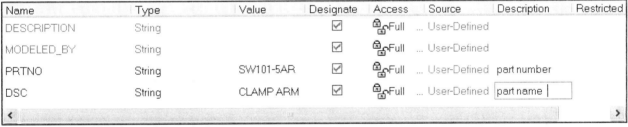

Figure 12.8(l) Add Value and Description

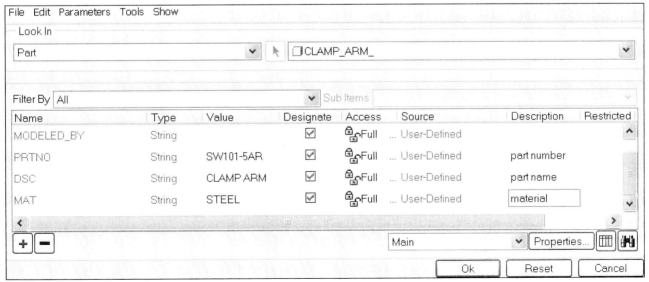

Figure 12.8(m) Completed Parameters

You can also access the parts parameters using the Relations dialog box. In the case of the Clamp_Arm, there was a relation created for controlling a features location.

Click: **Tools** ⇒ **Relations** ⇒ **Local Parameters** [Fig. 12.8(n)] ⇒ **Ok** ⇒ **File** ⇒ **Save** ⇒ **MMB** ⇒ **Window** ⇒ **Close**

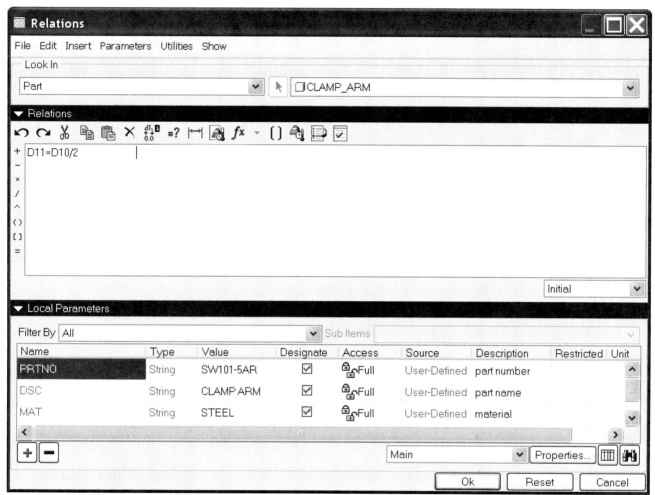

Figure 12.8(n) Relations Dialog Box

Retrieve the clamp swivel, click: **File** ⇒ **Open** ⇒ **clamp_swivel.prt** ⇒ **Open** [Fig. 12.9(a)] ⇒ **Tools** ⇒ **Parameters** ⇒ **Parameters** ⇒ **Add Parameter** ⇒ complete the parameters as shown [Fig. 12.9(b)] ⇒ **Ok** ⇒ **File** ⇒ **Save** ⇒ **MMB** ⇒ **File** ⇒ **Close Window**

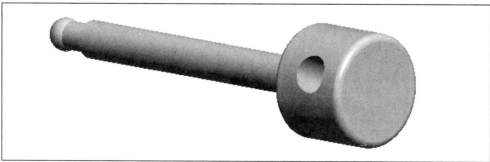

Figure 12.9(a) Clamp_Swivel

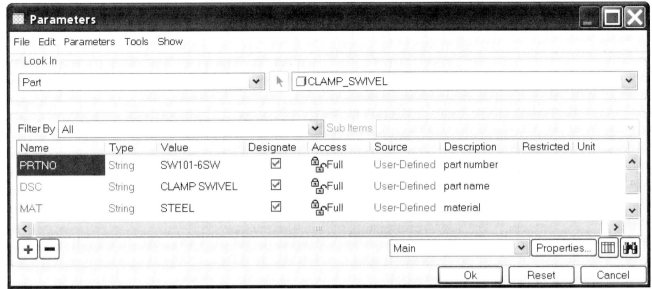

Figure 12.9(b) Clamp_Swivel Parameters: PRTNO SW101-6SW, DSC CLAMP SWIVEL, MAT STEEL

Retrieve the clamp ball, click: **File ⇒ Open ⇒ clamp_ball.prt ⇒ Open** [Fig. 12.10(a)] **⇒ Tools ⇒ Parameters ⇒ Parameters ⇒ Add Parameter ⇒** complete the parameters as shown [Fig. 12.10(b)] **⇒ Ok ⇒ File ⇒ Save ⇒ MMB ⇒ File ⇒ Close Window**

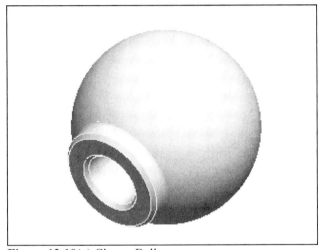

Figure 12.10(a) Clamp_Ball

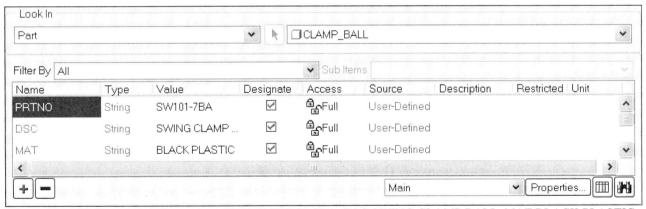

Figure 12.10(b) Clamp_Ball Parameters: PRTNO SW101-7BA, DSC SWING CLAMP BALL, MAT BLACK PLASTIC

Parameters can be added, deleted, and modified in Part Mode, Assembly Mode, or Drawing Mode. You can also add *parameter columns* to the Model Tree in Assembly Mode, and edit the parameter value. Use the following information to add parameters both to purchased components (standard parts) and to the remaining parts required for the subassembly and the assembly (Fig. 12.11). Use the Assembly Mode to input the information shown in Figures 12.12(a-e) to establish the part parameters.

Click: **File ⇒ Open ⇒ clamp_assembly.asm ⇒ Open**
(see commands on the following pages)

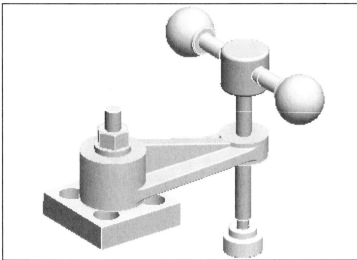

Figure 12.11 Clamp_Assembly

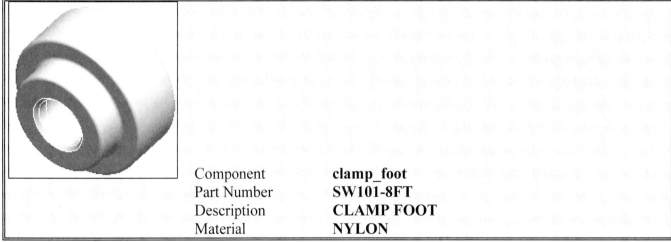

Component	**clamp_foot**
Part Number	**SW101-8FT**
Description	**CLAMP FOOT**
Material	**NYLON**

Figure 12.12(a) Clamp_Foot

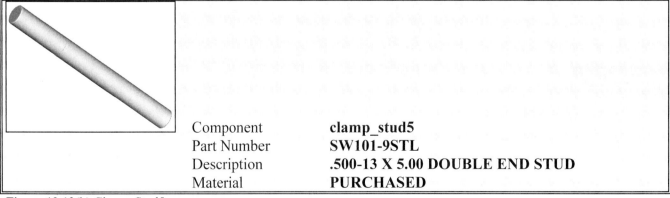

Component	**clamp_stud5**
Part Number	**SW101-9STL**
Description	**.500-13 X 5.00 DOUBLE END STUD**
Material	**PURCHASED**

Figure 12.12(b) Clamp_Stud5

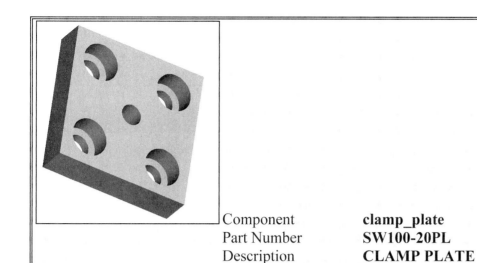

Component	**clamp_plate**
Part Number	**SW100-20PL**
Description	**CLAMP PLATE**
Material	**STEEL**

Figure 12.12(c) Clamp_Plate

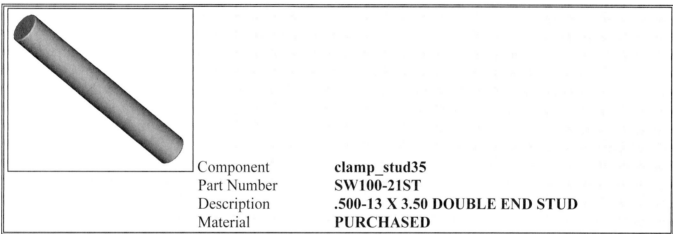

Component	**clamp_stud35**
Part Number	**SW100-21ST**
Description	**.500-13 X 3.50 DOUBLE END STUD**
Material	**PURCHASED**

Figure 12.12(d) Clamp_Stud35

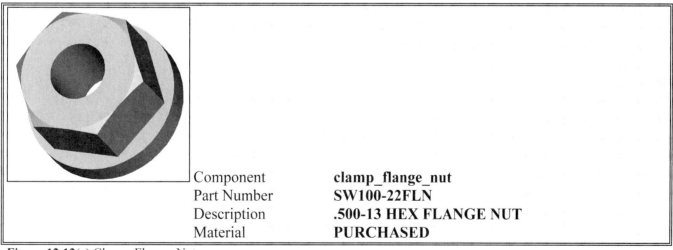

Component	**clamp_flange_nut**
Part Number	**SW100-22FLN**
Description	**.500-13 HEX FLANGE NUT**
Material	**PURCHASED**

Figure 12.12(e) Clamp_Flange_Nut

Click: **Settings** in the Model Tree ⇒ **Tree Columns** [Fig. 12.13(a)] ⇒ Type **Model Params** [Fig. 12.13(b)] ⇒ Name field, type **PRTNO** [Fig. 12.13(c)] ⇒ **Enter** ⇒ Name field, type **DSC** [Fig. 12.13(d)] ⇒ >> **Add column** [Fig. 12.13(e)] ⇒ Name field, type **MAT** ⇒ **Enter** [Fig. 12.13(f)] ⇒ **Apply** ⇒ **OK**

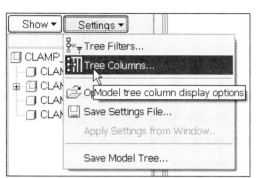

Figure 12.13(a) Tree Columns

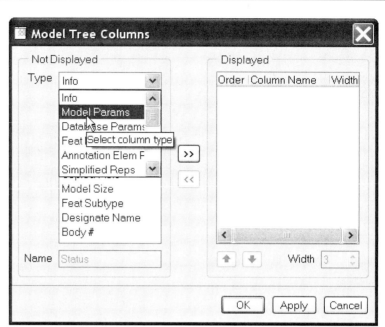

Figure 12.13(b) Model Params

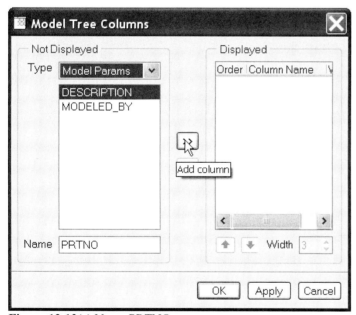

Figure 12.13(c) Name PRTNO

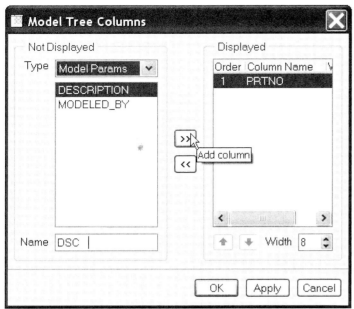

Figure 12.13(d) Name DSC

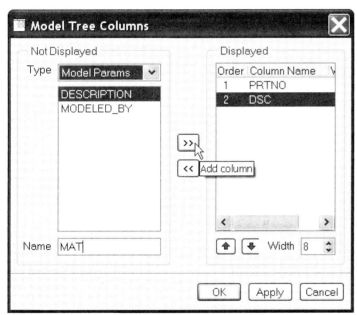

Figure 12.13(e) PRTNO and DSC

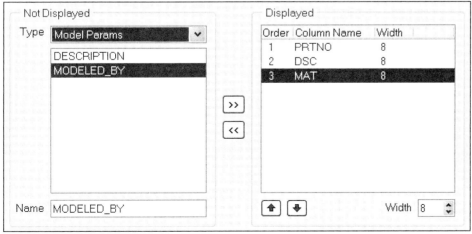

Figure 12.13(f) PRTNO, DSC and MAT

Resize the Model Tree and click on **CLAMP_PLATE.PRT** ⇒ click in **PRTNO** field [Fig. 12.14(a)] ⇒ ⬇ ⇒ **String** ⇒ in Value field, type **SW100-20PL** [Fig. 12.14(b)] ⇒ ☑Designated ⇒ in Description field, type **part number** ⇒ **Ok** [Fig. 12.14(c)] ⇒ click in **DSC** field ⇒ ⬇ ⇒ **String** ⇒ in Value field, type **CLAMP PLATE** [Fig. 12.14(d)] ⇒ ☑Designated ⇒ in Description field, type **part name** ⇒ **Ok** [Fig. 12.14(e)] ⇒ click in **MAT** field ⇒ ⬇ ⇒ **String** ⇒ in Value field, type **STEEL** [Fig. 12.14(f)] ⇒ ☑Designated ⇒ in Description field, type **material** ⇒ **Ok** ⇒ **File** ⇒ **Save** ⇒ **MMB**.

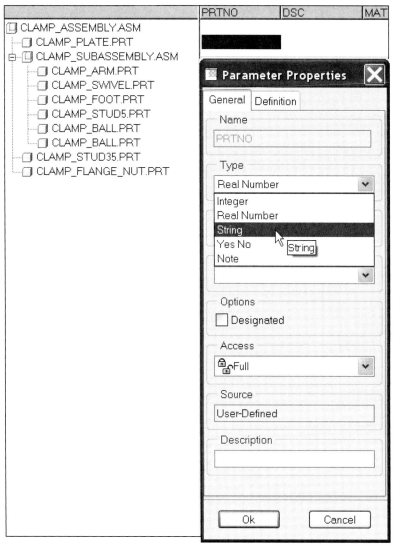

Figure 12.14(a) New Parameter String for PRTNO

Figure 12.14(b) Value SW100-20PL

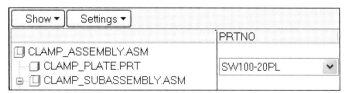

Figure 12.14(c) Part Number SW100-20PL

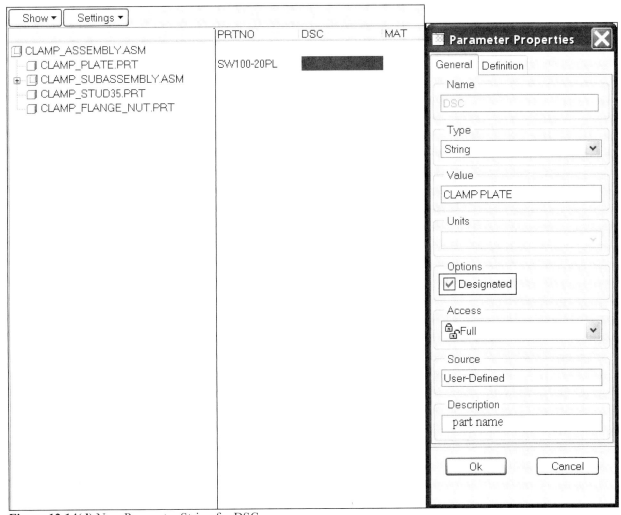

Figure 12.14(d) New Parameter String for DSC

Figure 12.14(e) Description CLAMP PLATE

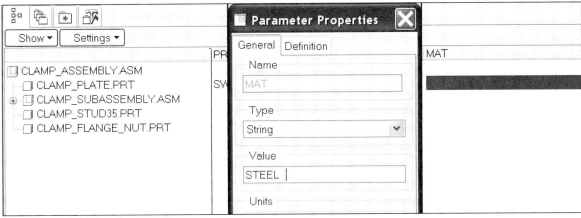

Figure 12.14(f) Material STEEL

Complete the parameters for the remaining components [Fig. 12.15(a)]

	PRTNO	DSC	MAT
CLAMP_ASSEMBLY.ASM			
CLAMP_PLATE.PRT	SW100-20PL	CLAMP PLATE	STEEL
CLAMP_SUBASSEMBLY.ASM			
CLAMP_ARM.PRT	SW101-5AR	CLAMP ARM	STEEL
CLAMP_SWIVEL.PRT	SW101-6SW	CLAMP SWIVEL	STEEL
CLAMP_FOOT.PRT	SW101-8FT	CLAMP FOOT	NYLON
CLAMP_STUD5.PRT	SW101-9STL	.500-13 X 5.00 DOUBLE END STUD	PURCHASED
CLAMP_BALL.PRT	SW101-7BA	SWING CLAMP BALL	BLACK PLASTIC
CLAMP_BALL.PRT	SW101-7BA	SWING CLAMP BALL	BLACK PLASTIC
CLAMP_STUD35.PRT	SW100-21ST	.500-13 X 3.50 DOUBLE END STUD	PURCHASED
CLAMP_FLANGE_NUT.PRT	SW100-22FLN	.500-13 X HEX FLANGE NUT	PURCHASED

Figure 12.15(a) Assembly Model Tree and Parameters

Click: **Tools** ⇒ **Parameters** ⇒ [▼] ⇒ **Part** [Fig. 12.15(b)] ⇒ click on **CLAMP_FOOT.PRT** [Fig. 12.15(c)] ⇒ Designate [☑] ⇒ add Parameter Descriptions ⇒ repeat for remaining components ⇒ **Ok**

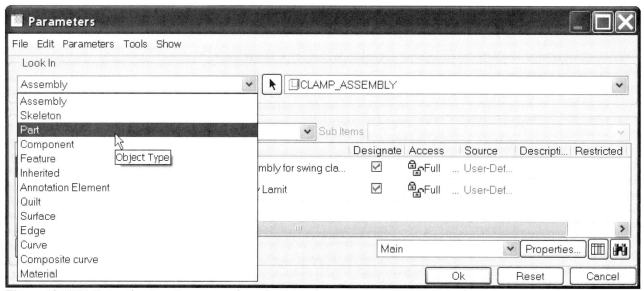

Figure 12.15(b) Extracting Part Parameters from the Assembly

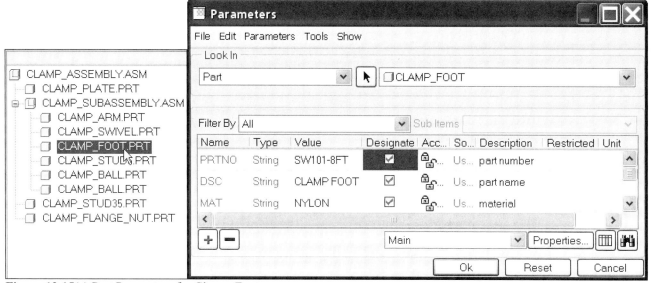

Figure 12.15(c) Part Parameters for Clamp_Foot

The Clamp_Flange_Nut, Clamp_Stud35 and the Clamp_Stud5 are standard parts that are copied from the PTC parts library; therefore they will have additional parameters [Fig. 12.16(a)] and relations [Figs. 12.16(b-c)].

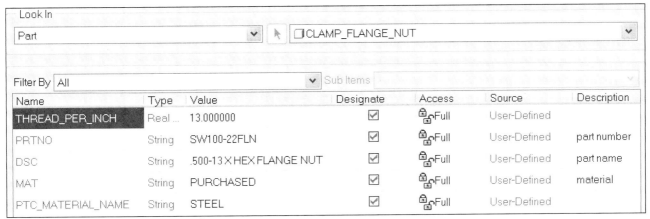

Figure 12.16(a) Clamp_Flange_Nut Parameters

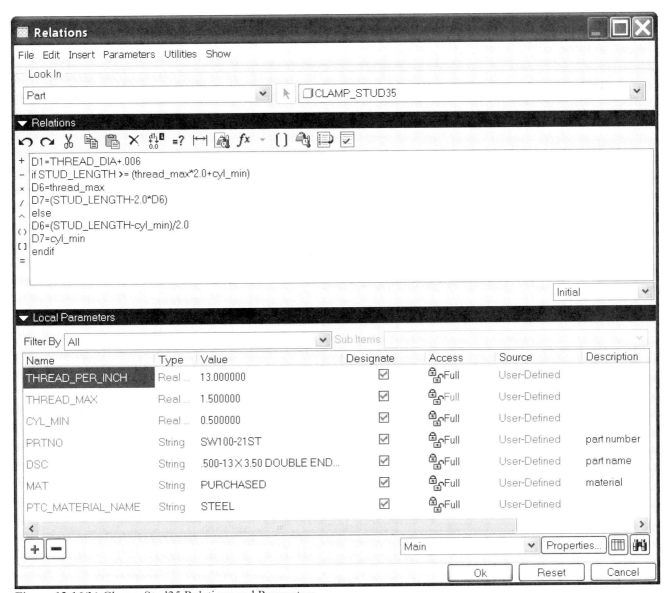

Figure 12.16(b) Clamp_Stud35 Relations and Parameters

Relations and parameters can be displayed together by clicking on the part name (here **CLAMP_STUD5**) in the CLAMP_ASSEMBLY Model Tree ⇒ **RMB** ⇒ **Open** ⇒ **Info** ⇒ **Relations and Parameters** [Fig. 12.16(c)] **File** ⇒ **Close Window**

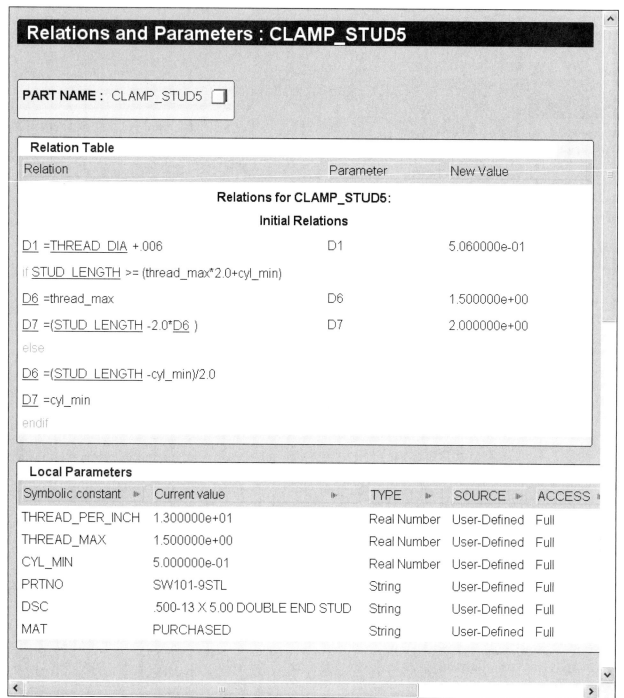

Figure 12.16(c) Clamp_Stud5 Relations and Parameters

Assembly Drawings

The parameters (and their values) have been established for each part. The assembly format with related parameters in a parts list table has been created and saved in your (format) directory. You can now create a drawing of the assembly, where the parts list will be generated automatically, and the assembly ballooned. The first assembly drawing will be of the Clamp_Subassembly [Figs. 12.17(a-b)].

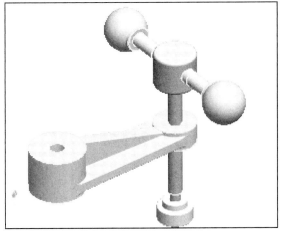

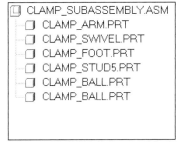

Figure 12.17(a) Clamp_Subassembly

Figure 12.17(b) Subassembly Model Tree

Click: **File** ⇒ **Set Working Directory** ⇒ select the working directory ⇒ **OK** ⇒ **Create a new object** ⇒ **Drawing** ⇒ Name **CLAMP_SUBASSEMBLY** ⇒ **Use default template** [Fig. 12.18(a)] ⇒ **OK** ⇒ Default Model **Browse** ⇒ **clamp_subassembly.asm** ⇒ **Open** ⇒ **Empty with format** ⇒ **Browse** ⇒ **Working Directory** ⇒ pick **asm_format_e. frm** ⇒ **Open** [Fig. 12.18(b)] ⇒ **OK**

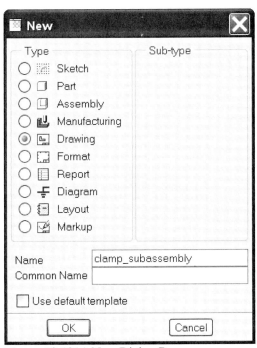

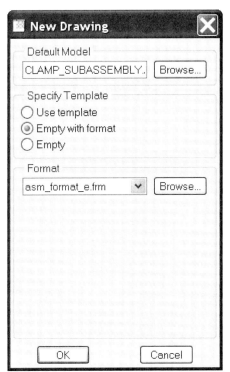

Figure 12.18(a) New Dialog Box

Figure 12.18(b) New Drawing Dialog Box

475

For the drawing to display with the correct style, modify the values of the Drawing Options.

Click: **Tools** ⇒ **Environment** ⇒ ☑ Snap to Grid ⇒ Display Style | Hidden Line ⇒ Tangent Edges | Dimmed ⇒ **OK** ⇒ **File** ⇒ **Properties** ⇒ **Drawing Options** ⇒ *default_font filled* ⇒ **Enter** ⇒ *draw_arrow_style filled* ⇒ **MMB** ⇒ *drawing_text_height* .50 ⇒ **MMB** ⇒ *crossec_arrow_length* .50 ⇒ **MMB** ⇒ *crossec_arrow_width* .17 ⇒ **MMB** ⇒ *max_balloon_radius* .50 ⇒ **MMB** ⇒ *min_balloon_radius* .50 ⇒ **MMB** ⇒ 💾 Save a copy of the currently displayed configuration file ⇒ Name **CLAMP_ASM** ⇒ **Ok** ⇒ **Apply** ⇒ **Close** ⇒ **MMB** ⇒ 🔍 ⇒ 🖼 ⇒ 💾 ⇒ **MMB** [Fig. 12.18(c)]

Note that the Repeat Region of the BOM [Fig. 12.18(d)] has been automatically filled in with the parameters read from the individual components.5

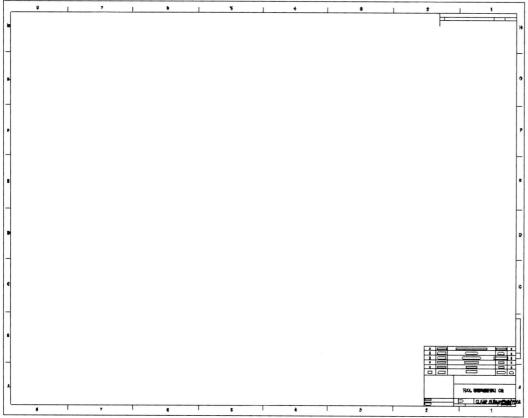

Figure 12.18(c) Drawing

5	SW101-9STL	.500-13 X 5.00 DOUBLE END STUD	PURCHASED	1
4	SW101-8FT	CLAMP FOOT	NYLON	1
3	SW101-7BA	SWING CLAMP BALL	BLACK PLASTIC	2
2	SW101-6SW	CLAMP SWIVEL	STEEL	1
1	SW101-5AR	CLAMP ARM	STEEL	1
ITEM	PT NUM	DESCRIPTION	MATERIAL	QTY

Figure 12.18(d) BOM

Click: **Datum axes** off ⇒ **Datum points** off ⇒ **Coordinate systems** off ⇒ double-click on **SCALE: 1.000** in the lower left-hand of the graphics window ⇒ Enter value for scale ⇒ type **1.50** ⇒ **MMB** ⇒ **Disallow the movement of drawing views with the mouse** unlock ⇒ **Create a general view** ⇒ **No Combined State** ⇒ **OK** ⇒ ⇒ (top view) [Fig. 12.19(a)] ⇒ **Geometry references** ⇒ [Fig. 12.19(b)] ⇒ Select Reference 1 **Front** [Fig. 12.19(c)]

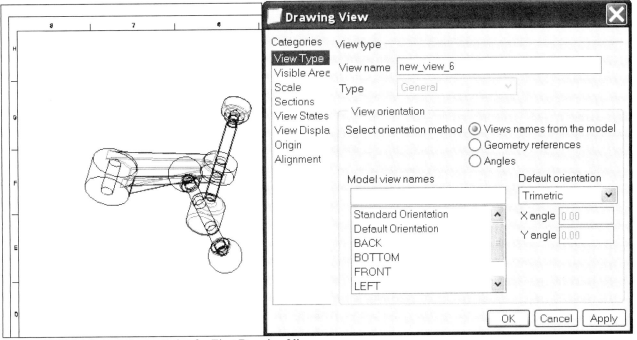

Figure 12.19(a) Select Center Point for First Drawing View

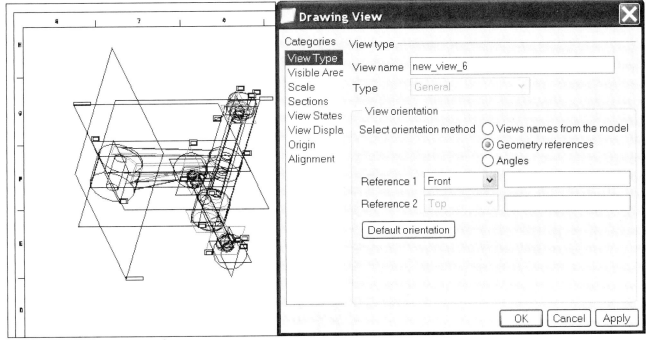

Figure 12.19(b) Geometry references

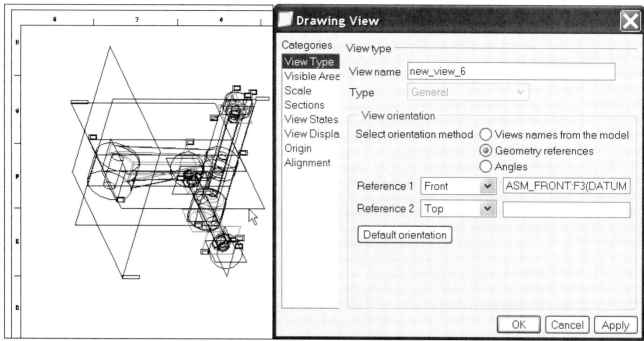

Figure 12.19(c) Reference 1 Front ASM_FRONT:F3(DATUM PLANE)

Select Reference 2 **Top** [Reference 2 | Top | ASM_TOP:F2(DATUM PLANE)] [Fig. 12.19(d)] ⇒ **OK** ⇒

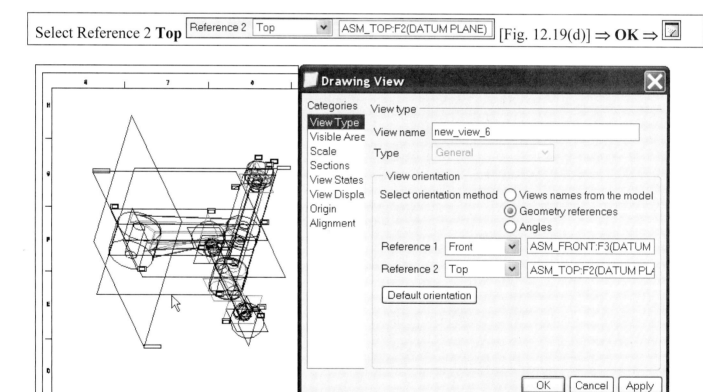

Figure 12.19(d) Reference 2 Top ASM_TOP:F2(DATUM PLANE)

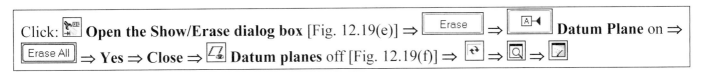

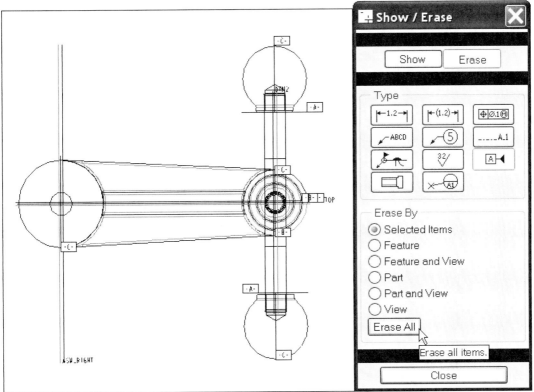

Figure 12.19(e) Show/Erase Dialog Box- Erase All Datum Planes

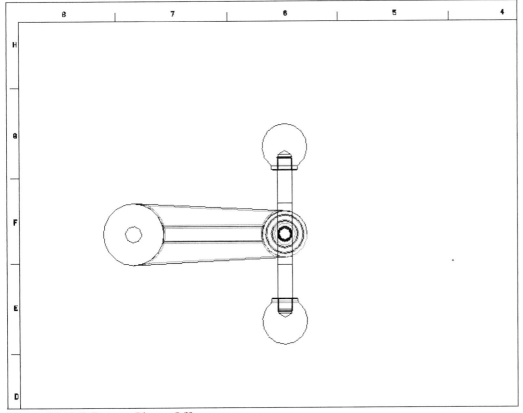

Figure 12.19(f) Datums Planes Off

The only other view needed to show the subassembly is a front section view.

Click: 🖫 ⇒ **MMB** ⇒ pick on the Top view to highlight ⇒ **RMB** ⇒ **Insert Projection View** [Fig. 12.20(a)] ⇒ `Select CENTER POINT for drawing view.` select below the view previously created [Fig. 12.20(b)]

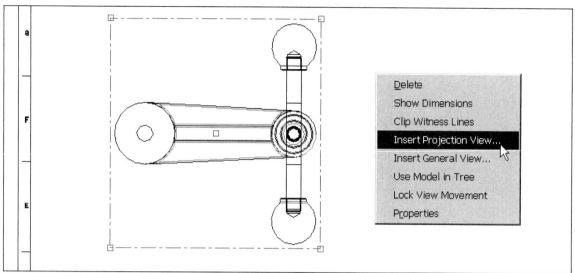

Figure 12.20(a) Place the Front View

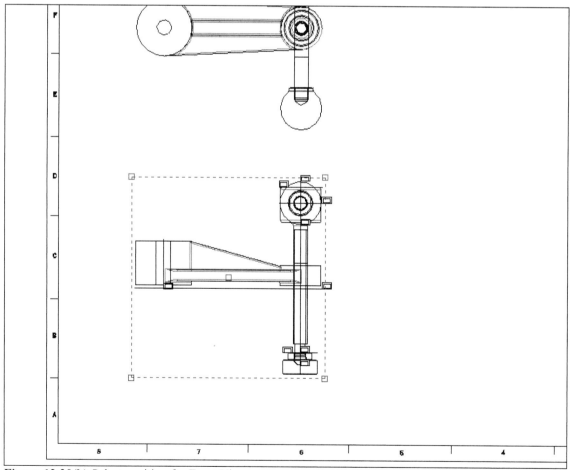

Figure 12.20(b) Select position for Front View

Click: 🖫 ⇒ **MMB** ⇒ with front view still selected/highlighted, click: **RMB** ⇒ **Properties** [Fig. 12.20(c)] ⇒ **Sections** ⇒ ⦿ 2D cross-section [Fig. 12.20(d)] ⇒ ➕ [Figs. 12.20(e-f)] ⇒ A ⇒ **OK**

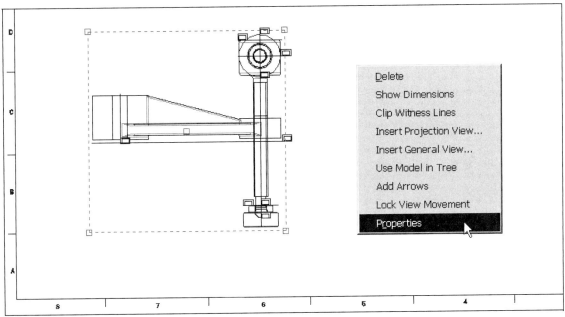

Figure 12.20(c) Properties

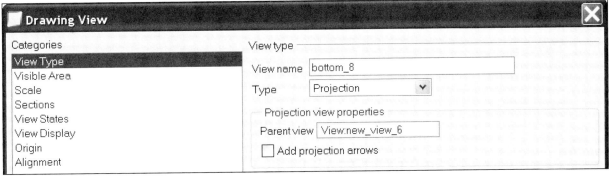

Figure 12.20(d) Drawing View Dialog Box

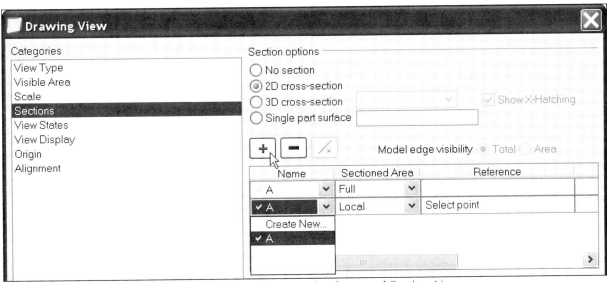

Figure 12.20(e) Drawing View 2D section (Must have previously created Section A)

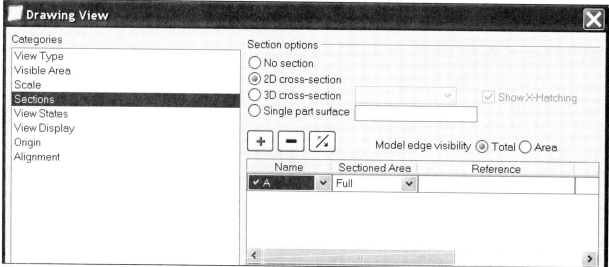

Figure 12.20(f) Section A, Full

Click: ⬚ **Datum planes** off ⇒ **LMB** to deselect ⇒ ⬚ ⇒ ⬚ ⇒ [Erase] ⇒ ⬚ **Datum Plane** on ⇒ [Erase All] ⇒ **Yes** ⇒ **Close** ⇒ reposition the views as needed ⇒ ⬚ ⇒ ⬚ ⇒ ⬚ ⇒ ⬚ ⇒ **MMB** ⇒ pick on the Front view to highlight/select ⇒ **RMB** ⇒ **Add Arrows** [Fig. 12.20(g)] ⇒ pick on the Top view [Figs. 12.20(h-i)] ⇒ **LMB** to deselect

Figure 12.20(g) Add Arrows

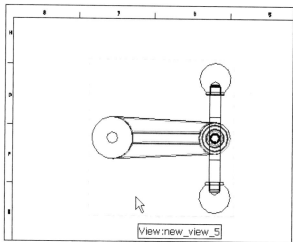
Figure 12.20(h) Top View

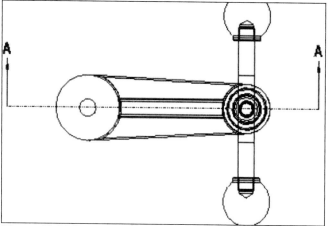
Figure 12.20(i) Top View with Section Cutting Plane Arrows

Pro/E provides tools to alter the display of the section views to comply with industry ASME standard practices. Most companies require that the crosshatching on parts in section views of assemblies be "clocked" such that parts that meet do not use the same section lining (crosshatching) spacing and angle. This makes the separation between parts more distinct. First, modify the visibility of the views to remove hidden lines and make the tangent edges dimmed. Next, show all centerlines and clip as needed.

> Press and hold down the **Ctrl** key and pick on both views ⇒ click **RMB** with cursor outside of the view outlines ⇒ **Properties** ⇒ Display Style ⇒ No Hidden [Fig. 12.21(a)] ⇒ Tangent edges display style ⇒ Dimmed [Fig. 12.21(b)] ⇒ **Apply** ⇒ **Close** ⇒ **LMB** to deselect

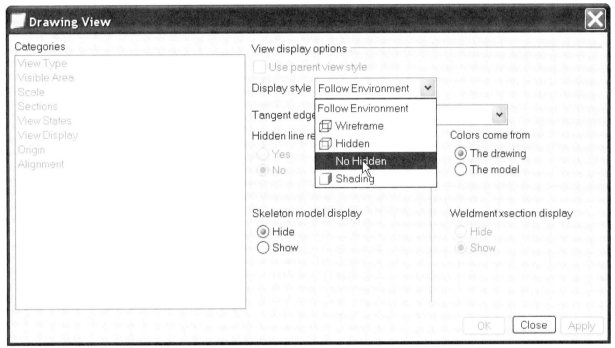

Figure 12.21(a) No Hidden

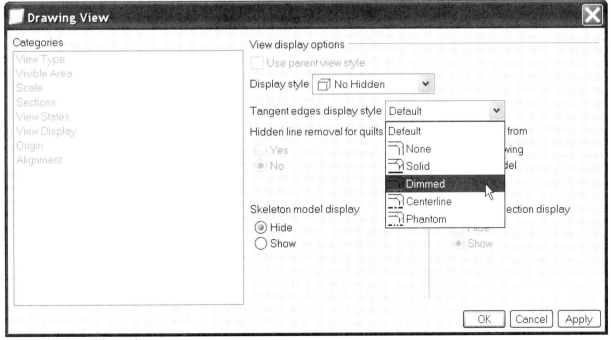

Figure 12.21(b) Dimmed

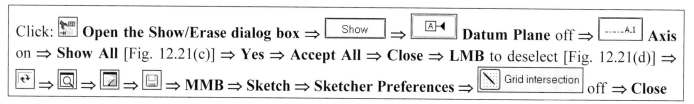

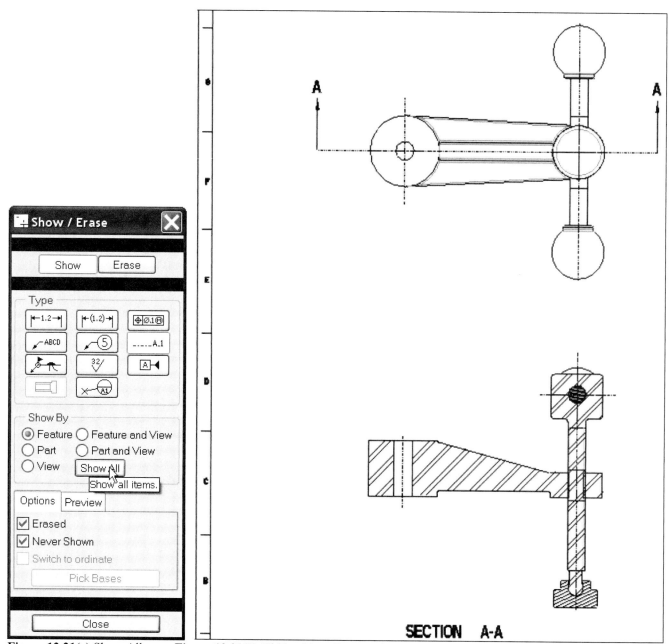

Figure 12.21(c) Show All **Figure 12.21(d)** Hidden Lines Removed, Tangent Edges Dimmed, Centerline Lines Displayed

Double-click on the crosshatching in the Front view (Clamp_Stud5 is now active) [Fig. 12.21(e)] ⇒ **Fill** [Fig. 12.21(f)] ⇒ click **Next Xsec** until Clamp_Foot is active [Fig. 12.21(g)] ⇒ **Hatch** ⇒ **Angle** ⇒ **135** [Fig. 12.21(h)]

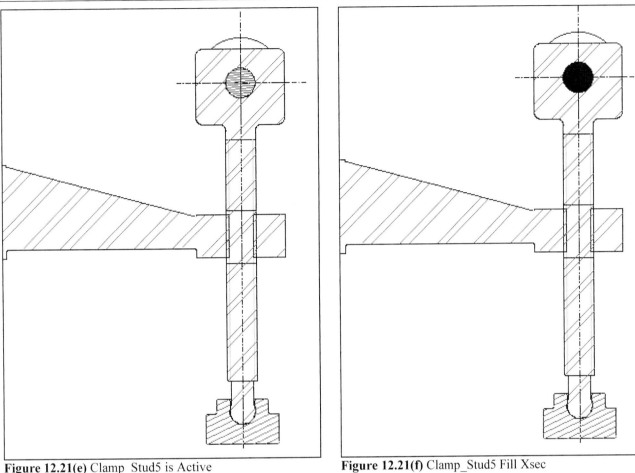

Figure 12.21(e) Clamp_Stud5 is Active   **Figure 12.21(f)** Clamp_Stud5 Fill Xsec

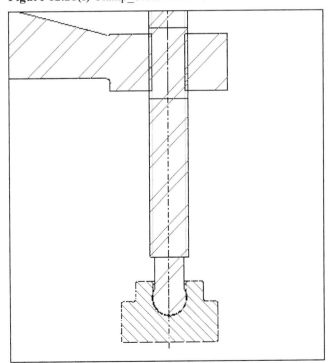

Figure 12.21(g) Clamp_Foot is Active   **Figure 12.21(h)** Clamp_Foot Hatch Angle **135**

Click: **Next Xsec** ⇒ **Next Xsec** Clamp_Arm is active ⇒ **Hatch** ⇒ **Delete line** (if necessary) ⇒ **Spacing** ⇒ **Half** ⇒ **Angle** ⇒ **120** [Fig. 12.21(i)] ⇒ **Prev Xsec** Clamp_Swivel is active ⇒ **Delete line** (if necessary) ⇒ **Spacing** ⇒ **Half** ⇒ **MMB** ⇒ **LMB** ⇒ [icon] ⇒ [icon] ⇒ [icon] [Fig. 12.21(j)] ⇒ [icon] ⇒ **MMB**

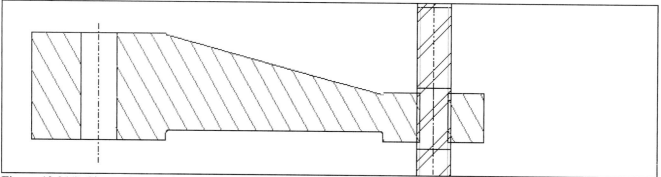

Figure 12.21(i) Clamp_Arm, Delete Line, Hatch Spacing Half, Angle **120**

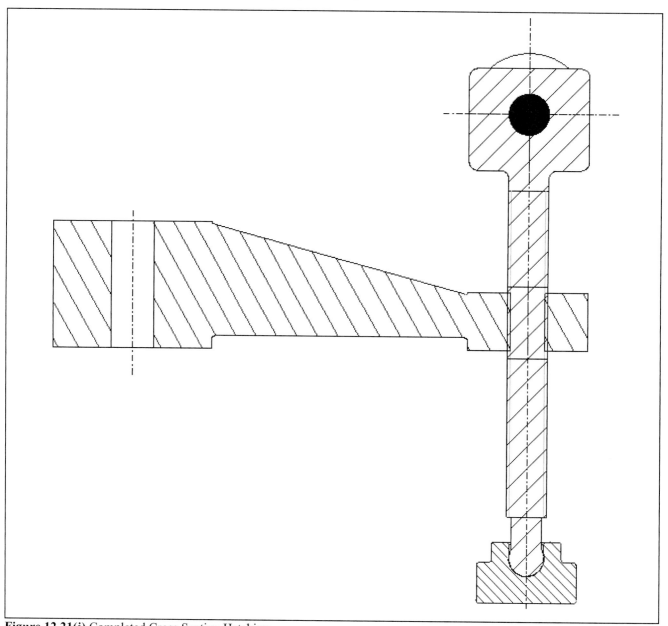

Figure 12.21(j) Completed Cross Section Hatching

To complete the drawing, the balloons must be displayed for each component. Balloons are displayed in the top view as the default, because it was the first view that was created.

Click: **Table** ⇒ **BOM Balloons** ⇒ **Set Region** ⇒ **Simple** ⇒ pick in the BOM field [Fig. 12.22(a)] ⇒ **Create Balloon** ⇒ **Show All** [Fig. 12.22(b)] ⇒ **MMB** ⇒ press and hold the **Ctrl** key and pick on the balloons for the Clamp_Arm (1), Clamp_Stud5 (5) and Clamp_Foot (4) ⇒ **RMB** ⇒ **Move Item to View** [Fig. 12.22(c)] ⇒ pick in the Front view [Fig. 12.22(d)] ⇒ pick on and reposition each balloon as needed [Fig. 12.22(e)] ⇒ **LMB** to deselect ⇒ 🔄 **Update** ⇒ 🔍 ⇒ 📄 ⇒ 💾 ⇒ **MMB**

5	SW101-9STL	.500-13 X 5.00 DOUBLE END STUD	PURCHASED	1
4	SW101-8FT	CLAMP FOOT	NYLON	1
3	SW101-7BA	SWING CLAMP BALL	BLACK PLASTIC	2
2	SW101-6SW	CLAMP SWIVEL	STEEL	1
1	SW101-5AR	CLAMP ARM	STEEL	1
ITEM	PT NUM	DESCRIPTION	MATERIAL	QTY

Figure 12.22(a) Set Region

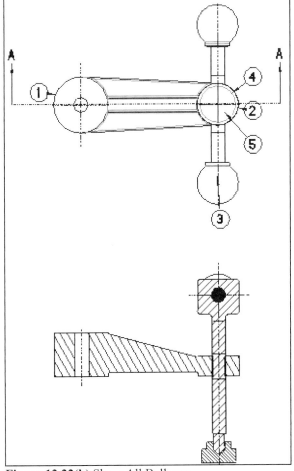

Figure 12.22(b) Show All Balloons

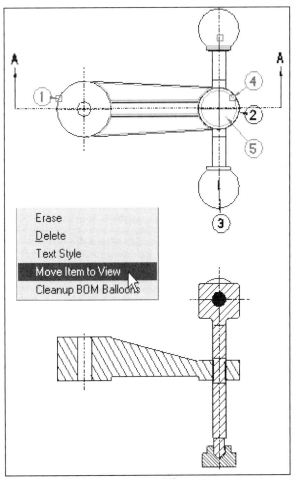

Figure 12.22(c) Move Item to View

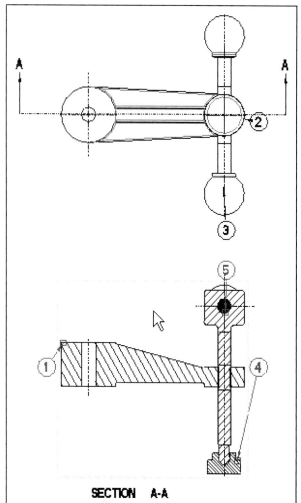

Figure 12.22(d) Balloons Moved to Front View

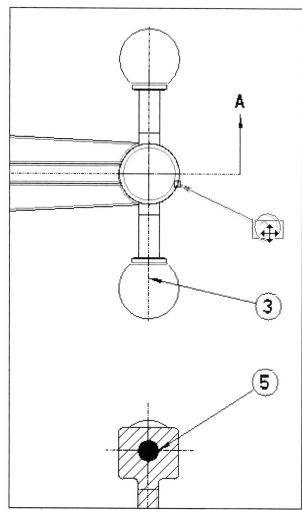

Figure 12.22(e) Reposition Balloons

Pick on balloon **3** (Clamp_Ball) ⇒ **RMB** [Fig. 12.22(f)] ⇒ **Edit Attachment** ⇒ **On Entity** ⇒ pick on the edge of the Clamp_Ball [Fig. 12.22(g)] ⇒ **MMB** ⇒ reposition balloon ⇒ **LMB** ⇒ [icon] ⇒ [icon] ⇒ [icon] ⇒ [icon] ⇒ **MMB** ⇒ **File** ⇒ **Delete** ⇒ **Old Versions** ⇒ **MMB** [Fig. 12.22(h)] ⇒ **Window** ⇒ **Close**

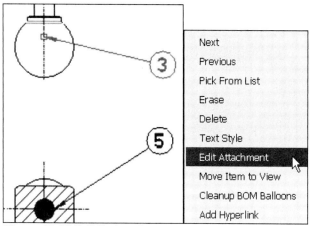

Figure 12.22(f) Edit Attachment

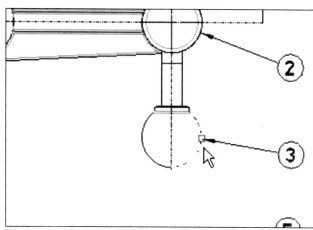

Figure 12.22(g) Pick the Edge of the Clamp_Ball

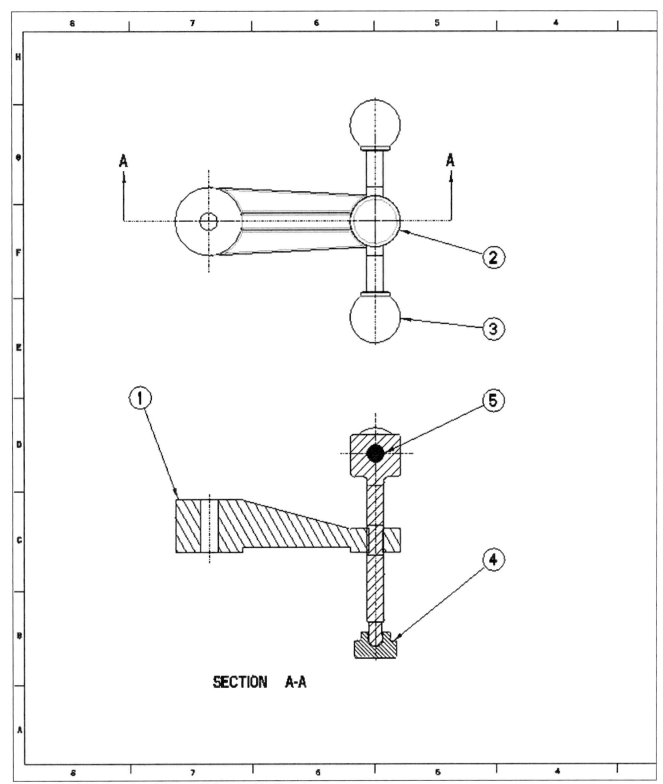

Figure 12.22(h) Completed Clamp_Subassembly Drawing

Create the drawing for the Swing Clamp Assembly. This assembly is composed of the subassembly (in the previous drawing), the plate, the short stud, and the nut. The drawing will use the same format created for the subassembly. Formats are read-only files that can be used as many times as needed.

Click: [icon] **Create a new object** ⇒ [◉ Drawing] ⇒ Name **CLAMP_ASSEMBLY** ⇒ [☐ Use default template] ⇒ **OK** ⇒ Default Model **Browse** ⇒ **clamp_assembly.asm** ⇒ **Open** ⇒ [◉ Empty with format] ⇒ Format **Browse** ⇒ [icon] **Working Directory** ⇒ select **asm_format_e.frm** ⇒ **Open** ⇒ **OK** ⇒ [icon] **Zoom In** on the title block [Fig. 12.23(a)] ⇒ [icon] ⇒ [icon] ⇒ **RMB** ⇒ **Properties** ⇒ **Drawing Options** ⇒ [icon] **Open a configuration file** ⇒ **clamp_asm.dtl** ⇒ **Open** ⇒ **Apply** ⇒ **Close** ⇒ **MMB**

ITEM	PT NUM	DESCRIPTION	MATERIAL	QTY
8	SW101-9STL	.500-13 X 5.00 DOUBLE END STUD	PURCHASED	1
7	SW101-8FT	CLAMP FOOT	NYLON	1
6	SW101-7BA	SWING CLAMP BALL	BLACK PLASTIC	2
5	SW101-6SW	CLAMP SWIVEL	STEEL	1
4	SW101-5AR	CLAMP ARM	STEEL	1
3	SW100-22FLN	.500-13 X HEX FLANGE NUT	PURCHASED	1
2	SW100-21ST	.500-13 X 3.50 DOUBLE END STUD	PURCHASED	1
1	SW100-20PL	CLAMP PLATE	STEEL	1

TOOL ENGINEERING CO.

DRAWN / ISSUED 1.000 CLAMP_ASSEMBLY A SHEET 1 OF 1

Figure 12.23(a) BOM and Title Block

Click: [icon] **Datum axes** off ⇒ [icon] **Datum points** off ⇒ [icon] **Coordinate systems** off ⇒ [icon] **Datum planes** on ⇒ double-click on **SCALE:1.000** in the lower left-hand of the graphics window [SCALE:1.000 TYPE:ASSEM] ⇒ [Enter value for scale 1.50] ⇒ **MMB** ⇒ [icon] **Disallow the movement of drawing views with the mouse** unlock ⇒ [icon] **Create a general view** ⇒ **No Combined State** ⇒ **OK** ⇒ [Select CENTER POINT for drawing view.] pick where you want a top view ⇒ [◉ Geometry references] ⇒ [icon] ⇒ select Reference 1 Front **CL_ASM_FRONT** ⇒ select Reference 2 Top **CL_ASM_TOP** ⇒ **OK** ⇒ with the view still highlighted, click **RMB** ⇒ **Insert Projection View** ⇒ pick in the general area where you want a front view located ⇒ **RMB** ⇒ **Properties** ⇒ Categories **Sections** ⇒ [◉ 2D cross-section] ⇒ [+] ⇒ [v] ⇒ **Create New** ⇒ **Planar** ⇒ **Single** ⇒ **Done** ⇒ type **A** ⇒ **MMB** ⇒ [icon] ⇒ [Select or create an assembly datum] pick on **CL_ASM_TOP** ⇒ **Apply** ⇒ **Close** ⇒ **RMB** ⇒ **Add Arrows** ⇒ [Pick a view for arrows where the section is perp.] pick in the Top view ⇒ [icon] **Datum planes** off ⇒ **LMB** to deselect

Click: 🗇 **Open the Show/Erase dialog box** ⇒ ⌈Erase⌉ ⇒ ⌈A⌉ **Datum Plane** on (everything else off) ⇒ ⌈Erase All⌉ ⇒ **Yes** ⇒ ⌈Show⌉ ⇒ ⌈A⌉ **Datum Plane** off ⇒ ⌈----A_1⌉ **Axis** on ⇒ **Show All** ⇒ **Yes** ⇒ **Accept All** ⇒ **Close** ⇒ **LMB** ⇒ press and hold down the **Ctrl** key and pick on both views ⇒ click **RMB** with cursor outside of the view outlines ⇒ **Properties** ⇒ Display Style **No Hidden** ⇒ Tangent edges **Dimmed** ⇒ **MMB** ⇒ **LMB** ⇒ 🗘 ⇒ 🔍 ⇒ 🗹 ⇒ 💾 ⇒ **MMB** [Fig. 12.23(b)]

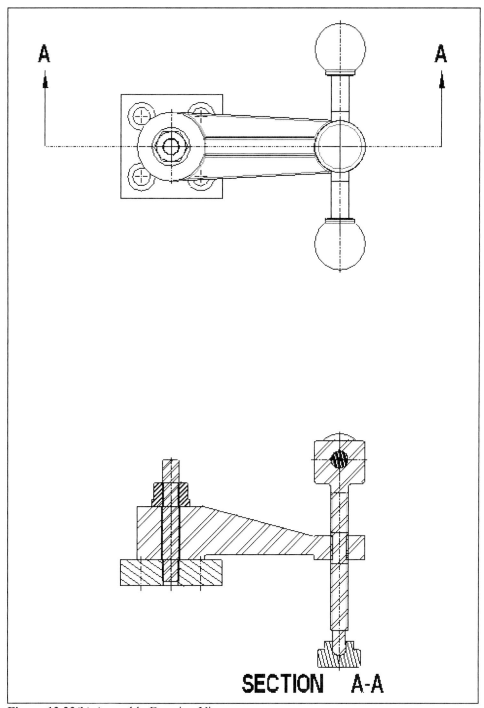

Figure 12.23(b) Assembly Drawing Views

Click: **Table** ⇒ **Repeat Region** ⇒ **Attributes** ⇒ pick in the BOM field [Fig. 12.24(a)] ⇒ **Flat** ⇒ **MMB** ⇒ **MMB** ⇒ **MMB** [Fig. 12.24(b)] ⇒ **Table** ⇒ **BOM Balloons** ⇒ **Set Region** ⇒ pick in the BOM field ⇒ **Create Balloon** ⇒ **Show All** ⇒ **MMB** [Fig. 12.24(c)]

ITEM	PT NUM	DESCRIPTION	MATERIAL	QTY
8	SW101-9STL	.500-13 X 5.00 DOUBLE END STUD	PURCHASED	1
7	SW101-8FT	CLAMP FOOT	NYLON	1
6	SW101-7BA	SWING CLAMP BALL	BLACK PLASTIC	2
5	SW101-6SW	CLAMP SWIVEL	STEEL	1
4	SW101-5AR	CLAMP ARM	STEEL	1
3	SW100-22FLN	.500-13 X HEX FLANGE NUT	PURCHASED	1
2	SW100-21ST	.500-13 X 3.50 DOUBLE END STUD	PURCHASED	1
1	SW100-20PL	CLAMP PLATE	STEEL	1

Figure 12.24(a) Showing with BOM Attribute Recursive

ITEM	PT NUM	DESCRIPTION	MATERIAL	QTY
3	SW100-22FLN	.500-13 X HEX FLANGE NUT	PURCHASED	1
2	SW100-21ST	.500-13 X 3.50 DOUBLE END STUD	PURCHASED	1
1	SW100-20PL	CLAMP PLATE	STEEL	1

TOOL ENGINEERING CO.

Figure 12.24(b) Showing with BOM Attribute Flat

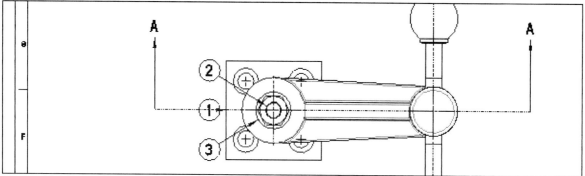

Figure 12.24(c) Balloons Displayed in Top View

While pressing the **Ctrl** key, pick on all three balloons ⇒ **RMB** ⇒ **Move Item to View** ⇒ pick in the Front view [Fig. 12.24(d)] ⇒ pick on and reposition each balloon as needed ⇒ pick on balloon **2** [Fig. 12.24(e)] ⇒ **RMB** ⇒ **Edit Attachment** ⇒ **On Entity** ⇒ pick on edge [Fig. 12.24(f)] ⇒ **MMB** ⇒ **LMB**

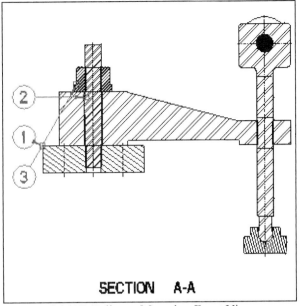

Figure 12.24(d) Balloons Moved to Front View

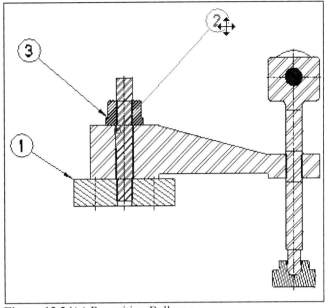

Figure 12.24(e) Reposition Balloons

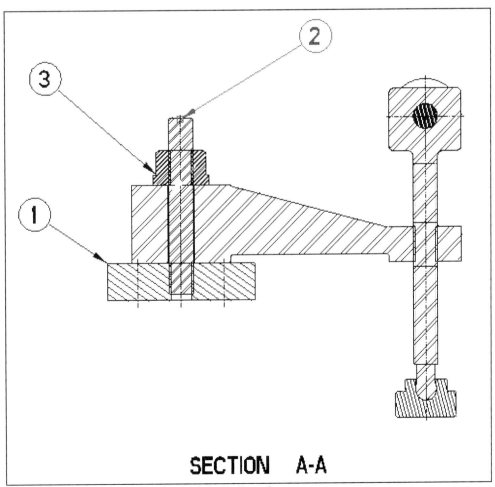

Figure 12.24(f) Attachment Changed for Balloon 2

Most companies (and as per drafting standards) require that round purchased items, such as nuts, bolts, studs, springs, and die pins be excluded from sectioning even when the section cutting plane passes through them. Remove the section lining (crosshatching) from the Clamp_Flange_Nut and the Clamp_Stud35 in the front section view.

Double-click on the crosshatching in the Front view [Fig. 12.25(a)]. The Clamp_Flange_Nut is now active [Fig. 12.25(b)]. ⇒ **Excl Comp** to eliminate Xsec of Clamp_Flange_Nut [Fig. 12.25(c)] ⇒ **Next Xsec** Clamp_Stud35 is now active ⇒ **Excl Comp** to eliminate Xsec of Clamp_Stud35 [Fig. 12.25(d)] ⇒ **MMB** ⇒ **LMB** ⇒ 🔄 ⇒ 🔍 ⇒ 📝 ⇒ 💾 ⇒ **MMB**

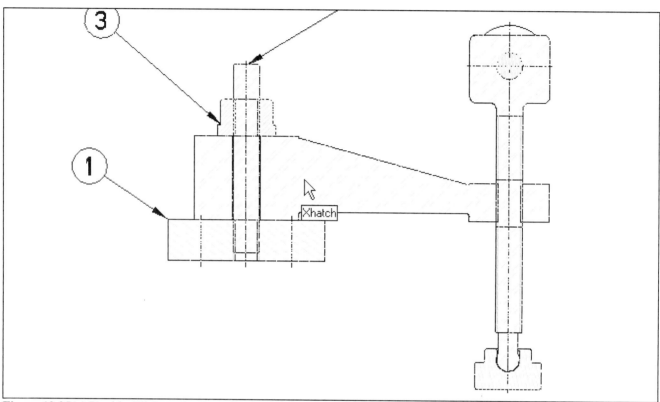

Figure 12.25(a) Double-Click on the Cross Section Lining

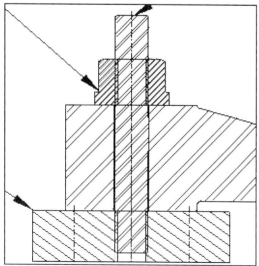

Figure 12.25(b) Clamp_Flange_Nut Active

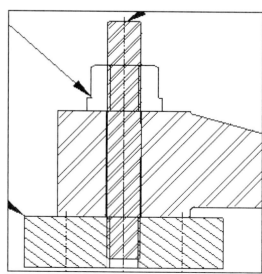

Figure 12.25(c) Section Lining Removed

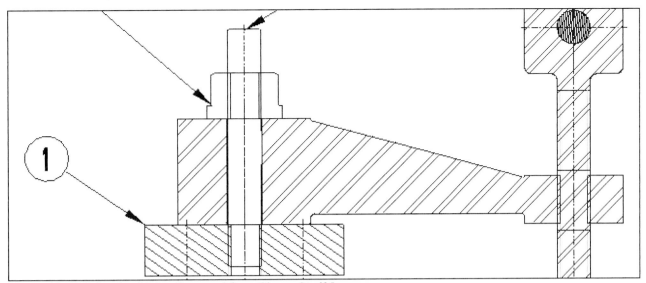

Figure 12.25(d) Section Lining Removed from Clamp_Stud35

Pick on the crosshatching ⇒ **RMB** ⇒ **Properties** the Clamp_Flange_Nut is now active ⇒ **Next Xsec** ⇒ **Next Xsec** Clamp_Stud5 is now active ⇒ **Fill** ⇒ **Next Xsec** ⇒ **Next Xsec** Clamp_Swivel is now active ⇒ **Delete line** (if necessary) ⇒ **Next Xsec** Clamp_Arm is now active ⇒ **Delete line** (if necessary) ⇒ **Angle** ⇒ **120** ⇒ **Next Xsec** Clamp_Plate is now active ⇒ **Angle** ⇒ **45** (Fig. 12.26) ⇒ **MMB** ⇒ **LMB** ⇒ **View** ⇒ **Drawing Display** ⇒ **Component Display** ⇒ with the Ctrl key pressed, pick the flange nut and the 3.50 stud in the front view ⇒ **OK** ⇒ **Standard** ⇒ **Done** ⇒ **MMB** ⇒ **MMB** ⇒ [icon] ⇒ [icon] ⇒ [icon] ⇒ [icon] ⇒ **MMB**

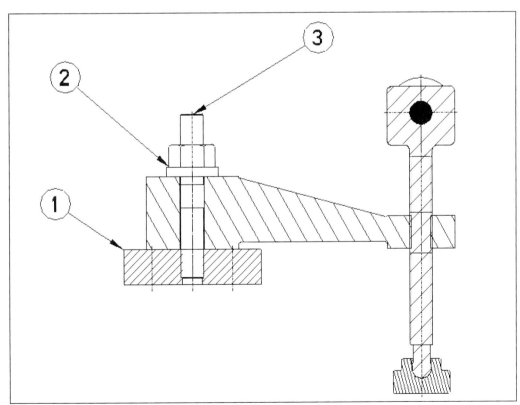

Figure 12.26 Clamp_Plate Section Lining Angle is now **45** Degrees

The numbering of the components in assemblies may need to be different from the default setting. To change the balloon numbering, you must use **Fix Index**.

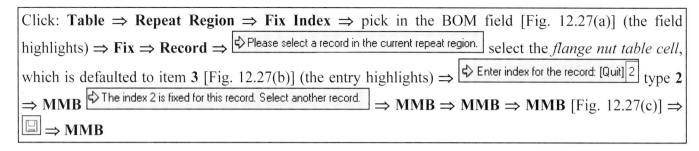

Click: **Table** ⇒ **Repeat Region** ⇒ **Fix Index** ⇒ pick in the BOM field [Fig. 12.27(a)] (the field highlights) ⇒ **Fix** ⇒ **Record** ⇒ [Please select a record in the current repeat region.] select the *flange nut table cell*, which is defaulted to item **3** [Fig. 12.27(b)] (the entry highlights) ⇒ [Enter index for the record: [Quit] 2] type **2** ⇒ **MMB** [The index 2 is fixed for this record. Select another record.] ⇒ **MMB** ⇒ **MMB** ⇒ **MMB** [Fig. 12.27(c)] ⇒ [💾] ⇒ **MMB**

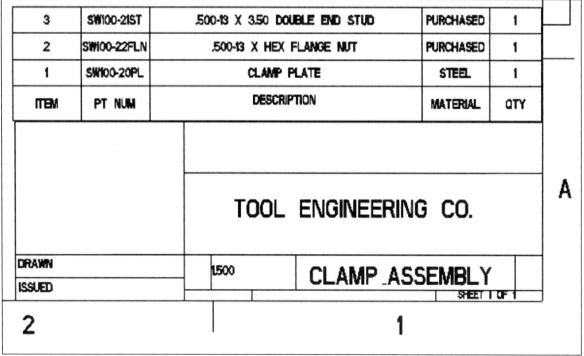

Figure 12.27(a) Pick in the BOM Field

Figure 12.27(b) Pick in the Clamp_Flange_Nut Table Cell

Figure 12.27(c) The Clamp_Flange_Nut is Now Listed Second

Exploded Swing Clamp Assembly Drawings

The process required to place an exploded view on a drawing is similar to adding assembly orthographic views. The BOM will display all components on this sheet. You will be required to fix the BOM sequence and manually create balloons.

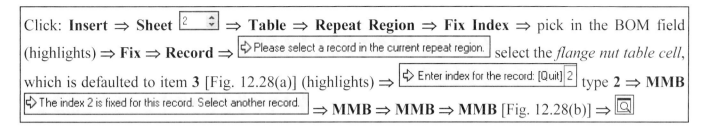

8	SW101-9STL	.500-13 X 5.00 DOUBLE END STUD	PURCHASED	1
7	SW101-8FT	CLAMP FOOT	NYLON	1
6	SW101-7BA	SWING CLAMP BALL	BLACK PLASTIC	2
5	SW101-6SW	CLAMP SWIVEL	STEEL	1
4	SW101-5AR	CLAMP ARM	STEEL	1
3	SW100-22FLN	.500-13 X HEX FLANGE NUT	PURCHASED	1
2	SW100-21ST	.500-13 X 3.50 DOUBLE END STUD	PURCHASED	1
1	SW100-20PL	CLAMP PLATE	STEEL	1
ITEM	PT NUM	DESCRIPTION	MATERIAL	QTY

Figure 12.28(a) Pick in the Clamp_Flange_Nut Table Cell on Sheet 2

8	SW101-9STL	.500-13 X 5.00 DOUBLE END STUD	PURCHASED	1
7	SW101-8FT	CLAMP FOOT	NYLON	1
6	SW101-7BA	SWING CLAMP BALL	BLACK PLASTIC	2
5	SW101-6SW	CLAMP SWIVEL	STEEL	1
4	SW101-5AR	CLAMP ARM	STEEL	1
3	SW100-21ST	.500-13 X 3.50 DOUBLE END STUD	PURCHASED	1
2	SW100-22FLN	.500-13 X HEX FLANGE NUT	PURCHASED	1
1	SW100-20PL	CLAMP PLATE	STEEL	1
ITEM	PT NUM	DESCRIPTION	MATERIAL	QTY

Figure 12.28(b) On Sheet 2 the Clamp_Flange_Nut is Now Listed Second

Click: **RMB** ⇒ **Insert General View** ⇒ [No Presentation / COMB0001 / DEFAULT ALL] ⇒ **OK** ⇒ Select CENTER POINT for drawing view ⇒ Categories- View Type- Model view names ⇒ [BACK / BOTTOM / EXPLODE1 / FRONT] ⇒ Default orientation ⇒ ⇒ Isometric [Fig. 12.29(a)] ⇒ OK / Cancel / Apply ⇒ View States ⇒ **Combined state** ⇒ COMB0001 ⇒ Explode view ☑ Explode components in view ⇒ Assembly explode state ⇒ ⇒ EXP0001 [Fig. 12.29(b)] ⇒ **Apply** ⇒ View Display ⇒ Display style ⇒ No Hidden ⇒ Tangent edges display style ⇒ Dimmed [Fig. 12.29(c)] ⇒ **Apply** ⇒ **Close**

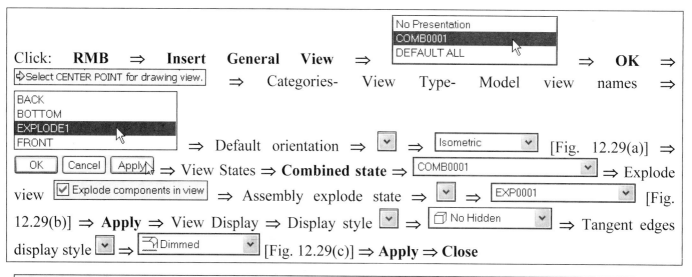

Figure 12.29(a) View Type

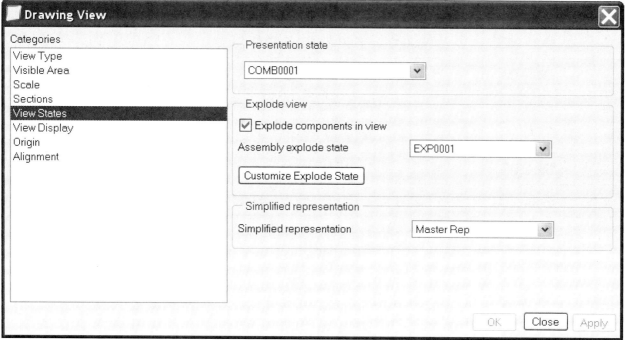

Figure 12.29(b) View States

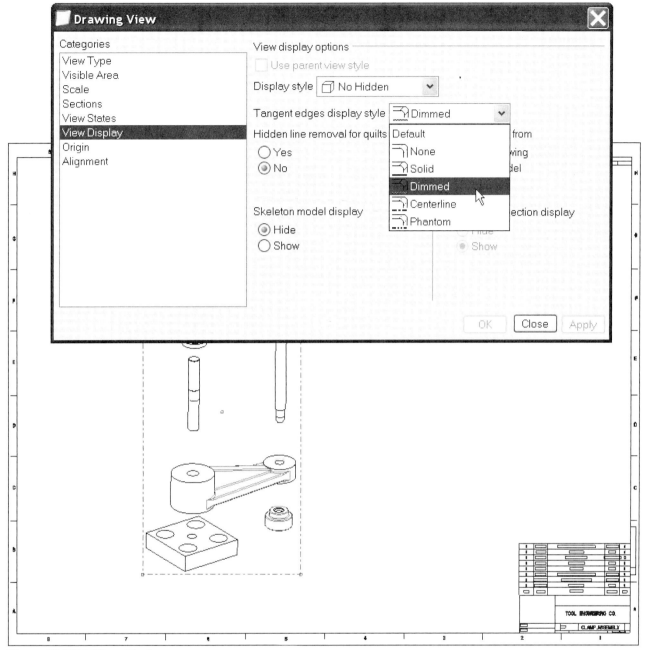

Figure 12.29(c) View Display

Click: **RMB** ⇒ **Lock View Movement** off (uncheck) ⇒ adjust and position the view as needed [Fig. 12.29(d)]

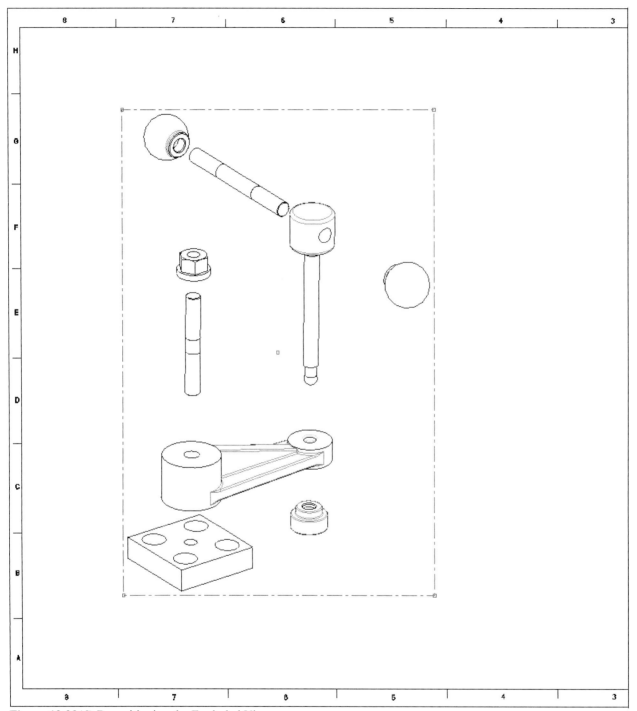

Figure 12.29(d) Repositioning the Exploded View

Click: 🔧 **Open the Show/Erase dialog box** ⇒ Show ⇒ ----A_1 **Axis** ⇒ **Show All** ⇒ **Yes** ⇒ **Accept All** ⇒ **Close** ⇒ **LMB** [Fig. 12.29(e)] ⇒ 🔄 ⇒ 🔍 ⇒ 📐 ⇒ 💾 ⇒ **MMB**

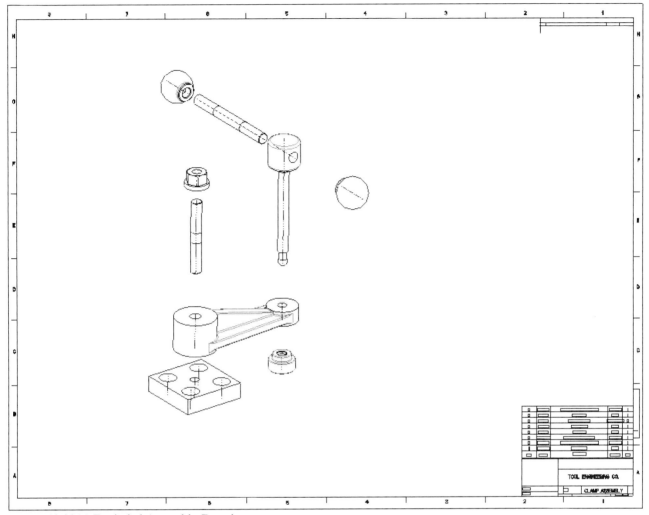

Figure 12.29(e) Exploded Assembly Drawing

Clip each centerline (axis) to extend between components that are in line [Fig. 12.29(f)]

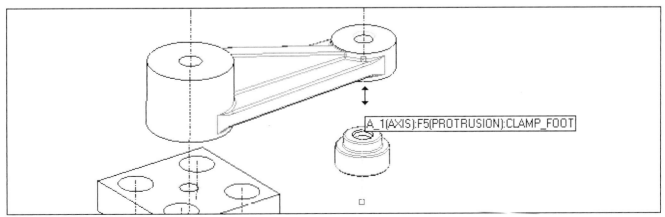

Figure 12.29(f) Clip the Centerlines Axes

To add balloons to a drawing sheet not using a parametric title block, you need to create each balloon separately. Create balloons for the components on the second sheet. The balloons added must correspond to the BOM.

Click: **Insert** ⇒ **Balloon** ⇒ **With Leader** ⇒ **Make Note** ⇒ **On Entity** ⇒ **Arrow Head** ⇒ [Select edges, entities,] pick edge of Clamp_Arm [Figs. 12.30(a-b)] ⇒ **OK** ⇒ **Done** ⇒ [Select LOCATION for note.] ⇒ [icon] ⇒ [Enter NOTE: 4] ⇒ **Enter** ⇒ **Enter** [Fig. 12.30(c)] ⇒ **Make Note** ⇒ continue until all *eight* components are ballooned ⇒ **Done/Return** ⇒ [icons] ⇒ **MMB** ⇒ **LMB**

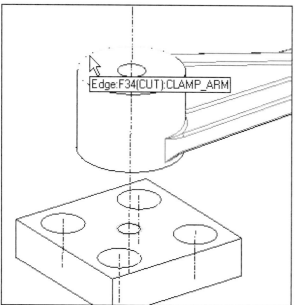

Figure 12.30(a) Highlight Edge of Clamp_Arm

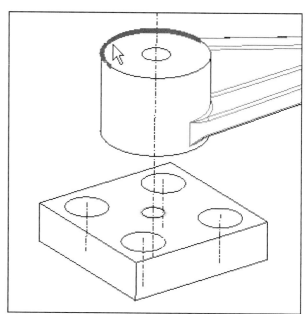

Figure 12.30(b) Select on the Edge

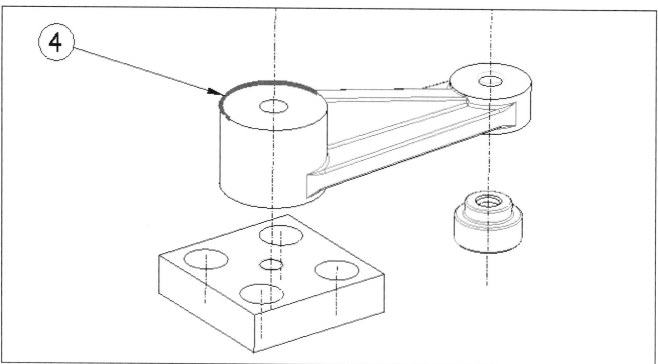

Figure 12.30(c) Balloon Placed

After the ballooning is complete, reposition the balloons and their attachment points as needed to clean up the drawing [Fig. 12.30(d)].

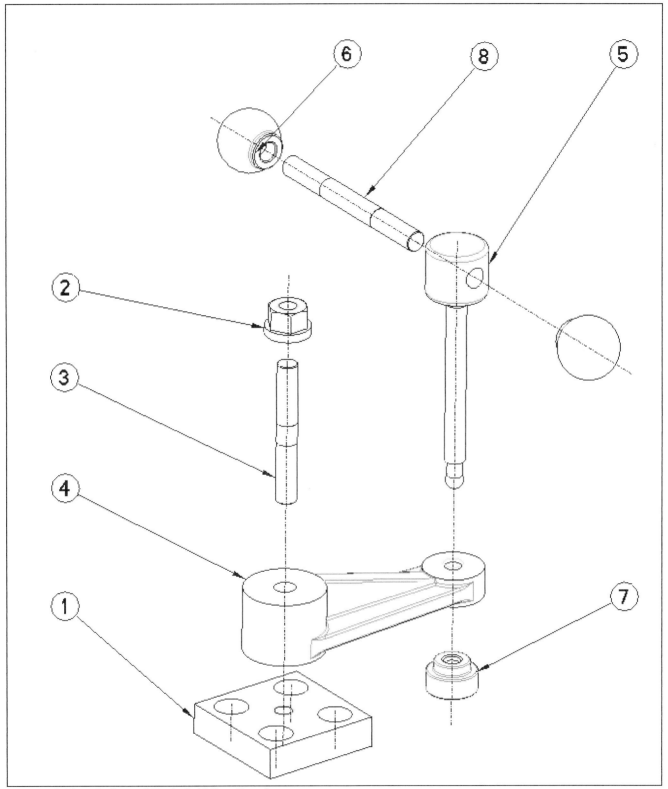

Figure 12.30(d) Ballooned Exploded View Drawing

Pick on pick balloon **6** ⇒ **RMB** [Fig. 12.30(e)] ⇒ **Edit Attachment** ⇒ **On Surface** ⇒ **Filled Dot** [Fig. 12.30(f)] ⇒ pick a place on the Clamp_Ball's surface ⇒ **MMB** ⇒ **LMB** [Fig. 12.30(g)] ⇒ ▨ ⇒ ▨ ⇒ ▨ ⇒ ▨ ⇒ **MMB** ⇒ **File** ⇒ **Delete** ⇒ **Old Versions** ⇒ **MMB** [Fig. 12.30(h)] ⇒ **Window** ⇒ **Close**

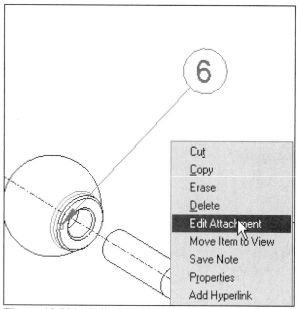

Figure 12.30(e) Edit Attachment

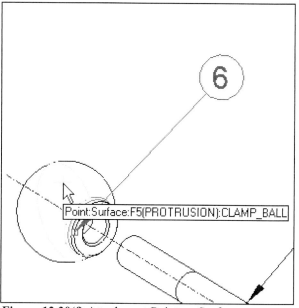

Figure 12.30(f) Attachment Point on Surface

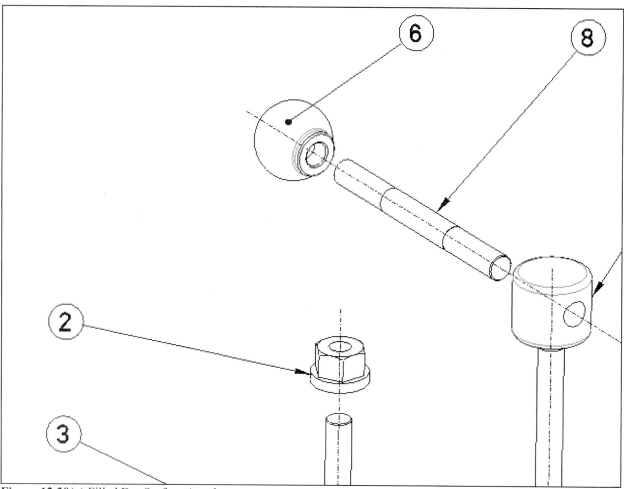

Figure 12.30(g) Filled Dot Surface Attachment

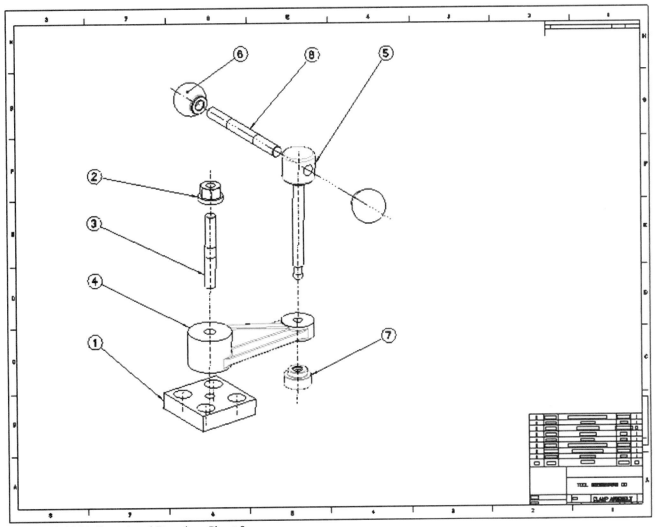

Figure 12.30(h) Completed Drawing, Sheet 2

Create a *documentation package* for the Clamp_Assembly. A complete documentation package contains all models and drawings required to manufacture the parts and assemble the components. Your instructor may change the requirements, but in general, create and plot/print the following:

- **Part Models** for all components

- **Detail Drawings** for each nonstandard component, such as the Clamp_Arm, Clamp_Swivel, Clamp_Foot, and Clamp_Ball *(do not detail the standard parts)*

- **Assembly Drawings** using standard orthographic ballooned views

- **Exploded Subassembly Drawing** of the ballooned subassembly

- **Exploded Assembly Drawing** of the ballooned assembly

A different assembly (Coupling) is available at ***www.cad-resources.com*** ⇒ ***Downloads***.

Notes:

Lessons 13-18 More Feature Tools

  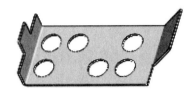

Lesson 13 Patterns

Lesson 14 Blends **Lesson 15 Sweeps**

Lesson 16 Helical Sweep **Lesson 17 Drafts** **Lesson 18 Shells**

Lesson 13 Patterns

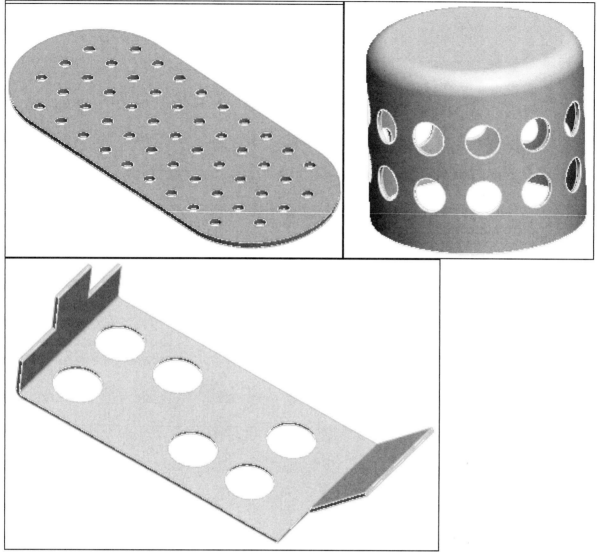

Figure 13.1 Parts with Patterns

OBJECTIVES

- Create **Patterns**
- Use **fill patterns**
- Create an **axial pattern**
- Model a **Sheetmetal** part and create a **directional pattern**

Patterns

Creating a pattern is a quick way to reproduce a feature. A pattern is parametrically controlled. Therefore, you can modify a pattern by changing pattern parameters, such as the number of instances, spacing between instances, and original feature dimensions. Modifying patterns is more efficient than modifying individual features. In a pattern, when you change dimensions of the original feature, Pro/E automatically updates the whole pattern.

This lesson will use direct modeling to introduce you to variations of the **Pattern Tool**.

Part Model (Plate- Fill Pattern)

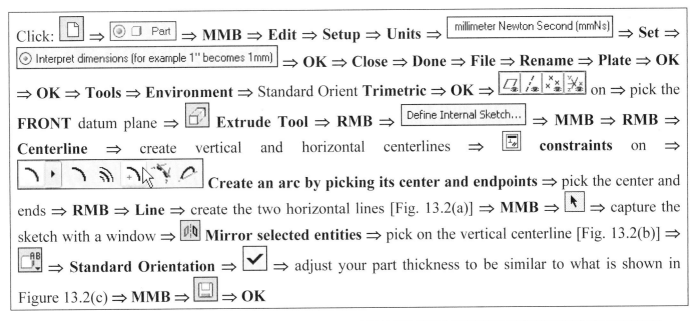

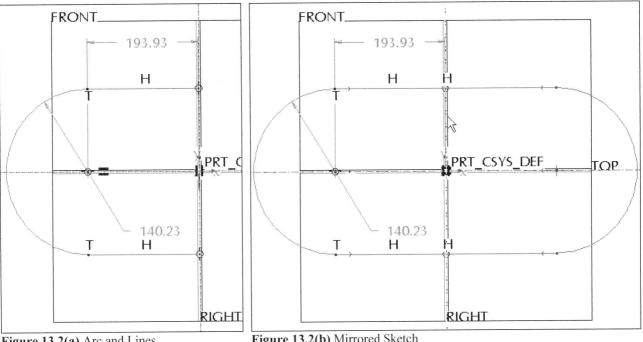

Figure 13.2(a) Arc and Lines   **Figure 13.2(b)** Mirrored Sketch

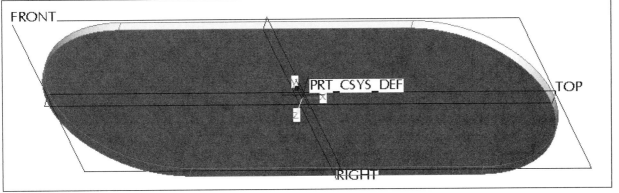

Figure 13.2(c) Extrusion

Click: **Hole Tool** ⇒ pick on the face of the part ⇒ **RMB** ⇒ [Secondary References Collector] ⇒ pick the **TOP** datum plane ⇒ press and hold **Ctrl** key ⇒ pick the **RIGHT** datum plane ⇒ **Placement** tab ⇒ make both Offset values **0.00** ⇒ [icon] ⇒ [icon] ⇒ change the diameter [⌀ 22.00] ⇒ **Enter** [Fig. 13.2(d)] ⇒ rotate the part [Fig. 13.2(e)] ⇒ [✓] from the dashboard ⇒ [icon] ⇒ **MMB** ⇒ with the Hole still selected in the Model Tree, click **RMB** ⇒ **Pattern** [Fig. 13.2(f)] ⇒ [Dimension] ⇒ [▼] ⇒ **Fill** [Fill] ⇒ **RMB** ⇒ [Define Internal Sketch...] ⇒ pick the upper face [Fig. 13.2(g)] ⇒ **Sketch** ⇒ [icon] **Create an entity from an edge** ⇒ [⊙ Loop] ⇒ pick the face [Fig. 13.2(h)] ⇒ [✓] [Fig. 13.2(i)]

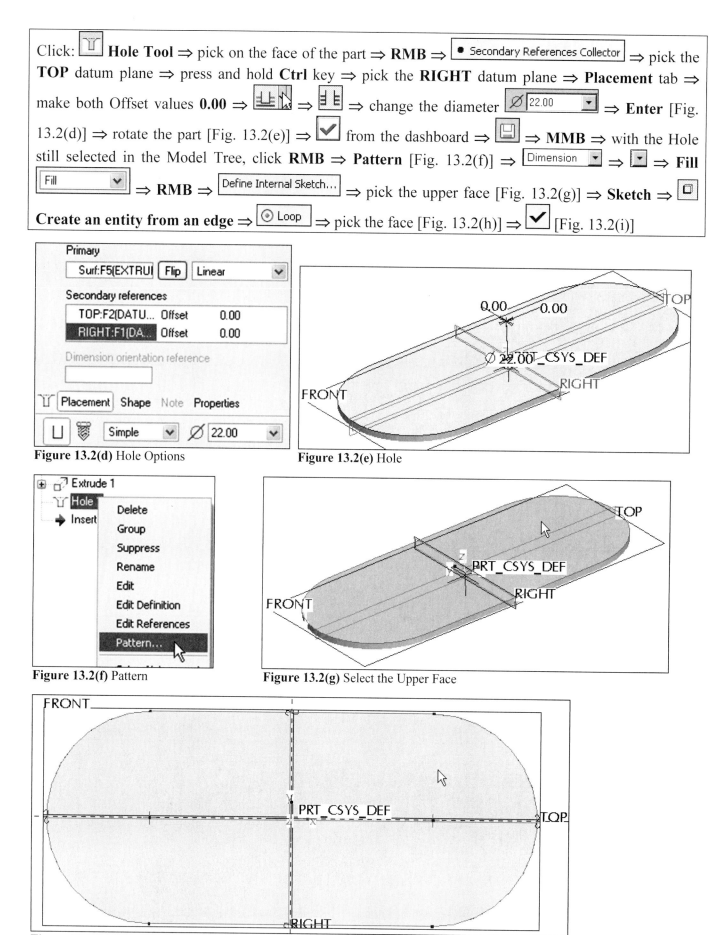

Figure 13.2(d) Hole Options   **Figure 13.2(e)** Hole

Figure 13.2(f) Pattern   **Figure 13.2(g)** Select the Upper Face

Figure 13.2(h) Select Surface to Specify as an Entity Loop

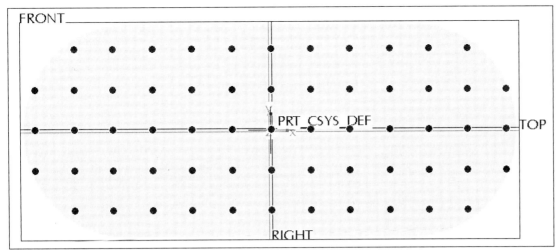
Figure 13.2(i) Pattern Preview

Pick on the unwanted copies that are near the parts edges [pick on a black dot and it changes to white (inactive)] [Fig. 13.2(j)] ⇒ [icon] ⇒ **Standard Orientation** ⇒ **MMB** to complete the command ⇒ **Edit** from the menu bar ⇒ **Scale Model** ⇒ .6 ⇒ **Enter** ⇒ **Yes** ⇒ [icon] ⇒ **MMB** ⇒ **File** ⇒ **Delete** ⇒ **Old Versions** ⇒ **MMB** ⇒ **View** ⇒ **Shade** [Fig. 13.2(k)] ⇒ **File** ⇒ **Close Window**

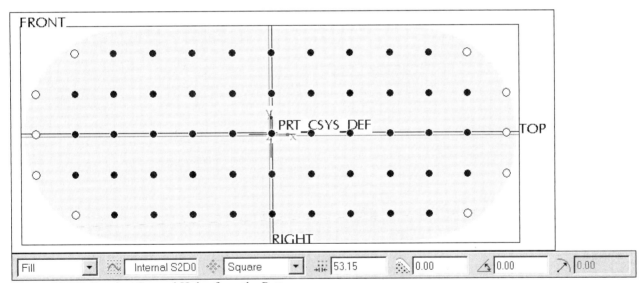
Figure 13.2(j) Deselect Several Holes from the Pattern

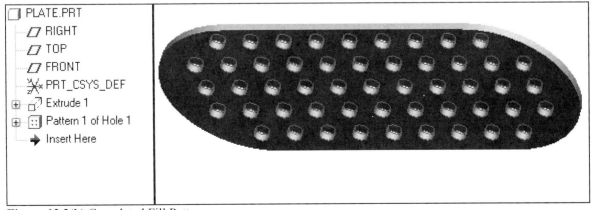
Figure 13.2(k) Completed Fill Pattern

Part Model (Cap- Axial Pattern)

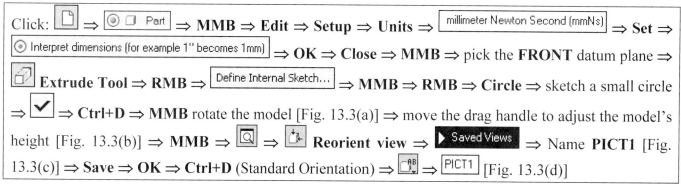

Click: ▢ ⇒ ⦿ ▢ Part ⇒ **MMB** ⇒ **Edit** ⇒ **Setup** ⇒ **Units** ⇒ [millimeter Newton Second (mmNs)] ⇒ **Set** ⇒ [⦿ Interpret dimensions (for example 1" becomes 1mm)] ⇒ **OK** ⇒ **Close** ⇒ **MMB** ⇒ pick the **FRONT** datum plane ⇒ [icon] **Extrude Tool** ⇒ **RMB** ⇒ [Define Internal Sketch...] ⇒ **MMB** ⇒ **RMB** ⇒ **Circle** ⇒ sketch a small circle ⇒ ✓ ⇒ **Ctrl+D** ⇒ **MMB** rotate the model [Fig. 13.3(a)] ⇒ move the drag handle to adjust the model's height [Fig. 13.3(b)] ⇒ **MMB** ⇒ [icon] ⇒ [icon] **Reorient view** ⇒ ▶ Saved Views ⇒ Name **PICT1** [Fig. 13.3(c)] ⇒ **Save** ⇒ **OK** ⇒ **Ctrl+D** (Standard Orientation) ⇒ [icon] ⇒ [PICT1] [Fig. 13.3(d)]

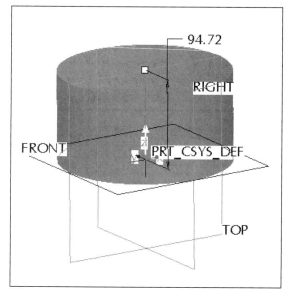

Figure 13.3(a) Circular Protrusion Preview

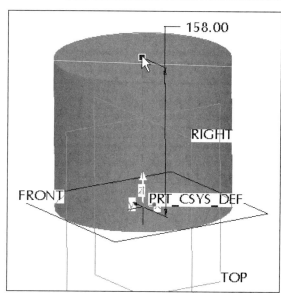

Figure 13.3(b) Change the Height

Figure 13.3(c) Orientation Dialog Box

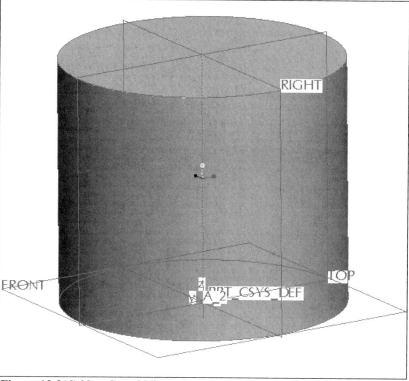

Figure 13.3(d) New Saved View

Pick on the top edge of the part ⇒ **RMB** ⇒ **Round Edges** [Fig. 13.4(a)] ⇒ **MMB** [Fig. 13.4(b)] ⇒
Hole Tool ⇒ pick on the vertical cylindrical face of the part ⇒ **RMB** ⇒ [Secondary References Collector] ⇒
pick the **RIGHT** datum plane [Fig. 13.5(a)] ⇒ press and hold **Ctrl** key ⇒ pick the **FRONT** datum plane
[Fig. 13.5(b)] ⇒ **Placement** tab ⇒ edit the values [Figs. 13.5(c-d)] ⇒ ⇒ **MMB** ⇒ **Ctrl+S** ⇒ **Enter**
⇒ **LMB** to deselect

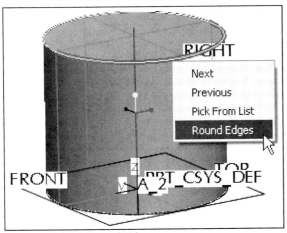

Figure 13.4(a) Round Edges

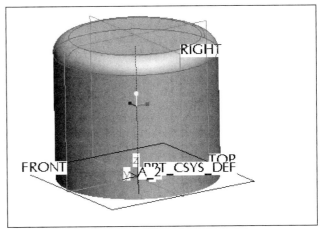

Figure 13.4(b) Completed Round

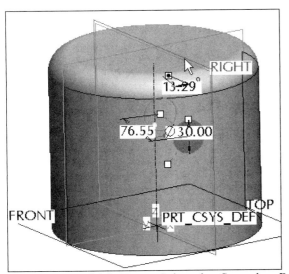

Figure 13.5(a) Right Datum Selected as Secondary Reference

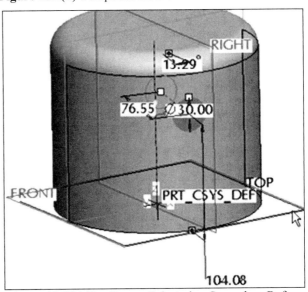

Figure 13.5(b) Top Datum Selected as Secondary Reference

Figure 13.5(c) Default Values

Figure 13.5(d) New Values and Depth

Pick on the hole ⇒ **Ctrl+C** ⇒ **Edit** ⇒ [Paste Special...] ⇒ check the options [Fig. 13.6(a)] ⇒ **OK** [Fig. 13.6(b)] ⇒ **Transformations** tab [Fig. 13.6(c)] ⇒ Direction reference- pick **Front** datum [Figs. 13.6(d-e)] ⇒ move the drag handle down [Fig. 13.6(f)] ⇒ **MMB** ⇒ **Ctrl+D** ⇒ **View** ⇒ **Orientation** ⇒ **Previous** ⇒ **LMB** to deselect ⇒ **Ctrl+S** ⇒ **Enter**

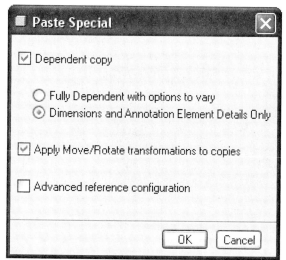

Figure 13.6(a) Paste Special Dialog Box

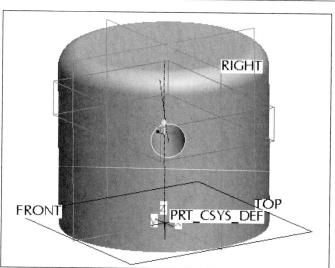

Figure 13.6(b) Paste Preview Box

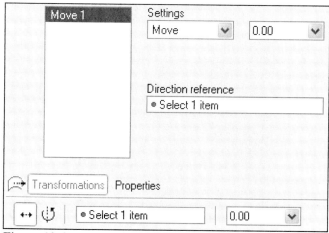

Figure 13.6(c) Transformations Tab

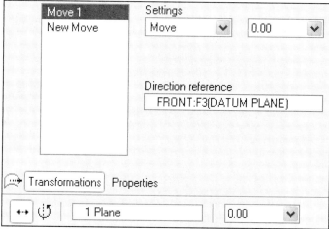

Figure 13.6(d) Direction reference FRONT

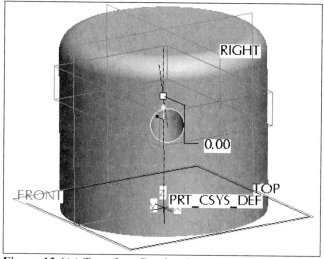

Figure 13.6(e) Transform Preview Box

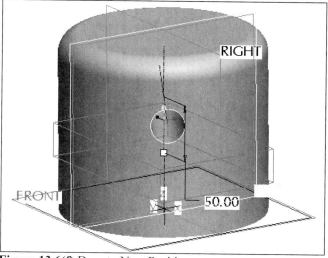

Figure 13.6(f) Drag to New Position

With the **Ctrl** key pressed, click on the **Hole** and the **Moved Copy** in the Model Tree ⇒ **RMB** ⇒ **Group** [Fig. 13.6(g)] ⇒ ⊞ ⇒ ⊞ [Fig. 13.6(h)] ⇒ **RMB** ⇒ **Pattern** [Fig. 13.6(i)] ⇒ **Dimensions** tab ⇒ **MMB** rotate the model to see the dimensions clearer ⇒ Direction 1- pick the **30°** dimension ⇒ pattern members, type **6** [1 | 6 | 1 item(s)] [Fig. 13.6(j)] ⇒ **Enter** ⇒ **MMB** [Fig. 13.6(k)] ⇒ [AB] ⇒ [PICT1] [Fig. 13.6(l)] ⇒ **LMB** to deselect ⇒ **Ctrl+S** ⇒ **Enter** [Fig. 13.6(m)]

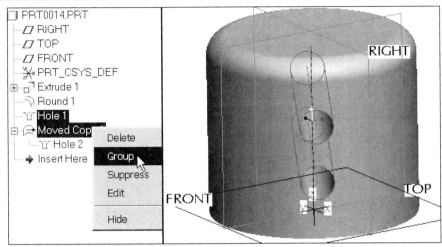

Figure 13.6(g) Grouping the Features

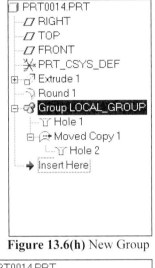

Figure 13.6(h) New Group

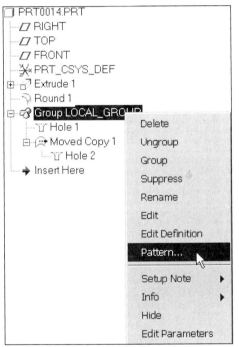

Figure 13.6(i) Pattern the Group

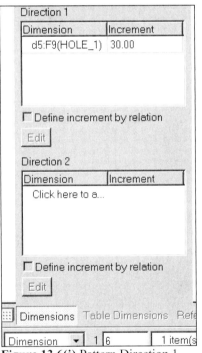

Figure 13.6(j) Pattern Direction 1

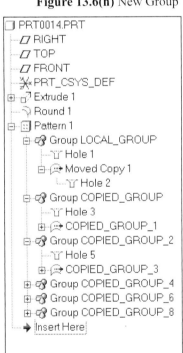

Figure 13.6(k) Pattern

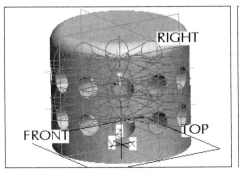
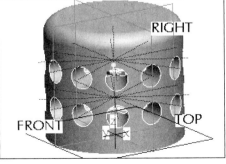

Figures 13.6(l-m) Patterned Holes

Pick on [Insert Here] in the Model Tree ⇒ drag **Insert Here** to a position before the **Pattern** [Pattern] and drop *(this will roll back the model to a state before the pattern was created and enter the Insert Mode)* (Fig. 13.7) ⇒ **MMB** rotate the model to see the bottom surface ⇒ pick the bottom surface [Fig. 13.8(a)] ⇒ **Shell Tool** [Fig. 13.8(b)] ⇒ [✓ 👓] [Fig. 13.8(c)] ⇒ [▶] ⇒ **MMB** ⇒ **Edit** ⇒ **Resume** ⇒ **Resume All** ⇒ **Ctrl+S** ⇒ **Enter** [Fig. 13.8(d)]

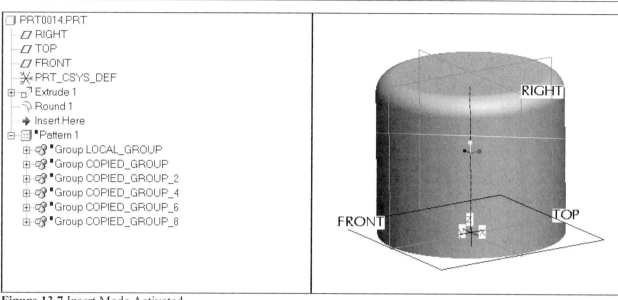

Figure 13.7 Insert Mode Activated

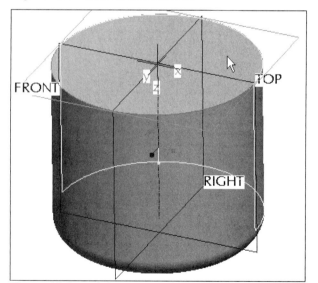

Figure 13.8(a) Pick on the Bottom Surface

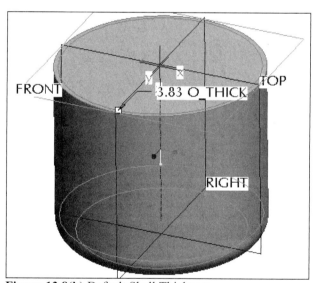

Figure 13.8(b) Default Shell Thickness

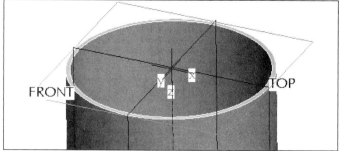

Figure 13.8(c) Shell Previewed

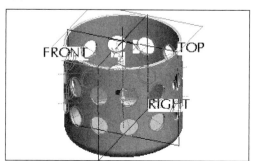

Figure 13.8(d) Resumed holes

Change the name and scale of the part: click: **File ⇒ Rename ⇒ Cap ⇒ OK ⇒ OK ⇒ Edit ⇒ Scale Model ⇒ .5** [Enter scale [1.0000]: .5] **⇒ Enter ⇒ Yes** [Fig. 13.9(a)] ⇒ in the Model Tree, pick on **Extrude ⇒ Shift** key ⇒ pick on **Shell ⇒ RMB ⇒ Edit** [Fig. 13.9(b)] ⇒ [AB] ⇒ [PICT1] [Fig. 13.9(c)] ⇒ [💾] ⇒ **MMB ⇒ File ⇒ Delete ⇒ Old Versions ⇒ MMB ⇒ File ⇒ Close Window ⇒ File ⇒ Erase ⇒ Not Displayed ⇒ OK**

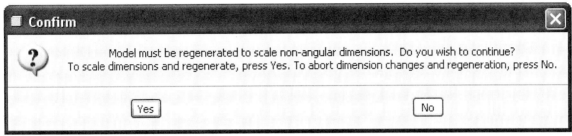

Figure 13.9(a) Confirm Dialog Box

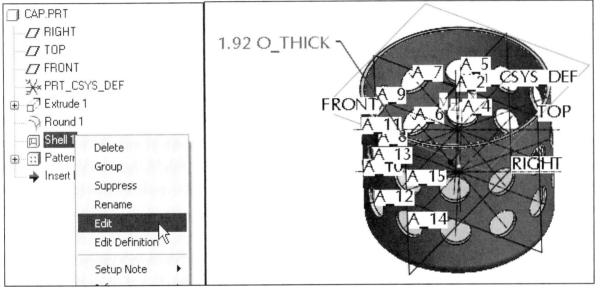

Figure 13.9(b) Edit

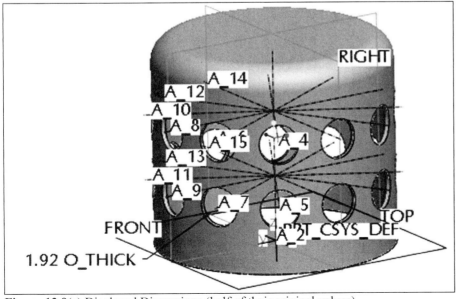

Figure 13.9(c) Displayed Dimensions (half of their original values)

Part Model (Sheetmetal- Directional Pattern)

The Pro/E part database is different when you create parts using Pro/SHEETMETAL. All sheet metal parts are by definition thin-walled constant-thickness parts. Because of this, sheet metal parts have some unique properties, which other Pro/E parts do no have. You may convert a solid part into a sheet metal part, but not a sheet metal part into a solid part.

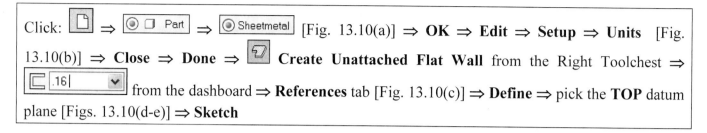

Click: ▢ ⇒ ● Part ⇒ ● Sheetmetal [Fig. 13.10(a)] ⇒ **OK** ⇒ **Edit** ⇒ **Setup** ⇒ **Units** [Fig. 13.10(b)] ⇒ **Close** ⇒ **Done** ⇒ **Create Unattached Flat Wall** from the Right Toolchest ⇒ .16 from the dashboard ⇒ **References** tab [Fig. 13.10(c)] ⇒ **Define** ⇒ pick the **TOP** datum plane [Figs. 13.10(d-e)] ⇒ **Sketch**

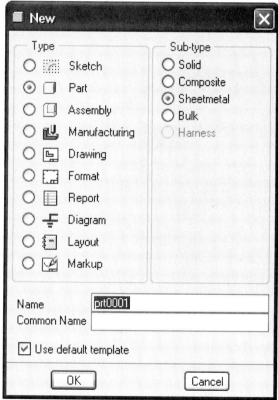

Figure 13.10(a) Sheetmetal

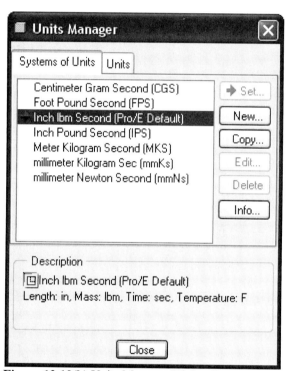

Figure 13.10(b) Units Manager Dialog Box

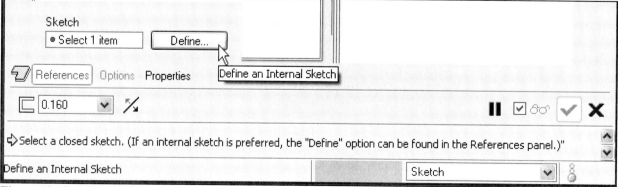

Figure 13.10(c) First Wall Dashboard

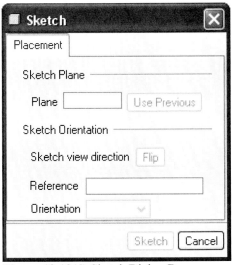

Figure 13.10(d) Sketch Dialog Box

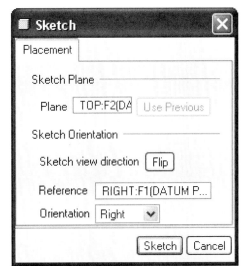

Figure 13.10(e) TOP Datum Selected for Sketch Plane

Click: **RMB** ⇒ **Rectangle** ⇒ sketch a rectangle [Fig. 13.10(f)] ⇒ ✓ ⇒ **Tools** ⇒ **Environment** ⇒ Standard Orient **Isometric** ⇒ **Apply** ⇒ **Close** ⇒ **Ctrl+D** ⇒ **Preview** ⇒ **MMB** ⇒ **MMB** ⇒ **Ctrl+S** ⇒ **Enter** [Fig. 13.10(g)] ⇒ **LMB** to deselect

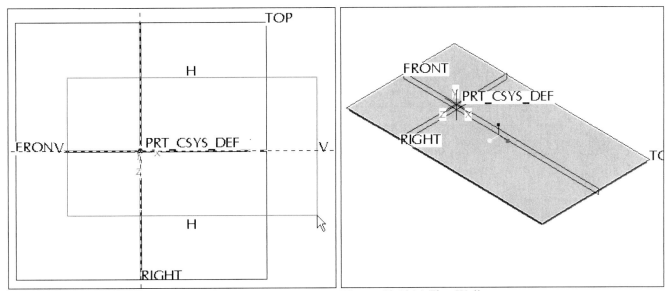

Figure 13.10(f) Sketch a Rectangle

Figure 13.10(g) First Wall

Change the scale of the part: click: **Edit** ⇒ **Scale Model** ⇒ **.02** [Enter scale [1.0000]: .02] ⇒ **Enter** ⇒ **Yes** ⇒ 🔍 ⇒ 💾 ⇒ **MMB** ⇒ double-click on the wall to see all of the dimensions [Fig. 13.10(h)] ⇒ 🔧 **Create Flat Wall** ⇒ pick on the upper edge of the first wall [Fig. 13.10(i)] ⇒ **Rectangle** ⇒ ▼ ⇒ **T** [Figs. 13.10(j-k)] ⇒ **Shape** tab [Fig. 13.10(l)] ⇒ **MMB** ⇒ 🔍 ⇒ 💾 ⇒ **MMB** [Fig. 13.10(m)]

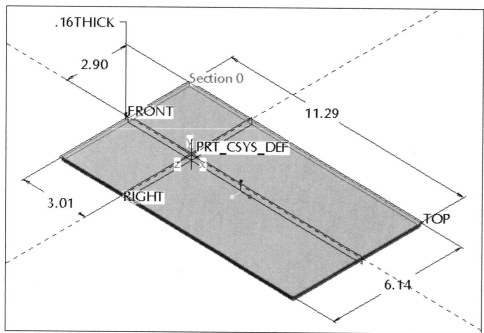

Figure 13.10(h) Walls Dimensions Displayed (yours will be different)

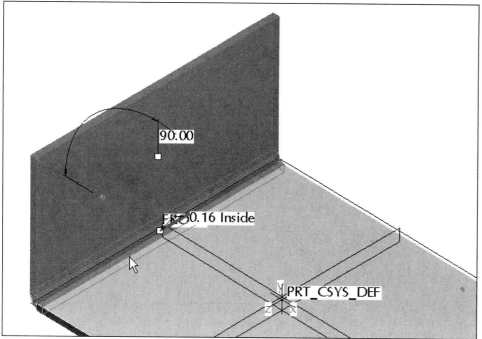

Figure 13.10(i) Pick on Upper Edge of First Wall

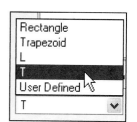

Figure 13.10(j) Select **T**

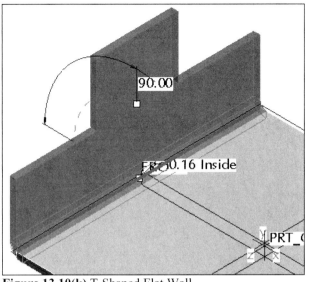

Figure 13.10(k) T-Shaped Flat Wall

Figure 13.10(l) Shape Tab

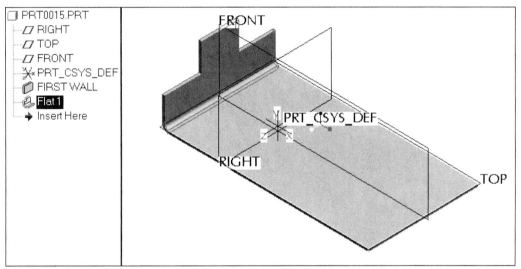

Figure 13.10(m) Flat Wall

Click: **Create Flange Wall** ⇒ pick on the opposite edge of the first wall [Fig. 13.10(n)]

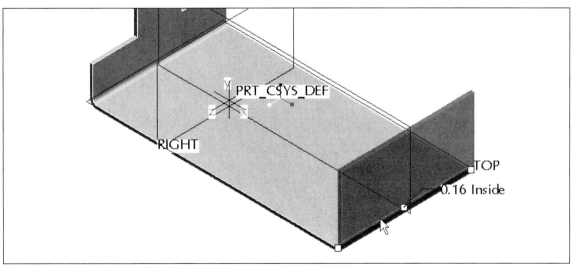

Figure 13.10(n) Flange Wall

Click: **Profile** tab [Fig. 13.10(o)] ⇒ **45** ⇒ **Enter** ⇒ **MMB** rotate model [Fig. 13.10(p)] ⇒ move a drag handle to shorten the wall [Fig. 13.10(q)] ⇒ **MMB**

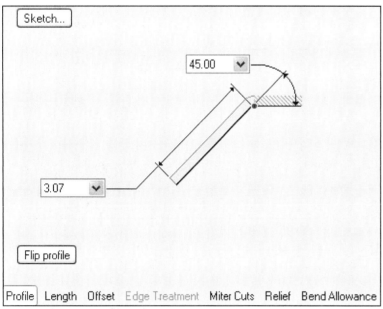
Figure 13.10(o) Profile Tab

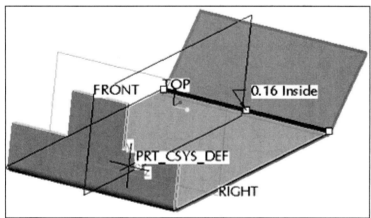
Figure 13.10(p) Profile Preview

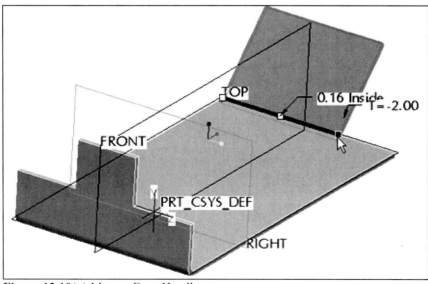
Figure 13.10(q) Move a Drag Handle

MMB rotate model ⇒ **Ctrl+S** ⇒ **Enter** ⇒ **File** ⇒ **Rename** ⇒ **Sheetmetal** ⇒ **OK** ⇒ **OK** ⇒ click on the three features in the Model Tree ⇒ **Edit** [Fig. 13.10(r)] ⇒ **MMB** rotate model

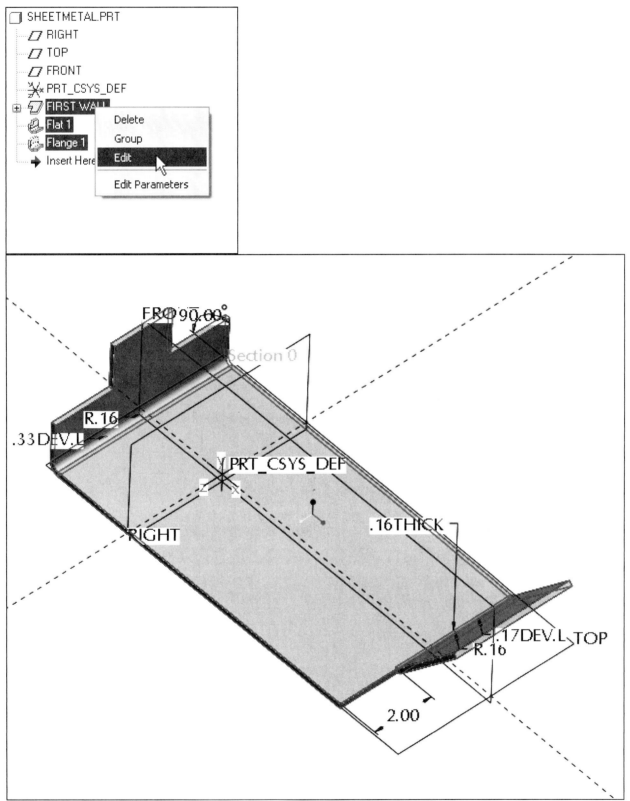

Figure 13.10(r) Edit

Click: **Insert** from the menu bar ⇒ [Hole...] ⇒ pick on the face of the part [Fig. 13.10(s)] ⇒ **RMB** ⇒ [Secondary References Collector] ⇒ pick the front edge [Fig. 13.10(t)] ⇒ press and hold **Ctrl** key ⇒ pick on the adjacent edge [Fig. 13.10(u)] ⇒ adjust dimensions ⇒ [icon] [Fig. 13.10(v)] ⇒ **MMB** [Fig. 13.10(w)]

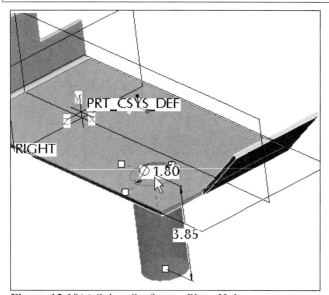
Figure 13.10(s) Select Surface to Place Hole

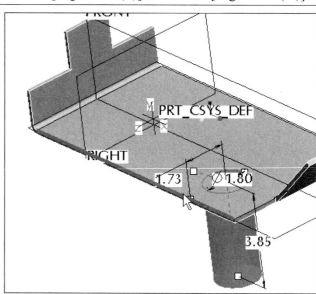
Figure 13.10(t) First Edge Reference

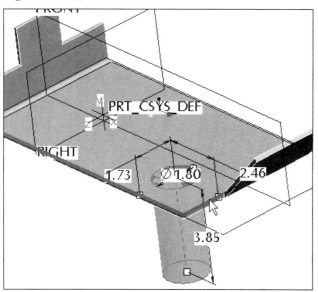
Figure 13.10(u) Second Edge Reference

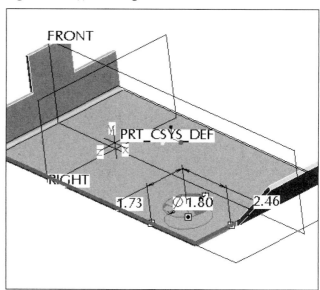
Figure 13.10(v) Through All

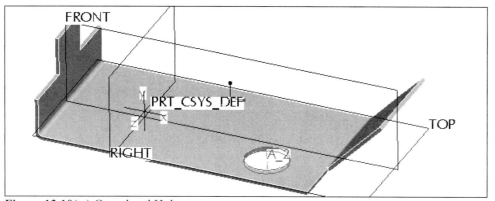
Figure 13.10(w) Completed Hole

With the Hole selected in the Model Tree, click: **RMB** ⇒ **Pattern** ⇒ **Ctrl+MMB** zoom in ⇒ **Dimension** ⇒ ⇒ **Direction** ⇒ pick on the same first edge used to locate the hole [Fig. 13.10(x)] ⇒ **Flip the first direction** ⇒ type **4** as the number of holes in this direction ⇒ **Enter** ⇒ move the drag handle to space the holes [Fig. 13.10(y)] ⇒ **Click here to add item** ⇒ **Select 1 item** ⇒ pick on the same second edge used to locate the hole [Fig. 13.10(z)] ⇒ move the second direction drag handle [Fig. 13.11(a)]

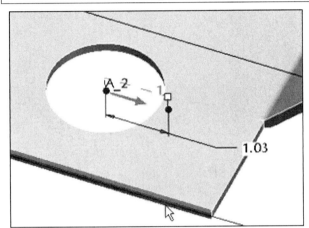

Figure 13.10(x) First Direction

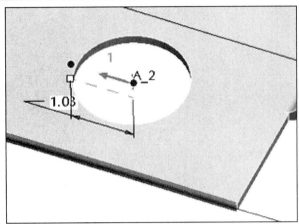

Figure 13.10(y) First Direction Reversed

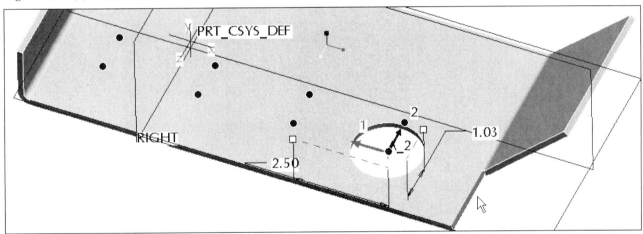

Figure 13.10(z) Second Direction

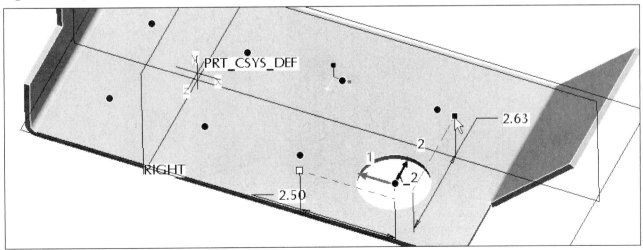

Figure 13.11(a) Second Direction Adjusted

Pick on the two pattern hole preview dots (black) to remove them from the pattern (dots turn white) [Fig. 13.11(b)] ⇒ **MMB** [Fig. 13.11(c)] ⇒ **LMB** to deselect

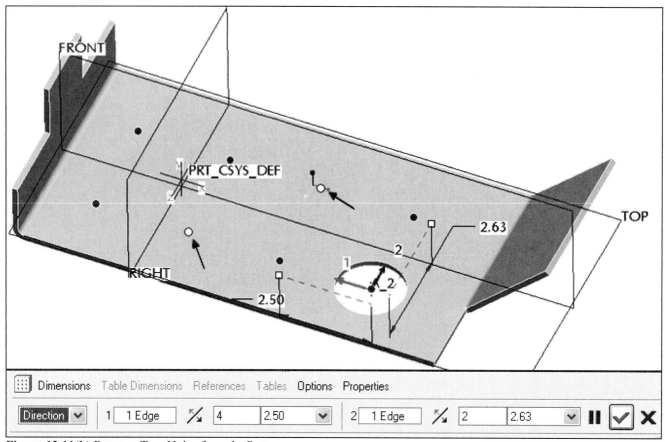

Figure 13.11(b) Remove Two Holes from the Pattern

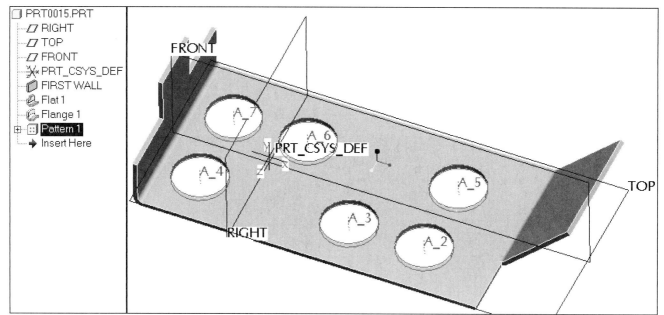

Figure 13.11(c) Completed Hole Pattern

Click: ▢ **Refit** ⇒ **View** ⇒ **Orientation** ⇒ **Standard Orientation** ⇒ ▢ **Create Flat Pattern** [Fig.13.12(a)] ⇒ pick on the upper surface [Fig. 13.12(b)] ⇒ **Edit** ⇒ **Undo: Flat Pattern** ⇒ **Info** ⇒ **Model Size** ⇒ Length of bounding box diagonal = 15.6217 INCH (your size will be different) [Fig. 13.12(c)] ⇒ **Ctrl+S** ⇒ **Enter** ⇒ **File** ⇒ **Close Window**

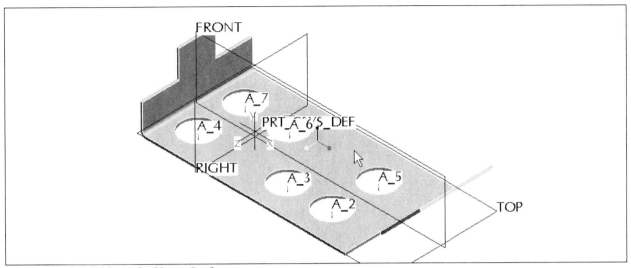

Figure 13.12(a) Pick on the Upper Surface

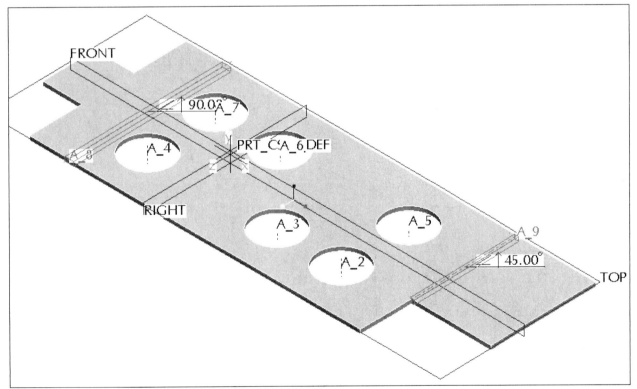

Figure 13.12(b) Completed Flat Pattern

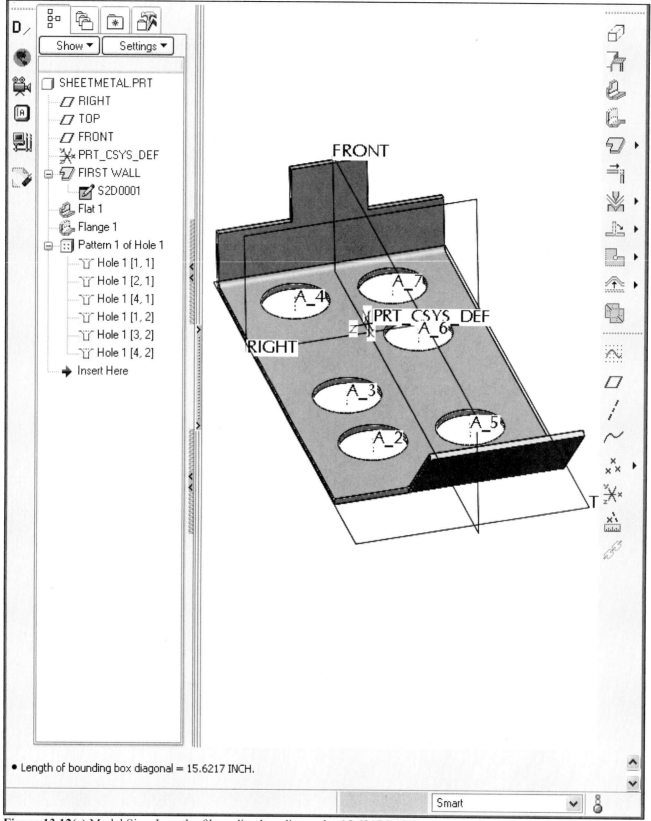

Figure 13.12(c) Model Size: Length of bounding box diagonal = 15.6217 INCH.

For those who wish additional information on surfaces, see ***www.cad-resources.com*** ⇒ ***Downloads***, for extra projects.

Lesson 14 Blends

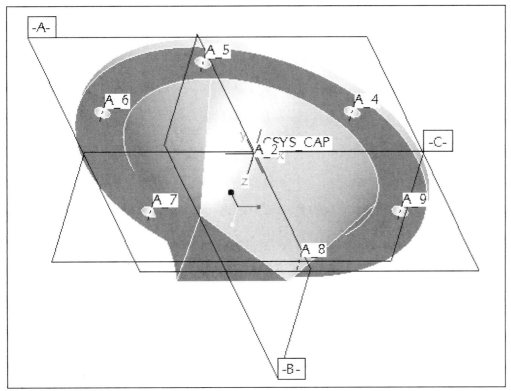

Figure 14.1 Cap

OBJECTIVES

- Create a **Parallel Blend** feature
- Use the **Shell Tool**
- Create a **Swept Blend**

BLENDS

A blended feature consists of a series of at least two planar sections that are joined together at their edges with transitional surfaces to form a continuous feature. The Cap (Figs. 14.1 and 14.2) uses a blend feature in its design. A Blend can be created as a **Parallel Blend**, or you can construct a **Swept Blend**.

Blend Sections

Blended surfaces are created between the corresponding sections. Figure 14.3 shows a parallel blend for which the *section* consists of several *subsections*. Each segment in the subsection is matched with a segment in the following subsection; to create the transitional surfaces; Pro/E connects the *starting points* of the subsections and continues to connect the vertices of the subsections in a clockwise manner. By changing the starting point of a blend subsection, you can create blended surfaces that twist between the subsections. The default starting point is the first point sketched in the subsection. You can position the starting point to the endpoint of another segment by choosing the option Start Point and selecting the new position.

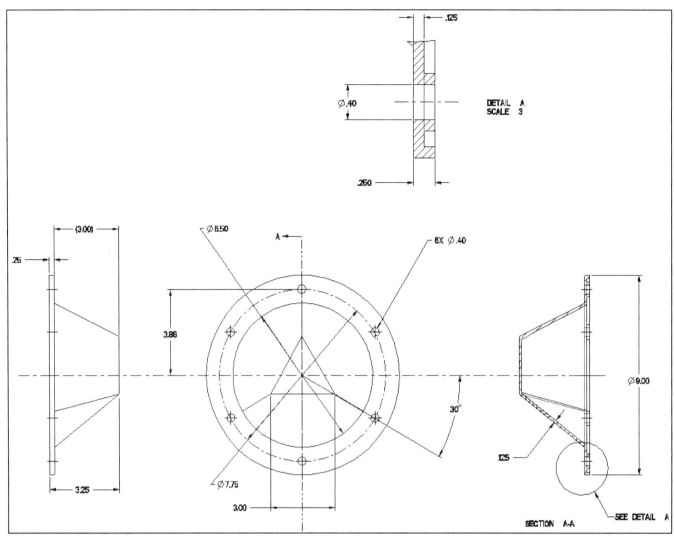

Figure 14.2 Cap Detail

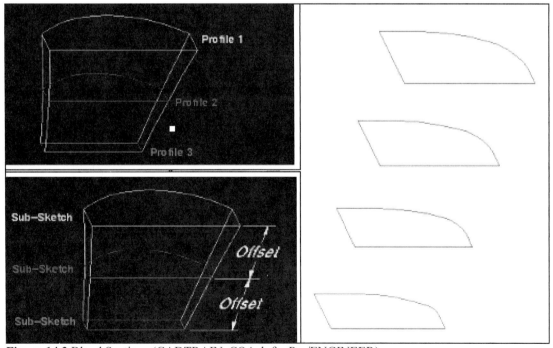

Figure 14.3 Blend Sections (CADTRAIN, COAch for Pro/ENGINEER)

Blend Options

Blends use one of the following transitional surface options:

- **Straight** Create a straight blend by connecting vertices of different subsections with straight lines. Edges of the sections are connected with ruled surfaces.
- **Smooth** Create a smooth blend by connecting vertices of different subsections with smooth curves. Edges of the sections are connected with ruled (spline) surfaces.
- **Parallel** All blend sections lie on parallel planes in one section sketch.
- **Rotational** The blend sections are rotated about the **Y** axis, up to a maximum of **120°**. Each section is sketched individually and aligned using the coordinate system of the section.
- **General** The sections of a general blend can be rotated about and translated along the **X**, **Y**, and **Z** axes. Sections are sketched individually and aligned using the coordinate system of the section.
- **Regular Sec** The feature will use the regular sketching plane.
- **Project Sec** The feature will use the projection of the section on the selected surface. This is used for parallel blends only.
- **Select Sec** Select section entities (not available for parallel blends).
- **Sketch Sec** Sketch section entities.

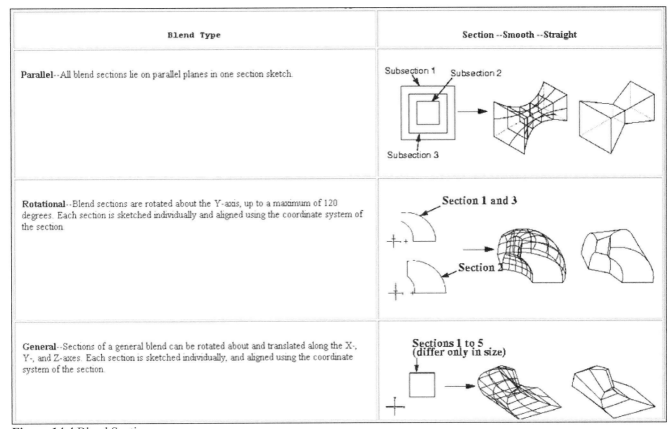

Figure 14.4 Blend Sections

Parallel Blends

A parallel blend (Fig. 14.5) is created from a single section that contains multiple sketches called *subsections*. A first or last subsection can be defined as a point resulting in a blend vertex. The starting point for each subsection must be carefully selected as per the design requirements (Fig. 14.6).

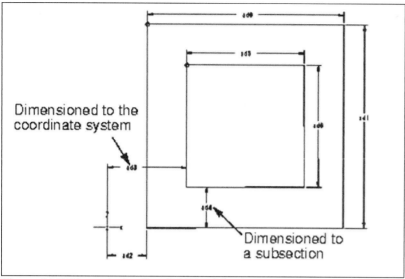

Figure 14.5 Dimensioning Parallel Blend Sections

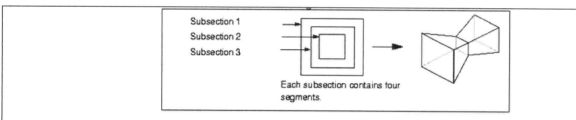

Starting Point of a Section

To create the transitional surfaces, Pro/ENGINEER connects the starting points of the sections and continues to connect the vertices of the sections in a clockwise manner. By changing the starting point of a blend subsection, you can create blended surfaces that twist between the sections (see the illustration Starting Points and Blend Shape).

The default starting point is the first point sketched in the subsection. You can place the starting point at the endpoint of another segment by choosing the option **Start Point** from the
SEC TOOLS menu and selecting the point.

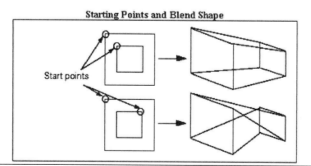

Figure 14.6 Starting Points

Rotational and General Blends

You create rotation blends (Fig. 14.7) by sketching a series of profiles which are offset by an angle. A rotational blend is created by sections that are rotated about the Y axis.

In a general blend (Fig. 14.8), sections can be rotated about and translated along the X, Y, and Z axes. Each section is sketched individually, and aligned using the coordinate system of the section.

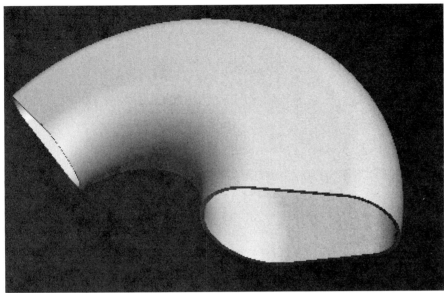

Figure 14.7 Blend Sections (CADTRAIN, COAch for Pro/ENGINEER)

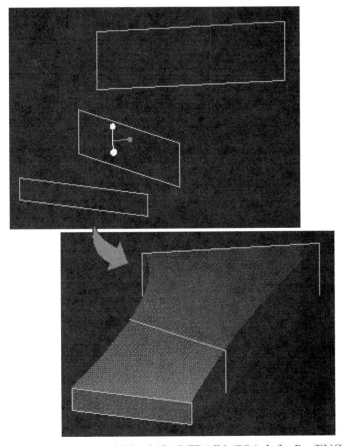

Figure 14.8 General Blend (CADTRAIN, COAch for Pro/ENGINEER)

Lesson 14 STEPS

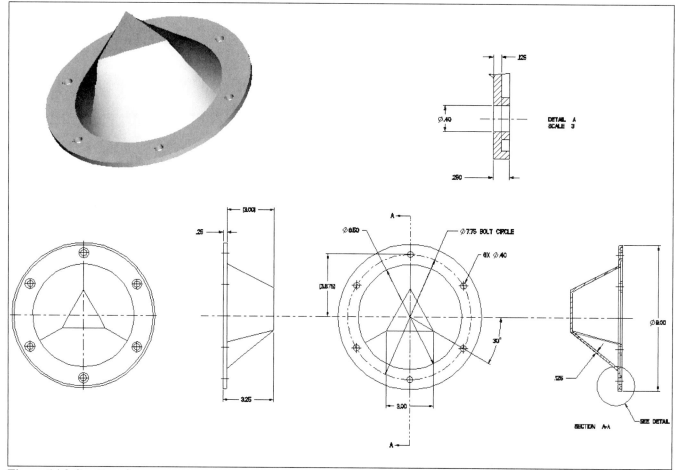

Figure 14.9 Cap Drawing

Cap

The Cap is a part created with a Parallel Blend (Figs. 14.9 through 14.16). The blend sections are a circle and a triangle. Because *the sections of a blend must have equal segments*, the "circle" is actually three equal arcs. The part is shelled as the last feature in its creation. The Shell Tool will create *bosses* around each hole as it hollows out the part. A cross section will be created in the Part mode to be used when you are detailing the Cap in the Drawing mode.

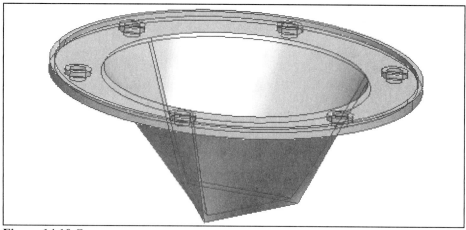

Figure 14.10 Cap

Start a new part. Click: **Create a new object** ⇒ Part ⇒ Name **CAP** ⇒ Use default template ⇒ **MMB** ⇒ **Tools** ⇒ **Environment** ⇒ Standard Orient **Trimetric** ⇒ **Apply** ⇒ **OK** ⇒ **Edit** ⇒ **Setup**

- **Material** = PVC (pvc.mtl)
- **Units** = Inch lbm Second

Set datums (using **Geom Tol**)
- Datum TOP = **C**
- Datum FRONT = **A**
- Datum RIGHT = **B**

Rename the coordinate system
- Coordinate System = **CSYS_CAP**

Tools ⇒ **Options**
- *sketcher_dec_places*   3
- *default_dec_places*   3
- *tol_mode*   plusminus

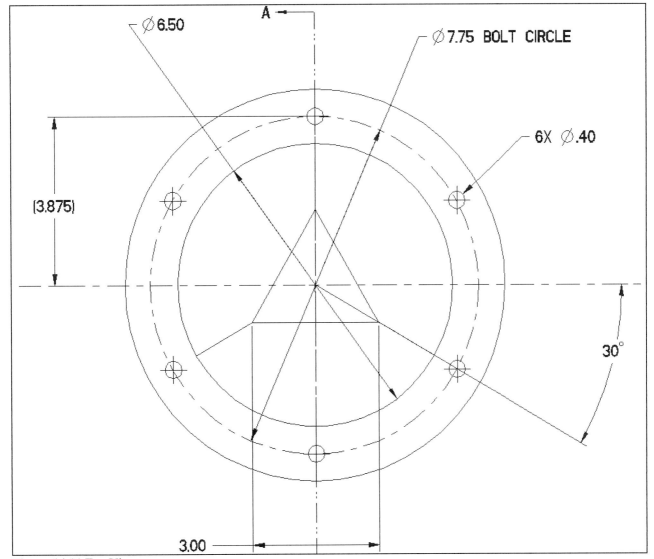

Figure 14.11 Top View

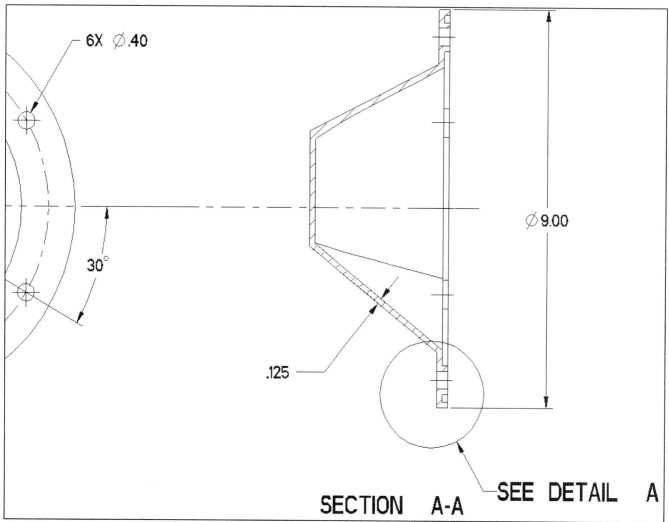

Figure 14.12 SECTION A-A

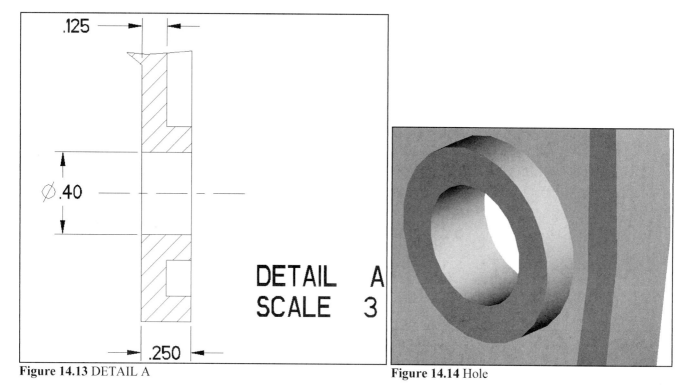

Figure 14.13 DETAIL A

Figure 14.14 Hole

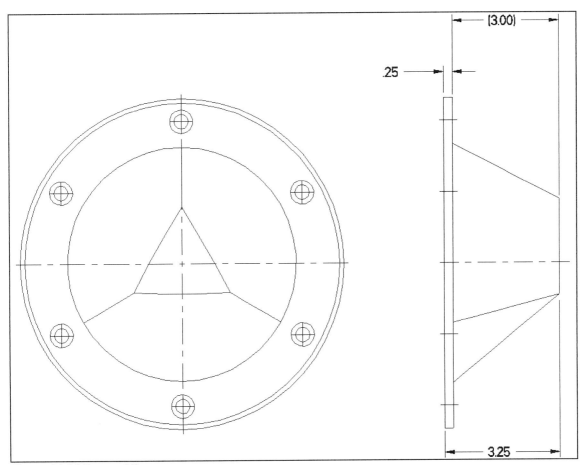

Figure 14.15 Bottom View

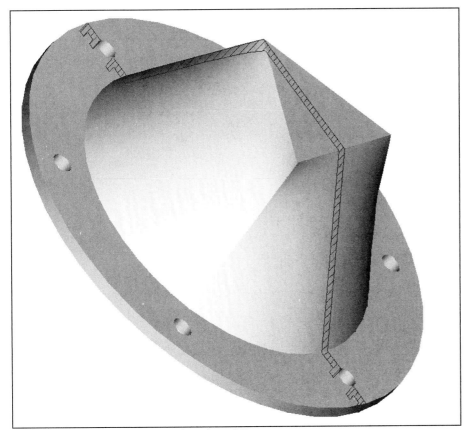

Figure 14.16(a) Section Through Hole and Boss

Figure 14.16(b) Hole and Boss

Model the circular protrusion that is ⌀**9.00** by **.25** thick shown in Figure 14.17. Sketch the first protrusion on datum **A (FRONT)** and centered on **B (RIGHT)** and **C (TOP)**.

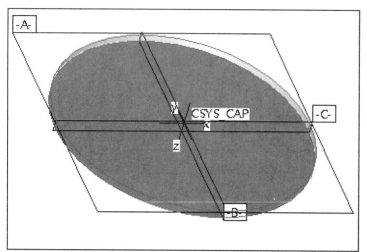

Figure 14.17 First Protrusion

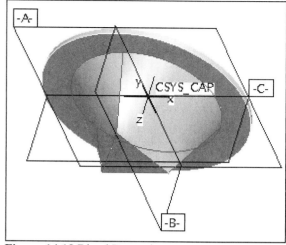

Figure 14.18 Blend Protrusion

Create the blend protrusion (Fig. 14.18), click: **Insert** ⇒ **Blend** ⇒ **Protrusion** ⇒ **Parallel** ⇒ **Regular Sec** ⇒ **Sketch Sec** ⇒ **MMB** ⇒ **Straight** ⇒ **MMB** ⇒ pick the top surface of the first protrusion (Fig. 14.19) ⇒ **MMB** to confirm direction of feature creation ⇒ **MMB** for default orientation ⇒ **Toggle the grid** on (Fig. 14.20)

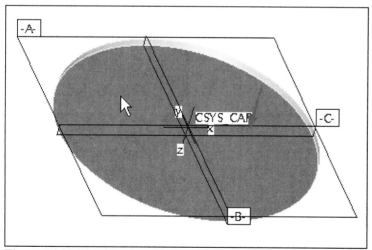

Figure 14.19 Blend Feature Starting Surface and Direction of Creation

The section grid would be better utilized if it were a Polar grid rather than a Cartesian grid.

Change the grid type and size, click: **Sketch** ⇒ **Options** ⇒ **Miscellaneous** tab ⇒ ☑ Snap to Grid ⇒ **Parameters** tab ⇒ ⊙ Polar ⇒ Grid Spacing **Manual** ⇒ Radial **.50** ⇒ Angular **30°** ⇒ ✓ (Fig. 14.20) ⇒ **Hidden Line** (Fig. 14.21) ⇒ add three centerlines (Figs. 14.22 to 14.25)

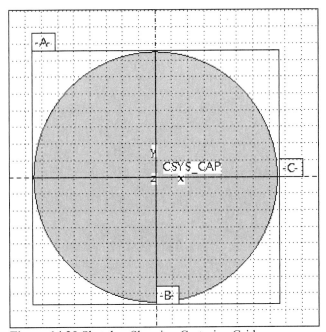
Figure 14.20 Sketcher Showing Cartesian Grid

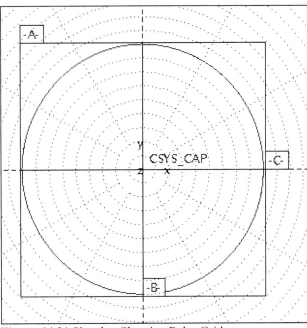
Figure 14.21 Sketcher Showing Polar Grid

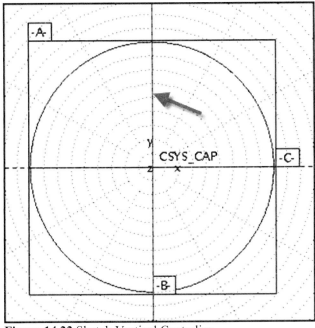
Figure 14.22 Sketch Vertical Centerline

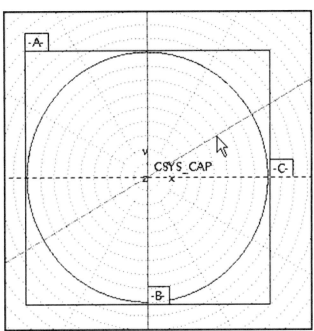
Figure 14.23 Sketch 30 Degree Centerline

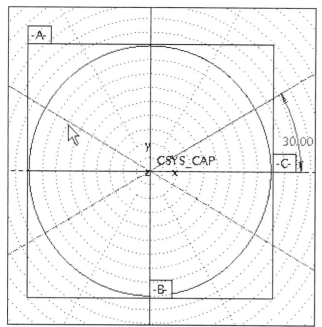
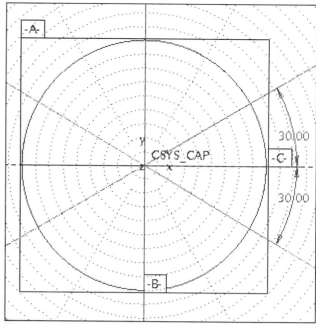

Figure 14.24 Sketch Second **30** Degree Centerline   **Figure 14.25** Three Centerlines

Sketch the *first section* of the blend by creating *three* equal **120°** arcs, click: [⌒] *flyout* ⇒ [⌒] **Create an arc by picking its center and endpoints** ⇒ Sketch each arc in a counterclockwise direction. Note the location of the start point for this section (Fig.14.26). ⇒ add a diameter dimension (Fig. 14.26)

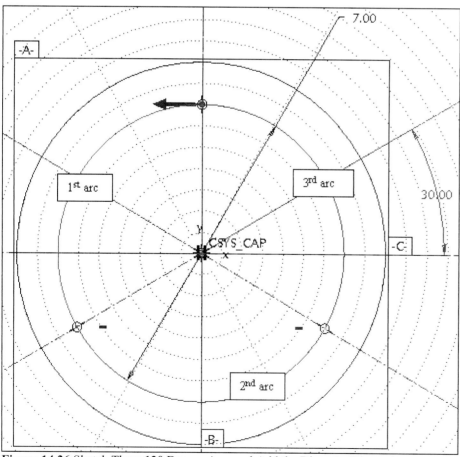

Figure 14.26 Sketch Three **120** Degree Arcs and Add the Diameter Dimension

Click: **RMB** ⇒ **Toggle Section** ⇒ sketch the second parallel section (the first section is *grayed* out) ⇒ ⬜ ⇒ ⬜ **Create 2 point lines** ⇒ sketch the three lines of the triangle starting at the top so that the start point is near the start point of the first section and picking points *in the same direction* in which the arcs were created (Fig. 14.27) ⇒ dimension and modify [Figs. 14.28(a-c)] ⇒ ✓ ⇒ add the two **120** degree dimensions (Fig. 14.29) ⇒ ✓ **Continue with the current section**

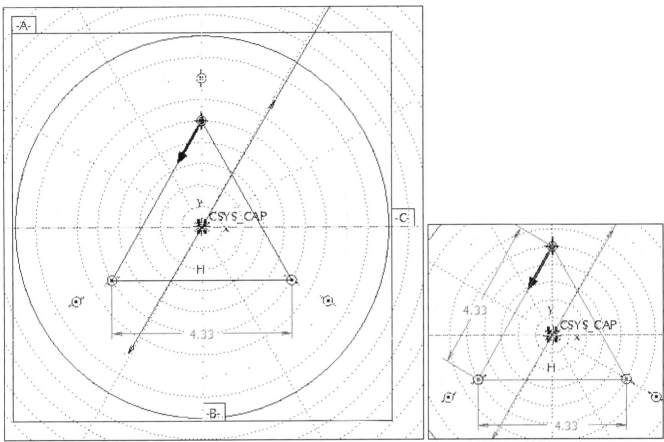

Figure 14.27 Sketch Three Lines

Figure 14.28(a) Add Dimensions

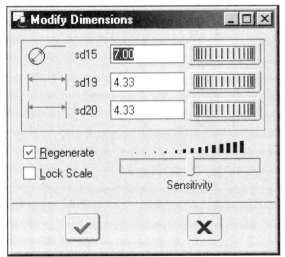

Figure 14.28(b) Modify the Dimensions

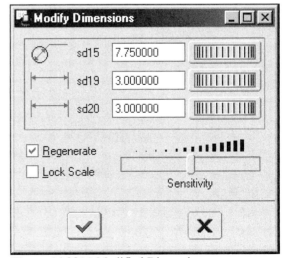

Figure 14.28(c) Modified Dimensions

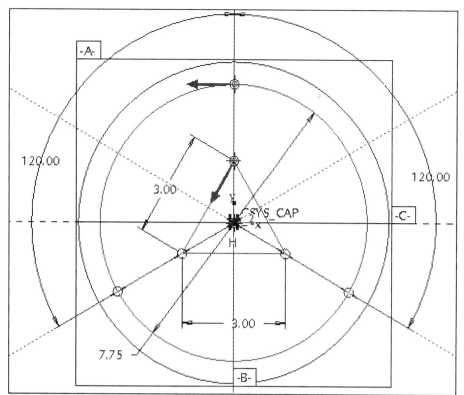

Figure 14.29 Add the **120** Degree Dimensions

Click: **Blind** ⇒ **MMB** ⇒ type **3.00** at the prompt ⇒ **MMB** ⇒ **Preview** ⇒ **MMB** ⇒ with the blend feature highlighted on the model and Model Tree, click **RMB** ⇒ **Edit** ⇒ double-click on the **7.75** dimension and change the diameter to **6.50** (Fig. 14.30) ⇒ **Enter** ⇒ [icon] **Regenerates Model** ⇒ **Ctrl+D** ⇒ [icon] ⇒ [icon] ⇒ **MMB** (Fig. 14.31)

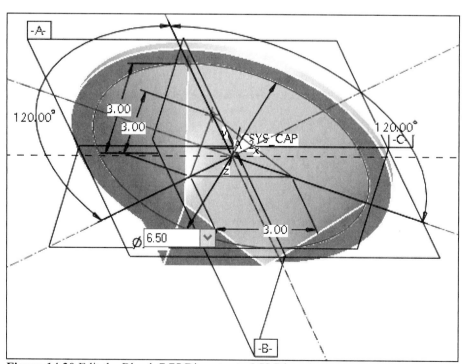

Figure 14.30 Edit the Blend, **7.75** Diameter to **6.50** Diameter

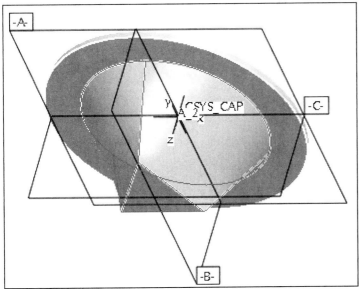

Figure 14.31 Completed Blend

Create and pattern the six equally spaced holes Ø**.400** on a Ø**7.75** bolt circle, click: **Hole Tool** ⇒ complete the hole with the options and references provided in Figure 14.32 ⇒ **LMB** to deselect

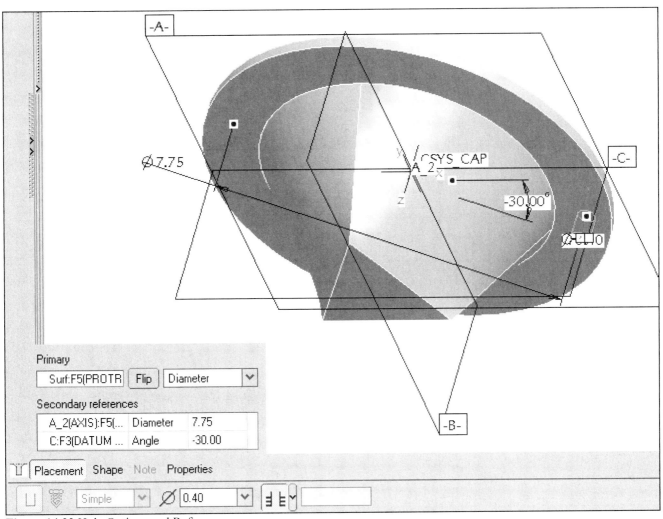

Figure 14.32 Hole Options and References

Click on the hole (on the model or in the Model Tree) ⇒ **RMB** ⇒ **Pattern** ⇒ Complete the pattern with the options and references provided in Figure 14.33 and Figure 14.34. (Hint: select the **30** degree dimension to pattern). ⇒ **LMB** to deselect

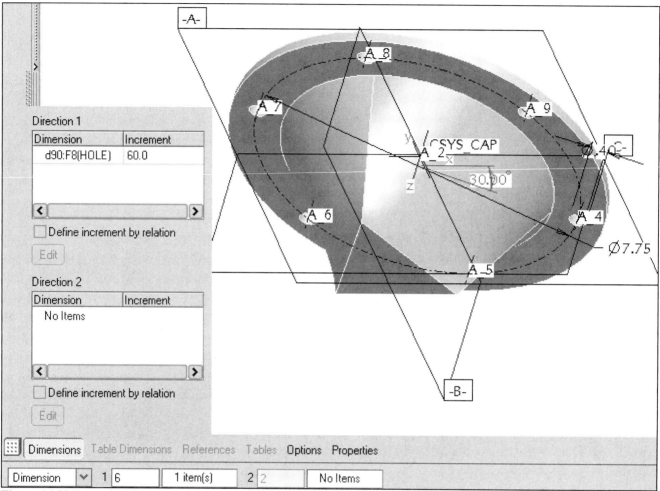

Figure 14.33 Patterning the Hole

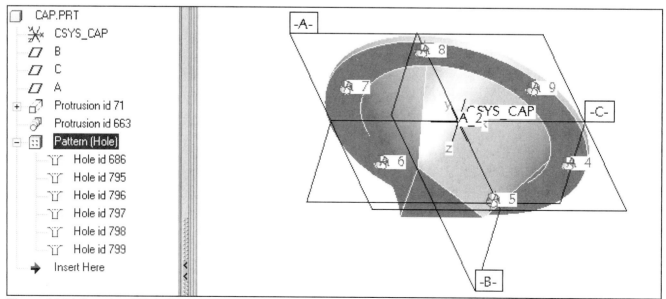

Figure 14.34 Completed Hole Pattern

Shell the part, click: ▭ **Shell Tool** ⇒ pick the bottom surface of the part as the surface to remove [Fig. 14.35(a)] ⇒ Thickness **.125** ⇒ **Enter** ⇒ ☑👓 [Fig. 14.35(b)] ⇒ ▶ ⇒ ✓ (Fig. 14.36)

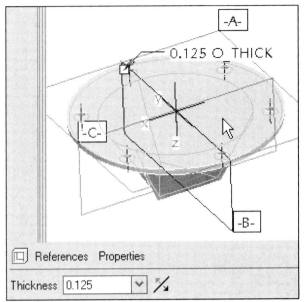

Figure 14.35(a) Shell Tool

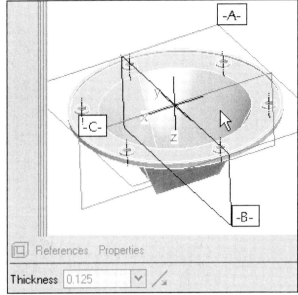

Figure 14.35(b) Shell Preview

The shell will automatically create the *bosses* (Fig. 14.36), because the **.125** thickness is left around all previously created features. The *bosses* around the holes are created automatically at **.250** larger than the holes: (**.125** + **.125** + **.400** = ∅**.650**). If the *bosses* were not desired, you would simply reorder the holes to come after the shell feature.

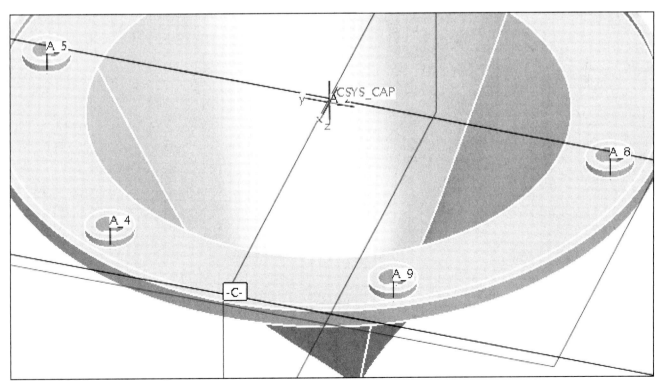

Figure 14.36 Completed Shell

Measure the size of the boss, click: **Analysis** ⇒ **Measure** ⇒ **Diameter** [Figs. 14.37(a-c)] ⇒ pick the vertical surface of a boss ⇒ ✓

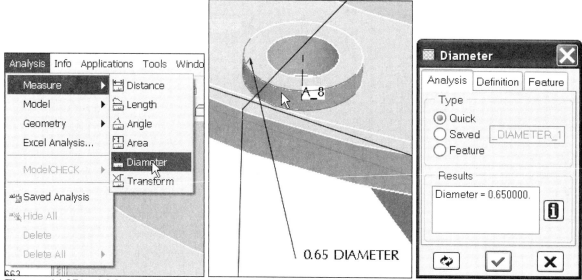

Figures 14.37(a-c) Measure (Results **Diameter = 0.650000**)

Now create a cross section to be used in the Drawing mode when detailing the Cap.

Click: 📷 **Start the view manager** ⇒ Xsec **Create Cross Sections** ⇒ **New** ⇒ type **A** ⇒ **Enter** ⇒ **Planar** ⇒ **Single** ⇒ **MMB** ⇒ pick datum **B** ⇒ **Display** ⇒ **Xhatching** (Fig. 14.38) ⇒ **Close** ⇒ 📋 ⇒ **Standard Orientation** ⇒ 🔍 ⇒ 💾 ⇒ **MMB**

Figure 14.38 SECTION A-A

Save the Cap under a new name, and complete the **ECO** [Figs. 14.39(a-b)] to establish a second part with different dimensional sizes and features. Add **R.20** rounds to the edges. Insert the rounds to be affected by the shell.

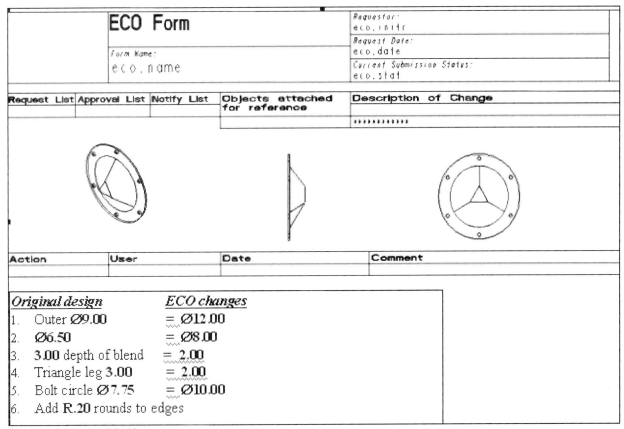

Figure 14.39(a) ECO Changes

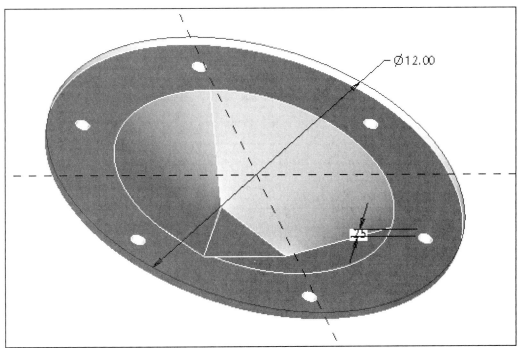

Figure 14.39(b) Completed ECO Part

For other projects see *www.cad-resources.com* ⇒ **Downloads**.

Notes:

Lesson 15 Sweeps

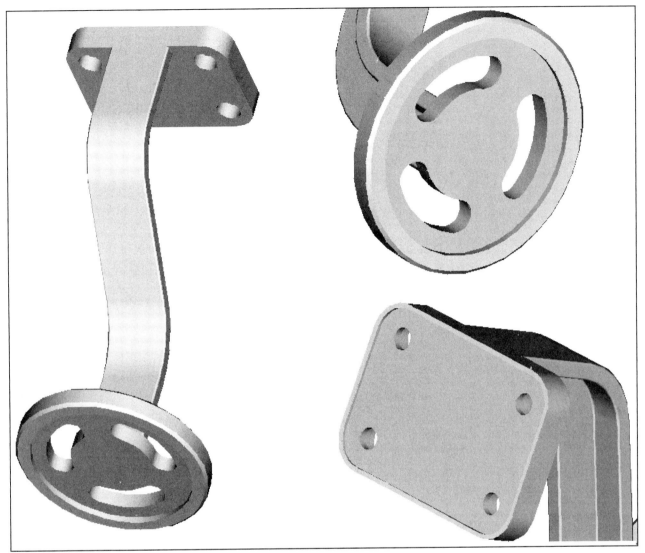

Figure 15.1 Bracket

OBJECTIVES

- Create a **constant-section sweep** feature
- Sketch a **Trajectory** for a sweep
- Sketch and locate a **Sweep section**
- Understand the difference between adding and not adding **Inner Faces**
- Be able to **Edit** a sweep
- Understand the difference between a **Sketched Trajectory** and a **Selected Trajectory**

SWEEPS

A Sweep is created by sketching or selecting a *trajectory* and then sketching a *section* to follow along it. The Bracket, shown in Figures 15.1 and 15.2, uses a simple sweep in its design.

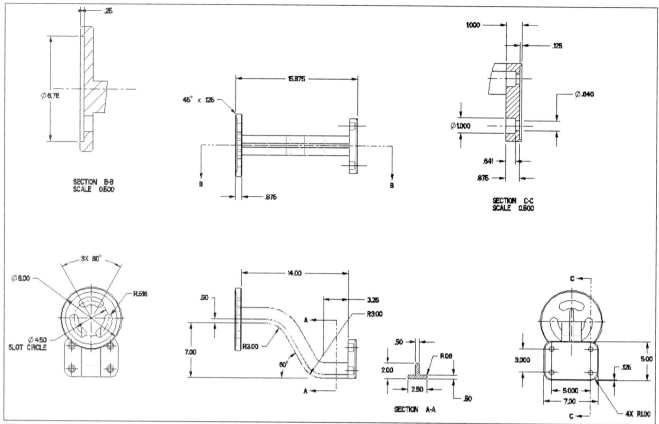

Figure 15.2 Bracket Detail

A *constant-section sweep* (Fig. 15.3) can use either trajectory geometry sketched at the time of feature creation or a trajectory made up of selected datum curves or edges. The trajectory [Figs. 15.4 (a-b)] must have adjacent reference surfaces or be planar (Fig. 15.5). When defining a sweep, Pro/E checks the specified trajectory for validity and establishes normal surfaces. When ambiguity exists, Pro/E prompts you to select a normal surface.

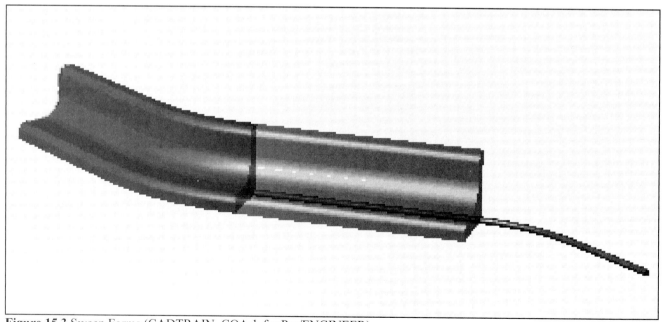

Figure 15.3 Sweep Forms (CADTRAIN, COAch for Pro/ENGINEER)

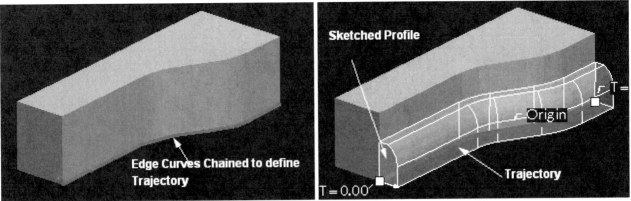

Figures 15.4(a-b) Sweep Trajectory and Section (CADTRAIN, COAch for Pro/ENGINEER)

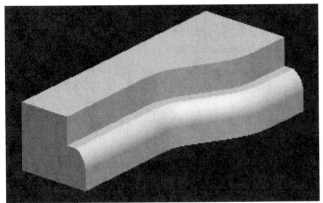

Figure 15.5 Completed Sweep (CADTRAIN, COAch for Pro/ENGINEER)

The following options are available for sweeps:

- **Sketch Traj** Sketch the sweep trajectory using Sketcher mode
- **Select Traj** Select a chain of existing curves or edges as the sweep trajectory
- **Merge Ends** Merge the ends of the sweep, if possible, into the adjacent solid
- **Free Ends** Do not attach the sweep end to adjacent geometry
- **No Inn Fcs** For closed sections, does not add top and bottom faces [Fig. 15.6(a)]
- **Add Inn Fcs** For open sections, add top and bottom faces to close the swept solid [Fig. 15.6(b)]

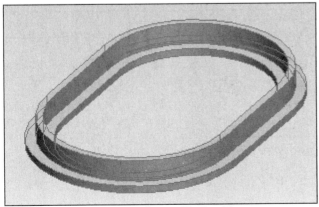

Figure 15.6(a) Sweep Attributes, **No Inn Fcs**

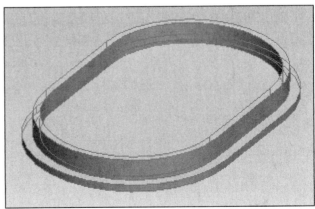

Figure 15.6(b) Sweep Attributes, **Add Inn Fcs**

Lesson 15 STEPS

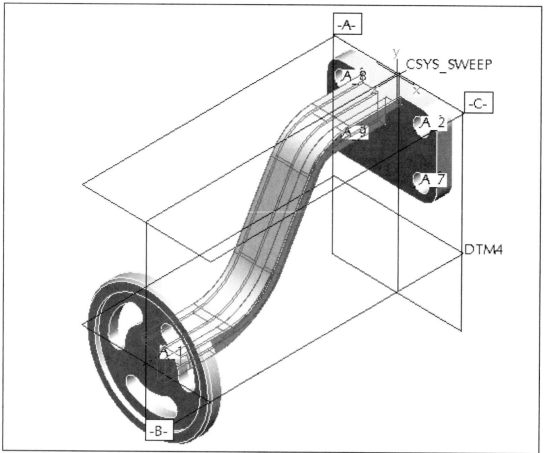

Figure 15.7 Bracket Showing Set Datum Planes and Coordinate System

Bracket

The Bracket (Fig. 15.7) requires the use of the Sweep command. The T-shaped section is swept along the sketched *trajectory*. The protrusions on both sides of the swept feature are to be created with the dimensions shown in Figures 15.8 through 15.14. Systematic commands are provided only for the sweep trajectory and its cross section.

Start a new part. Click: [icon] ⇒ [⊙ ☐ Part] ⇒ Name **Bracket** ⇒ [☑ Use default template] ⇒ **MMB** ⇒ **Edit** ⇒ **Setup**

- **Material** = al6061.mtl
- **Units** = Inch lbm Second

Set Datum and **Rename** the default datum planes and coordinate system:

- Datum **TOP** = **C**
- Datum **FRONT** = **A**
- Datum **RIGHT** = **B**
- Coordinate System = **CSYS_SWEEP**

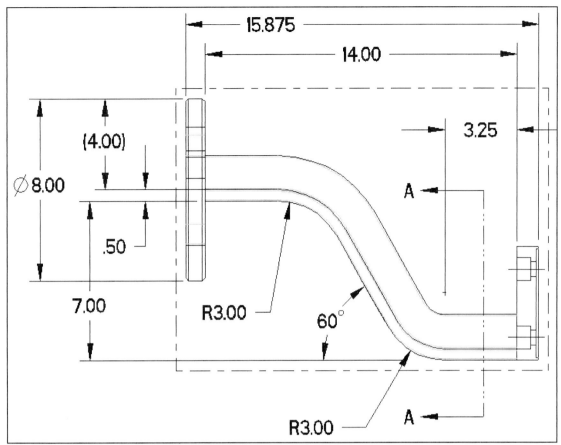

Figure 15.8 Bracket Drawing, Front View

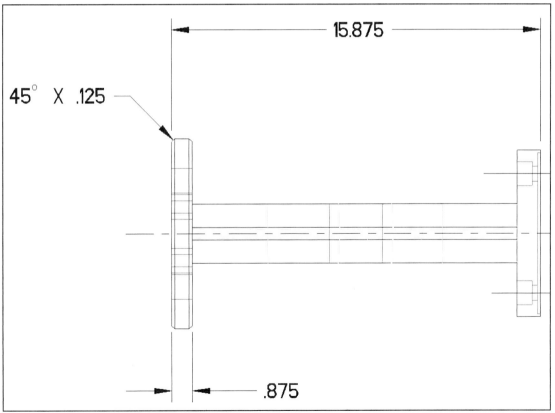

Figure 15.9 Bracket Drawing, Top View

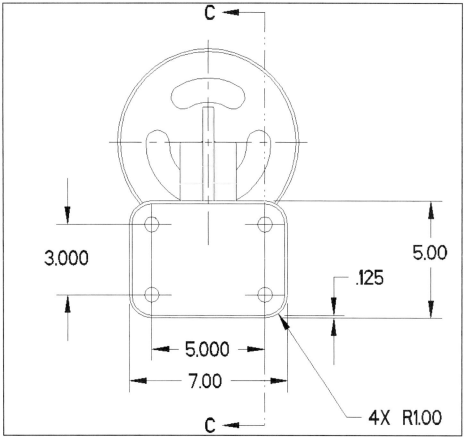

Figure 15.10 Bracket Drawing, Right Side View

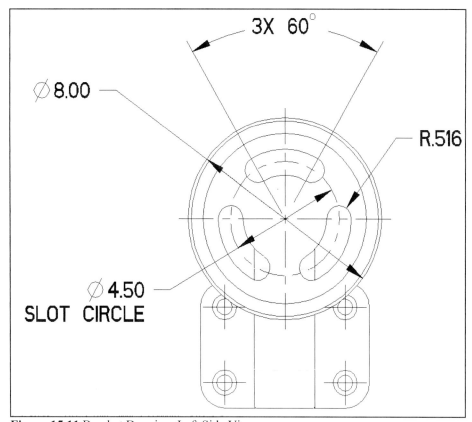

Figure 15.11 Bracket Drawing, Left Side View

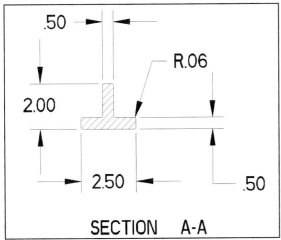
Figure 15.12(a) SECTION A-A

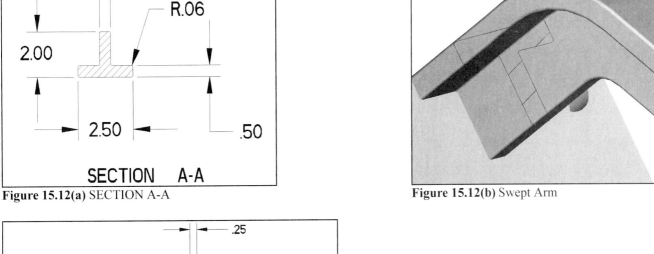
Figure 15.12(b) Swept Arm

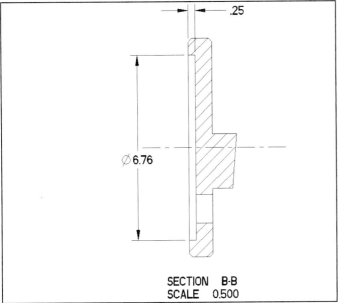
Figure 15.13(a) SECTION B-B

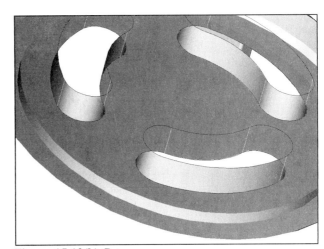
Figure 15.13(b) Cut

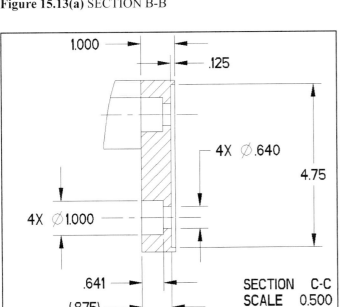
Figure 15.14(a) SECTION C-C

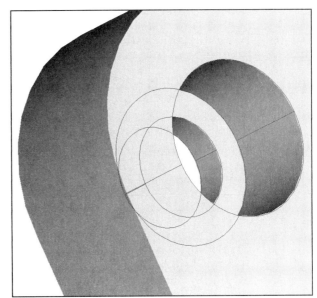
Figure 15.14(b) Counterbore

Start the Bracket by modeling the protrusion shown in Figure 15.15. This protrusion will be used to establish the sweep's position in space. Sketch the protrusion on datum **A**. The second protrusion is a sweep and is shown in Figure 15.16.

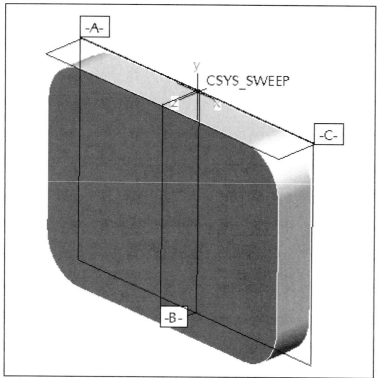

Figure 15.15 Bracket's First Protrusion

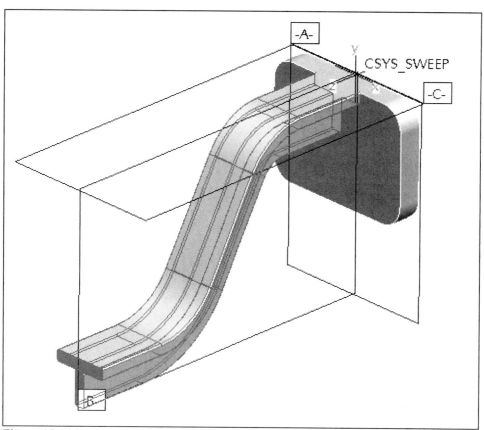

Figure 15.16 Swept Protrusion

Create the sweep. From the menu bar, click: **Insert** ⇒ **Sweep** ⇒ **Protrusion** ⇒ **Sketch Traj** ⇒ pick datum **B** as the sketching plane for the trajectory (Fig. 15.17) ⇒ **MMB** ⇒ **Top** ⇒ pick datum **C** as the orientation plane ⇒ **Sketch** from menu bar ⇒ **References** ⇒ delete datum **A** and add the front face of the protrusion ⇒ **Close** ⇒ sketch, dimension, and modify the trajectory (Figs. 15.18 through 15.21) ⇒ ✓ (Fig. 15.22)

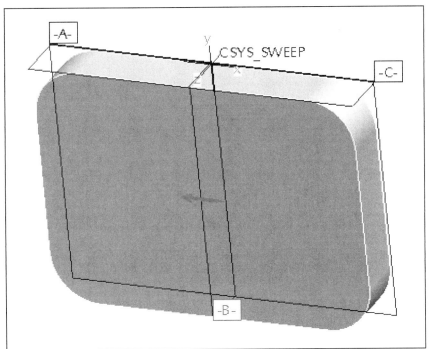

Figure 15.17 Select Datum B as the Trajectory Sketching Plane

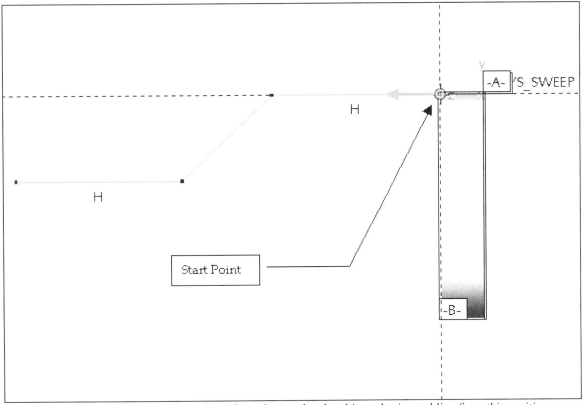

Figure 15.18 Sketch the Three Lines. Start the trajectory by sketching a horizontal line from this position.

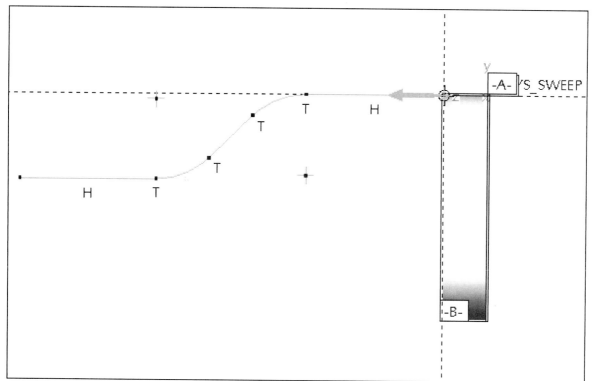

Figure 15.19 Add the Arc Fillets

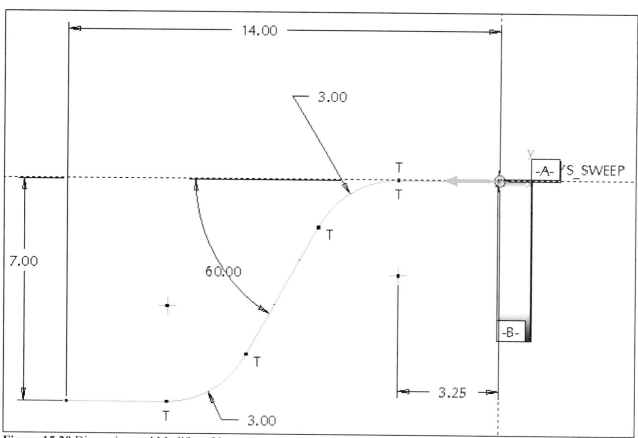

Figure 15.20 Dimension and Modify as Necessary

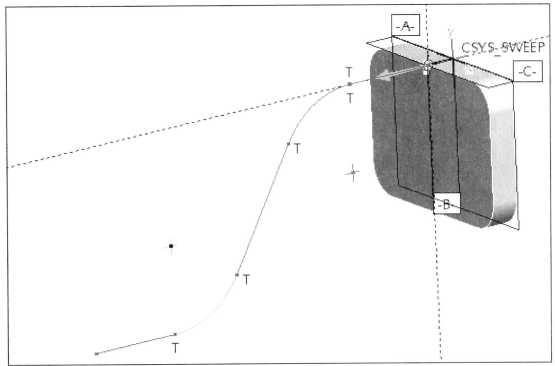

Figure 15.21 Trajectory Sketch

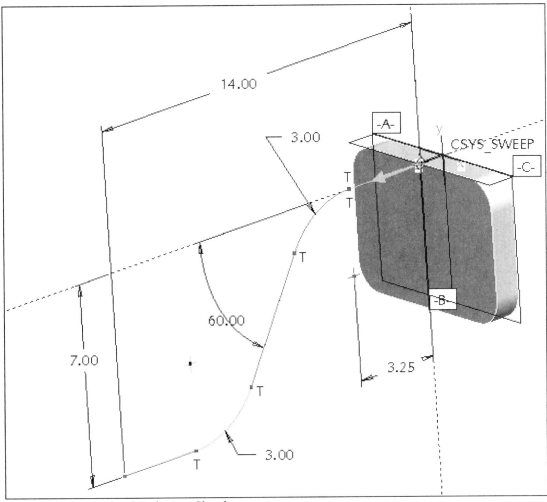

Figure 15.22 Completed Trajectory Sketch

Complete the sweep, click: **Free Ends** ⇒ **MMB** ⇒ **Sketch** (from menu bar) ⇒ **Options** ⇒ **Miscellaneous** tab ⇒ ☑ Grid ⇒ ☑ Snap To Grid ⇒ **Parameters** tab ⇒ Grid Spacing: **Manual** ⇒ ☑ Equal Spacing ⇒ **X = .25** ⇒ **Enter** ⇒ ✓ ⇒ sketch the *eight* lines of the section [Figs. 15.23(a-b)] ⇒ add a horizontal centerline and sketch *eight* fillets (Fig. 15.24) ⇒ add and reposition dimensions as needed (Figures 15.25 and 15.26) ⇒ modify the section [Figs. 15.27(a-b) and 15.28] ⇒ ✓ ⇒ **Preview** (Fig. 15.29) ⇒ **OK** ⇒ 💾 ⇒ **MMB**

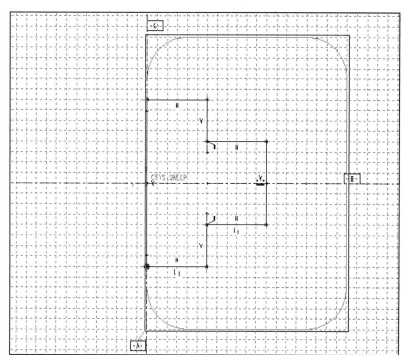
Figure 15.23(a) Sketch the Eight Lines

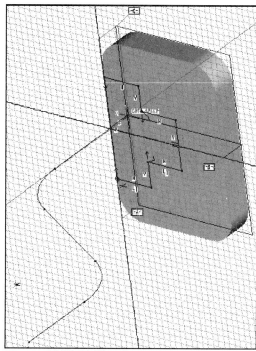
Figure 15.23(b) Sketch

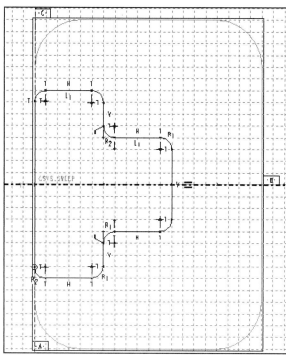
Figure 15.24 Sketch the Centerline and Arc Fillets

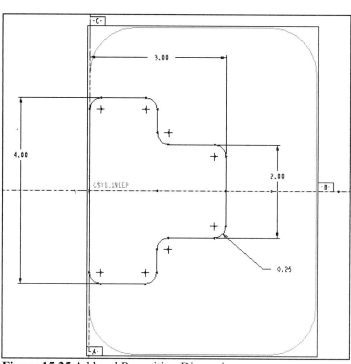
Figure 15.25 Add and Reposition Dimensions

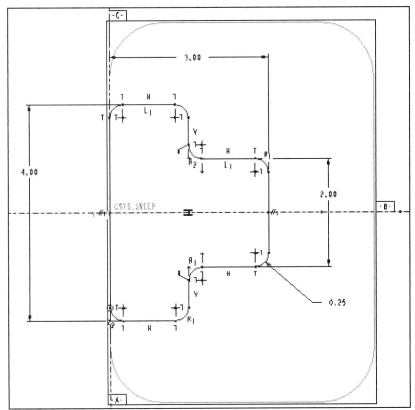

Figure 15.26 Constraints On

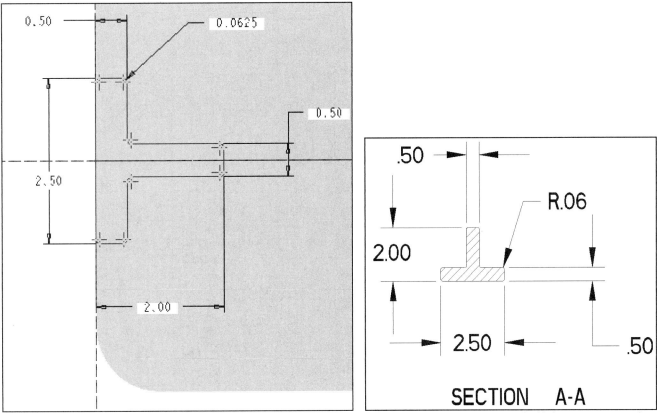

Figures 15.27(a-b) Modify the second **.50** dimension is optional. Pro/E will assume that the two web thicknesses are equal. (As per design intent, if they are both to be displayed on a drawing, then both should be on the section sketch.)

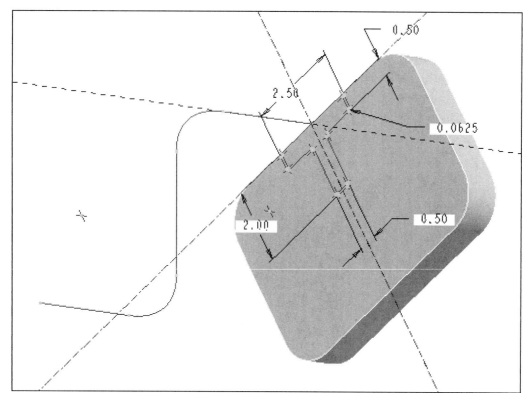

Figure 15.28 Section Rotated

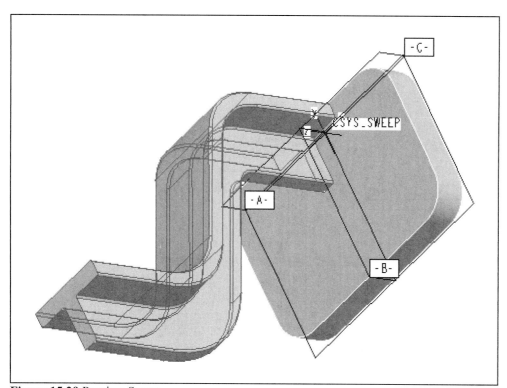

Figure 15.29 Preview Sweep

Complete the part by modeling the remaining features (Figs. 15.30 through 15.34).

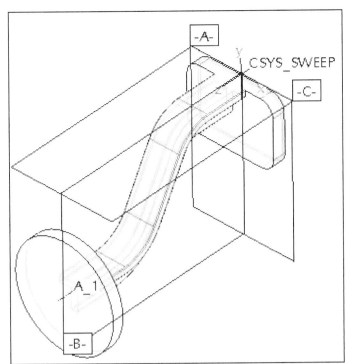
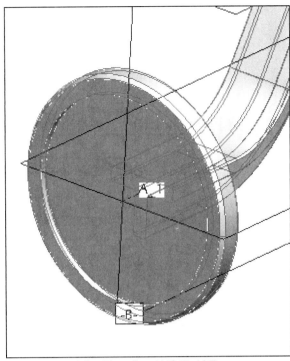

Figures 15.30(a-b) Add the Third Protrusion; Add the Cut (⌀**6.76** by **.250** deep) and the Chamfers (**45° X .125**)

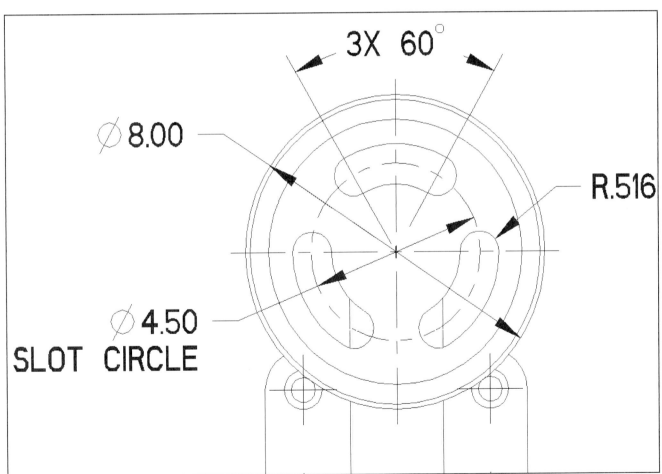

Figure 15.31(a) In order to pattern the slot feature, create a new datum through A_1 and at an angle, to use as the orientation (reference) plane.

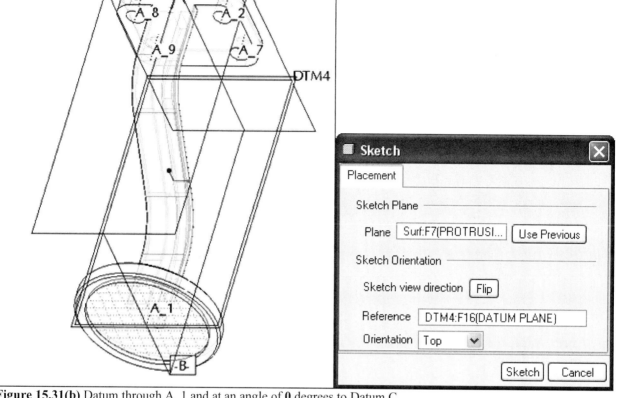

Figure 15.31(b) Datum through A_1 and at an angle of **0** degrees to Datum C

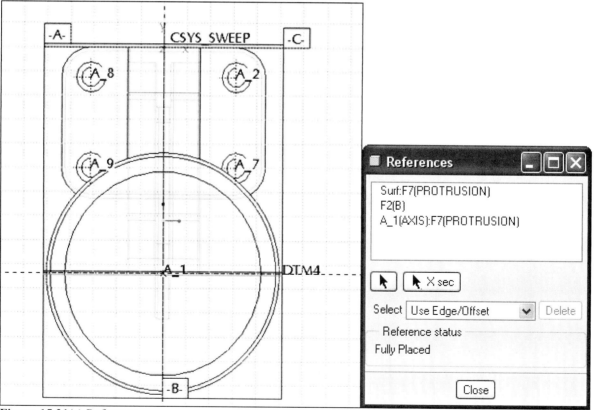

Figure 15.31(c) References

564

Add a Sketcher Point at the center of the round protrusion and then set the options as shown [Fig. 15.31(d)] **Sketch** ⇒ **Options** ⇒ **Miscellaneous** tab ⇒ set as displayed ⇒ **Constraints** tab ⇒ set as displayed ⇒ **Parameters** tab ⇒ Use the sketcher point as the **Origin** of the polar grid ⇒ Sketch the section [Figs. 15.31(e-g)] ⇒ complete the feature ⇒ pattern the slot [Figs. 15.32(a-c)] ⇒ create the cut and pattern the counterbore holes [Figs. 15.33(a-b)] ⇒ complete the part (Fig. 15.34)

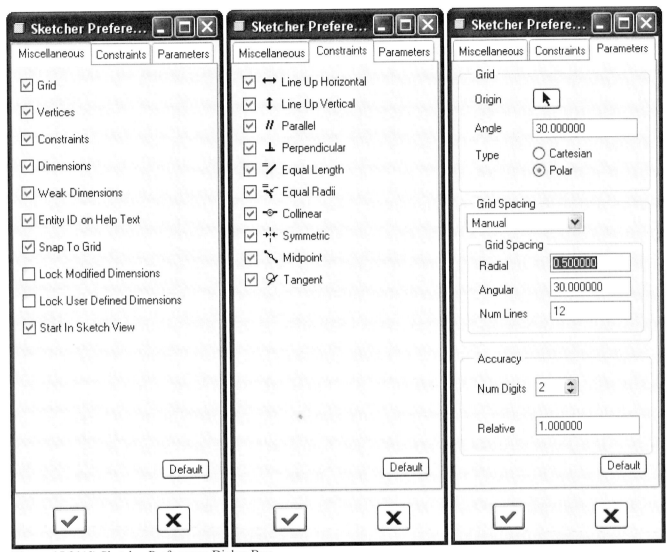

Figure 15.31(d) Sketcher Preferences Dialog Box

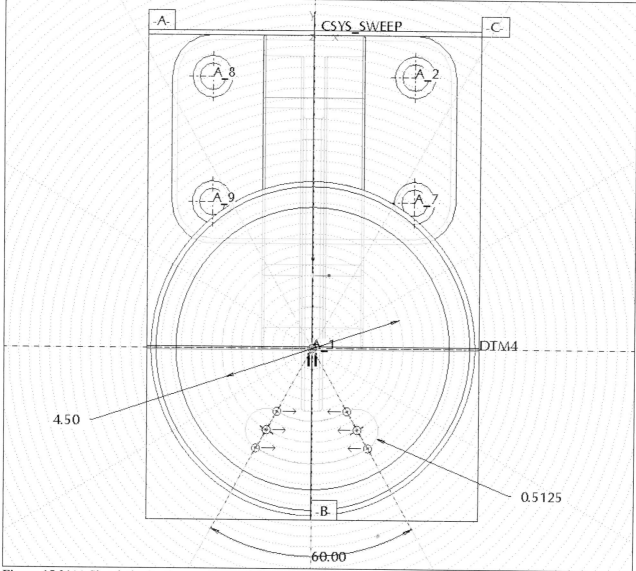

Figure 15.31(e) Sketch the Section

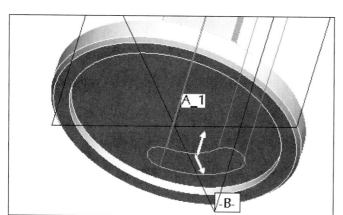

Figure 15.31(f) Cut Direction

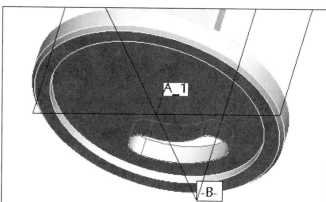

Figure 15.31(g) Completed Cut

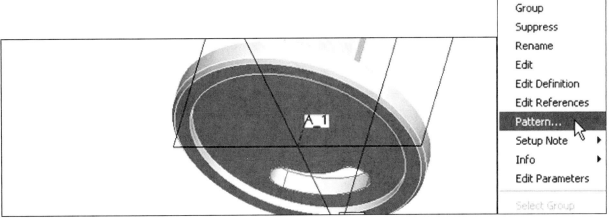

Figure 15.32(a) Pattern the Slot

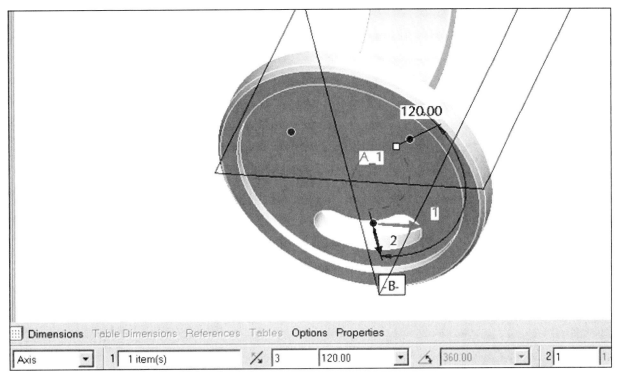

Figure 15.32(b) Pattern the Slot Using the Axis

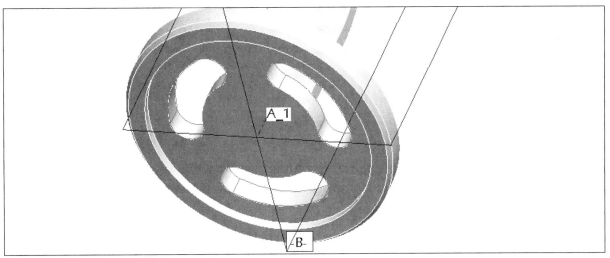

Figure 15.32(c) Patterned Slot

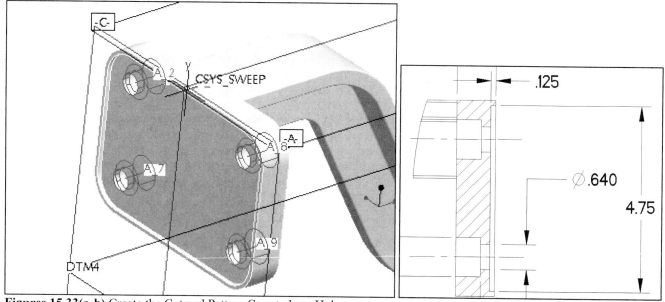

Figures 15.33(a-b) Create the Cut and Pattern Counterbore Holes

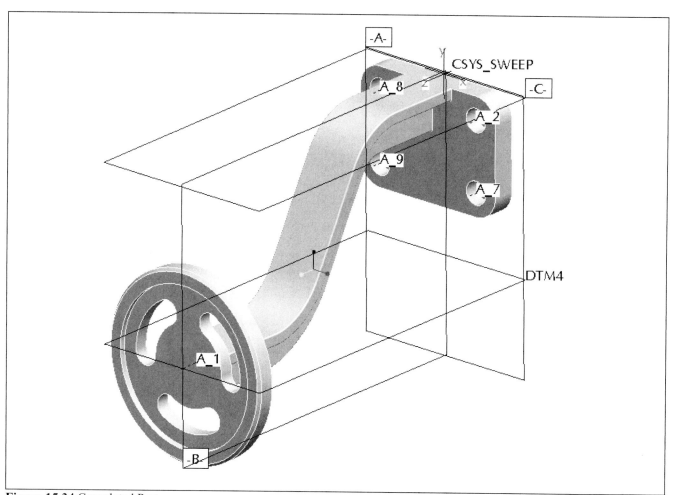

Figure 15.34 Completed Part

Download projects from ***www.cad-resources.com* ⇒ *Downloads.***

Lesson 16 Helical Sweeps and 3D Model Notes

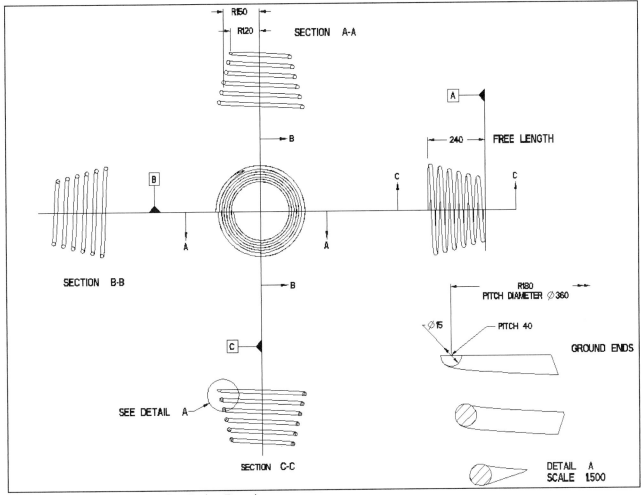

Figure 16.1 Helical Compression Spring Drawing

OBJECTIVES

- Create springs with a **Helical Sweep**
- Model a **helical compression spring**
- Use sweeps to create **hooks** on **extension springs**
- Design an **extension spring** with a **machine hook**
- Create **plain ground ends** on a spring
- Model a **convex spring**
- Create **3D Notes**

HELICAL SWEEPS AND 3D MODEL NOTES

A **helical sweep** (Fig. 16.1) is created by sweeping a section along a helical *trajectory*. The trajectory is defined by both the *profile* of the *surface of revolution* (which defines the distance from the section origin of the helical feature to its *axis of revolution*) and the *pitch* (the distance between coils). The trajectory and the surface of revolution are construction tools and do not appear in the resulting geometry. **3D Model Notes** are pieces of text, which can contain links (URL's) to World Wide Web pages, which you can attach to objects in Pro/E. Model notes, increase the amount of information that you can attach to any object in your model.

Helical Sweeps

The Helical Sweep command is available (Fig. 16.2) for both solid and surface features. Use the following ATTRIBUTES menu options, presented in mutually exclusive pairs, to define the helical sweep feature:

- **Constant** The pitch is constant
- **Variable** The pitch is variable and defined by a graph
- **Thru Axis** The section lies in a plane that passes through the axis of revolution
- **Norm To Traj** The section is oriented normal to the trajectory (or surface of revolution)
- **Right Handed** The trajectory is defined by the right-hand rule
- **Left Handed** The trajectory is defined by the left-hand rule

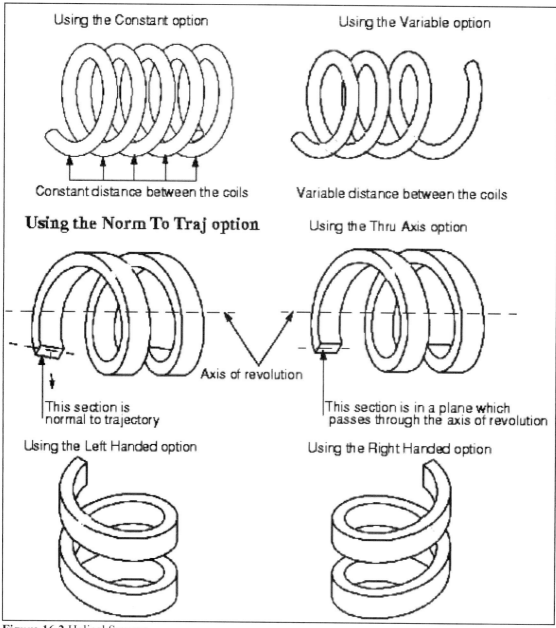

Figure 16.2 Helical Sweeps

3D Model Notes

Model notes are text strings that you can attach to objects [Fig. 16.3(a)]. You can attach any number of notes to any object in your model. When you attach a note to an object, the object is considered the parent of the note. When you delete the parent object, all child notes are deleted with it. The attachment is at the end of the model note leader line. Notes do not have to be attached to a parent. You can also allocate a URL to each model note [Fig. 16.3(b)]. You can use model notes to:

- Communicate with members of your workgroup as to how to review or use a model
- Explain how you approached or solved a design problem when modeling
- Explain changes that you have made to the features of a model over time

Modifying Note Text Style

You can modify the following text style attributes of an existing model note:

- Style name, Font, width factor, and slant angle of text
- Line spacing and angle
- Horizontal and vertical justification
- Note orientation. That is, you can make the note a mirror image of itself
- System and user-defined colors for note text

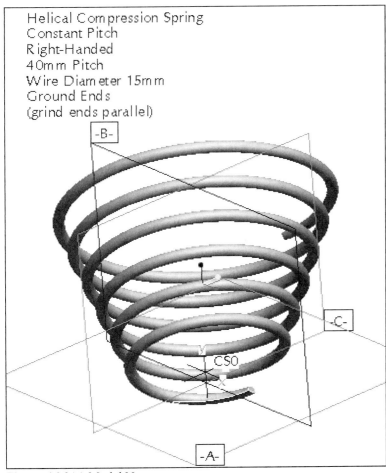

Figure 16.3(a) Model Notes

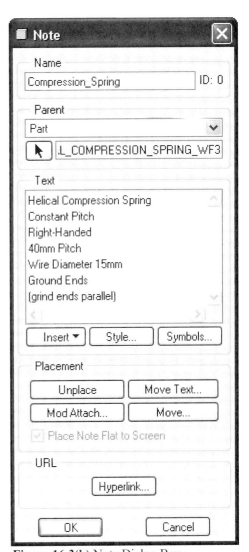

Figure 16.3(b) Note Dialog Box

Lesson 16 STEPS

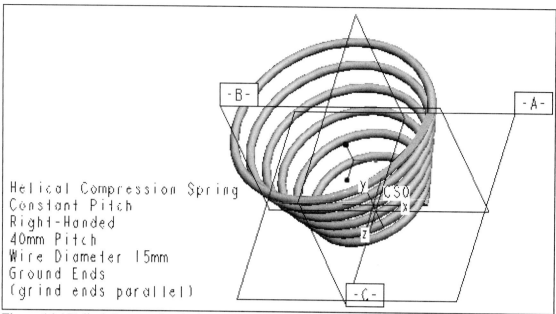

Figure 16.4 Helical Compression Spring with Datum Planes and Model Note

Helical Compression Spring

Springs (Fig. 16.4) and other helical features are created with the Helical Sweep command. A helical sweep is created by sweeping a *section* along a *trajectory* that lies in the *surface of revolution:* the trajectory is defined by both the *profile* of the surface of revolution and the distance between coils. The model for this lesson is a *constant-pitch right-handed helical compression spring with ground ends, a pitch of 40 mm, and a wire diameter of 15 mm* (Figs. 16.4 through 16.8).

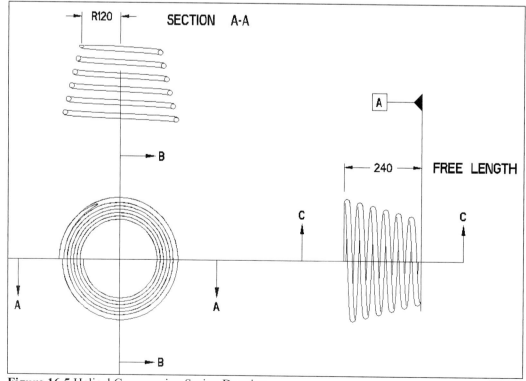

Figure 16.5 Helical Compression Spring Drawing

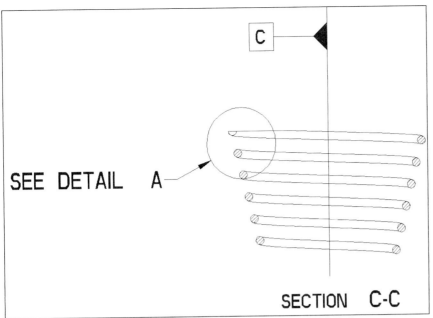

Figure 16.6 Helical Compression Spring Drawing, SECTION C-C

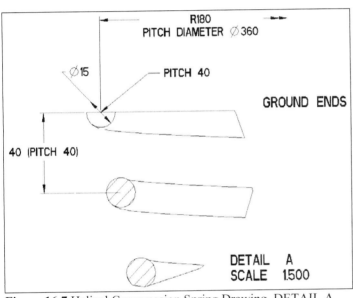

Figure 16.7 Helical Compression Spring Drawing, DETAIL A

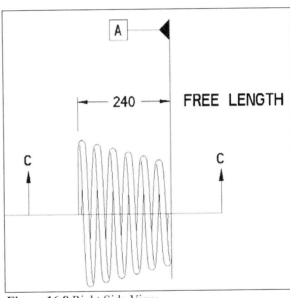

Figure 16.8 Right Side View

Start a new part. Click: **Create a new object** ⇒ Part ⇒ Name **Helical_Compression_Spring** ⇒ Use default template ⇒ **MMB** ⇒ **Edit** ⇒ **Setup** ⇒ setup up your project as follows:

- **Material** = ss.mtl
- **Units** = millimeter Newton Second

Set Datum and **Rename** the default datum planes and coordinate system:

- Datum TOP = **A**
- Datum FRONT = **B**
- Datum RIGHT = **C**
- Coordinate System = **CS0**

Create the first protrusion: **Insert** ⇒ **Helical Sweep** ⇒ **Protrusion** ⇒ **Constant** ⇒ **Thru Axis** ⇒ **Right Handed** ⇒ **MMB** ⇒ pick datum **B (FRONT)** ⇒ **MMB** ⇒ **MMB** ⇒ sketch a line [(Figs. 16.9(a-b)] ⇒ **Create 2 point centerlines** add a vertical centerline along datum **C (RIGHT)**

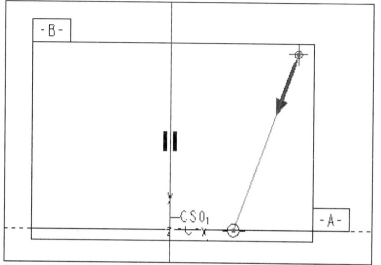
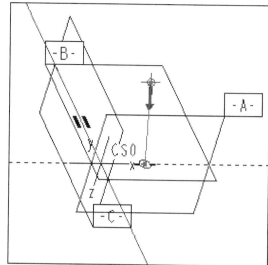

Figure 16.9(a) Sketch the Profile Line and a Vertical Centerline   **Figure 16.9(b)** Sketch in 3D

Click: create all required dimensions ⇒ **Modify the values of dimensions** change the values to the design sizes [Figs. 16.10(a-b)] ⇒ ✓ ⇒ enter the pitch value **40** at the prompt ⇒ **MMB**

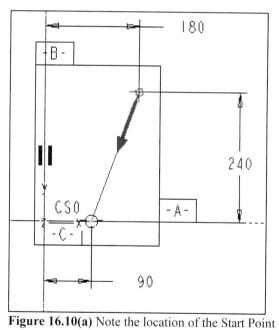
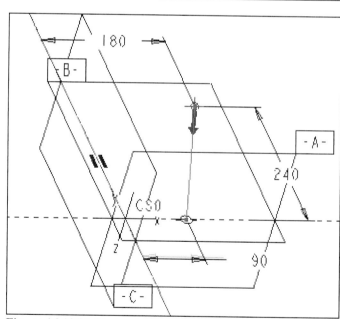

Figure 16.10(a) Note the location of the Start Point   **Figure 16.10(b)** Start point in 3D

Sketch the section geometry of the spring at the crosshairs (a circle), click: **Create circle** [Figs. 16.11(a-b)] ⇒ **Modify the values of dimensions** (**15**) [Figs. 16.12(a-b)] ⇒ **MMB** ⇒ ✓ ⇒ **MMB** ⇒ **Shading** ⇒ ⇒ **MMB** ⇒ ⇒ **Standard Orientation** [Figs. 16.13(a-b)]

574

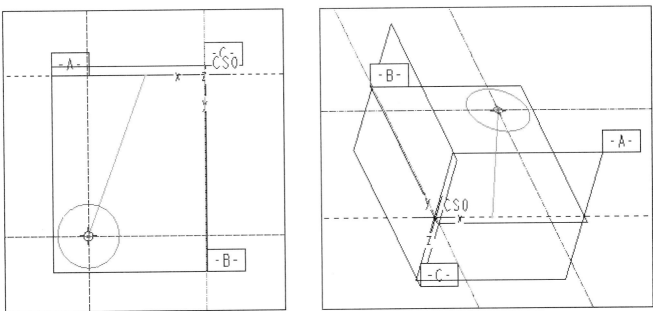

Figures 16.11(a-b) Sketching the Circle, Sketch a Circle as the Section Geometry (wire diameter)

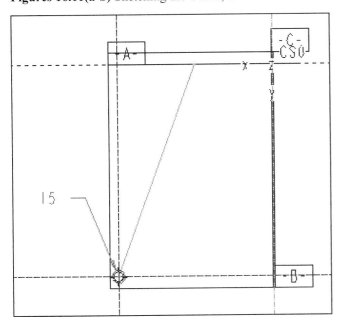

Figure 16.12(a) Wire Diameter 15

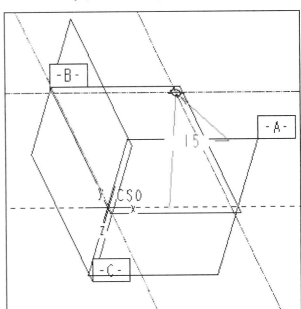

Figure 16.12(b) Wire Diameter Sketch

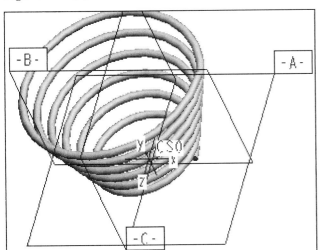

Figure 16.13(a) Swept Protrusion

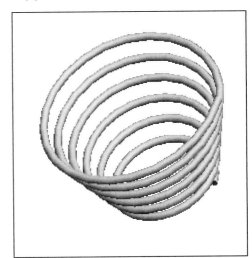

Figure 16.13(b) Shaded Spring

Create the *ground ends*, click: ⬚ **Extrude Tool** ⇒ ⬚ **Extrude on both sides** ⇒ **RMB** ⇒ **Remove Material** ⇒ **RMB** ⇒ **Define Internal Sketch** ⇒ Plane: pick datum **C** ⇒ Reference: pick datum **A** ⇒ Orientation: **Bottom** [Figs. 16.14(a-b)] ⇒ **MMB** ⇒ **RMB** ⇒ **Line** [Figs. 16.15(a-b)] create the line ⇒ modify the dimensions (Fig. 16.16) ⇒ spin the model as needed ⇒ ✓ (Fig. 16.17)

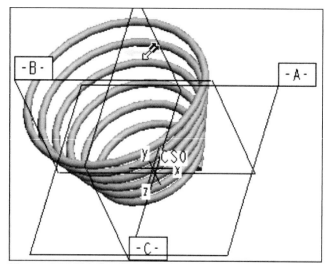

Figure 16.14(a) Cut Sketch Orientation

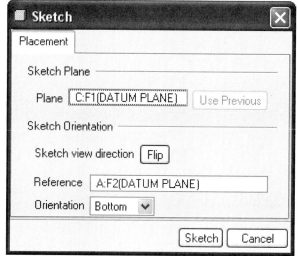

Figure 16.14(b) Sketch Dialog

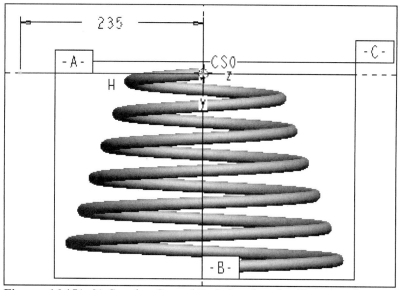

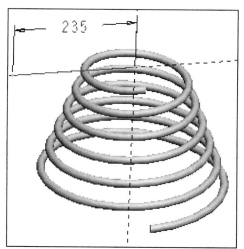

Figures 16.15(a-b) Creating Ground Ends (any length will work as long as it goes beyond the spring)

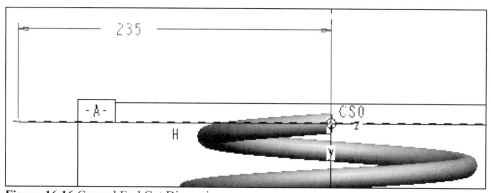

Figure 16.16 Ground End Cut Dimension

Extend a depth handle to include the spring (Fig. 16.17 and Fig. 16.18) ⇒ ☑👓 (Fig. 16.19) ⇒ verify material direction ⇒ 👓 ⇒ (if necessary) ⚋ **Change material direction** ⇒ ✓ ⇒ 🔍 ⇒ ▨ ⇒ 💾 ⇒ MMB ⇒ File ⇒ Delete ⇒ Old Versions ⇒ MMB

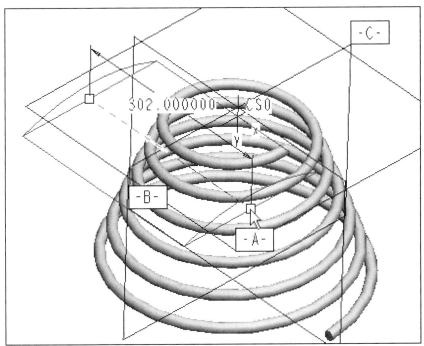

Figure 16.17 Depth Handle

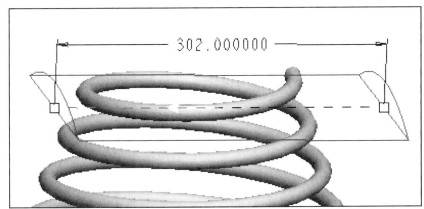

Figure 16.18 Ground End Depth

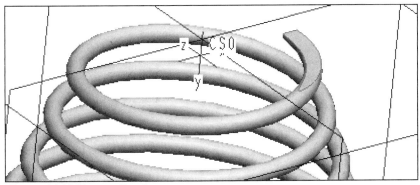

Figure 16.19 Feature Preview

The second ground end (Figs. 16.20 and 16.21) is created using similar commands. ⇒ complete the spring ⇒ **Ctrl+D** ⇒ **Ctrl+S** ⇒ **Enter** (Fig. 16.22)

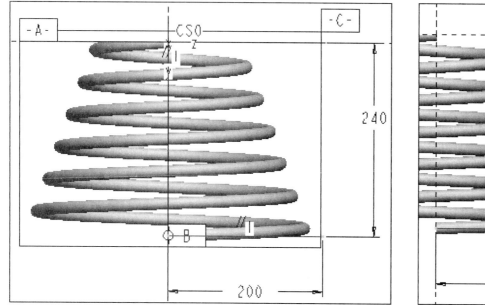

Figure 16.20 Creating the Second Ground End

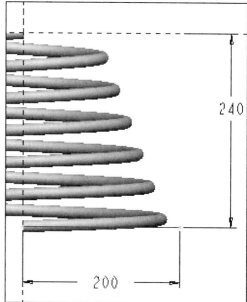

Figure 16.21 Shaded Sketch

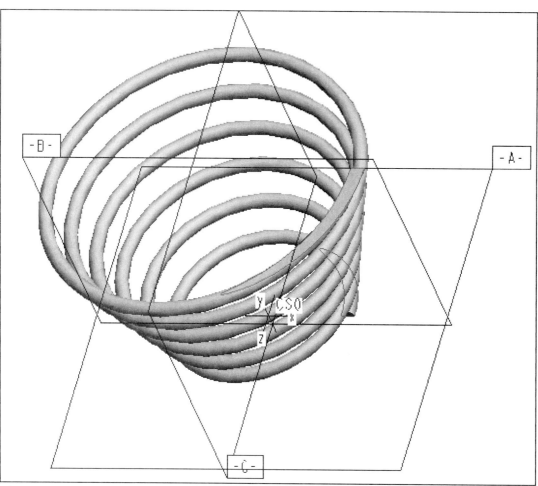

Figure 16.22 Completed Helical Compression Spring

Save a copy the Helical Compression Spring by clicking: **File** ⇒ **Save a Copy** ⇒ Type a different name-- **HEL_COMP_SPR_GRND_ENDS**. Figure 16.23 provides an **ECO** for the new spring. Rename the file you are working on by clicking: **File** ⇒ **Rename** ⇒ provide a unique name such as **HEL_EXT_SPR_MACH_ENDS** ⇒ delete the existing ground ends ⇒ modify the pitch to **10 mm** ⇒ change the wire diameter to **7.5 mm** ⇒ Complete the extension spring (Figs. 16.24 through 16.33). The free length is to be **120 mm**. The large radius will now be **90 mm** (*was 180mm*), and the small radius will be **60 mm** (*was 90mm*).

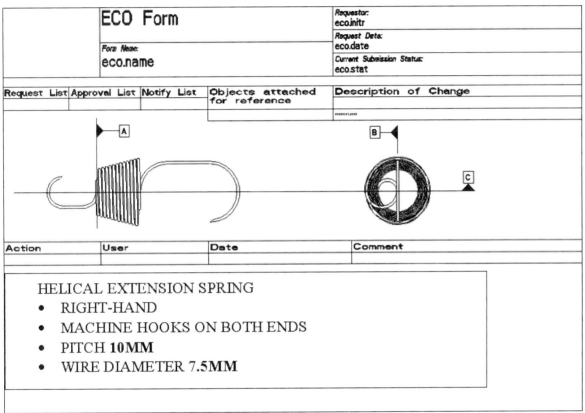

Figure 16.23 ECO to Create a Helical Extension Spring

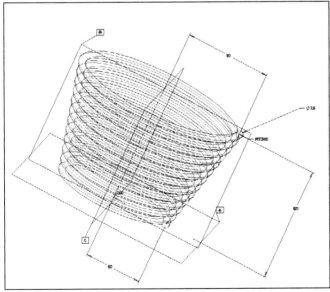

Figure 16.24 ECO Changes

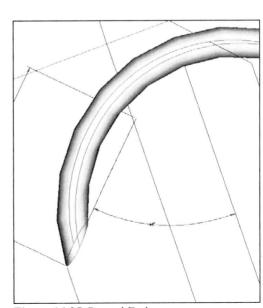

Figure 16.25 Ground End

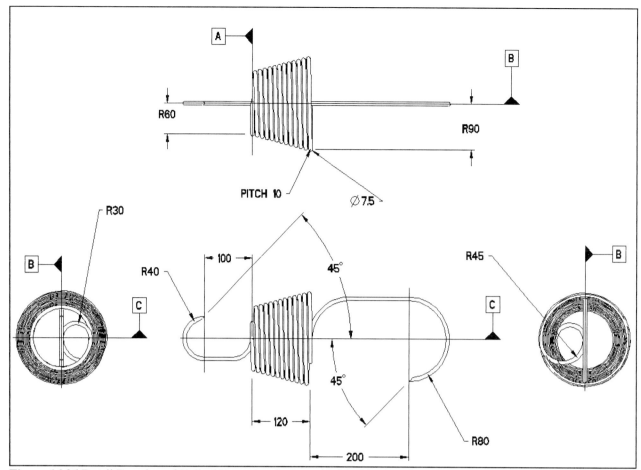

Figure 16.26 Detail Drawing of Helical Extension Spring with Machine Hook Ends

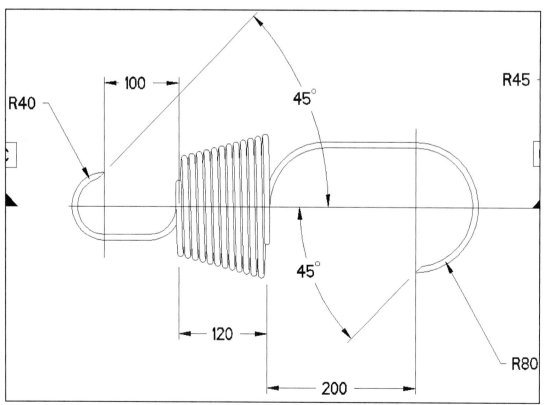

Figure 16.27 Front View

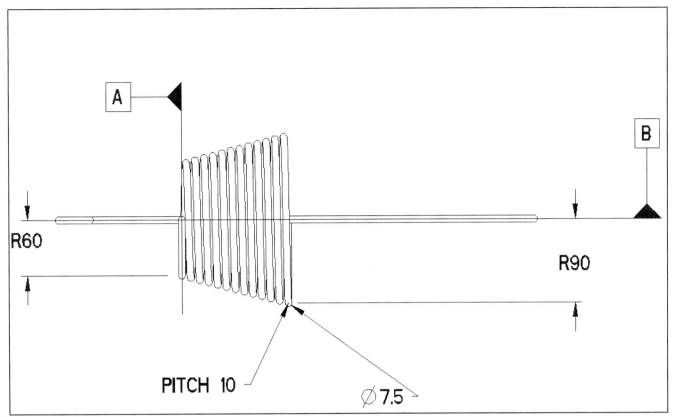

Figure 16.28 Top View

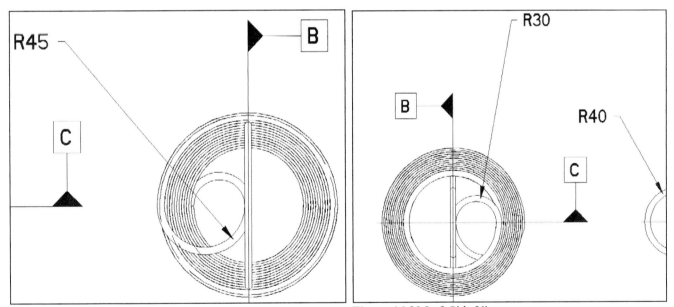

Figure 16.29 Right Side View

Figure 16.30 Left Side View

Create the machine hooks using simple sweeps and cuts, as shown in Figures 16.31 through 16.33.

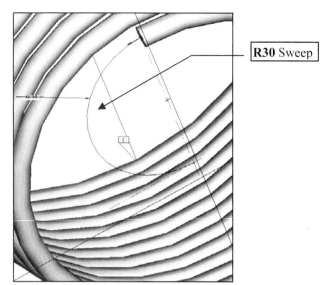

Figure 16.31(a) Sweep **R30**

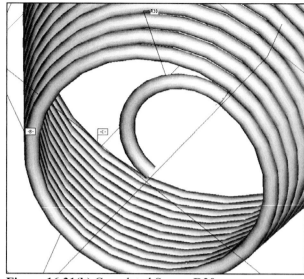

Figure 16.31(b) Completed Sweep **R30**

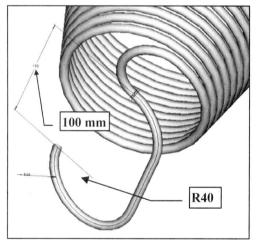

Figure 16.32(a) Small Hook End Sweep

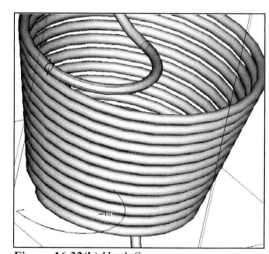

Figure 16.32(b) Hook Sweep

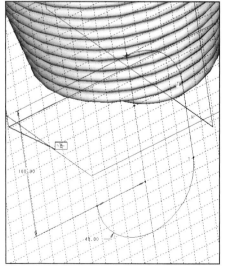

Figure 16.33(a) Large Hook End Sweep

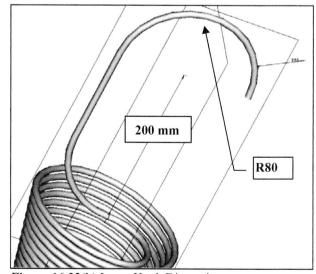

Figure 16.33(b) Large Hook Dimensions

582

Model Notes

When you attach a note to an object, the object is considered the "parent" of the note. Deleting the parent deletes all of the notes of the parent. You can attach model notes anywhere in the model; they do not have to be attached to a parent. Here we will add a note to the part and describe the spring.

Open the saved spring file that has the ground ends. Choose the following commands, click: **Edit** ⇒ **Setup** ⇒ **Notes** ⇒ **New** ⇒ type **Compression_Spring** as the name of the note; no spaces are allowed in the name ⇒ pick in the Text area and type the note (Fig. 16.34):

Helical Compression Spring
Constant Pitch
Right-handed
40 mm Pitch
Wire Diameter 15 mm
Ground Ends
(grind ends parallel)

Click: ☑ Place Note Flat to Screen ⇒ **Place** ⇒ **No Leader** ⇒ **Standard** ⇒ **MMB** ⇒ pick a place on the screen to place the note [Figs. 16.35(a-b)] ⇒ **OK** ⇒ **Done/Return** ⇒ **Done**

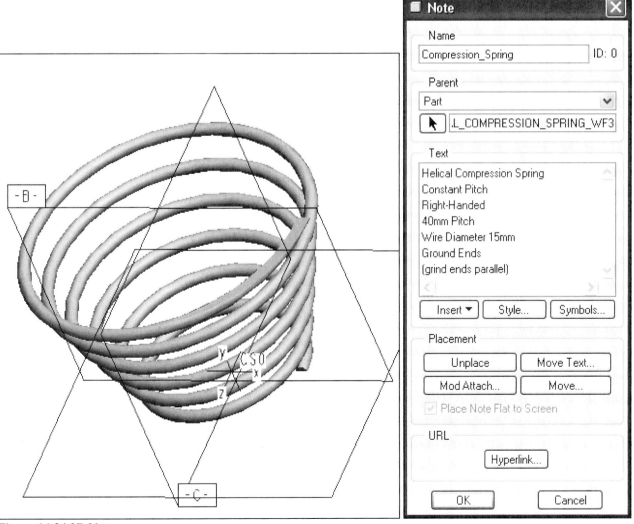

Figure 16.34 3D Notes

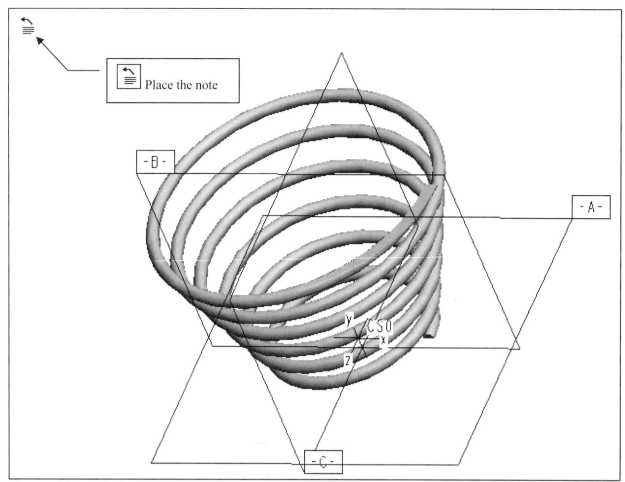

Figure 16.35(a) Placing the Note

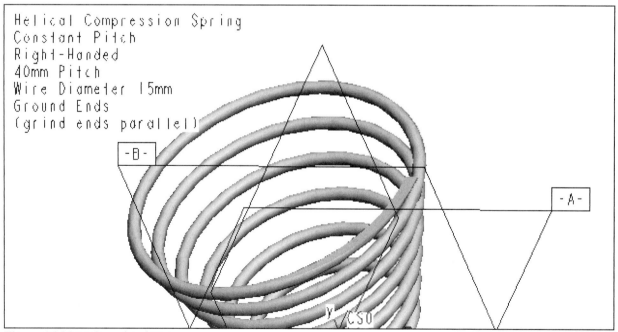

Figure 16.35(b) Completed Note

You can toggle the note ☑ 3D Notes and ☐ 3D Notes in the Environment dialog box (Fig. 16.36). Display the note in the Model Tree by clicking: **Settings** ⇒ **Tree Filters** ⇒ ☑ Annotations (Fig. 16.37) ⇒ **Apply** ⇒ **OK**

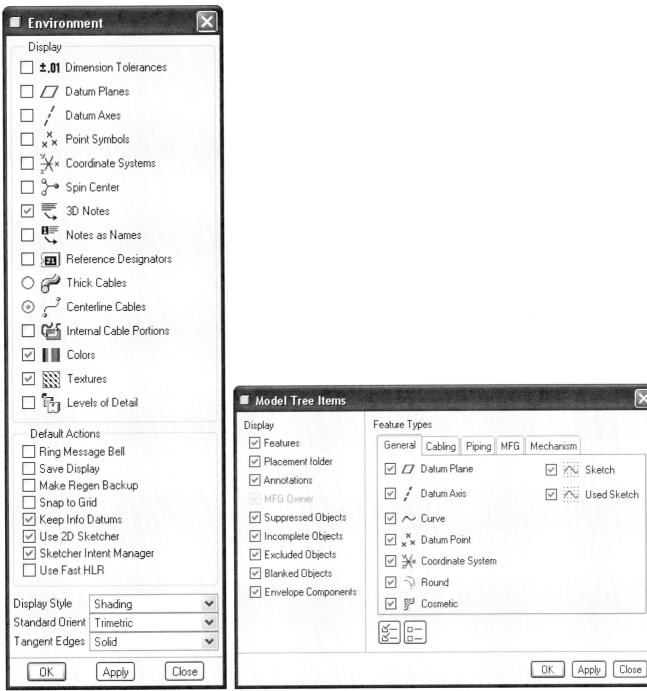

Figure 16.36 Displaying 3D Notes   **Figure 16.37** Displaying Notes in the Model Tree

You can also perform a variety of functions directly from the Model Tree. Click on [Compression_Spring icon] in the Model Tree ⇒ **RMB** [Fig. 16.38(a)] ⇒ **Properties** [Fig. 16.38(b)] ⇒ URL click: **Hyperlink** (Fig. 16.39) ⇒ type the URL or internal link: **www.americanprecspring.com** (Fig. 16.40)

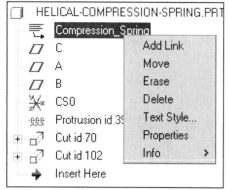

Figure 16.38(a) Model Tree (RMB)

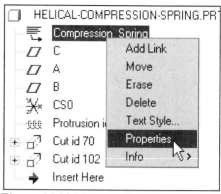

Figure 16.38(b) Properties

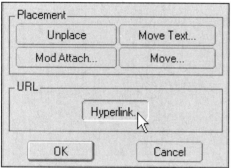

Figure 16.39 Hyperlink

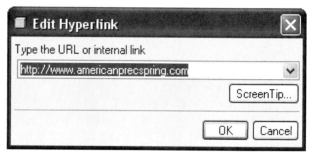

Figure 16.40 URL

Create a screen tip that will display as your cursor passes over the note, click: [ScreenTip..] ⇒ type **SPRING COMPANY** [Figs. 16.41(a-b)] ⇒ **OK** ⇒ **MMB** ⇒ **MMB** ⇒ [icon] ⇒ [icon] ⇒ [icon] ⇒ **MMB** ⇒ **File** ⇒ **Delete** ⇒ **Old Versions** ⇒ **MMB**

Figure 16.41(a) Screen Tip

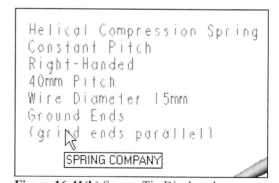

Figure 16.41(b) Screen Tip Displayed

Open the URL, click: **Compression_Spring** ⇒ **RMB** ⇒ Open URL [Figs. 16.42(a-b)] ⇒ pick in the browser window [Fig. 16.42(c)] ⇒ close **Browser** ⇒ **Ctrl+D** ⇒ **Ctrl+S** ⇒ **Enter**

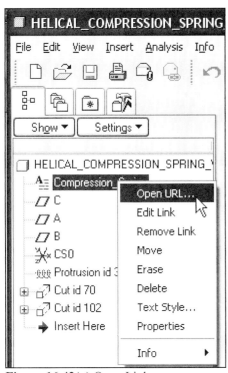

Figure 16.42(a) Open Link

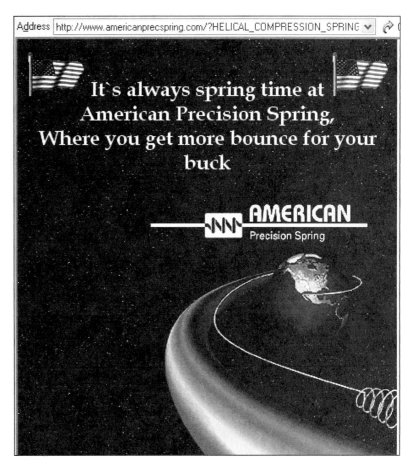

Figure 16.42(b) American Precision Spring Website

...simply be the best.

Quality Springs Since 1979

American Precision Spring Corporation has provided customers worldwide with custom precision springs and metal parts since 1979. We manufacture component parts for a variety of industries, including aerospace, computing, electronic instruments, and medical equipment. We produce a wide array of quality parts including torsion springs, compression springs, extension springs, flat springs, wire forms, leaf springs, battery contacts, four slide parts, punch press parts and more. Please give us a call or follow the links below for more infomation.

Overview of all Departments - Our history, commitment to quality and overview of our products
Four Slide Parts - Tooling, Wire forms, and Sheet Metal Parts
Punch Press Parts - Single station, Compound, and Progressive dies
Coiling - Compression Springs, Extension Springs, and Torsion Springs
Short run - Sheet Metal and Wire Forms
Inspection - MIL-I-45208A Spec
Employment - Employment Opportunities at American Precision Spring
Shipping-department - Track your shipment
Map - How to find us.
Future - What`s ahead
Faq - Common Questions

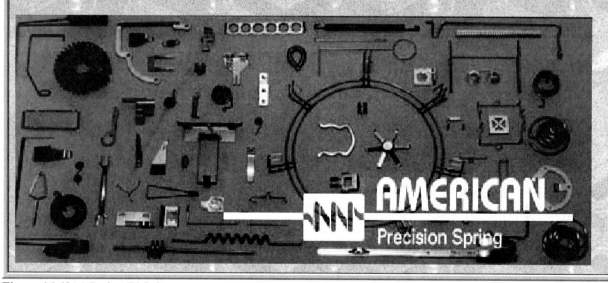

Figure 16.42(c) Spring Website

Download projects from ***www.cad-resources.com*** $\Rightarrow$ ***Downloads***.

Lesson 17 Shell, Reorder, and Insert Mode

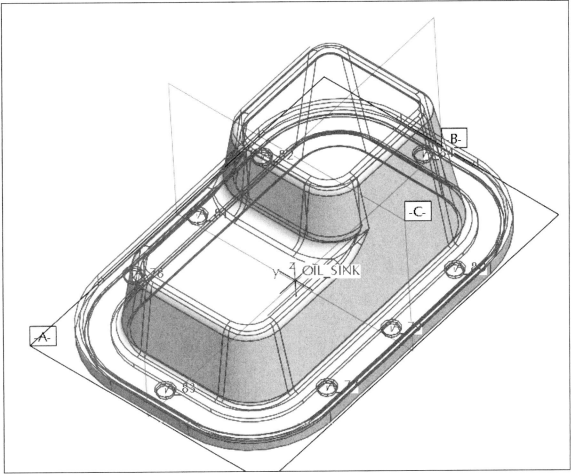

Figure 17.1 Oil Sink

OBJECTIVES

- Master the use of the **Shell Tool**
- Alter the creation sequence with **Reorder**
- **Insert** a feature at a specific point in the design order
- Create a **Hole Pattern** using a **Table**

SHELL, REORDER, AND INSERT MODE

The **Shell Tool** removes a surface or surfaces from the solid and then hollows out the inside of the solid, leaving a shell of a specified wall thickness, as in the Oil Sink (Fig. 17.1). When Pro/E makes the shell, all the features that were added to the solid before you chose the Shell Tool are hollowed out. Therefore, the *order of feature creation* is very important when you use the Shell Tool. You can alter the feature creation order by using the **Reorder** option. Another method of placing a feature at a specific place in the feature/design creation order is to use the **Insert Mode** option.

Creating Shells

The Shell Tool [Figs. 17.2(a-c)] enables you to remove a surface or surfaces from the solid, then hollows out the inside of the solid, leaving a shell of a specified wall thickness. If you flip the thickness side by entering a negative value, dragging a handle, or using the ⌧ **Change thickness direction** icon, the shell thickness is added to the outside of the part. If you do not select a surface to remove, a "closed" shell is created, with the whole inside of the part hollowed out and no access to the inside. In this case, you can add the necessary cuts or holes to achieve proper geometry at a later time.

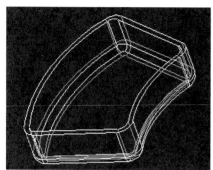 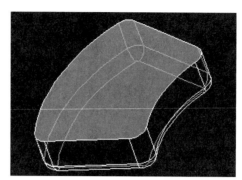 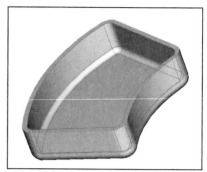

Figures 17.2(a-c) Shell

When defining a shell, you can also select surfaces where you want to assign a different thickness. You can specify independent thickness values for each such surface. However, you cannot enter negative thickness values, or flip the thickness side, for these surfaces. The thickness side is determined by the default thickness of the shell. When Pro/E makes the shell, all the features that were added to the solid before you started the Shell Tool are hollowed out. Therefore, the order of feature creation is very important when you use the Shell Tool. To access the Shell Tool, click 🗆 icon in the Right Toolchest, or click Insert ⇒ Shell on the top menu bar. The Thickness box lets you change the value for the default shell thickness. You can type the new value, or select a recently used value from the drop-down list.

In the graphics window, you can use the shortcut menu (**RMB**) to access the following options:

- **Remove Surfaces** Activates the collector of surfaces. You can select any number of surfaces
- **Non Default Thickness** Activates the collector of surfaces with a different thickness
- **Excluded Surfaces** Activates the collector of excluded surfaces
- **Clear** Remove all references from the collector that is currently active
- **Flip** Change the shell side direction

The Shell Dashboard displays the following slide-up/down panels:

- **References** Contains the collector of references used in the Shell feature
- **Options** Contains the collector of Excluded surfaces
- **Properties** Contains the feature name and an icon to access feature information

The [References] slide-up/down panel contains the following elements:

- The **Removed surfaces** collector lets you select the surfaces to be removed. If you do not select any surfaces, a "closed" shell is created.
- The **Non-default thickness** collector lets you select surfaces where you want to assign a different thickness. For each surface included in this collector, you can specify an individual thickness value.

The [Properties] panel contains the Name text box [Name SHELL_ID_200 ⓘ], where you can type a custom name for the shell feature, to replace the automatically generated name. It also contains the ⓘ icon that you can click to display information about this feature in the Browser.

Reordering Features

You can move features forward or backward in the feature creation (regeneration) order list, thus changing the order in which features are regenerated [Figs. 17.3(a-b)]. Use Edit ⇒ Feature Operations ⇒ Reorder to activate the command.

You can reorder multiple features in one operation, as long as these features appear in *consecutive* order. Feature reorder *cannot* occur under the following conditions:

- **Parents** Cannot be moved so that their regeneration occurs after the regeneration of their children
- **Children** Cannot be moved so that their regeneration occurs before the regeneration of their parents

You can select the features to be reordered by choosing an option:

- **Select** Select features to reorder by picking on the screen and/or from the Model Tree
- **Layer** Select all features from a layer by selecting the layer
- **Range** Specify the range of features by entering the regeneration numbers of the starting and ending features

You can reorder features in the Model Tree by dragging one or more features to a new location in the feature list. If you try to move a child feature to a higher position than its parent feature, the parent feature moves with the child feature in context, so that the parent/child relationship is maintained.

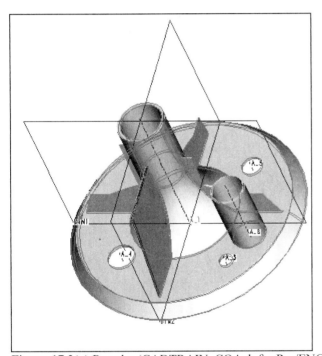

Figure 17.3(a) Reorder (CADTRAIN, COAch for Pro/ENGINEER)

Figure 17.3(b) Reorder

Inserting Features

Normally, Pro/E adds a new feature after the last existing feature in the part, including suppressed features. Insert Mode allows you to add new features at any point in the feature sequence, except before the base feature or after the last feature. You can also Insert features using the Model Tree. There is an arrow-shaped icon on the Model Tree that indicates where features will be inserted upon creation. By default, it is always at the end of the Model Tree. You may drag the location of the *arrow* higher or lower in the tree to insert features at a different point. When the *arrow* is dropped at a new location, the model is rolled backward or forward in response to the insertion *arrow* being moved higher or lower in the tree.

Lesson 17 STEPS

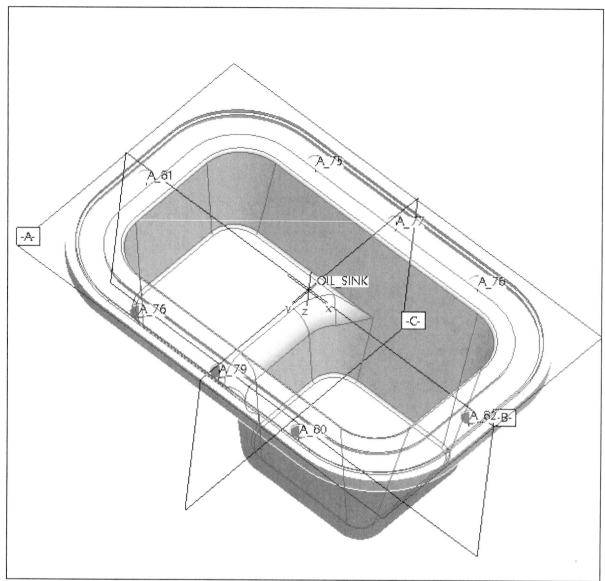

Figure 17.4 Oil Sink

Oil Sink

The Oil Sink (Fig. 17.4) requires the use of the **Shell Tool**. The shelling of a part should be done after the desired protrusions and most rounds have been modeled. This lesson part will have you create a protrusion, a cut, and a set of rounds. Some of the required rounds will be left off the part model on purpose. Pro/E's **Insert Mode** option enables you to insert a set of features at an earlier stage in the design of the part. In other words, you can create a feature after or before a selected existing feature even if the whole model has been completed. You can also *move the order in which a feature was created* and therefore have subsequent features affect the reordered feature. A round created after a shell operation can be reordered to appear before the shell, to have the shell be affected by the round.

In this lesson, you will also insert a round or two before the existing shell feature using Insert Mode. The rounds will be shelled after the **Resume** option is picked, because the rounds now appear before the shell feature. The details shown in Figures 17.5(a) through (m) provide the design dimensions.

592

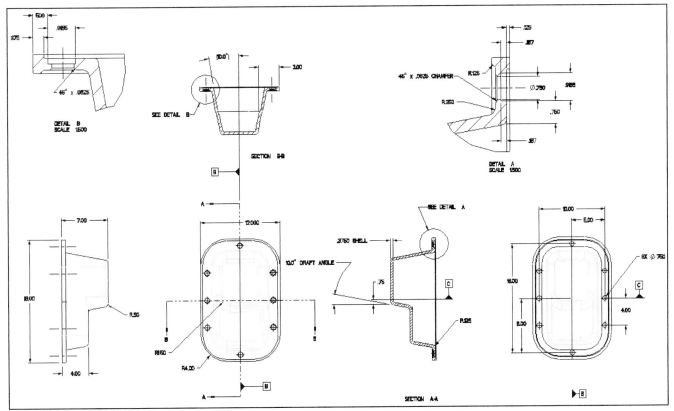

Figure 17.5(a) Oil Sink Detail Drawing

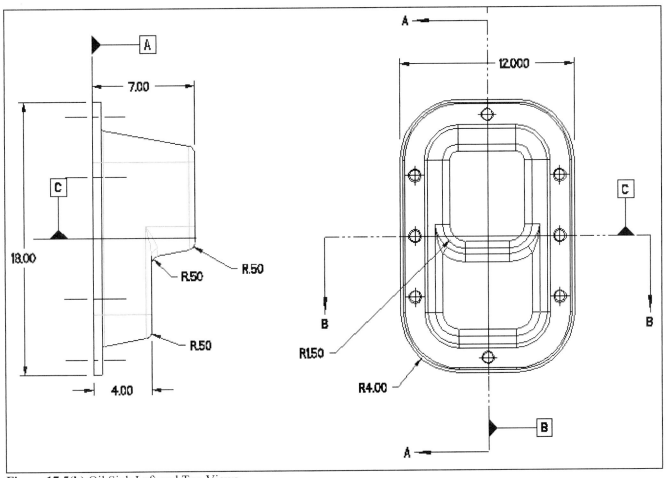

Figure 17.5(b) Oil Sink Left and Top Views

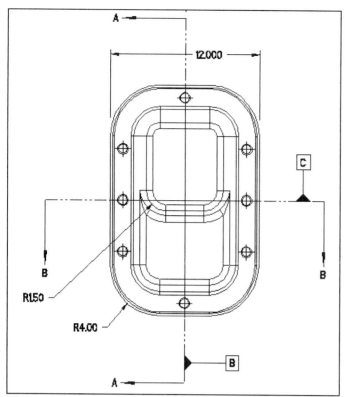

Figure 17.5(c) Oil Sink Top View Dimensions

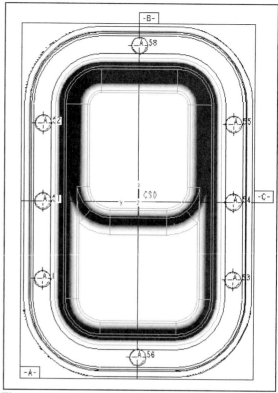

Figure 17.5(d) Oil Sink Top View

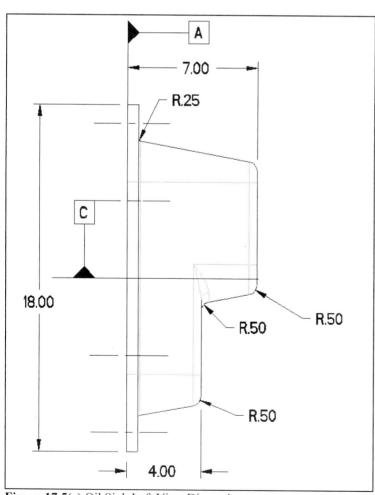

Figure 17.5(e) Oil Sink Left View Dimensions

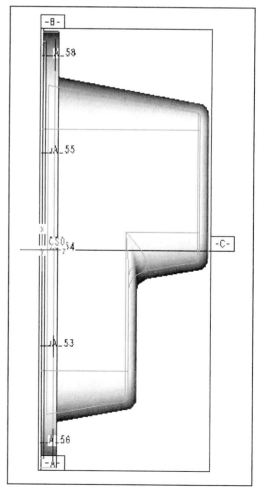

Figure 17.5(f) Oil Sink Left View

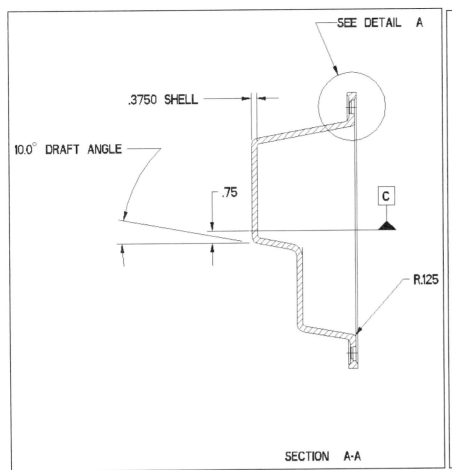

Figure 17.5(g) Oil Sink SECTION A-A

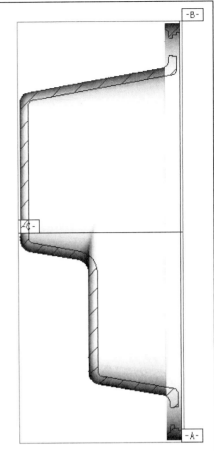

Figure 17.5(h) Oil Sink Right View

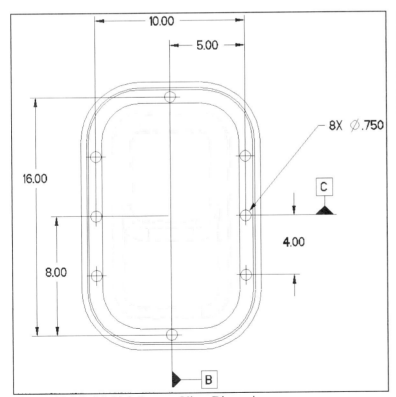

Figure 17.5(i) Oil Sink Bottom View Dimensions

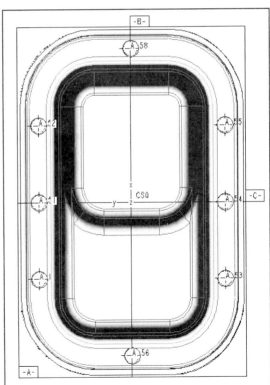

Figure 17.5(j) Oil Sink Bottom View

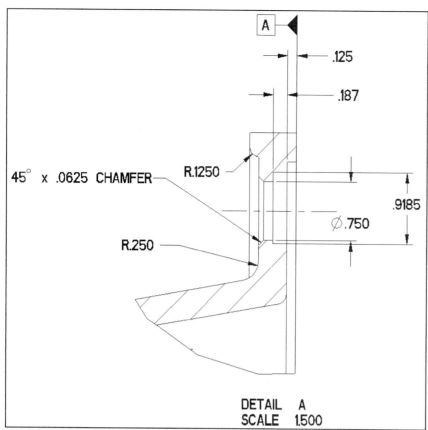

Figure 17.5(k) Oil Sink DETAIL A

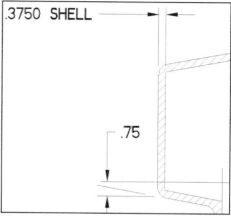

Figure 17.5(l) Oil Sink Shell

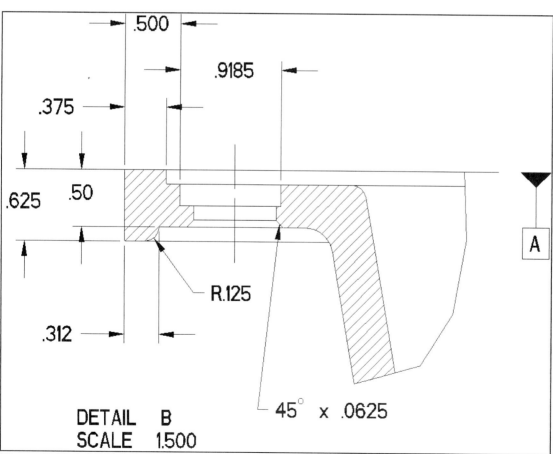

Figure 17.5(m) Oil Sink DETAIL B

Click: **File** ⇒ **Set Working Directory** select the working directory ⇒ **OK** ⇒ [icon] ⇒ [Part] ⇒ **OIL_SINK** ⇒ [✓] ⇒ **OK** ⇒ **Tools** ⇒ **Environment** ⇒ [✓ Snap to Grid] ⇒ [✓ Use 2D Sketcher] ⇒ [Display Style | Hidden Line] ⇒ [Tangent Edges | Dimmed] ⇒ **Apply** ⇒ **OK**

Set up the working environment and defaults: **Tools** ⇒ **Options** ⇒ Showing: **Current Session** ⇒ Option: *default_dec_places* ⇒ Value: **3** ⇒ **Enter** ⇒ Option: *sketcher_dec_places* ⇒ Value: **3** ⇒ **Enter** ⇒ **Apply** ⇒ **Close** ⇒ load your saved customization file ⇒ **Tools** ⇒ **Customize Screen** ⇒ **File** ⇒ **Open Settings** ⇒ click on your saved file ⇒ **Open** ⇒ **OK**

Set and Assign the material: **Edit** ⇒ **Setup** ⇒ **Units** ⇒ Units Manager **Inch lbm Second** ⇒ **Close** ⇒ **Material** ⇒ **steel.mtl** ⇒ **>>>** ⇒ **OK** ⇒ **Done**

Change the coordinate system name and set the datums: double-click on the default coordinate system name in the Model Tree-- **PRT_CSYS_DEF** ⇒ **OIL_SINK** ⇒ **MMB** ⇒ **Edit** ⇒ **Setup** ⇒ **Geom Tol** ⇒ **Set Datum** ⇒ pick **TOP** from the model ⇒ Name- **B** ⇒ **OK** ⇒ pick **FRONT** ⇒ Name- **A** ⇒ **OK** ⇒ pick **RIGHT** ⇒ Name- **C** ⇒ **MMB** ⇒ **MMB** ⇒ **MMB** ⇒ **MMB** ⇒ [icon] ⇒ **MMB** (Fig. 17.6)

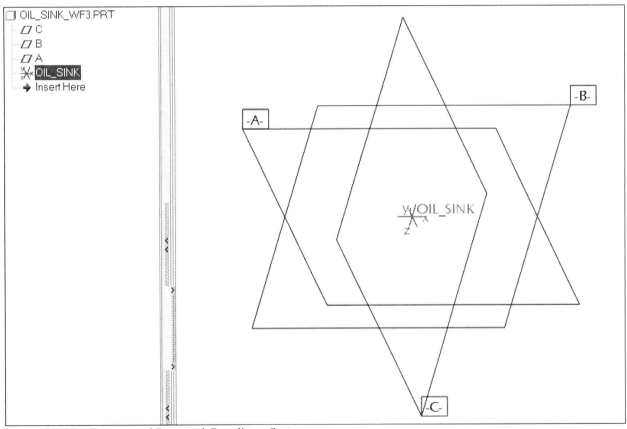

Figure 17.6 Set Datums and Renamed Coordinate System

Make the first protrusion **.50** (thickness) **X 12.00** (height) **X 18.00** (length), with **R4.00** rounds (add the fillets to the sketch). Sketch on datum plane **A**, and center the first protrusion horizontally on datum **B** and vertically on datum **C** [Figs. 17.7(a-b)].

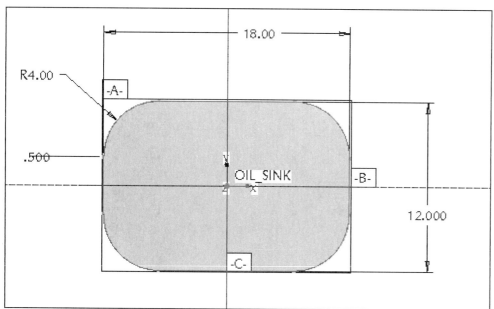

Figure 17.7(a) Dimensions for First Protrusion

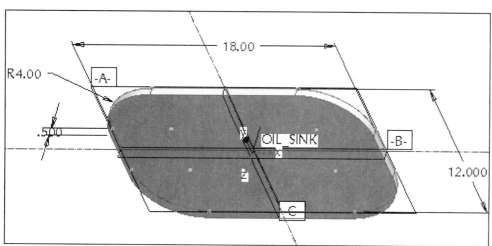

Figure 17.7(b) Standard Orientation

Make the second protrusion offset from the edge of the first protrusion **3.00**, with a height of **7.00** [Figs. 17.8(a-b)]. Sketch on top surface of the first protrusion. Then, create the cut [Figs. 17.9(a-b)].

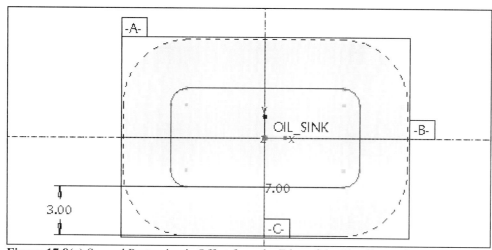

Figure 17.8(a) Second Protrusion is Offset from the Edge of the First Protrusion

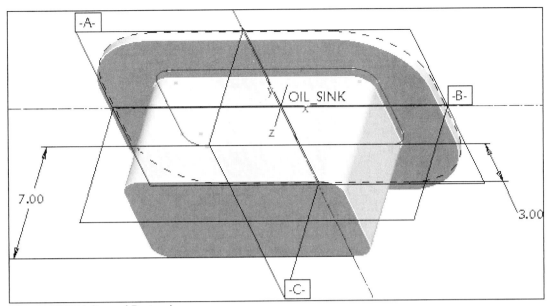

Figure 17.8(b) Second Protrusion

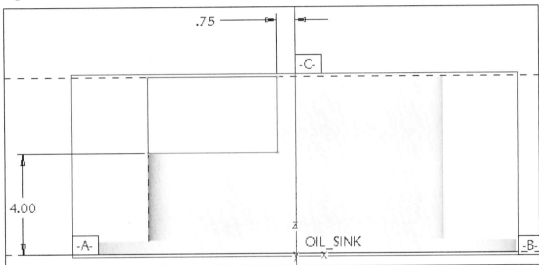

Figure 17.9(a) Cut

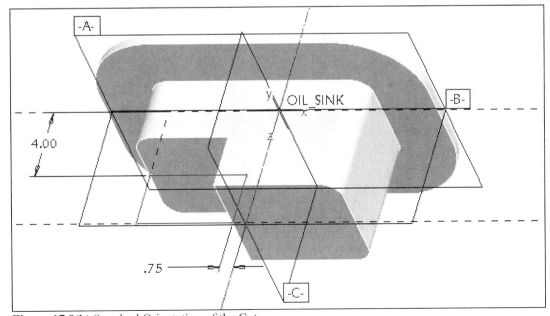

Figure 17.9(b) Standard Orientation of the Cut

Add the **R1.50** rounds [Figs. 17.10(a-b)]. Draft all vertical surfaces of the second protrusion **10** degrees. Use the top surface as the Draft hinge [Figs. 17.11(a-b)].

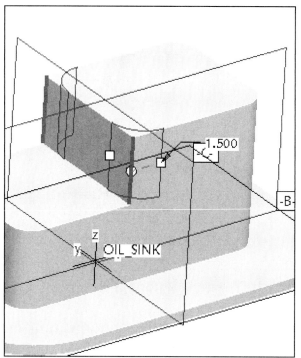

Figure 17.10(a) Create the **R1.50** Rounds

Figure 17.10(b) Completed Rounds

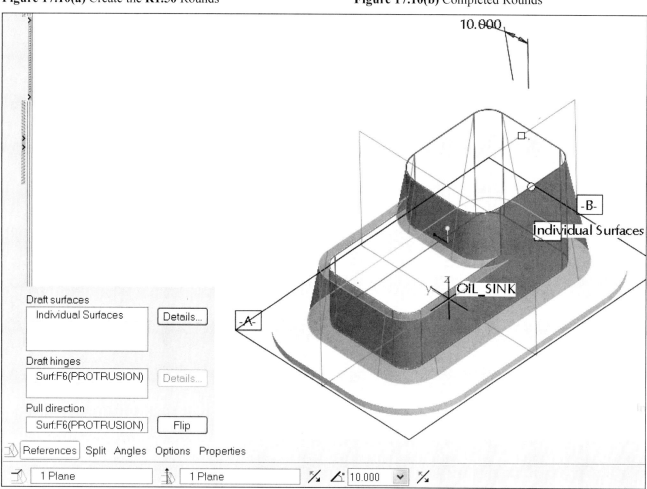

Figure 17.11(a) Draft References

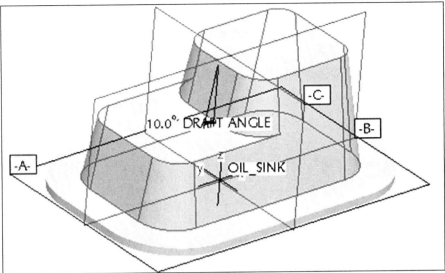

Figure 17.11(b) Drafted Sides

Click: **Context sensitive help** ⇒ ⇒ read about the Shell Tool ⇒ **Close** ⇒ **Shell Tool** ⇒ Thickness **.375** ⇒ **Enter** ⇒ spin the model ⇒ **References** tab ⇒ Removed surfaces-- select the bottom surface of the part [Fig. 17.12(a)] ⇒ ⇒ ⇒ **MMB** ⇒ **LMB** ⇒ ⇒ ⇒ ⇒ **MMB** ⇒ **File** ⇒ **Delete** ⇒ **Old Versions** ⇒ **MMB** [Fig. 17.12(b)]

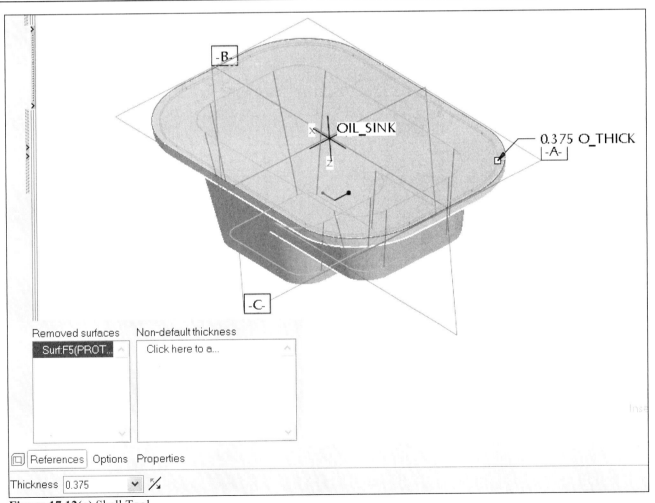

Figure 17.12(a) Shell Tool

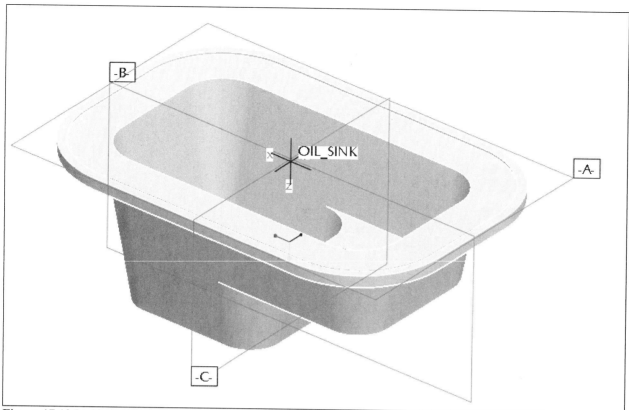

Figure 17.12(b) Shelled Part

The next feature you need to create is a *"lip"* around the part using a protrusion ⇒ Sketch two closed loops [Fig. 17.13(a)]. Use the edge of the first protrusion for the first loop and then create an offset edge (**-.3125**) for the second loop [Fig. 17.13(b)]. ⇒ The depth of the lip protrusion is **.125** [Figs. 17.13(c-d)].

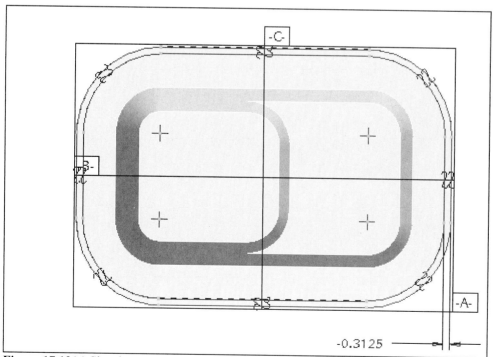

Figure 17.13(a) Sketch

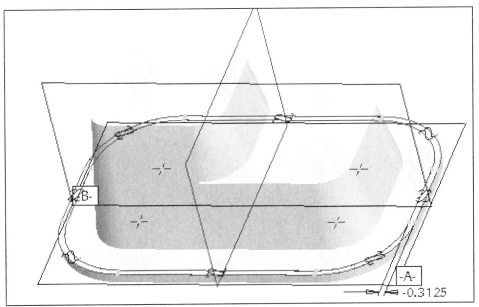

Figure 17.13(b) Standard Orientation of Sketch

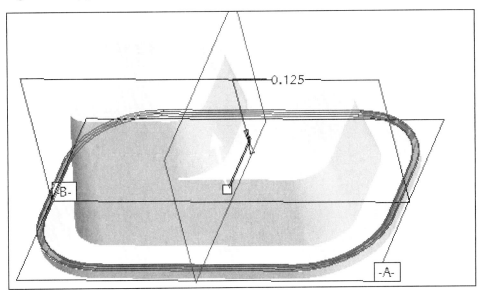

Figure 17.13(c) Depth **.125**

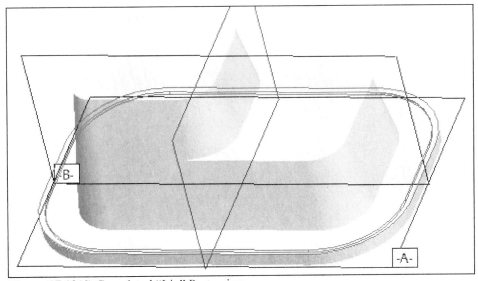

Figure 17.13(d) Completed "Lip" Protrusion

Add the **R.125** round to the inside of the *"lip"* (Fig. 17.14) ⇒ 💾 ⇒ **MMB**

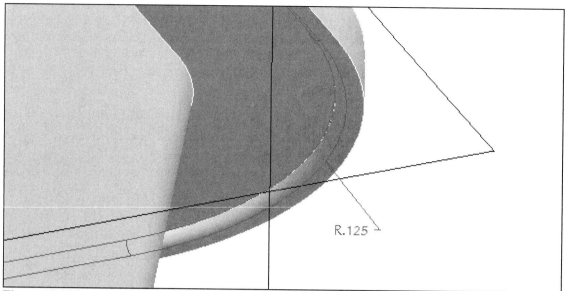

Figure 17.14 Round **R.125**

The next feature is a cut measuring **.9185** wide by **.187** deep [Figs. 17.15(a-b)]. The sketch will be composed of two closed loops as with the lip-like protrusion created previously [Figs. 17.16(a-c)].

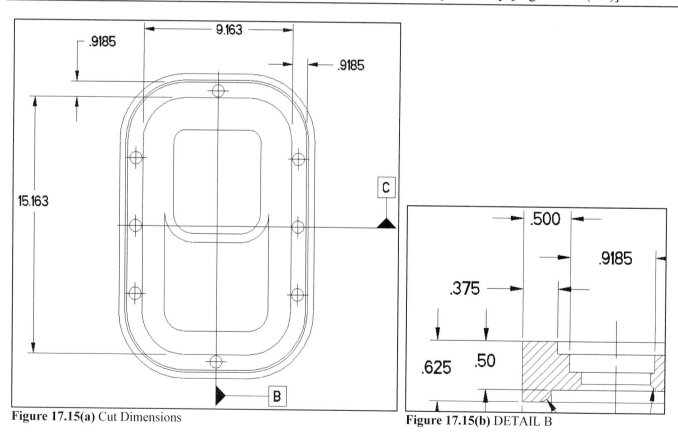

Figure 17.15(a) Cut Dimensions

Figure 17.15(b) DETAIL B

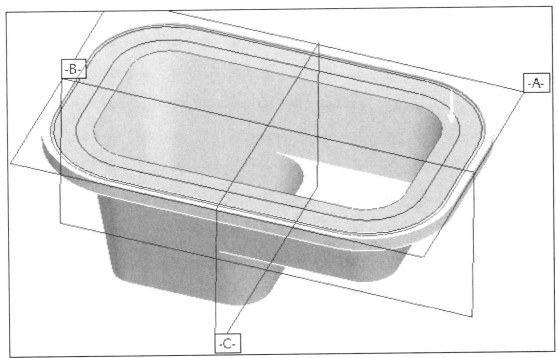
Figure 17.16(a) Cut Surface

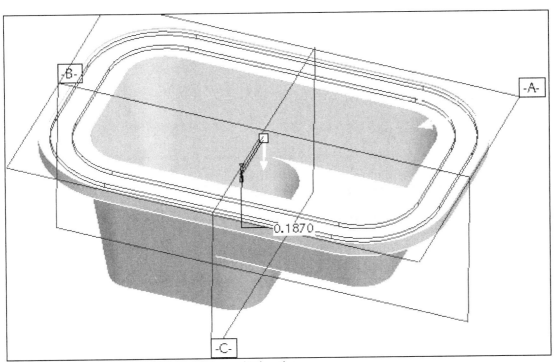
Figure 17.16(b) Depth and Material Removal Direction

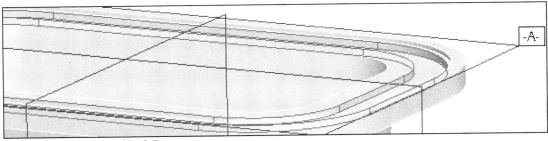
Figure 17.16(c) Completed Cut

605

Add another **R.125** round to the inside edge [Figs. 17.17(a-b)] ⇒ 🖫 ⇒ **MMB**

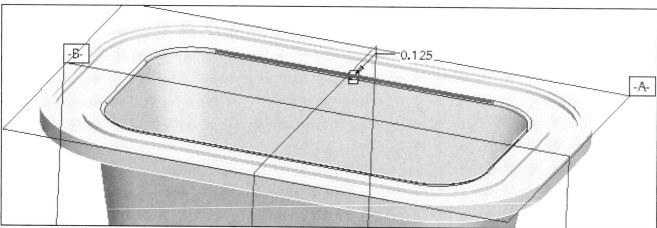

Figure 17.17(a) Edge Round **R.125**

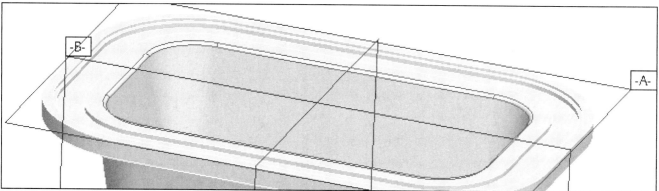

Figure 17.17(b) Completed Round

Create a **R.250** round as shown in Figure 17.18 ⇒ **Ctrl+S** ⇒ **MMB**

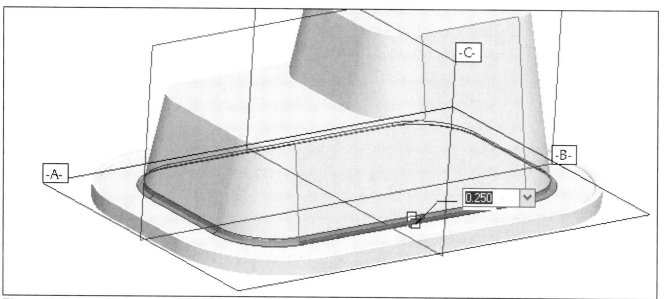

Figure 17.18 Round **R.250**

The countersunk holes will be added next [Figs. 17.19(a-b)].

Click: **Hole Tool** from Right Toolchest ⇒ spin the part ⇒ change the diameter to **.750** ⇒ **Placement** tab ⇒ select the surface for placement [Fig. 17.19(c)] ⇒ **Drill to intersect with all surfaces**

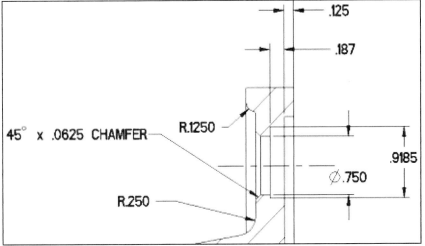

Figure 17.19(a) Hole Dimensions

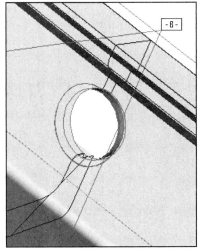

Figure 17.19(b) X-Section of Hole

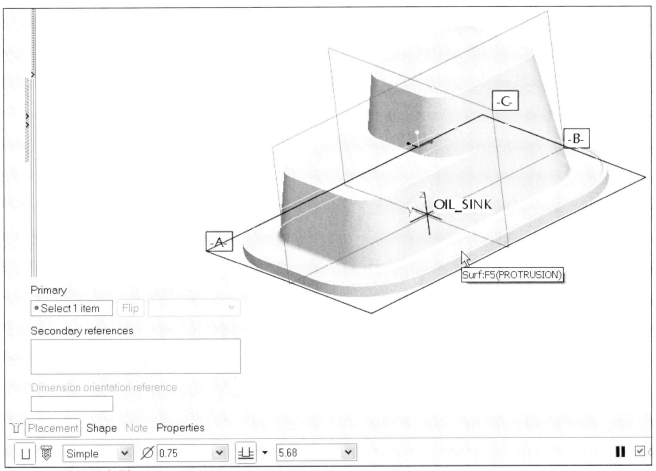

Figure 17.19(c) Hole Placement

Move one drag handle [Fig. 17.19(d)] to datum **C** and the other to datum **B** [Fig. 17.19(e)]

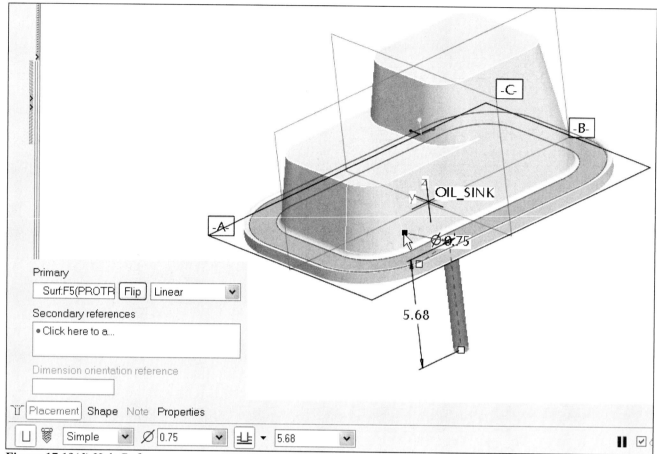

Figure 17.19(d) Hole References

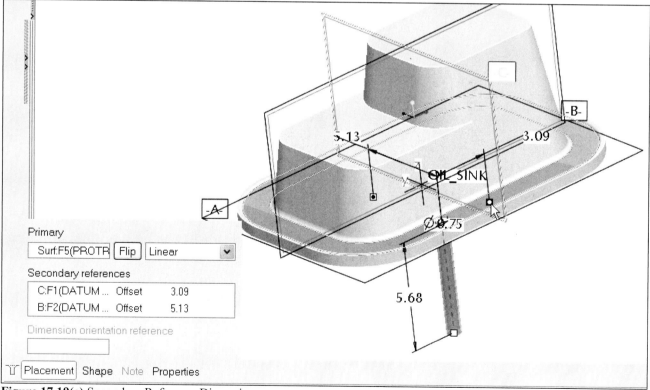

Figure 17.19(e) Secondary Reference Dimensions

Modify the values to be **4.00** from datum **C** and **5.00** from datum **B** [Fig. 17.19(f)]

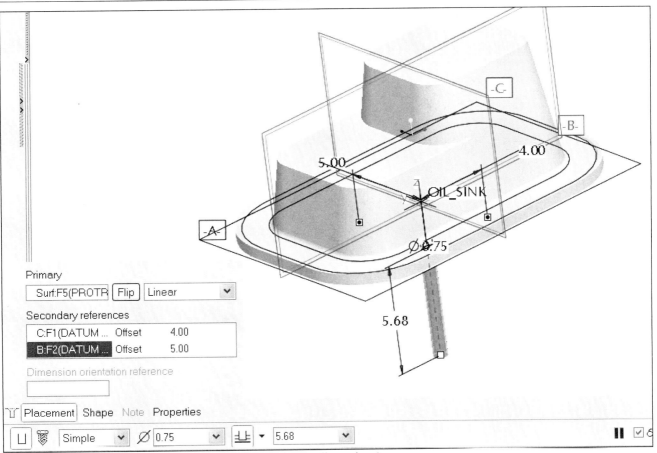

Figure 17.19(f) Secondary Reference Dimensions (position dimensions)

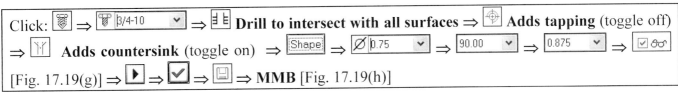

Click: [icon] ⇒ [3/4-10] ⇒ [icon] **Drill to intersect with all surfaces** ⇒ [icon] **Adds tapping** (toggle off) ⇒ [icon] **Adds countersink** (toggle on) ⇒ [Shape] ⇒ [Ø 0.75] ⇒ [90.00] ⇒ [0.875] ⇒ [✓ 👓] [Fig. 17.19(g)] ⇒ [▶] ⇒ [✓] ⇒ [💾] ⇒ **MMB** [Fig. 17.19(h)]

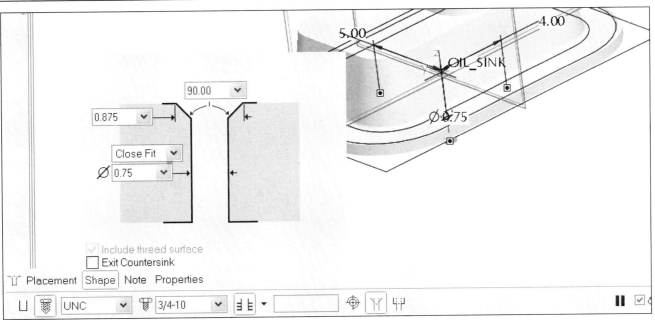

Figure 17.19(g) Shape

609

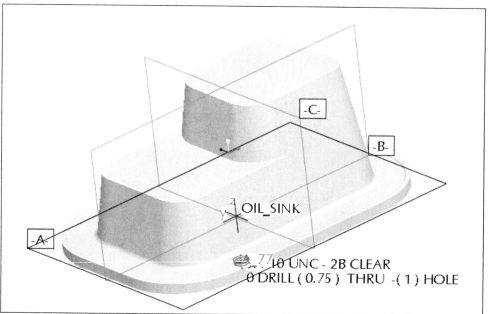

Figure 17.19(h) Completed Countersunk Hole

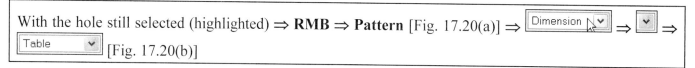

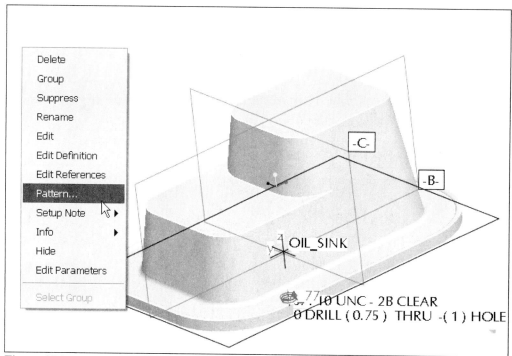

Figure 17.20(a) Pattern the Hole

Figure 17.20(b) Pattern Members Defined by Table

With the **Ctrl** key pressed, click on **4.000** and then on **5.000** [Fig. 17.20(c)]

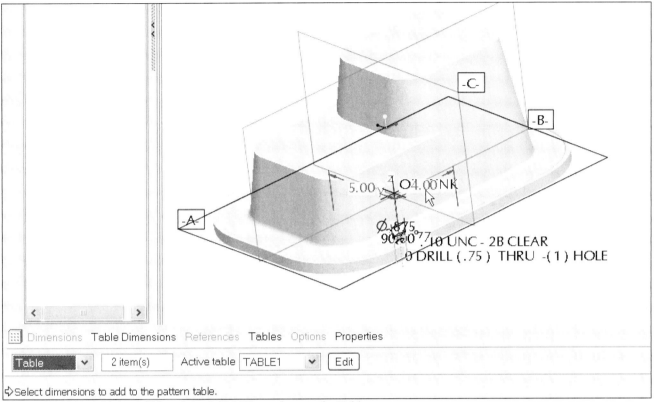

Figure 17.20(c) Add the **4.000** and the **5.000** Dimensions to the Table

Click: [Edit] [Fig. 17.20(d)] ⇒ add the information [Figs. 17.20(e-g)] ⇒ **File** ⇒ **Exit** [Fig. 17.20(h)]

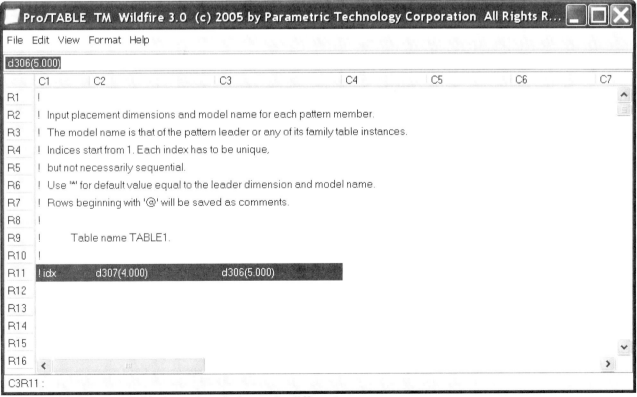

Figure 17.20(d) Pattern Table (your d symbols may be different)

R8	!	
R9	!	Tabl
R10	!	
R11	!idx	
R12		1
R13		2
R14		3
R15		4
R16		5
R17		6
R18		7
R19		

Figure 17.20(e) Add numbers 1-7

R8	!		
R9	!	Table name TABLE1.	
R10	!		
R11	!idx	d307(4.000)	
R12		1	*
R13		2	0.000
R14		3	-4.000
R15		4	0.000
R16		5	-4.000
R17		6	8.000
R18		7	-8.000
R19			

Figure 17.20(f) Add Values in the Second Column (* means identical value)

R9	!	Table name TABLE1.	
R10	!		
R11	!idx	d307(4.000)	d306(5.000)
R12	1	*	-5.000
R13	2	0.000	*
R14	3	-4.000	*
R15	4	0.000	-5.000
R16	5	-4.000	-5.000
R17	6	8.000	0.000
R18	7	-8.000	0.000

Figure 17.20(g) Add Values in the Third Column

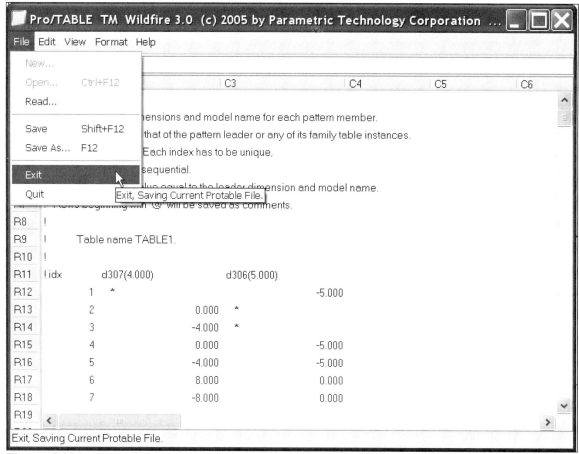

Figure 17.20(h) Completed Table

Click: ☑ [Fig. 17.20(i)] ⇒ 🖫 ⇒ **MMB** [Fig. 17.20(j)] ⇒ **LMB** ⇒ check your settings in the Navigator ⇒ **Settings** ⇒ **Tree Filters** ⇒ ☑ Suppressed Objects ⇒ **OK**

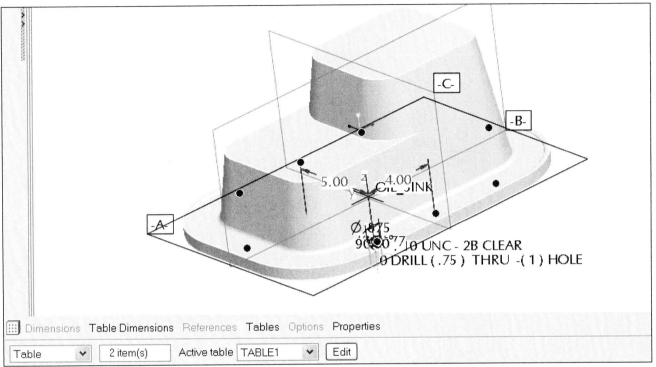

Figure 17.20(i) Previewed Pattern

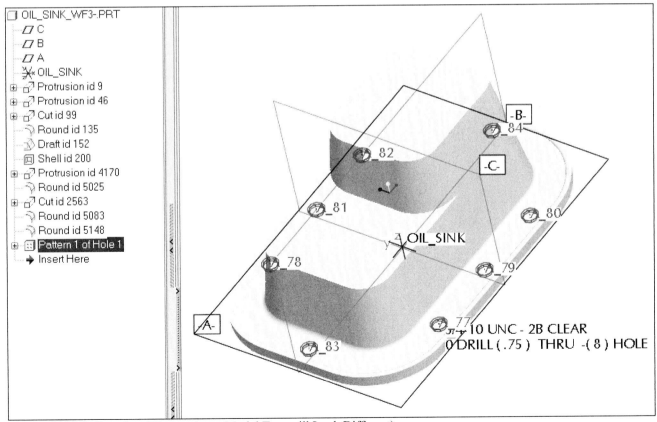

Figure 17.20(j) Completed Pattern (your Model Tree will Look Different)

The next series of features will be created purposely at the wrong stage in this project. You will now create the **R.50** round [Figs. 17.21(a-b)]. Because the design intent is to have a constant thickness for the part, the round should have been created before the shell. The Reorder capability will be used to change the position of this round in the design sequence. Using the Model Tree, you can pick and drag the round to a new location in the feature list. *(The Reorder capability can also be completed using Edit ⇒ Feature Operations ⇒ Reorder ⇒ pick the feature to reorder ⇒ OK ⇒ Done ⇒ Select the new position).*

Pick on the top edge of the part ⇒ **RMB** ⇒ **Round Edges** [Fig. 17.21(a)] ⇒ move drag handle to **.500** [Fig. 17.21(b)] ⇒ **MMB** [Fig. 17.21(c)] ⇒ Spin the part, and then pick on the inner surface (Fig. 17.22). Notice that the rounds do not propagate on the internal edges.

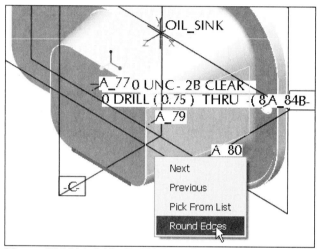

Figure 17.21(a) Add a **R.500** Round

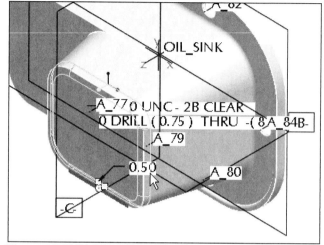

Figure 17.21(b) Move Drag Handle to **500**

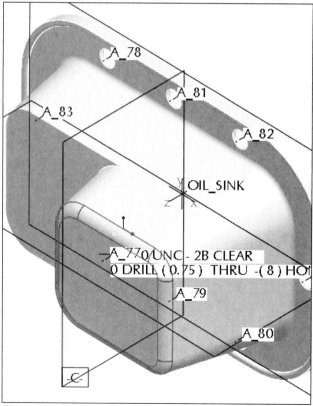

Figure 17.21(c) Completed Round

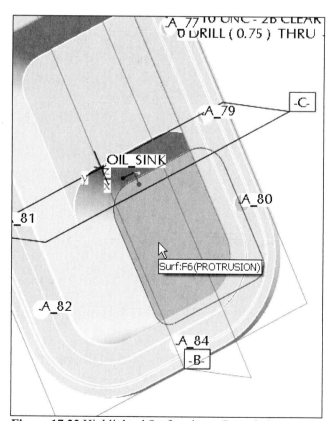

Figure 17.22 Highlighted Surface is not Rounded

Reorder the round to appear before the Shell: click on the **Round** [Fig. 17.23(a)] and drag [Round 1] to a position before/above [Shell id 200] [Fig. 17.23(b)] and drop [Fig. 17.23(c)] ⇒ 🖫 ⇒ **MMB**

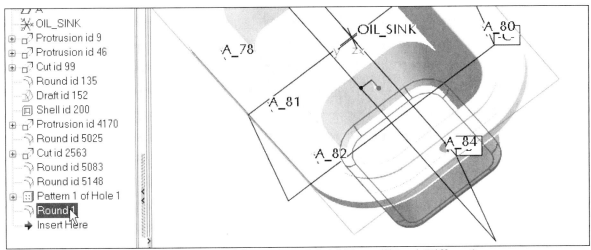

Figure 17.23(a) Click on Round in the Model Tree (your Model Tree will Look Different)

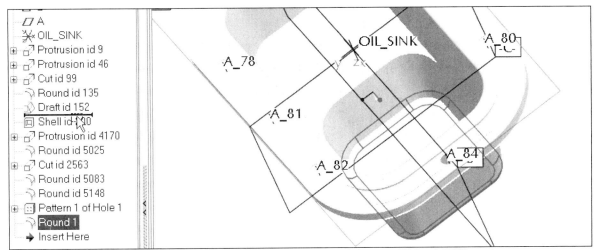

Figure 17.23(b) Move Cursor above the Shell Feature (your Model Tree will Look Different)

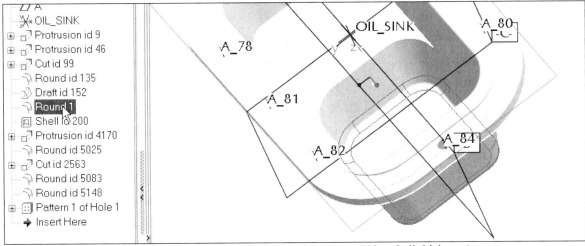

Figure 17.23(c) Reordered Round Shows on the Inside of the Part (.500 – shell thickness)

You can also Insert new features using the Model Tree. The arrow-shaped icon [→ Insert Here] in the Model Tree indicates where features will be inserted upon creation and is by default at the end (or bottom) of the Model Tree.

The Insert capability can also be completed using Edit ⇒ Feature Operations ⇒ Insert Mode ⇒ Activate ⇒ Select a feature to insert after ⇒ Done.

By dragging the location of the insert node higher, so that its position is before existing features, you can insert a new feature at that stage of the model history. When the *insert node* is dropped at a new location, the model is rolled backward (suppressed) or forward in response to the insertion node being moved higher or lower. The Model Tree displays a small square (■) next to the features that are not active (suppressed).

The previous round was created at the wrong stage in the design sequence and then reordered. To eliminate the reordering of a feature, the remaining **R.50** rounds will be created using Insert Mode with the Model Tree.

Insert Mode allows you to insert a feature at a previous stage of the design sequence. This is like going back into the past and doing something you wish you had done before--not possible with life, but with Pro/E less of a problem. Add the additional **R.50** rounds.

In the Model Tree click on [→ Insert Here] [Fig. 17.24(a)] and drag it to a position before/above [Shell id 200] and drop [Fig. 17.24(b)] ⇒ [🔍] **Refit**

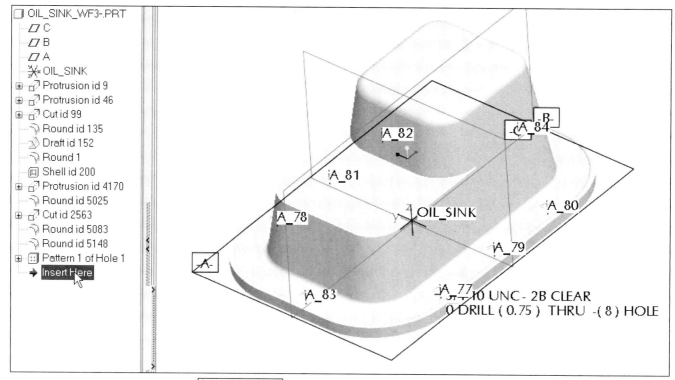

Figure 17.24(a) Insert Here Pointer [→ Insert Here] (your Model Tree will Look Different)

616

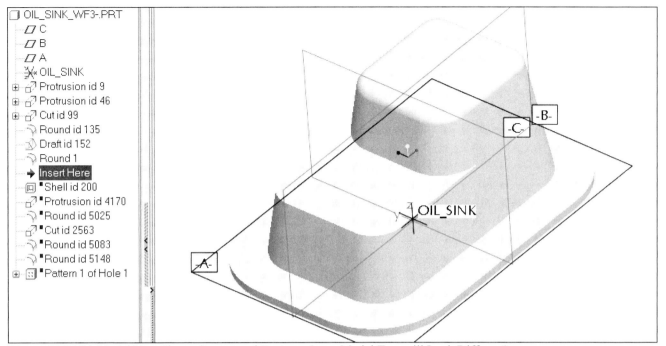

Figure 17.24(b) Model Tree Shows Suppressed Features (your Model Tree will Look Different)

Create two sets of rounds ⇒ **Round Tool** ⇒ **Sets** tab ⇒ pick the front edge ⇒ **RMB** ⇒ **Add set** ⇒ pick the second edge [Fig. 17.25(a)] ⇒ **MMB** [Fig. 17.25(b)]

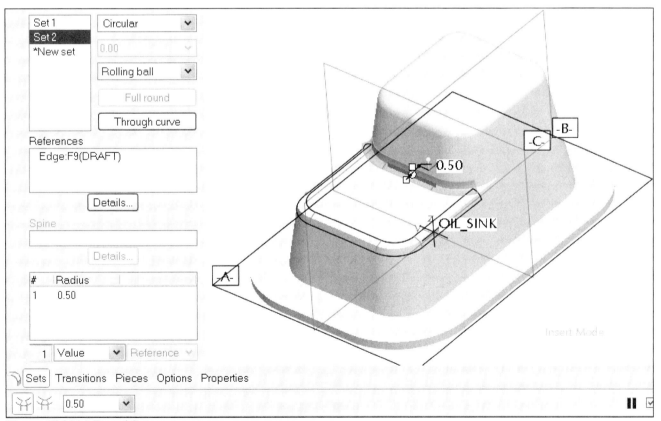

Figure 17.25(a) Round Sets

Rotate the model [Fig. 17.25(c)] ⇒ click on [→ Insert Here] [Fig. 17.25(c)]

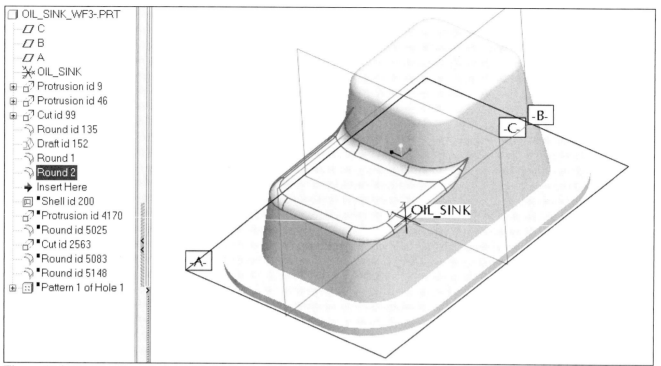

Figure 17.25(b) New Rounds (your Model Tree will Look Different)

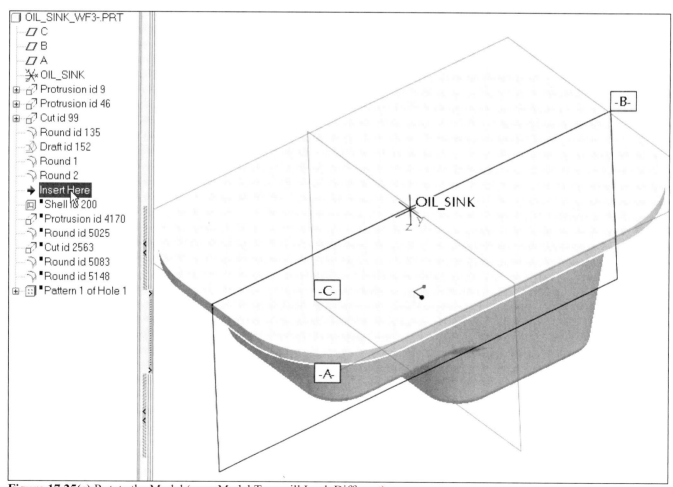

Figure 17.25(c) Rotate the Model (your Model Tree will Look Different)

Drag it to the bottom of the Model Tree list and drop [Fig. 17.26(a)] ⇒ **View** ⇒ **Shade** ⇒ click on the new rounds in the **Model Tree** ⇒ 💾 ⇒ **MMB** [Fig. 17.26(b)] ⇒ **File** ⇒ **Delete** ⇒ **Old Versions** ⇒ **MMB** [Fig. 17.26(c)]

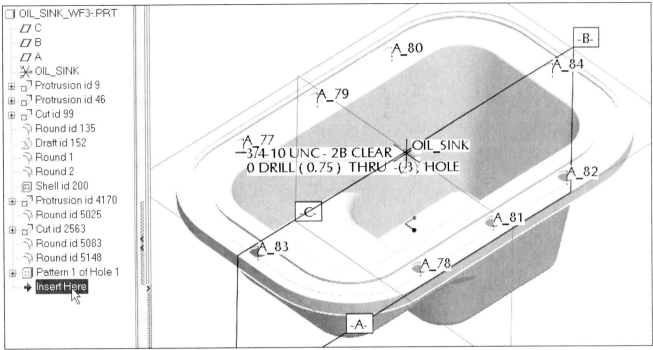

Figure 17.26(a) Drag and Drop Insert Node, All Features are Resumed

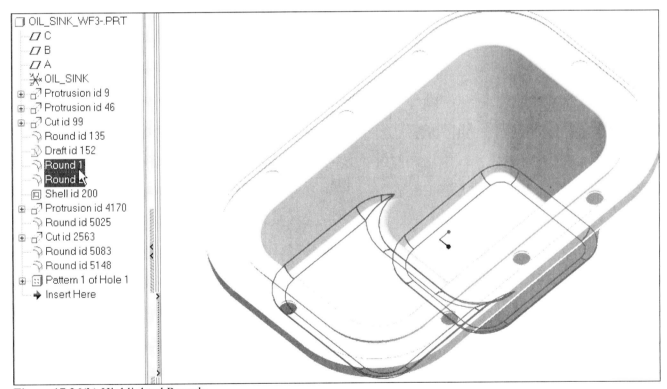

Figure 17.26(b) Highlighted Rounds

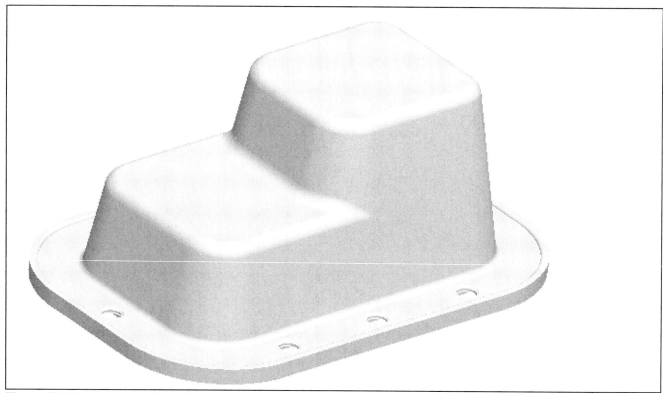

Figure 17.26(c) Completed Oil Sink

The rounds are now in the proper design sequence for the shell feature to have a constant (shell) thickness. Save the part with a new name (File ⇒ Save a Copy) and then complete the **ECO** (Fig. 17.27) using the current part (file) name. See ***www.cad-resources.com*** ⇒ ***Downloads*** for more part projects.

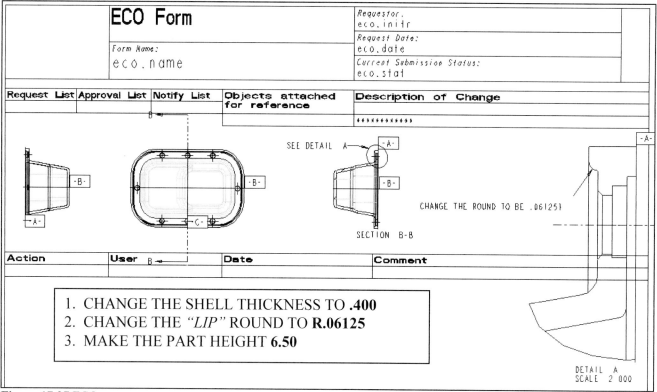

Figure 17.27 ECO

Lesson 18 Drafts, Suppress, and Text Extrusions

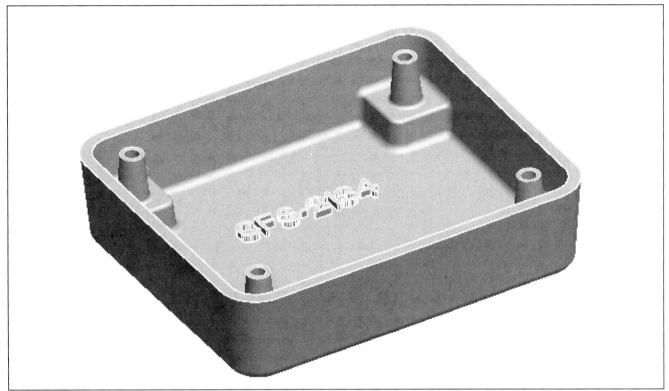

Figure 18.1 Enclosure

OBJECTIVES

- Create **Draft** features
- **Shell** a part
- **Suppress** features to decrease regeneration time
- **Resume** a set of suppressed features
- Create **Text** features on parts

DRAFTS, SUPPRESS, AND TEXT EXTRUSIONS

The **Draft** feature adds a draft angle between surfaces. A wide range of parts incorporate drafts into their design. Casting, injection mold, and die parts normally have drafted surfaces. The ENCLOSURE in Figure 18.1 and the object in Figure 18.2 are plastic injection-molded parts.

Suppressing features by using the **Suppress** command temporarily removes them from regeneration. Suppressed features can be "unsuppressed" (**Resume**) at any time. It is sometimes convenient to suppress text extrusions and rounds to speed up regeneration of the model. Suppressing removes the item from regeneration and requires you to resume the item later.

Hide is another option. Pro/E allows you to hide and unhide some types of model entities. When you hide an item, Pro/E removes the item from the graphics window. The hidden item remains in the Model Tree list, and its icon dims to reveal its hidden status. When you unhide an item, its icon returns to normal display (undimmed) and the item is redisplayed in the graphics window. The hidden status of items is saved with the model. Unlike the suppression of items, hidden items are regenerated.

Text can be included in a sketch for extruded extrusions and cuts, trimming surfaces, and cosmetic features. To decrease regeneration time of the model, text can be suppressed after it has been created. Text can also be drafted.

Drafts

The **Draft Tool** adds a draft angle between two individual surfaces or to a series of selected planar surfaces [Fig. 18.2(a)].

During draft creation, remember the following:

- You can draft only the surfaces that are formed by tabulated cylinders or planes [Fig. 18.2(b)].
- The draft direction must be normal to the neutral plane if a draft surface is cylindrical, [Figs. 18.2(c-d)].
- You cannot draft surfaces with fillets around the edge boundary. However, you can draft the surfaces first, and then fillet the edges [Fig. 18.2(e)].

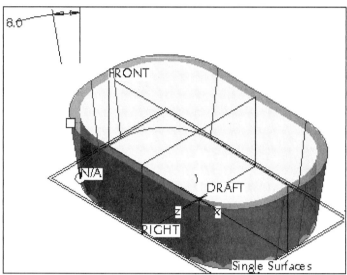

Figure 18.2(a) Draft

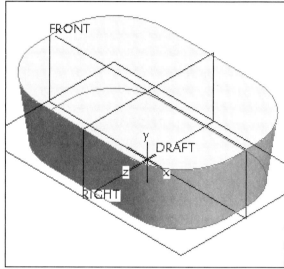

Figure 18.2(b) Drafted Extrusion

The following table lists the terminology used in drafts.

TERM	DEFINITION
Draft surfaces	Model surfaces selected for drafting.
Draft Hinges	Draft surfaces are pivoted about the intersection of the neutral plane with the draft surfaces.
Pull direction	Direction that is used to measure the draft angle. It is defined as normal to the reference plane.
Draft angle	Angle between the draft direction and the resulting drafted surfaces. If the draft surfaces are split, you can define two independent angles for each portion of the draft.
Direction of rotation	Direction that defines how draft surfaces are rotated with respect to the neutral plane or neutral curve.
Split areas	Areas of the draft surfaces to which you can apply different draft angles. Split object is also a choice.

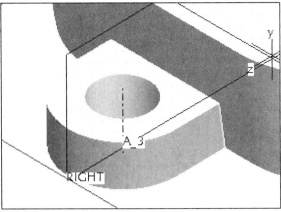

Figure 18.2(c) Second Extrusion

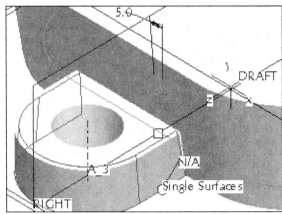

Figure 18.2(d) Drafted Extrusion

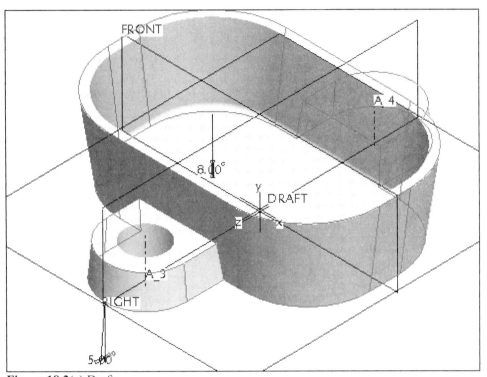

Figure 18.2(e) Drafts

Suppressing and Resuming Features

Suppressing a feature is similar to removing the feature from regeneration temporarily. You can "unsuppress" (**Resume**) suppressed features at any time. Features on a part can be suppressed to simplify the part model and decrease regeneration time. For example, while you work on one end of a shaft, it may be desirable to suppress features on the other end of the shaft. Similarly, while working on a complex assembly, you can suppress some of the features and components for which the detail is not essential to the current assembly process.

Unlike other features, the base feature cannot be suppressed. If you are not satisfied with your base feature, you can redefine the section of the feature, or you can delete it and start over again. Select feature(s) to suppress by: picking on it, selecting from the Model Tree [Figs. 18.3(a-b)], specifying a *range*, entering its *feature number* or *identifier*, or using *layers*.

You can use **Suppress** and **Resume** to simplify the part before inserting features such as text extrusions. In addition, you may wish to suppress the text extrusion if there is other work to be done on the part. Text extrusions take time to regenerate, and increase the file size considerably.

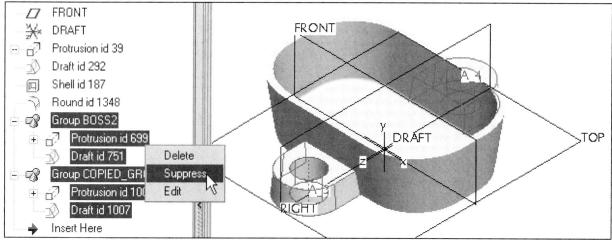

Figure 18.3(a) Suppress Grouped Features

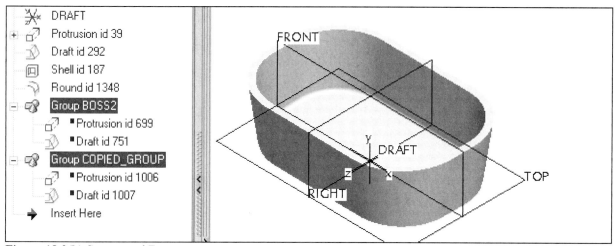

Figure 18.3(b) Suppressed Features

Text Extrusions

When you are modeling, **Text** can be included in a sketch for extruded extrusions and cuts, trimming surfaces, and cosmetic features (Fig. 18.4). The characters that are in an extruded feature use the font **font3d** as the default. Other fonts are available.

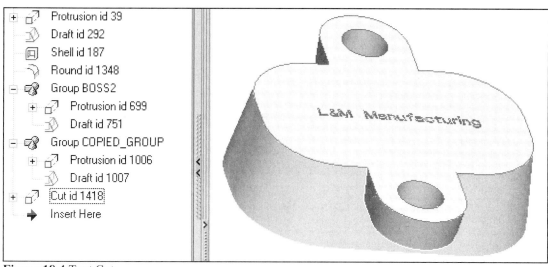

Figure 18.4 Text Cut

Lesson 18 STEPS

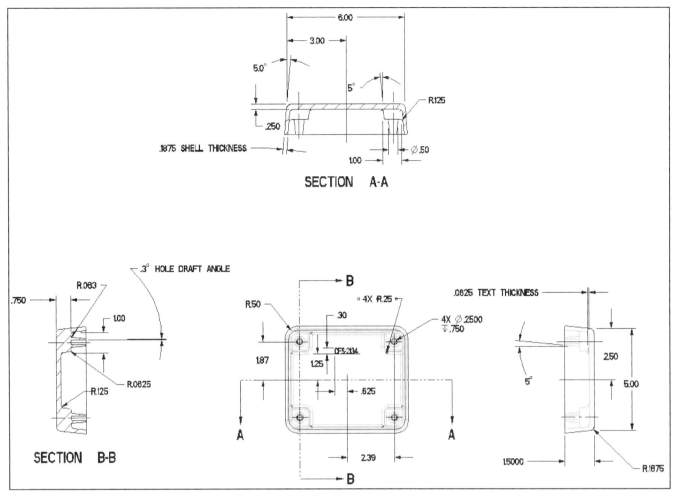

Figure 18.5 Enclosure

Enclosure

The Enclosure is a plastic injection-molded part. A variety of drafts will be used in the design of this part. A *raised text extrusion* will be modeled on the inside of the Enclosure, as shown in Figure 18.5. The dimensions for the part are provided in Figures 18.5 through 18.9.

Click: **File** ⇒ **Set Working Directory** select the working directory ⇒ **OK** ⇒ **Create a new object** ⇒ **Part** ⇒ **ENCLOSURE** ⇒ Use default template ⇒ **OK** ⇒ **Tools** ⇒ **Environment** ⇒ **Snap to Grid** ⇒ **Use 2D Sketcher** ⇒ **Tangent Edges Dimmed** ⇒ **OK** ⇒ **Tools** ⇒ **Options** ⇒ Showing: **Current Session** ⇒ Option: *default_dec_places* ⇒ Value: **3** ⇒ **Enter** ⇒ Option: *sketcher_dec_places* ⇒ Value: **3** ⇒ **Enter** ⇒ **Apply** ⇒ **Close** ⇒ **Edit** ⇒ **Setup** ⇒ **Units** ⇒ Units Manager **Inch lbm Second** ⇒ **Close** ⇒ **Material** ⇒ **fe20.mtl** (plastic) ⇒ **>>>** ⇒ **OK** ⇒ **Done** ⇒ double-click on the default coordinate system name in the Model Tree-- **PRT_CSYS_DEF** ⇒ type **CSYS_ENCLOSURE** ⇒ **Enter** ⇒ ⇒ **MMB** ⇒ load your customization by clicking **Tools** ⇒ **Customize Screen** ⇒ **File** ⇒ **Open Settings** ⇒ click on your saved file (if you made one and saved it previously) ⇒ **Open** ⇒ **OK**

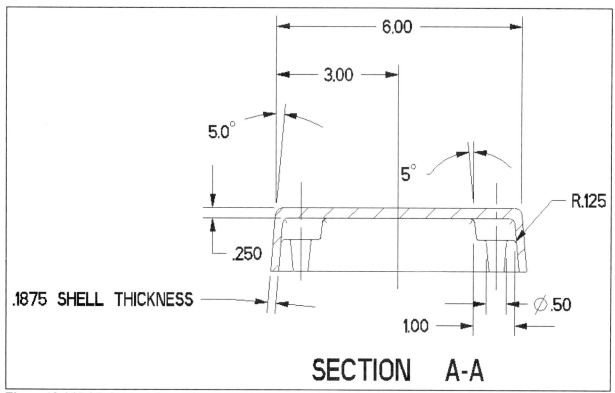

Figure 18.6 SECTION A-A (Top View)

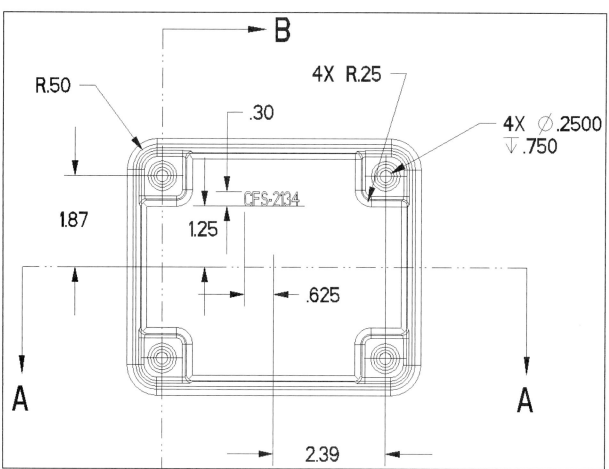

Figure 18.7 Front View

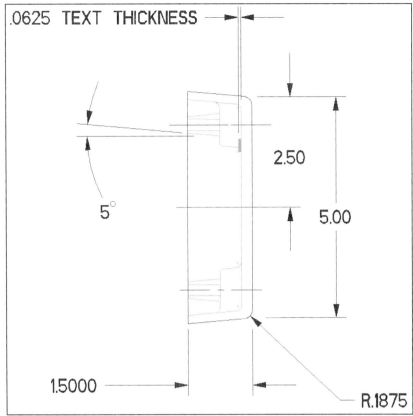

Figure 18.8 Right Side View

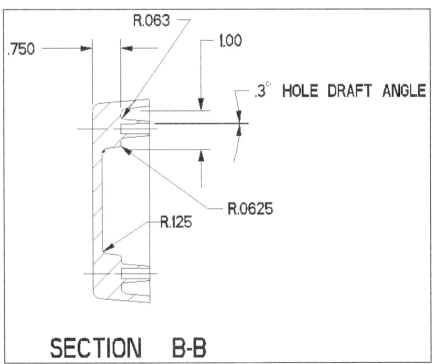

Figure 18.9 SECTION B-B (Left Side View)

Make first extrusion **6.00** (width) **X 5.00** (height) **X 1.50** (depth), with **R.50** rounds (add the fillets in the sketch rather than after the first extrusion is complete) [Figs. 18.18(a-b)]. Sketch on datum **FRONT** [Fig. 18.18(c)]. Center the first extrusion horizontally on datum **TOP** and vertically on datum **RIGHT** [Fig. 18.18(d)]. Add constraints if needed to control your sketch geometry.

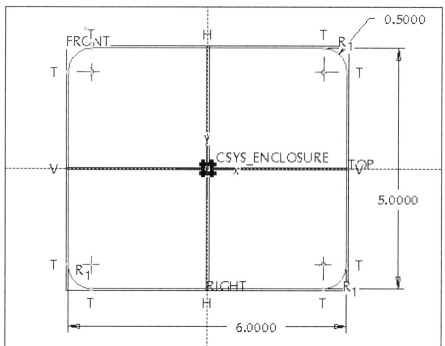

Figure 18.10(a) Sketch

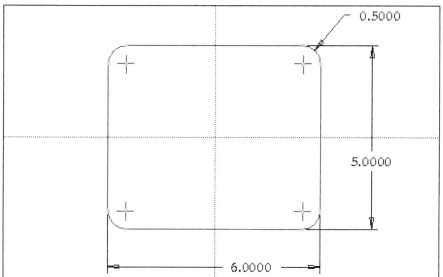

Figure 18.10(b) Dimensions

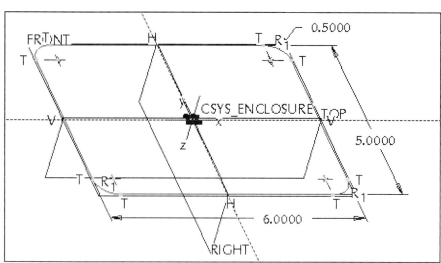

Figure 18.10(c) Standard Orientation

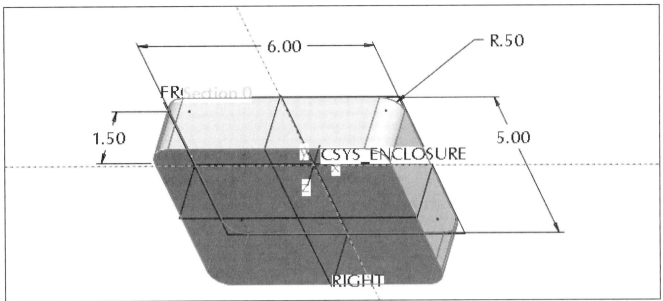

Figure 18.10(d) First Extrusion, **6.00 X 5.00 X 1.50**, and **R.50** rounds

Create the draft for the lateral surfaces of the extrusion, click: **Draft Tool** ⇒ **References** tab ⇒ select one of the lateral surfaces [Fig. 18.11(a)]

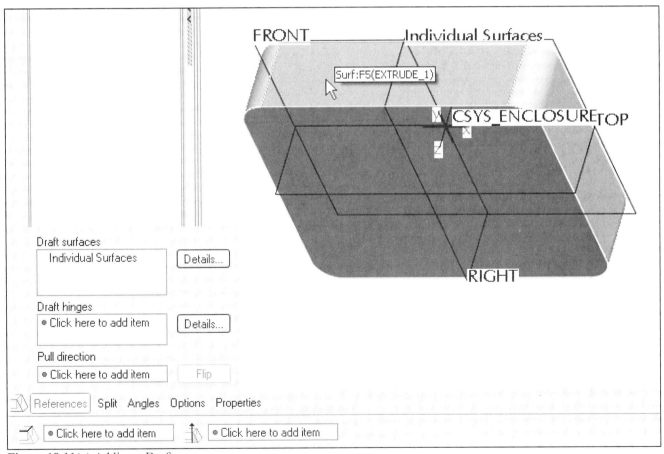

Figure 18.11(a) Adding a Draft

Click in the Draft hinges collector ⇒ pick **FRONT** [Fig. 18.11(b)] ⇒ in the Angle field type **5** ⇒ **Enter** ⇒ **MMB** ⇒ with Draft highlighted in the Model Tree, click **RMB** ⇒ **Edit** [Fig. 18.11(c)] ⇒ pick on **5.000** degrees ⇒ **RMB** ⇒ **Properties** ⇒ **Flip Arrows** ⇒ Number of decimal places **0** ⇒ **Enter** ⇒ **OK**

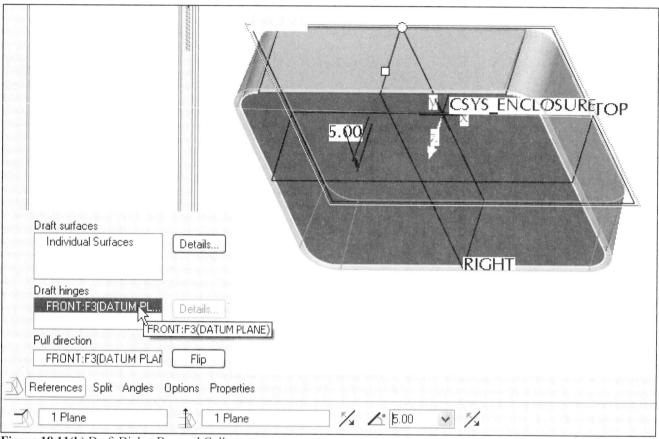

Figure 18.11(b) Draft Dialog Box and Collectors

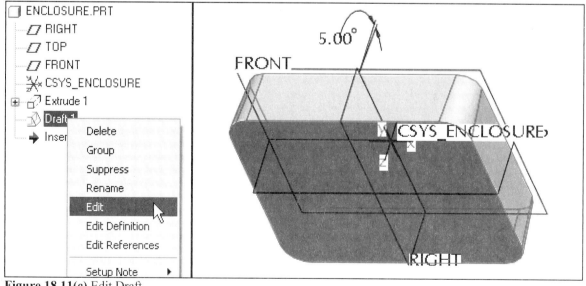

Figure 18.11(c) Edit Draft

Next, use the **Shell Tool** to create a part with a specific thickness.

Click: **Shell Tool** ⇒ pick the face to be removed [Fig. 18.12(a)] ⇒ type **.1875** in Thickness field Thickness 0.1875 ⇒ **Enter** ⇒ [Fig. 18.12(b)] ⇒

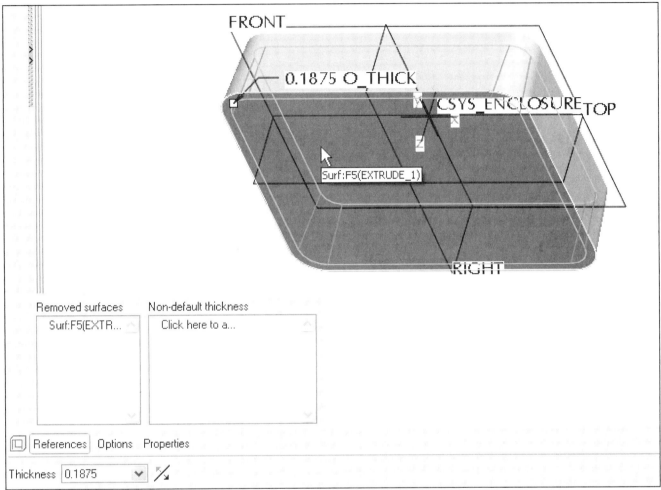

Figure 18.12(a) Using the Shell Tool

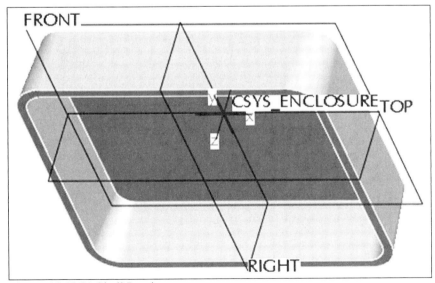

Figure 18.12(b) Shell Preview

Change the thickness of the enclosure to be **.25**, the walls will remain **.1875**. Click: **References** tab ⇒ click in the Non-default thickness collector ⇒ pick the face [Fig. 18.12(c)] (*highlights*) ⇒ type **.250** in the Dimension field ⇒ **Enter** ⇒ **MMB** [Fig. 18.12(d)] ⇒ **LMB**

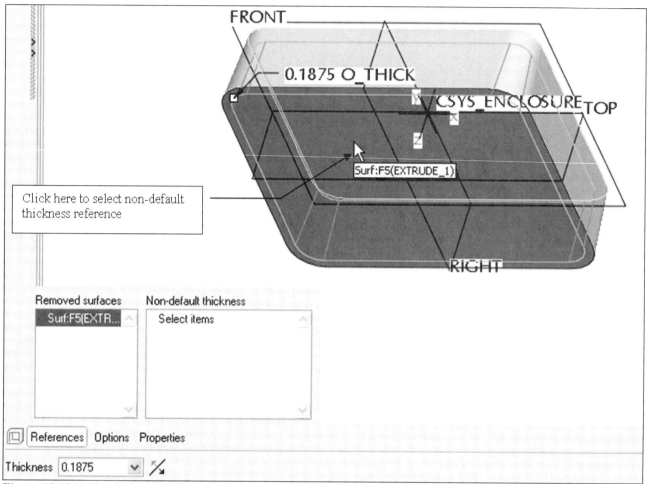

Figure 18.12(c) Non Default Thickness

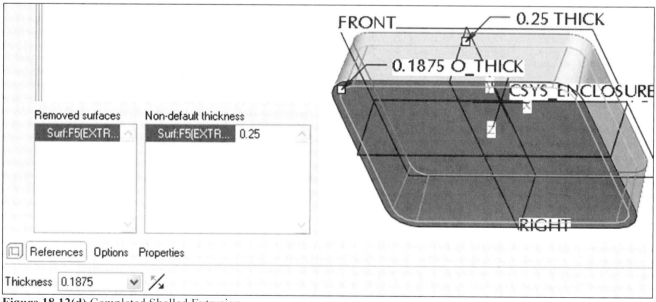

Figure 18.12(d) Completed Shelled Extrusion

Click: **File** ⇒ **Save** ⇒ **MMB** ⇒ Create a raised pedestal-like extrusion on the inside surface. Start by modeling a datum plane offset from datum **FRONT** by .75 [Fig. 18.13(a)]. **DTM1** will be used to control the height of the pedestal. ⇒ **Insert** ⇒ **Extrude** ⇒ select only the References shown in Figure 18.13(b) ⇒ use an existing *internal_shelled_edge* to start the section geometry [Fig. 18.13(c)] ⇒ add four lines and a fillet to complete the section [Fig. 18.13(d)] ⇒ use the dimensioning scheme in Figure 18.13(e) ⇒ ✓ [Fig. 18.13(f)]

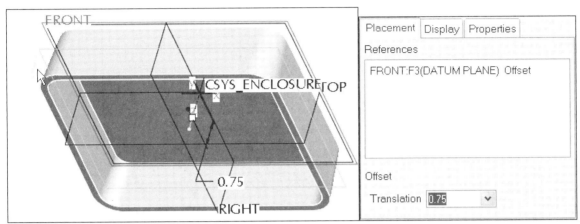

Figure 18.13(a) Offset Datum DTM1

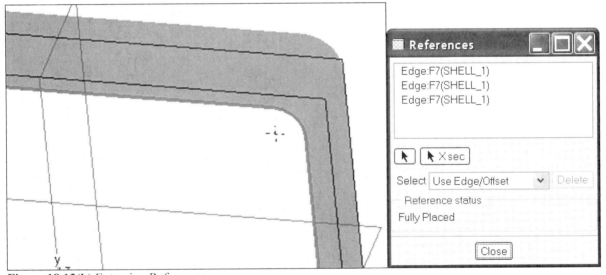

Figure 18.13(b) Extrusion References

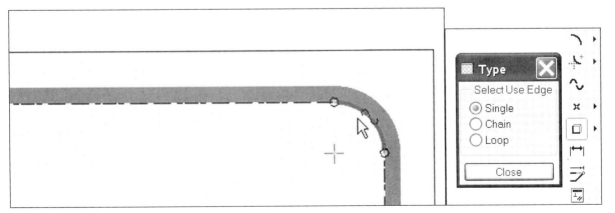

Figure 18.13(c) Create the First Entity using: ▫ **Create an entity from an edge**

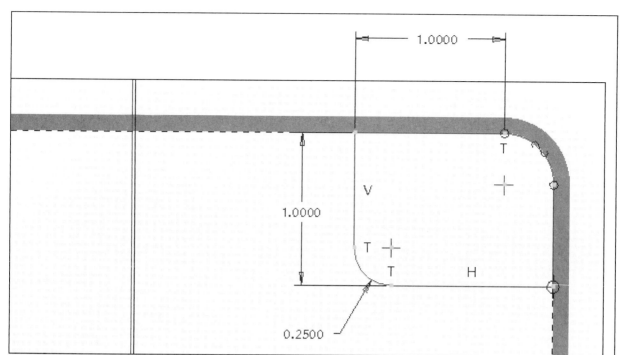

Figure 18.13(d) Section Sketch

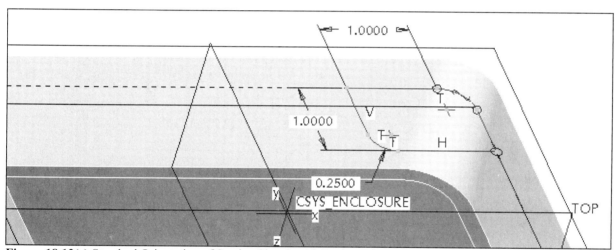

Figure 18.13(e) Standard Orientation of Section Sketch with Design Dimensions

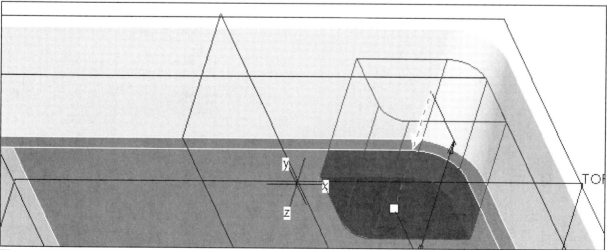

Figure 18.13(f) Previewed Extrusion Depth

Pick on the depth dimension ⇒ **RMB** ⇒ **To Selected** [Fig. 18.13(g)] ⇒ pick **DTM1** [Fig. 18.13(h)] ⇒ **MMB** ⇒ spin the model [Fig. 18.13(i)] ⇒ 🖫 ⇒ **MMB** ⇒ **LMB**

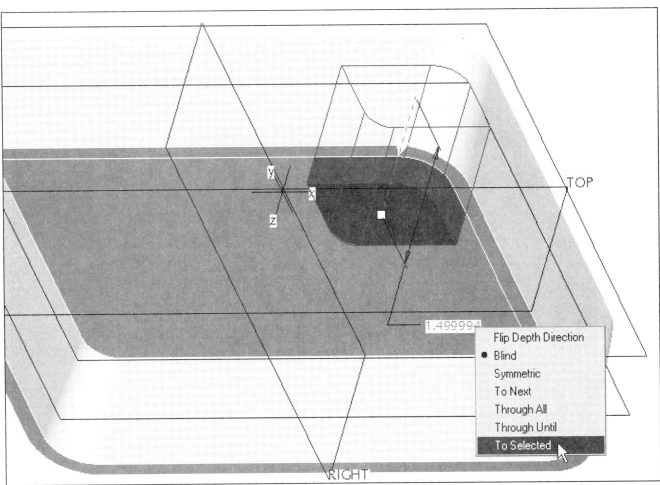

Figure 18.13(g) Determining the Depth Option

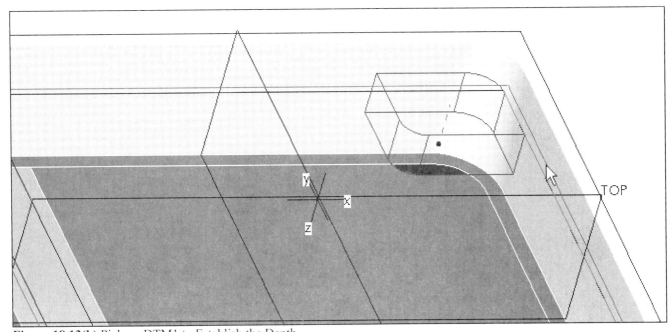

Figure 18.13(h) Pick on DTM1 to Establish the Depth

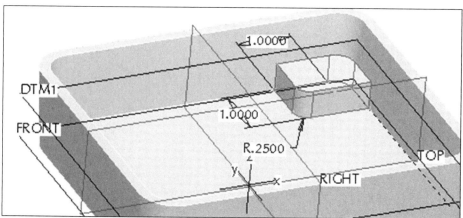

Figure 18.13(i) Completed Pedestal Extrusion

Click: 🗇 **Draft Tool** ⇒ **References** tab ⇒ select one vertical surface of the pedestal ⇒ in the Draft hinges collector, click in the field labeled: [• Click here to add item] (field changes to [× Select 1 item]) ⇒ pick **FRONT** ⇒ type **5** in Dimension field ⇒ **Enter** ⇒ 🗇 **Reverse pull direction** (Fig. 18.14) ⇒ **MMB**

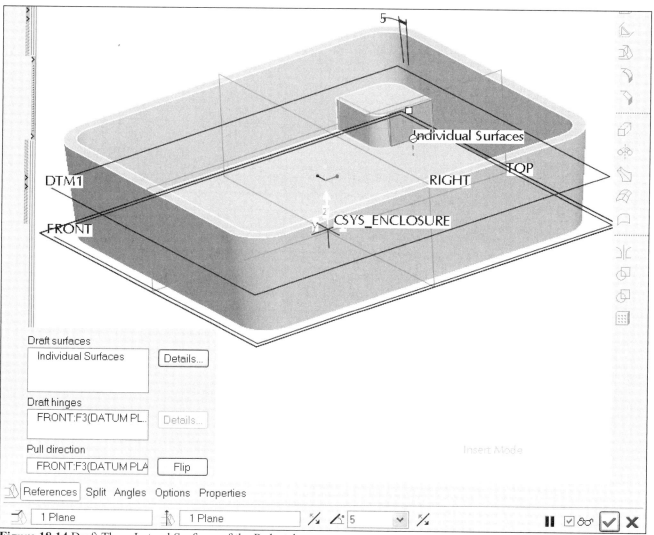

Figure 18.14 Draft Three Lateral Surfaces of the Pedestal

Model the circular extrusion ⇒ use the top surface of the pedestal as the sketching plane ⇒ Keep the default references. The section consists of one circle [Fig. 18.15(a)]. ⇒ ✓ ⇒ **MMB** ⇒ rotate the model [Fig. 18.15(b)] ⇒ click on the depth drag handle ⇒ **RMB** ⇒ **To Selected** [Fig. 18.15(c)] ⇒ pick on the surface [Fig. 18.15(d)] ⇒ **MMB** [Fig. 18.15(e)] ⇒ **LMB** to deselect

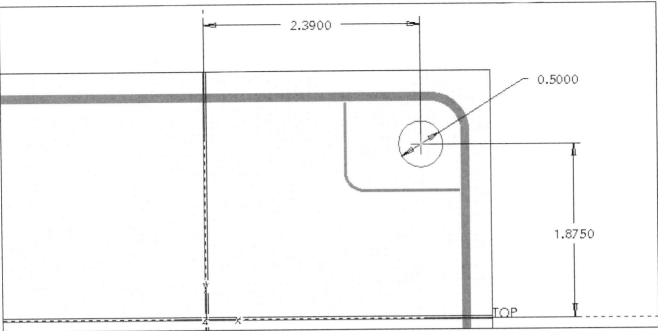

Figure 18.15(a) Section Sketch for Circular Extrusion

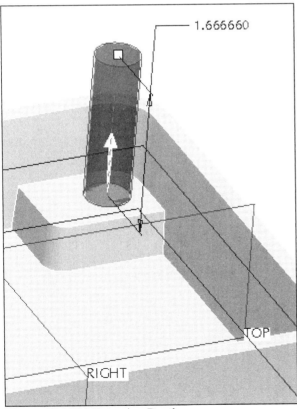

Figure 18.15(b) Extrusion Depth

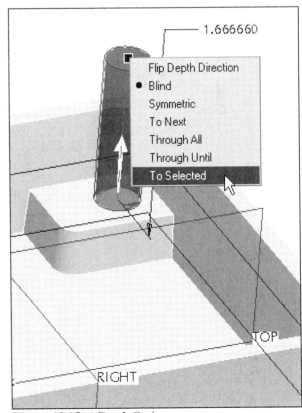

Figure 18.15(c) Depth Options

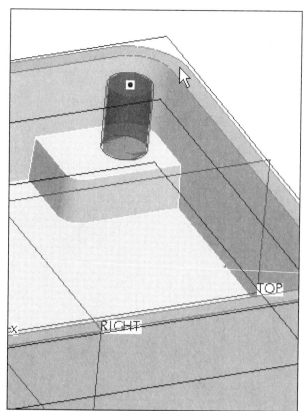

Figure 18.15(d) Select Surface

Figure 18.15(e) Completed Circular Extrusion

The circular feature looks correct, but there seems to be a problem with the pedestal.

Pick on the pedestal extrusion in the graphics window ⇒ **RMB** ⇒ **Edit Definition** [Fig. 18.16(a)]

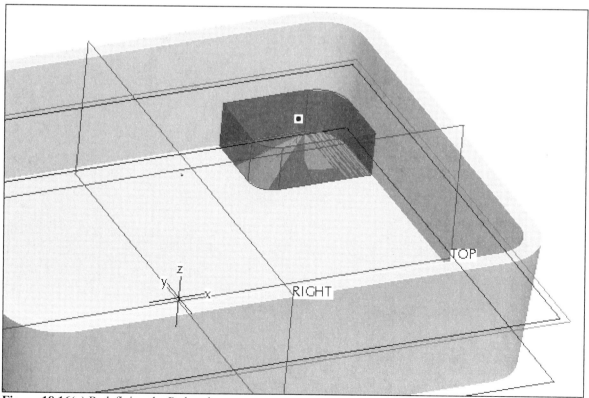

Figure 18.16(a) Redefining the Pedestal

Click: **RMB** ⇒ **Edit Internal Sketch** [Fig. 18.16(b)] ⇒ The dimension is referencing the end of the arc instead of the edge. Create a new defining dimension and modify the new dimension to **1.000** [Fig. 18.16(c)] ⇒ ✓ ⇒ ✓ ⇒ [Figs. 18.16(d-e)] ⇒ **LMB** to deselect

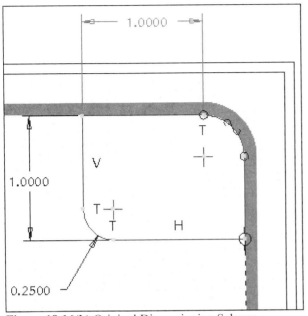
Figure 18.16(b) Original Dimensioning Scheme

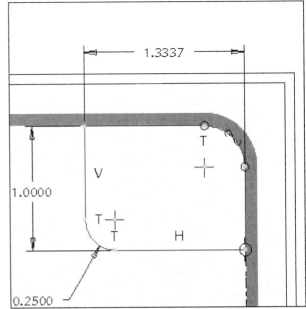
Figure 18.16(c) New Defining Dimension

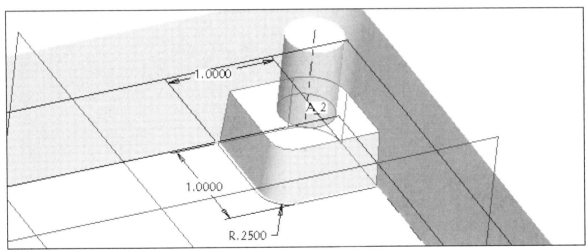
Figure 18.16(d) Redefined Pedestal

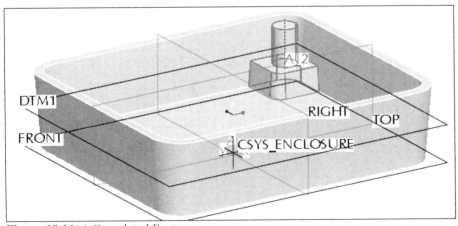
Figure 18.16(e) Completed Features

Draft the circular extrusion at **5°**. Use the upper surface of the circular extrusion as the draft hinge [Fig. 18.17(a)]. Select the drag handle. ⇒ **RMB** ⇒ **Flip Angle** ⇒ **MMB** ⇒ **RMB** ⇒ **Edit** ⇒ change the angle dimension properties to **Flip Arrows** and display **0** number of decimal places [Fig. 18.17(b)].

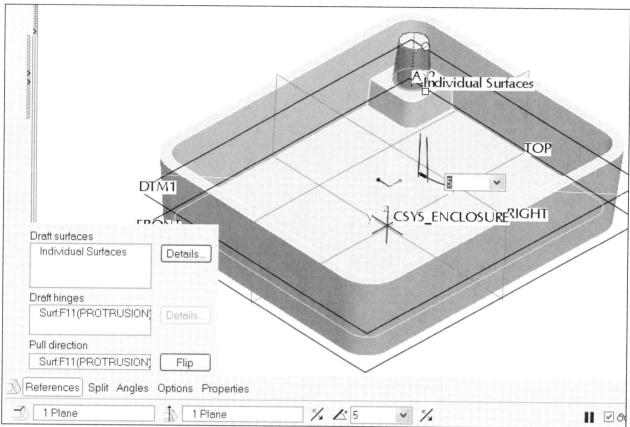

Figure 18.17(a) Draft the Circular Extrusion

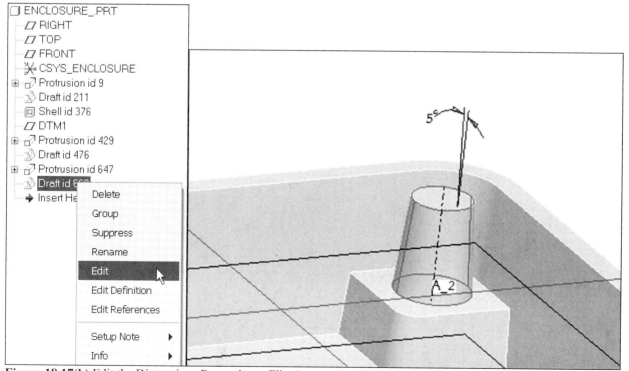

Figure 18.17(b) Edit the Dimensions Properties to Flip Arrows (properties set to 0 decimal places)

Create a **.250** diameter coaxial hole on the upper surface of the circular extrusion [Fig. 18.18(a)]. Use "To Selected" to establish the hole's depth to the top surface of the pedestal [Fig. 18.18(b)]. Complete the feature and save the part.

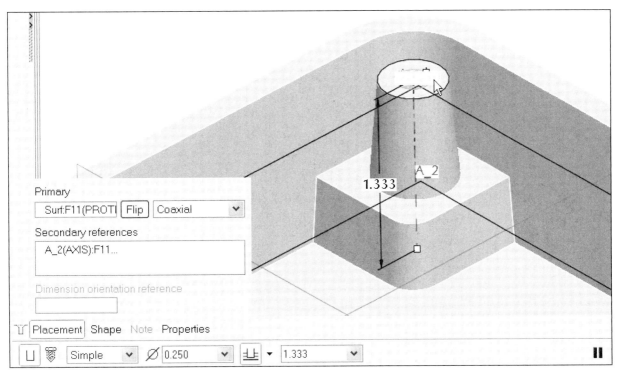

Figure 18.18(a) Coaxial Hole

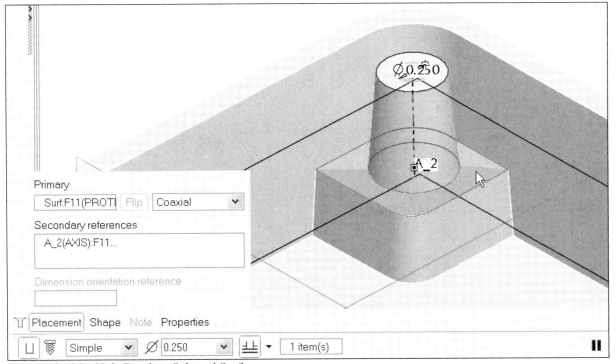

Figure 18.18(b) Hole Depth to Selected Surface

Add an internal draft of **.3°** to the coaxial hole (use the top surface of the cylinder as the draft hinge) [Fig. 18.18(c)].

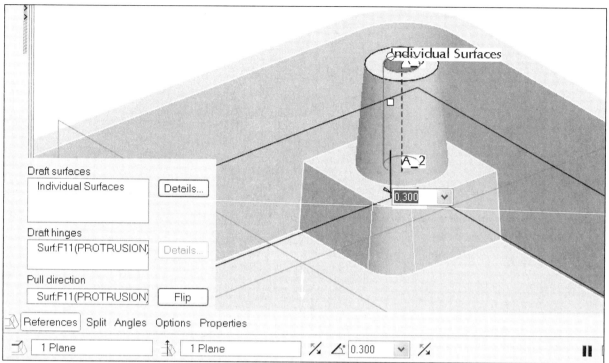

Figure 18.18(c) Draft the Coaxial Hole

Create the **.0625** and **.125** rounds [Figs. 18.19 (a-c)].

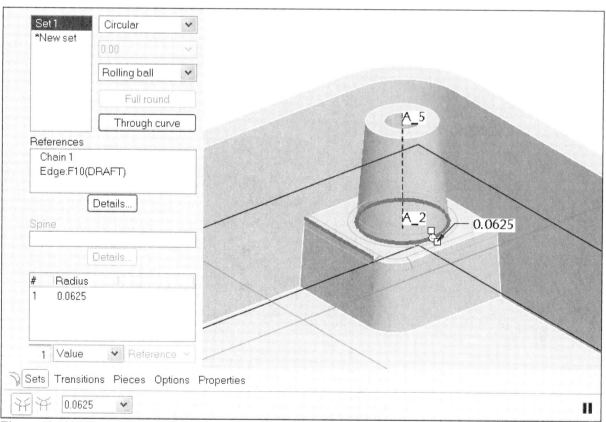

Figure 18.19(a) Round Set 1 (**R.0625**)

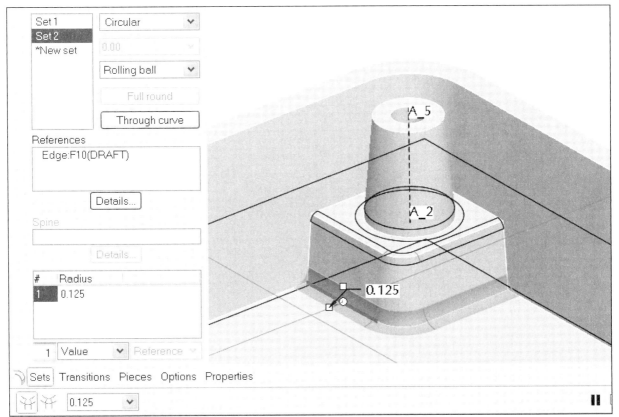

Figure 18.19(b) Round Set 2 (**R.125**)

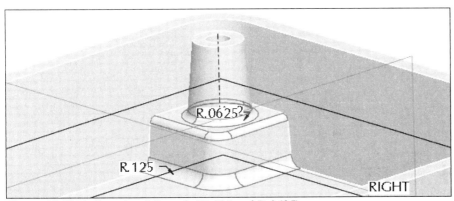

Figure 18.19(c) Completed Rounds (**R.125** and **R.0625**)

Group the extrusions, hole and rounds, click: select features (Fig. 18.20) ⇒ **RMB** ⇒ **Group** ⇒ click on the group name ⇒ type **PED_GRP** [PED_GRP] ⇒ **Enter**

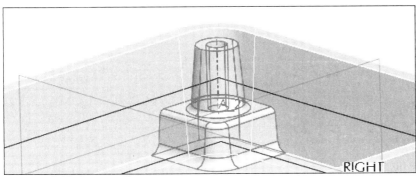

Figure 18.20 Create a Group

Create three identical grouped features, click: **Edit ⇒ Feature Operations ⇒ Copy ⇒ Mirror ⇒ Dependent ⇒ Done ⇒** from the Model Tree, select: Group **PED_GRP ⇒ MMB ⇒ MMB ⇒** select datum **RIGHT** [Fig. 18.21(a)] **⇒ Copy ⇒ Mirror ⇒ Dependent ⇒ Done ⇒** with the **Ctrl** key pressed, select: Group **PED_GRP** and Group **COPIED_GROUP** from the Model Tree **⇒ MMB ⇒ MMB ⇒** select datum **TOP** [Fig. 18.21(b)] **⇒ MMB ⇒ Ctrl+S ⇒ MMB ⇒ File ⇒ Delete ⇒ Old Versions ⇒ MMB**

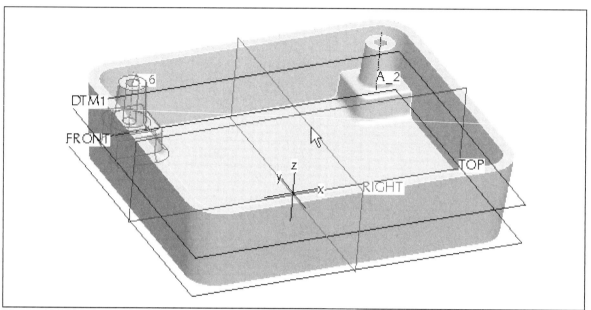

Figure 18.21(a) Group Copied and Mirrored about Datum RIGHT

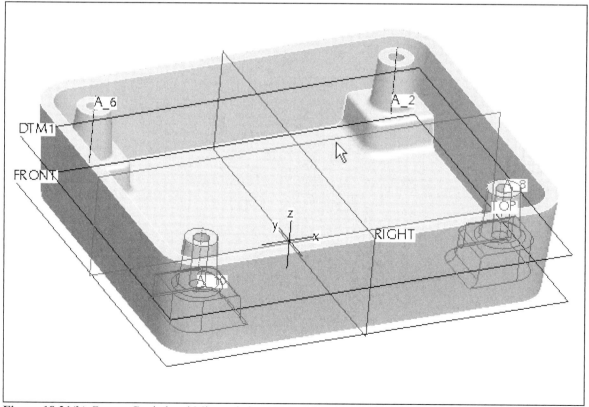

Figure 18.21(b) Groups Copied and Mirrored about Datum TOP

Create the internal round, click: **Round Tool** ⇒ type **.125** ⇒ select the inside of the shelled wall as the first reference [Fig. 18.22(a)] ⇒ **Sets** tab ⇒ press and hold the **Ctrl** key ⇒ select the top surface of the pedestal as the second reference [Fig. 18.22(b)] ⇒ **MMB**

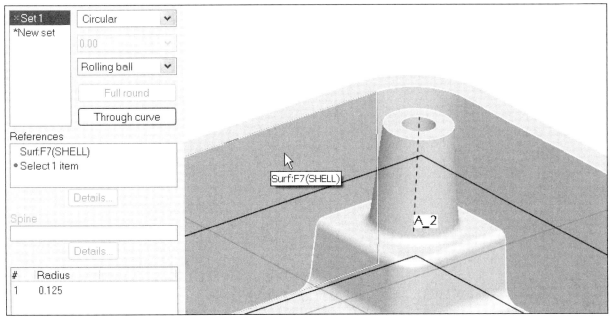

Figure 18.22(a) Select First Reference

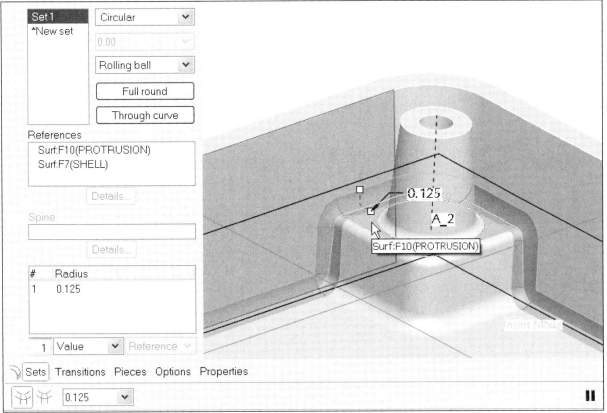

Figure 18.22(b) Select Second Reference

Click: [Item 2] [Fig. 18.22(c)] ⇒ Note: [Item 2 ☑] ⇒ **OK** ⇒ [icon] from dashboard [Fig. 18.22(d)] ⇒ **Quick Fix** from RESOLVE FEAT menu ⇒ **Redefine** ⇒ **Confirm** ⇒ **Options** tab ⇒ **Surface** [Fig. 18.22(e)] ⇒ [☑ 👓] **Preview** ⇒ [▶] **Resume** ⇒ [✓] ⇒ [🖫] ⇒ **MMB**

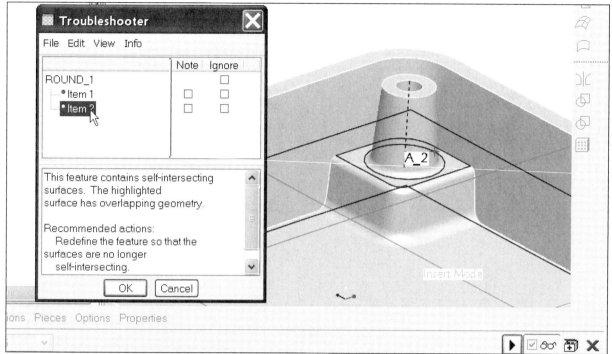

Figure 18.22(c) Select Troubleshooter Option

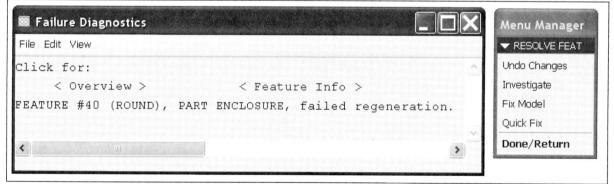

Figure 18.22(d) Failure Diagnostics Dialog Box

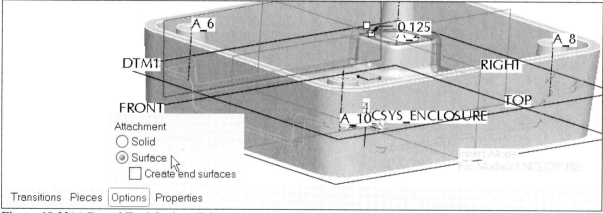

Figure 18.22(e) Round Tool Options Tab

The part's internal round may look correct, but it is not a solid round. Next, model a datum plane through the axis of two pedestals and use the datum plane to create a cross section.

Click: ⌗ **Datum Plane Tool** ⇒ References: pick axis **A_2** [Fig. 18.23(a)] ⇒ press and hold the **Ctrl** key ⇒ References: pick axis **A_6** *(your id's may be different)* [Fig. 18.23(b)] ⇒ **OK** (creates DTM2)

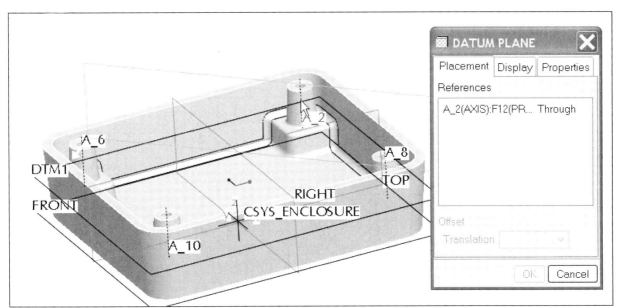

Figure 18.23(a) Select Axis A_2

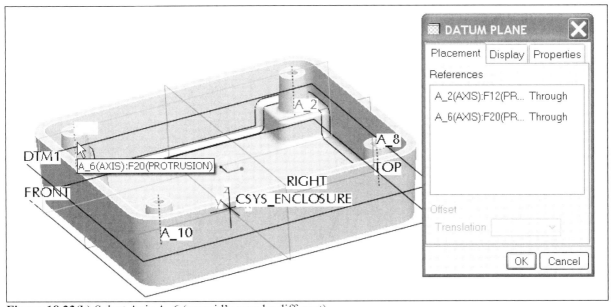

Figure 18.23(b) Select Axis A_6 (your id's may be different)

Create a cross section through the part, using datum DTM2.

Click: **Start the view manager** from Top Toolchest (View Manager dialog box displays) ⇒ **Xsec** tab ⇒ **New** ⇒ type name **A** ⇒ **Enter** ⇒ **Model** ⇒ **Planar** ⇒ **Single** ⇒ **MMB** ⇒ **Plane** ⇒ pick **DTM2** ⇒ click on **A** ⇒ **RMB** ⇒ **Visibility** [Fig. 18.24(a)] ⇒ click on section **A** ⇒ **RMB** ⇒ **Redefine** [Fig. 18.24(b)] ⇒ **Hatching** ⇒ **Fill** ⇒ **Done** ⇒ **Done/Return**

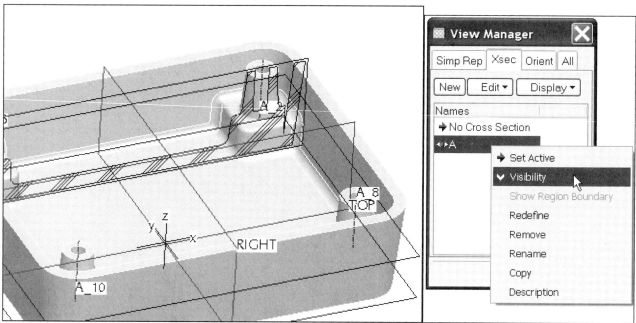

Figure 18.24(a) Show X-Section

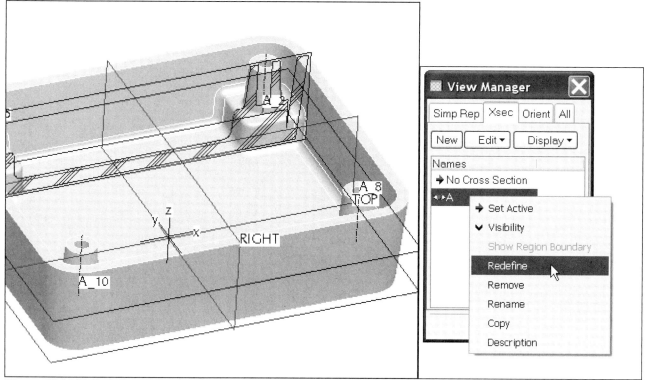

Figure 18.24(b) Redefine X-Section A

The gaps between the pedestals and the shelled walls were not filled by the Round Tool because the round was a surface round.

> Click on **A** [Fig. 18.24(c)] ⇒ **RMB** ⇒ **Set Active** [Fig. 18.24(d)] ⇒ spin the model ⇒ **RMB** on No Cross Section ⇒ **Set Active** ⇒ click on **A** ⇒ **RMB** ⇒ **Visibility** (uncheck) ⇒ **Close**

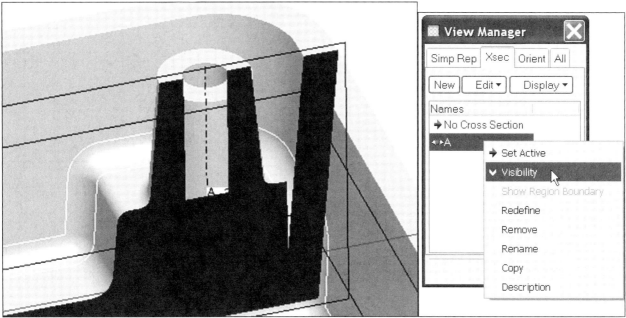

Figure 18.24(c) Filled X-Section

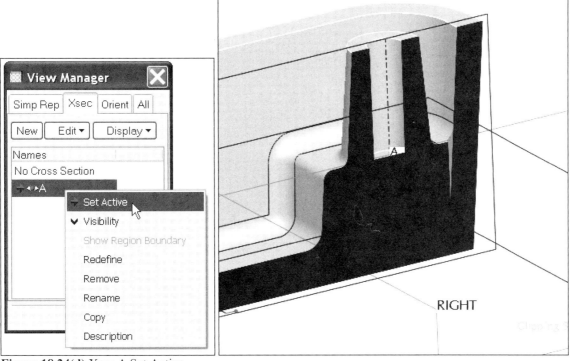

Figure 18.24(d) Xsec A Set Active

Click: on the surface round in the Model Tree (it will highlight on the model) ⇒ **Edit** from the menu bar ⇒ **Solidify** [Fig. 18.25(a)] ⇒ **References** tab [Fig. 18.25(b)]

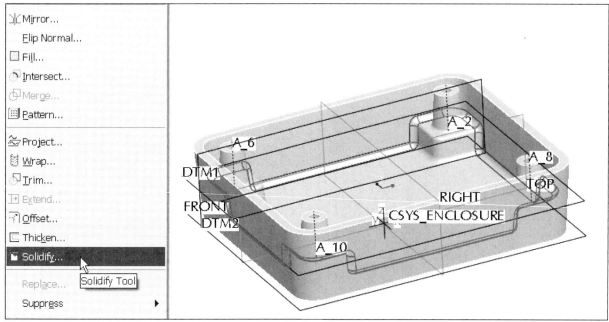

Figure 18.25(a) Solidify Tool

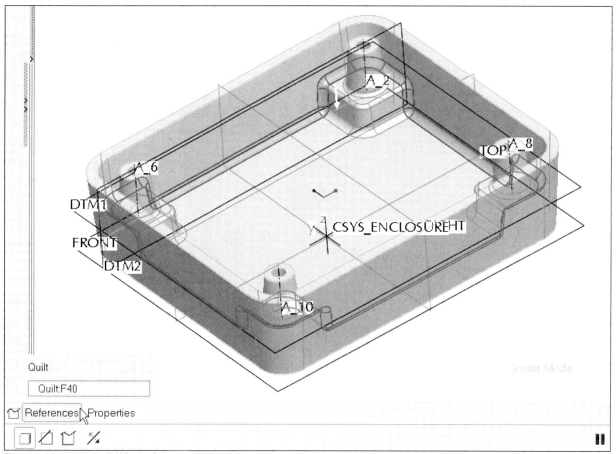

Figure 18.25(b) Solidify References Tab

Click: ⬜ **Fills volume delimited by quilt with solid material** [Fig. 18.25(c)] ⇒ **MMB** [Fig. 18.25(d)]

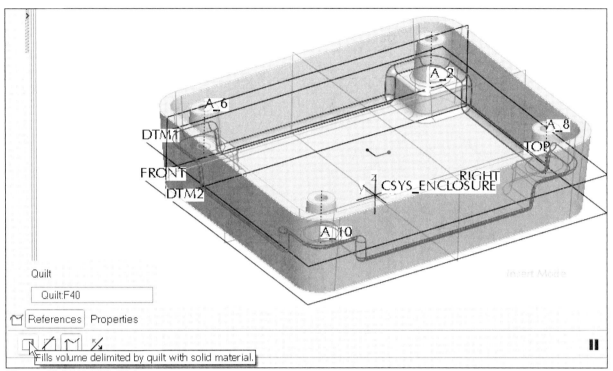

Figure 18.25(c) Fills Volume

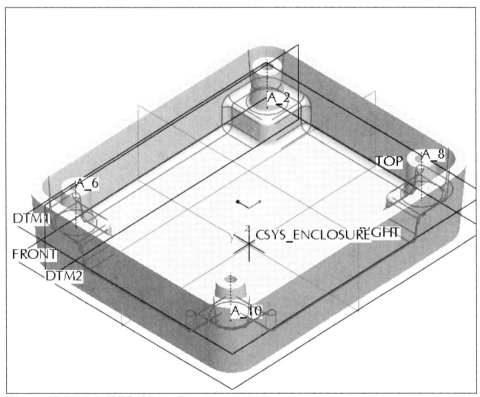

Figure 18.25(d) Solidified Round

Click: ▦ **Start the view manager** from Top Toolchest (View Manager dialog box displays) ⇒ **Xsec** tab ⇒ click on **A** ⇒ **Display** ⇒ **Visibility** (Fig. 18.26) ⇒ **RMB** ⇒ **Visibility** (uncheck) ⇒ **Close** ⇒ click on the last feature in the Model Tree to see it highlighted in the graphics window (Fig. 18.27) ⇒ 🖫 ⇒ **MMB**

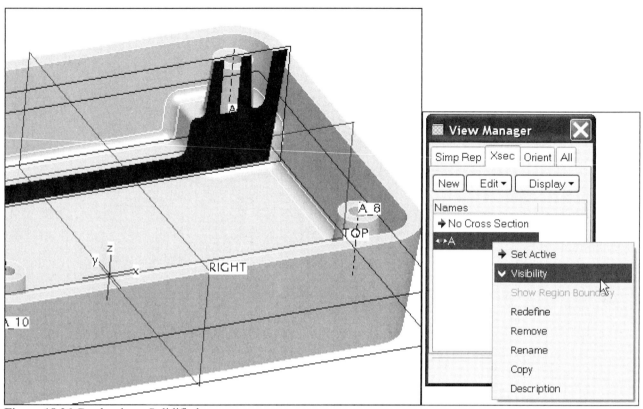

Figure 18.26 Gap has been Solidified

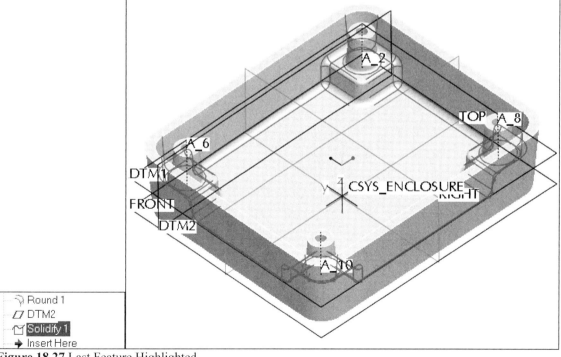

Figure 18.27 Last Feature Highlighted

Before creating the text extrusion, **Suppress** all the features after the shell command. Expand the Model Tree to include the feature number and status.

Click: **Settings** ⇒ **Tree Filters** ⇒ toggle on all options (Fig. 18.28) ⇒ **Apply** ⇒ **OK**

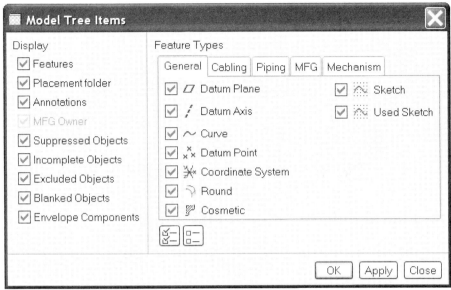

Figure 18.28 Model Tree Items Dialog Box

Click on **Group PED_GRP** in the model tree ⇒ press and hold the **Shift** key ⇒ click on the last feature in the Model Tree [Fig. 18.29(a)] ⇒ **RMB** ⇒ **Suppress** [Fig. 18.29(b)] ⇒ **OK** [Fig. 18.29(c)]

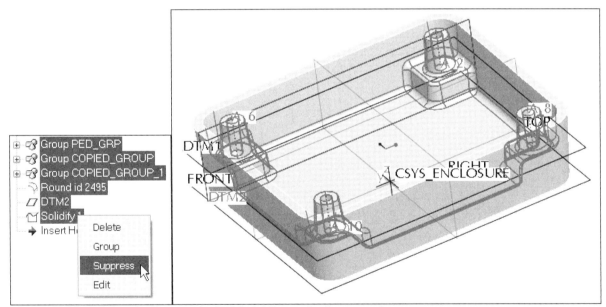

Figure 18.29(a) Select the Features in the Model Tree to be Suppressed

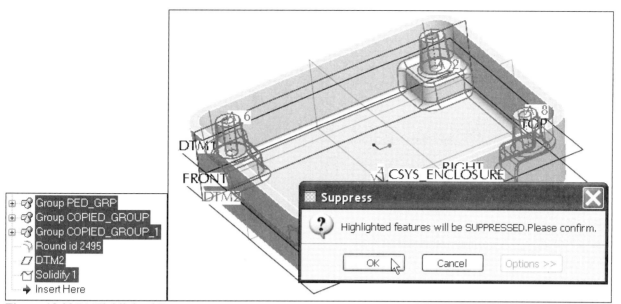

Figure 18.29(b) Highlighted Features

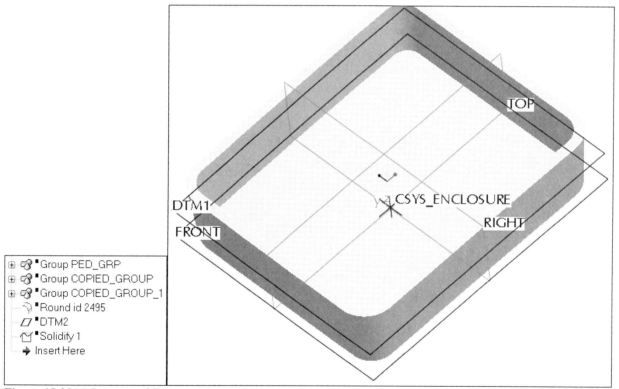

Figure 18.29(c) Suppressed Features

The regeneration time for your model will now be much shorter. Next, add the text extrusion.

Click: [AB] ⇒ **Standard Orientation** ⇒ [icon] **Extrude Tool** ⇒ **RMB** ⇒ **Define Internal Sketch** ⇒ Sketch Plane--- Plane: pick the inside surface of the enclosure for the sketching plane ⇒ Reference: **TOP** ⇒ Orientation: **Top** [Fig. 18.30(a)] ⇒ **Sketch** ⇒ **Tools** ⇒ **Environment** ⇒ [☑ Snap to Grid] ⇒ **Apply** ⇒ **OK** ⇒ [icon] **Toggle the grid** on ⇒ [icon] **Create text as a part of a section** ⇒ select *start point* of line to determine text starting position [Fig. 18.30(b)] ⇒ select *second point* of line to determine text height and orientation [Fig. 18.30(c)]

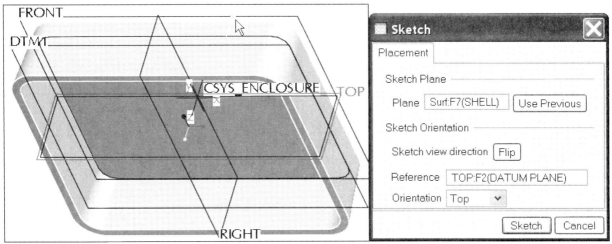

Figure 18.30(a) Sketching Plane, Inside Surface

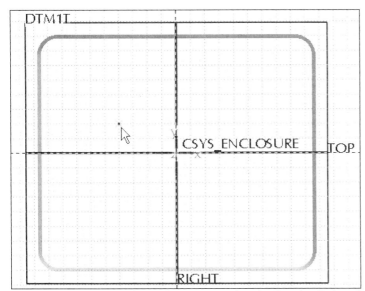

Figure 18.30(b) Pick First Point to Determine the Starting Point of the Lettering

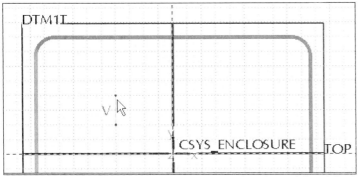

Figure 18.30(c) Pick Second Point to Determine the Height of the Lettering

Text Line- type **CFS-2134** [Fig. 18.30(d)] ⇒ **OK** ⇒ **Sketch** from menu bar ⇒ **Options** ⇒ **Parameters** tab ⇒ Num Digits ⇒ **4** ⇒ **MMB** ⇒ [cursor] ⇒ window-in the sketch to capture all dimensions ⇒ **RMB** ⇒ **Modify** ⇒ modify the dimensions [Fig. 18.30(e)] ⇒ [✓] from dialog box ⇒ [✓] **Continue** from Right Toolchest

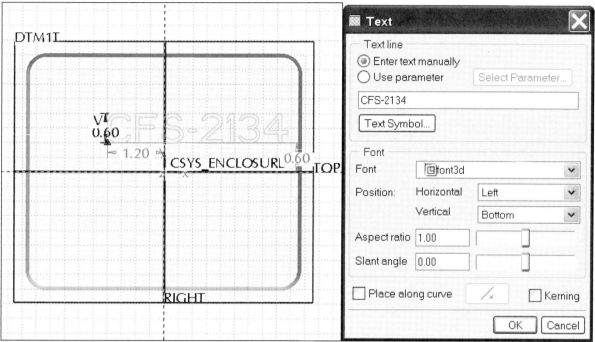

Figure 18.30(d) Type the Text "CFS-2134" (case sensitive)

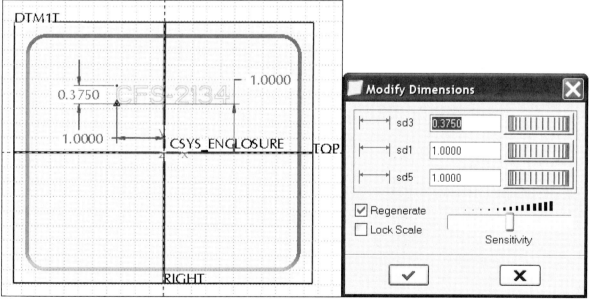

Figure 18.30(e) Modified Dimensions (dimensions are slightly different than those shown in Figure 18.7)

Press and hold **MMB** to spin the model [Fig. 18.30(f)] ⇒ double-click on the height dimension and modify it to **.0625** ⇒ **Enter** [Fig. 18.30(g)] ⇒ **MMB** [Fig. 18.30(h)] ⇒ 🖫 ⇒ **MMB**

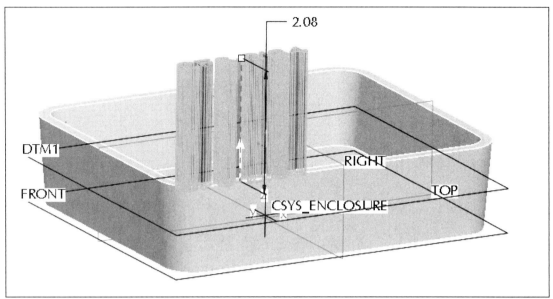

Figure 18.30(f) Dynamic Preview

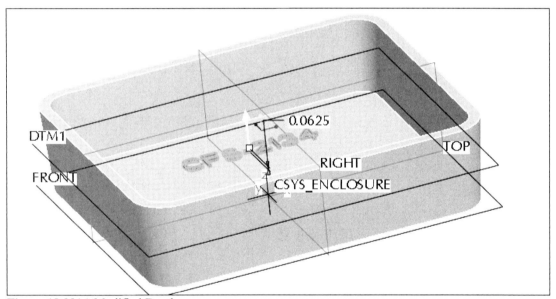

Figure 18.30(g) Modified Depth

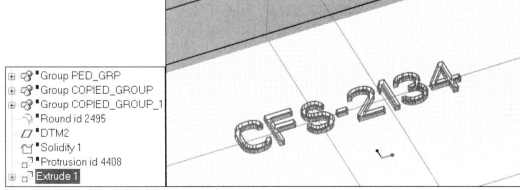

Figure 18.30(h) Completed Text Extrusion

Click: **Edit** ⇒ **Resume** ⇒ **Resume All** [Figs. 18.31(a-b)] ⇒ **Ctrl+D** [Fig. 18.31(c)] ⇒ **Ctrl+R** ⇒ **Ctrl+S** ⇒ **MMB** ⇒ **LMB**

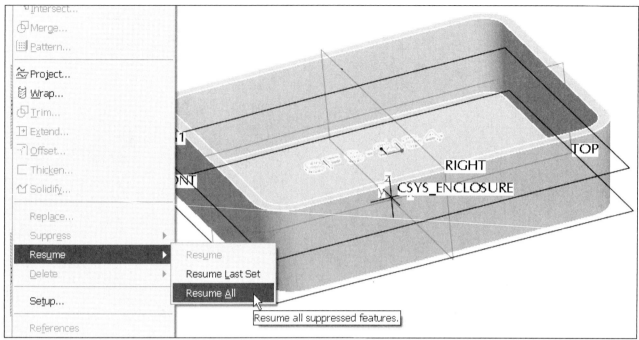

Figure 18.31(a) Resume All

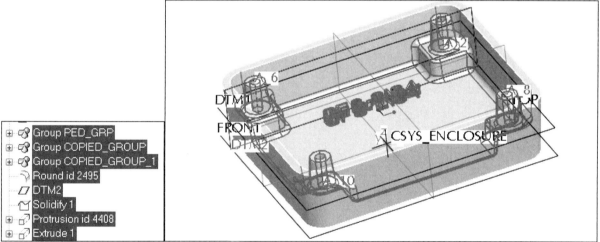

Figure 18.31(b) Suppressed Features Resumed

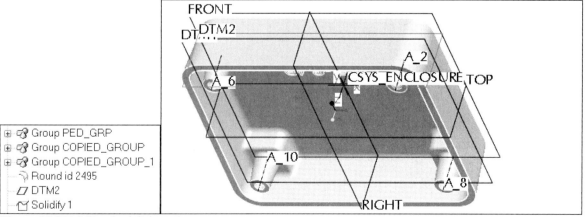

Figure 18.31(c) Standard Orientation

Spin the part ⇒ **View** ⇒ **Shade** ⇒ slowly pick on the parts' edge until it highlights [Fig. 18.32(a)] ⇒ depress and hold **Shift** key ⇒ pick on the surface [Fig. 18.32(b)] *[(Top surface edges will highlight, as the loop was selected. Note that this method was not necessary for this round since all the edges were tangent, but this was used to demonstrate another process of selection)]* ⇒ **RMB** ⇒ **Round Edges** [Fig. 18.32(c)] ⇒ **Ctrl+R** ⇒ modify the edge round to **.1875** [Fig. 18.32(d)] ⇒ **Enter** ⇒ **MMB** ⇒ **View** ⇒ **Shade** [Fig. 18.32(e)]

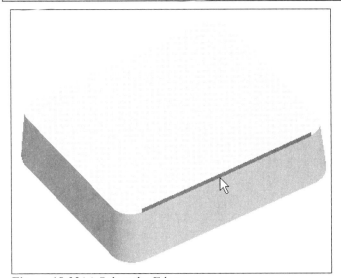

Figure 18.32(a) Select the Edge

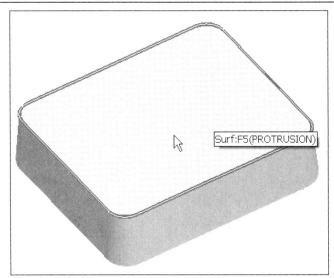

Figure 18.32(b) Shift + Select the Surface

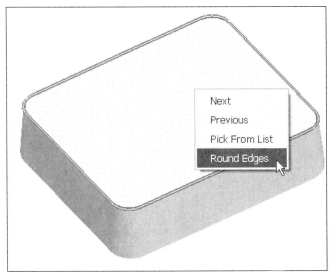

Figure 18.32(c) Round Edges

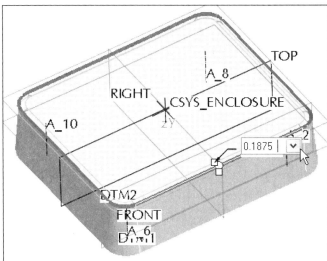

Figure 18.32(d) Modify to **.1875**

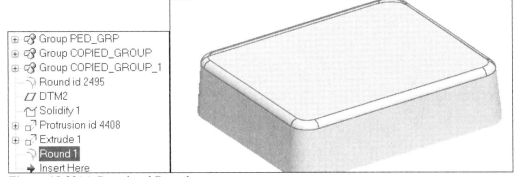

Figure 18.32(e) Completed Round

Click: **Ctrl+D** ⇒ select the text extrusion in the **Model Tree** ⇒ 🔍 **Orient Mode** on ⇒ **RMB** ⇒ **Velocity** [Fig. 18.33(a)] ⇒ hold down **MMB** and move the cursor about the screen to orbit the model [Fig. 18.33(b)] ⇒ **RMB** ⇒ **Exit Orient Mode** ⇒ **LMB** to deselect

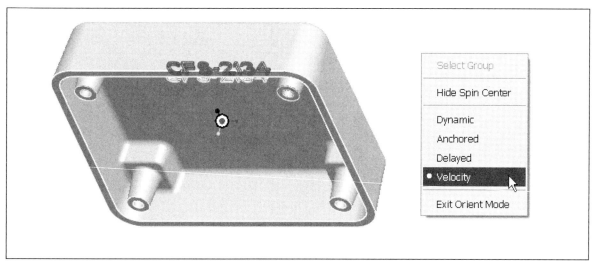

Figure 18.33(a) Orient Mode Velocity

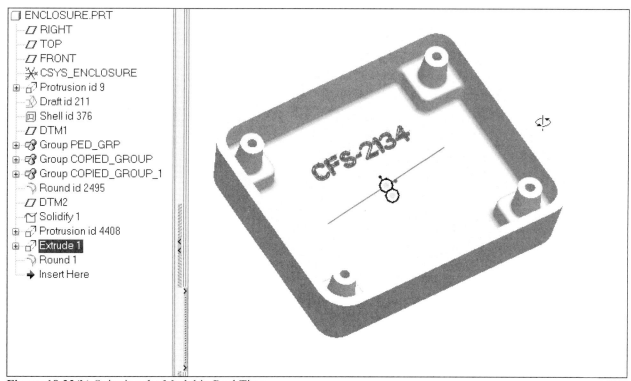

Figure 18.33(b) Spinning the Model in Real Time

Click: 🖉 ⇒ double-click on the text extrusion ⇒ modify the text depth to be **.250** [(Figs. 18.34(a-b)] ⇒ **Enter ⇒ Ctrl+G ⇒ Ctrl+S ⇒ MMB ⇒ File ⇒ Delete ⇒ Old Versions ⇒ MMB** (Fig. 18.35)

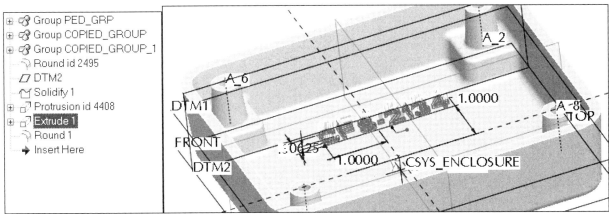

Figure 18.34(a) Edit the Text Extrusion

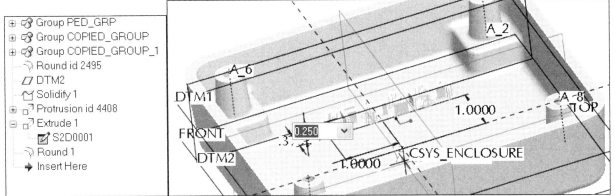

Figure 18.34(b) Modify the Text Extrusion to have a **.250** Depth (dimension is modified from that shown in Figure 18.8)

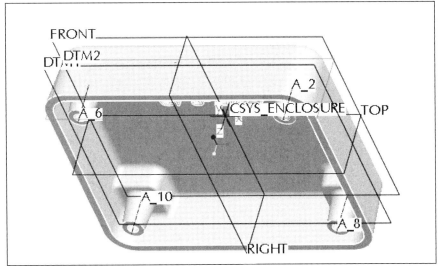

Figure 18.35 Completed Enclosure

Click: **Analysis** ⇒ **ModelCHECK** ⇒ **ModelCHECK Geometry Check** [Fig. 18.36(a)] ⇒ **OK** [Figs. 18.36(b-c)] ⇒ **Ctrl+S** ⇒ **MMB** ⇒ **Window** ⇒ **Close** ⇒ **File** ⇒ **Erase** ⇒ **Not Displayed** ⇒ **OK**

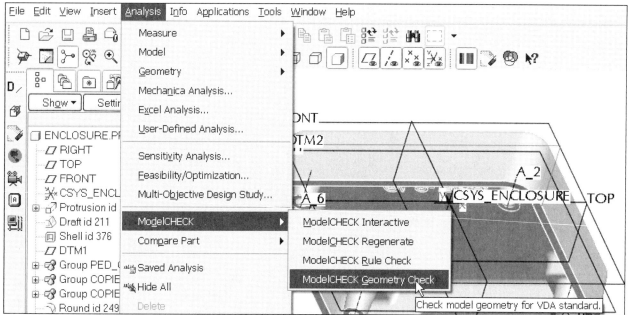

Figure 18.36(a) ModelCHECK

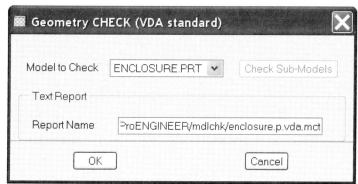

Figure 18.36(b) Geometry CHECK Dialog Box

Check	Result
1. F14: Small Distance between Edges	29
2. F17: High Edge Segment Concentration	127
3. M2: Identical Element - surfaces	17
4. M3b: Tangential Discontinuity - surface boundaries	257
5. M3c: Curvature Discontinuity - surface boundaries	114
6. SU10: Small Angle between Edges	31
7. SU8: Small Edge Segment	278
8. SU9: Small Radius of Curvature	104
9. T18: More than Two Surfaces per Edge	8

Figure 18.36(c) ModelCHECK List Shown in the Embedded Web Browser

See *www.cad-resources.com* ⇒ *Downloads* for more part projects.

Appendix

Customizing the User Interface (UI)

Customize your user interface to increase your efficiency in modeling. You can customize the Pro/E user interface, according to your needs or the needs of your group or company, to include the following:

- Create keyboard macros, called *mapkeys*, and add them to the menus and toolbars
- Add or remove existing toolbars
- Add split buttons to the toolbars (split buttons contain multiple closely-related commands and save toolbar space by hiding all but the first active command button)
- Move or remove commands from the menus or toolbars
- Change the location of the message area
- Add options to the Menu Manager
- Blank (make unavailable) options in the Menu Manager
- Set default command choices for Menu Manager menus

With a Pro/E file active, click: **Tools** ⇒ **Customize Screen** (Fig. A.1) ⇒ Customize dialog box opens with the Commands tab active (Fig. A.2) ⇒ Categories, click: **View** (Fig. A.3) ⇒ drag **Model Colors** and drop into the Top Toolchest (Fig. A.4)

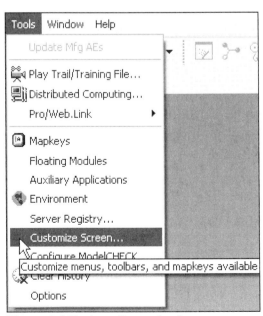

Figure A.1 Tools ⇒ Customize Screen

Figure A.2 Customize Dialog Box, Commands Tab

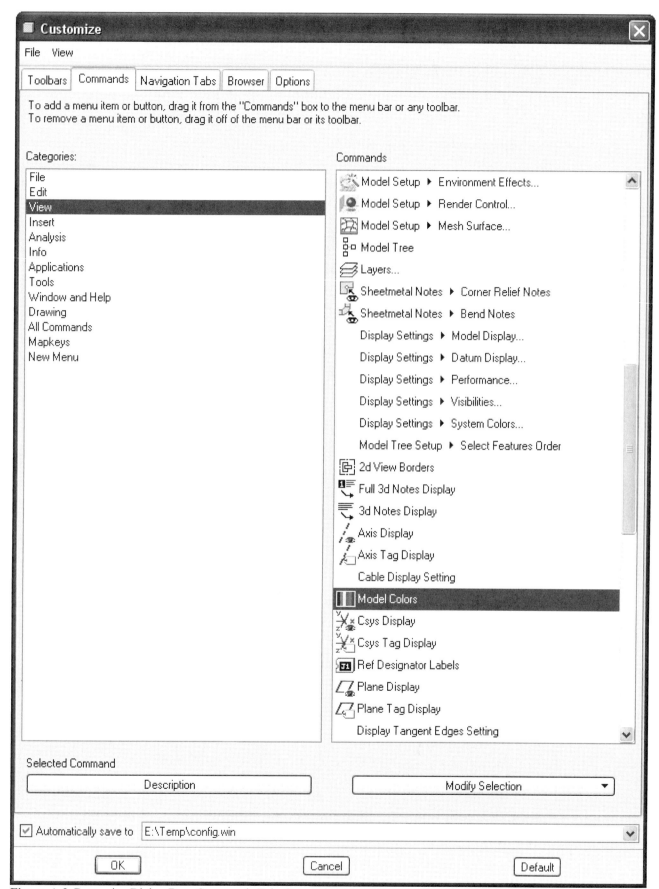

Figure A.3 Customize Dialog Box, Commands Tab, Categories View, Commands– Model Colors

Figure A.4 Top Toolchest with Newly Added Command Button

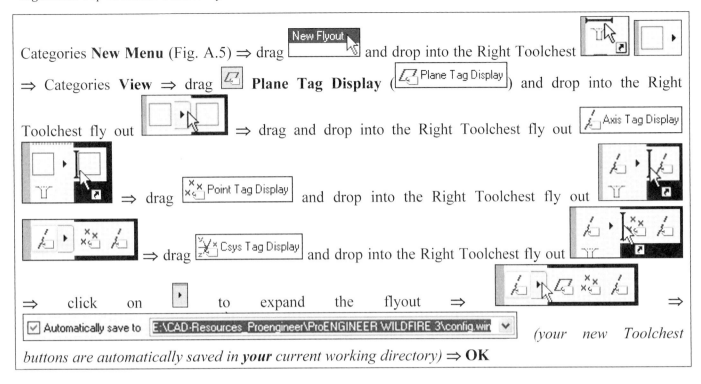

Figure A.5 New Menu Category

You can recall saved settings by clicking: *Tools ⇒ Customize Screen ⇒ File ⇒ Open Settings ⇒ select the file ⇒ Open ⇒ OK*. Buttons can be removed from the Toolchest using the exact same method, except, drag the buttons away from the Toolchest and release the mouse button. Customize your Pro/E screen with the buttons of commands you use frequently such as:

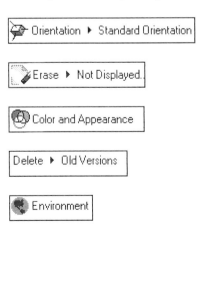

Next, add a Toolbar to the left side of the Navigator Window (Left Toolchest), click: **Tools** ⇒ **Customize Screen** ⇒ click **Toolbars** tab ⇒ ☑ Tools ⇒ **Left** (Fig. A.6) ⇒ **OK**

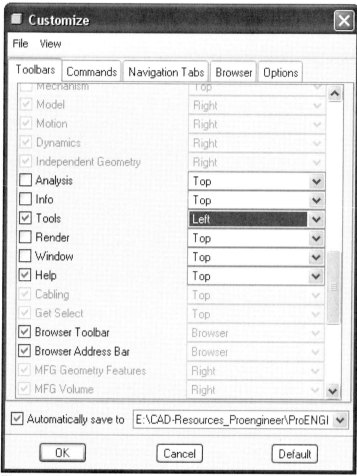

Figure A.6 Tools, Left

The Tools toolbar (Fig. A.7) includes: 🌐 **Set various environment options**, 📷 **Run trail or training file**, 🅰 **Create macros**, and 🖥 **Select hosts for distributed computing**.

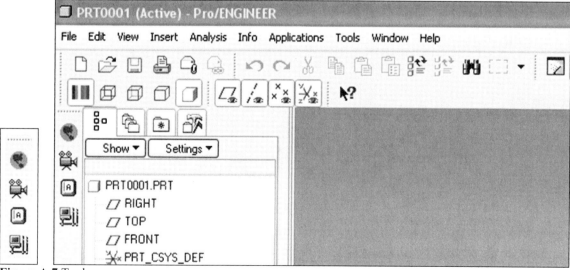

Figure A.7 Tools

The Navigation Tabs tab provides options for controlling the location of the Navigator (left or right), its width setting, and its placement in relation to the Model Tree settings.

Click: **Tools** ⇒ **Customize Screen** ⇒ **Navigation Tabs** tab and explore the settings (Fig. A.8)

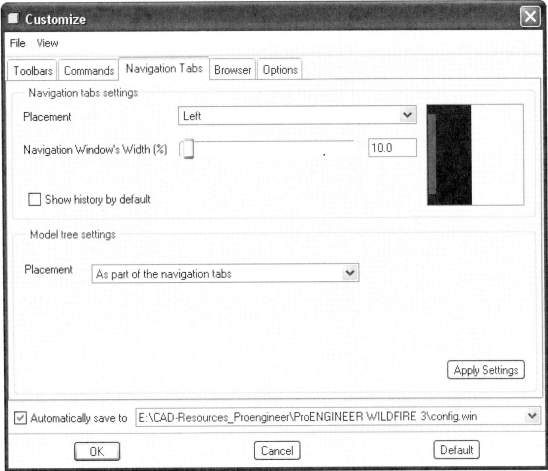

Figure A.8 Customize Dialog Box, Navigation Tabs

The Browser tab is used to control its window width and animation option. The Options tab provides settings to locate the Dashboard, Secondary Window size, and Menu display

Click: **Browser** tab and explore the options (Fig. A.9) ⇒ **Options** tab and explore its capabilities (Fig. A.10) ⇒ **OK**

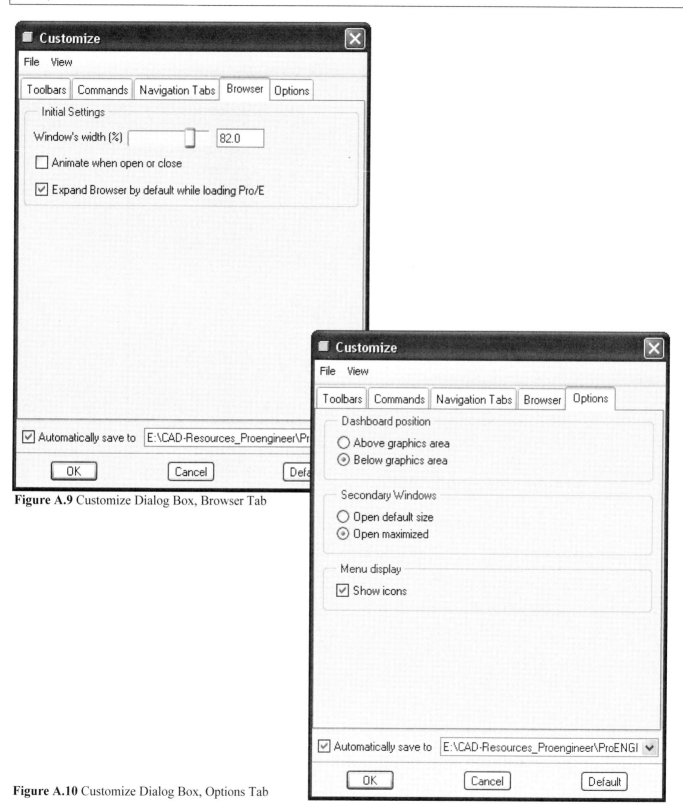

Figure A.9 Customize Dialog Box, Browser Tab

Figure A.10 Customize Dialog Box, Options Tab

Mapkeys

In Pro/E, a **Mapkey** is a macro that maps frequently used command sequences to certain keyboard keys or sets of keys. Mapkeys are saved in the configuration file, and are identified with the option *mapkey*, followed by the identifier and then the macro. You can define a unique key or combination of keys which, when pressed, executes the mapkey macro (for example, **F6** on your keyboard). You can create a mapkey for virtually any task you perform frequently within Pro/E.

By adding mapkeys to your Toolchest or menu bar, you can use mapkeys with a single mouse click or menu command and thus streamline your workflow in a visible way.

To create a mapkey, you can use the configuration file option *mapkey*, or, on the Pro/E menu bar, click Tools ⇒ Mapkeys, and then in the Mapkeys dialog box, you click New and record your mapkey in the Record Mapkey dialog box. Pro/E records your mapkey as you step through the sequence of keystrokes or command executions to define it. After you define the mapkey, Pro/E creates a corresponding icon and places it in the Customize dialog box under the Mapkeys category. To open the Customize dialog box, click: Tools ⇒ Customize Screen. On the Commands tabbed page, select the Mapkeys category. You can then drag the visible mapkey icon onto the Pro/E Toolchests. You can also create a label for the new mapkey.

You can also nest one mapkey within another, so that one mapkey initiates another. To do so, you include the mapkey name in the sequence of commands of the new mapkey you are defining.

Mapkeys include the ability to do the following:

- Pause for user interaction.
- Handle message window input more flexibly.
- Run operating system scripts and commands. The Record Mapkey dialog box contains the OS Script tabbed page, whose options allow you to run OS commands instead of Pro/E commands.

When you define a mapkey, Pro/E automatically records a pause when you make screen selections, so that you can make new selections while the mapkey is running. In addition, you can record a pause (at any place in the mapkey) along with a user-specified dialog prompt, which will appear at the corresponding point while the mapkey is running.

If you create a new mapkey that contains actions that open and make selections from dialog boxes, then when you run the mapkey, it does not pause for user input when it opens the dialog box. To set the mapkey to pause for user input when opening dialog boxes, you must select *Pause for keyboard input* on the Pro/E tab in the Record Mapkey dialog box before you create the new mapkey.

Mapkeys Dialog Box

You use the Mapkeys dialog box (Fig. A.11), to define new mapkeys, modify, and delete existing mapkeys, run a mapkey chosen from the list, and save mapkeys to a configuration file. To open the Mapkeys dialog box, click Tools ⇒ Mapkeys. The following defines each command option on the dialog box:

- **New** Allows you to define a new mapkey and opens the Record Mapkey dialog box
- **Modify** Allows you to modify the selected mapkey
- **Run** Allows you to run the selected mapkey
- **Delete** Allows you to delete the selected mapkey
- **Save** Allows you to save the selected mapkey to a configuration file
- **Changed** Allows you to save only the mapkeys changed in the current session
- **All** Save all the mapkeys

In the Record Mapkey dialog box (Fig. A.12), you can type the key sequence that is to be used to execute the mapkey in the Key Sequence text box. To use a function key, precede its name with a dollar sign ($). For example, to map to **F7**, type **$F7**. Type the Name and Description of the mapkey in the appropriate text boxes. On the Pro/E tab, specify how Pro/E will handle the prompts when running the mapkey by selecting one of the following commands:

- **Record keyboard input** (default selection) Record the keyboard input for prompts when defining the mapkey, and use it when running the macro
- **Accept system defaults** Accept the system defaults when running the macro
- **Pause for keyboard input** Pause for user input when running the macro

Figure A.11 Mapkey Dialog Box

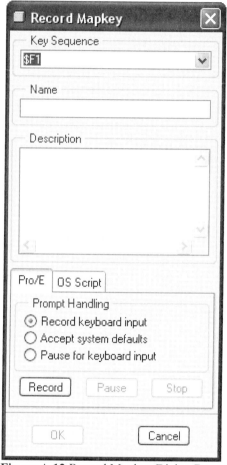

Figure A.12 Record Mapkey Dialog Box

Use Record to start recording the macro by selecting menu commands in the appropriate order. Use Pause to indicate where to pause while running the mapkey. Type the prompt in the Resume Prompt dialog box. Use Resume and continue recording the mapkey. When you run the macro, Pro/E will pause, display the prompt you typed, and give you the options to Resume running the macro or Cancel. Use Stop when you are finished recording the macro.

After you define the mapkey, a corresponding button appears in the Customize dialog box (Categories: Mapkeys). You can then drag the mapkey onto the Toolchest just like the Pro/E- supplied buttons.

Mapkeys include the ability to pause for user interaction, handle message window input flexibly, and run operating system commands.

Create a mapkey, click: **Tools** ⇒ **Mapkeys** Mapkeys dialog box opens (Fig. A.13) ⇒ **New** Record Mapkey dialog box opens (Fig. A.14) ⇒ Key Sequence- **$F5** ⇒ Name **COLORS** ⇒ Description **Opens Appearance Editor and starts the New Color Definition** (Fig. A.14) ⇒ **Record** ⇒ **View** from menu bar ⇒ **Color and Appearance** ⇒ **+** Add new appearance ⇒ Color: **Color** button (Fig. A.15) ⇒ **adjust for a new color** ⇒ **Stop** ⇒ **OK** ⇒ Save Mapkeys ⇒ Name my_config.pro (Fig. A.16) ⇒ **Ok** ⇒ **Close** (Fig. A.17) ⇒ **Close** Color Editor ⇒ **Close** Appearance Editor

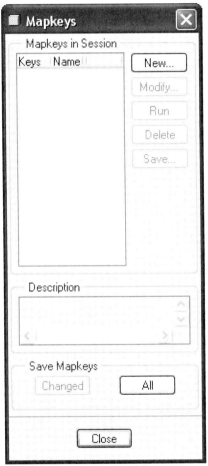

Figure A.13 Mapkeys Dialog Box

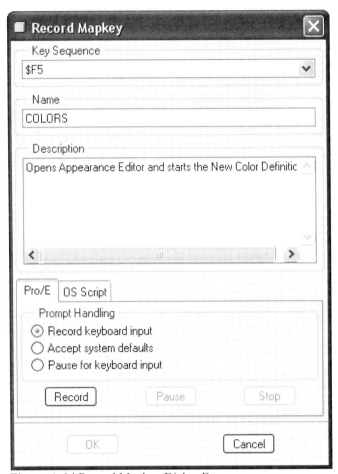

Figure A.14 Record Mapkey Dialog Box

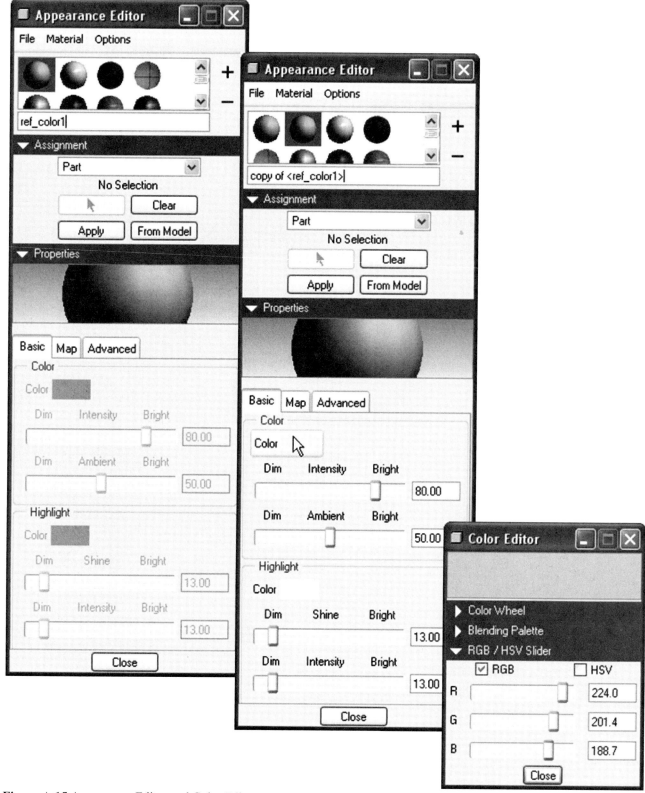

Figure A.15 Appearance Editor and Color Editor

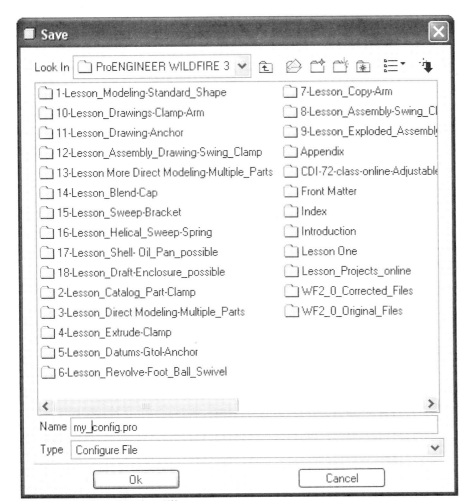

Figure A.16 Save Mapkey File **Figure A.17** Close Mapkeys

Click: **Tools** ⇒ **Customize Screen** ⇒ **Commands** tab ⇒ **Mapkeys** from the Categories list ⇒ click on the new mapkey **COLORS** (Fig. A.18) ⇒ **Modify Selection** ⇒ **Choose Button Image** (Fig. A.19) ⇒ select (Fig. A.20) ⇒ **RMB** (Fig. A.21) ⇒ **Edit Button Image** Button Editor Opens (Fig. A.22) ⇒ click on a color block and edit the picture as you wish (Fig. A.23) ⇒ **OK**

Figure A.18 Customize Dialog Box

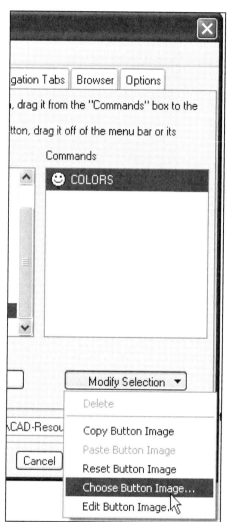

Figure A.19 Choose Button Image

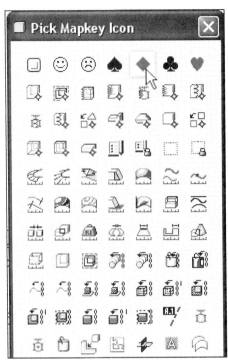

Figure A.20 Select an Icon

Figure A.21 Edit Button Image

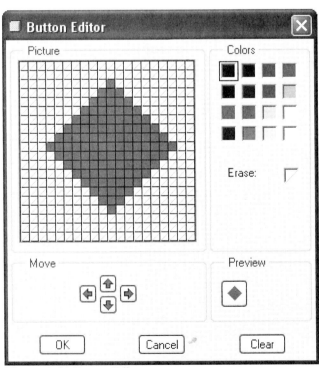

Figure A.22 Button Editor

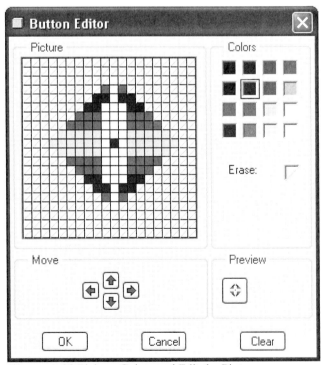

Figure A.23 Pick on Colors and Edit the Picture

Pick the new mapkey: **COLORS** (Fig. A.24) ⇒ **Description** ⇒ **COLORS Opens Appearance Editor and starts the New Color Definition** (Fig.A.25) ⇒ click on the button **COLORS** and drag to the Top Toolchest and drop (Fig. A.26) ⇒ **OK** ⇒ test the new button, click: **Opens Appearance Editor and starts the New Color Definition** (Fig. A.27), Color Editor dialog box opens ⇒ **Close** ⇒ **Close**

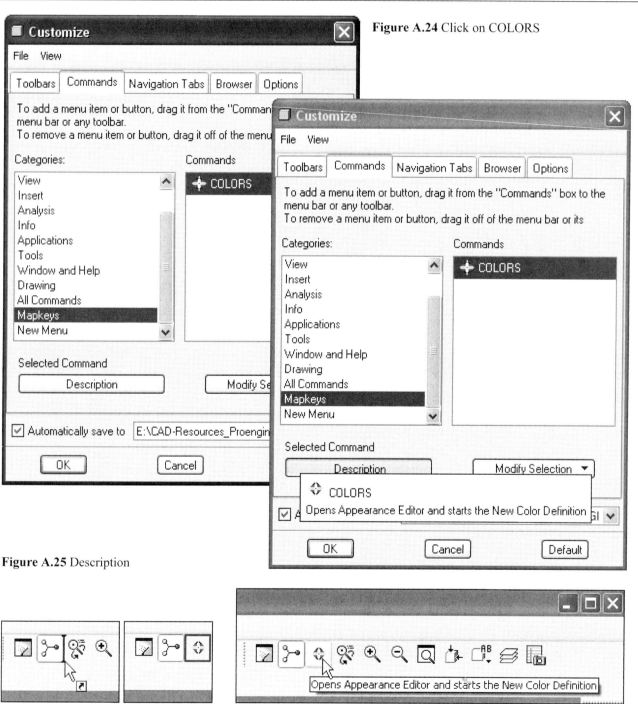

Figure A.24 Click on COLORS

Figure A.25 Description

Figure A.26 Drag and Drop in Toolchest   **Figure A.27** Color Editor Opens

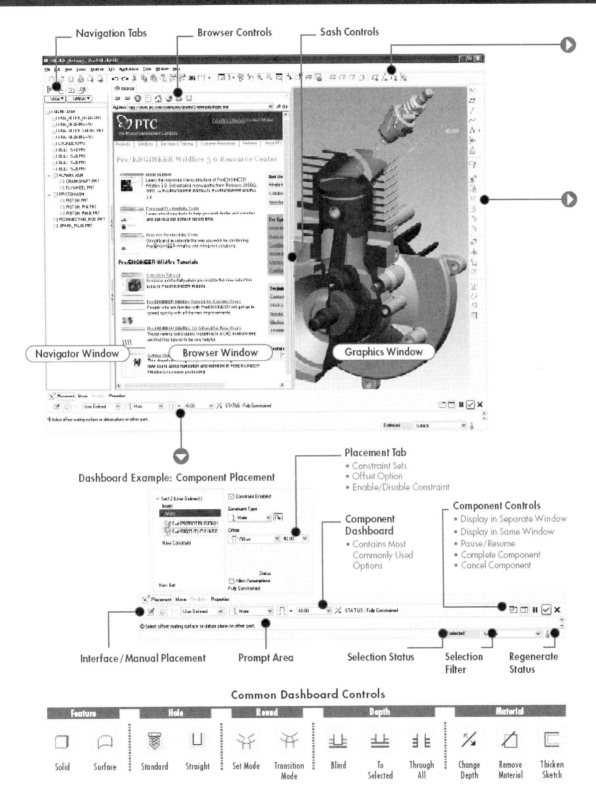

See www.ptc.com for PDF download of Quick Reference Cards.

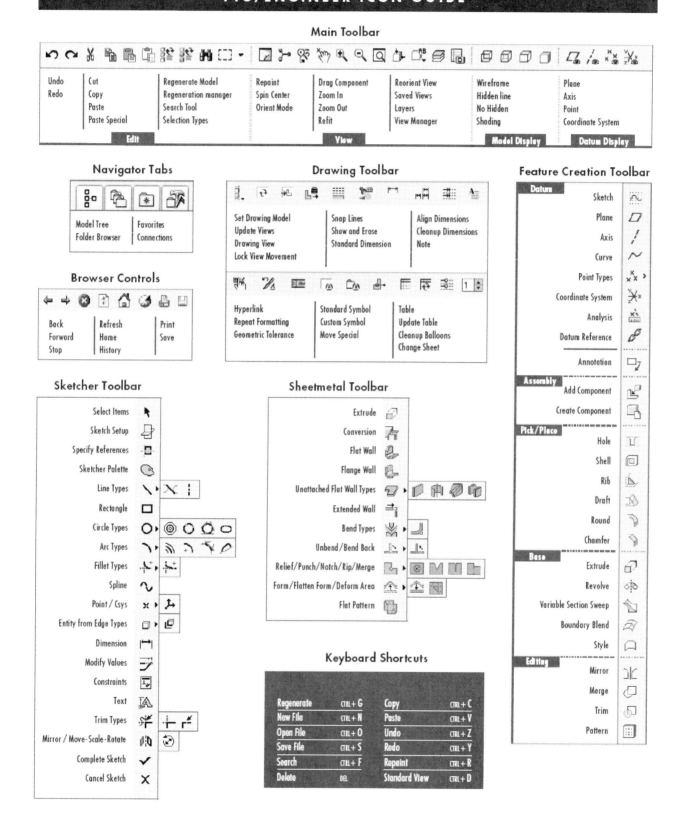

ORIENTING THE MODEL

DYNAMIC VIEWING

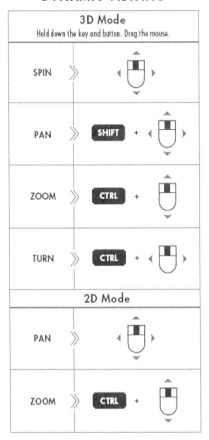

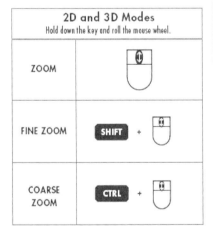

Using the Spin Center

Click the icon in the Main Toolbar to enable the Spin Center.
- Enabled – The model spins about the location of the spin center
- Disabled – The model spins about the location of the mouse pointer

Using Orient Mode

Click the icon in the Main Toolbar to enable Orient mode.
- Provides enhanced Spin/Pan/Zoom Control
- Disables selection and highlighting
- Right-click to access additional orient options
- Use the shortcut: CTRL + SHIFT + Middle-click

Using Component Drag Mode in an Assembly

Click the icon in the Main Toolbar to enable Component Drag mode.
- Allows movement of components based on their kinematic constraints or connections
- Click a location on a component, move the mouse, click again to stop motion.
- Middle-click to disable Component Drag mode

COMPONENT PLACEMENT CONTROLS
Allows reorientation of components during placement

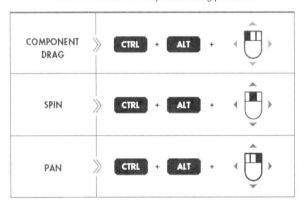

Object Mode
Provides enhanced Spin/Pan/Zoom Control:
1. Enable Orient mode
2. Right-click to enable Orient Object mode
3. Use Dynamic Viewing controls to orient the component
4. Right-click and select Exit Orient mode

DOC-RC60388-EN-350

See www.ptc.com for PDF download of Quick Reference Cards.

MAKING SELECTIONS

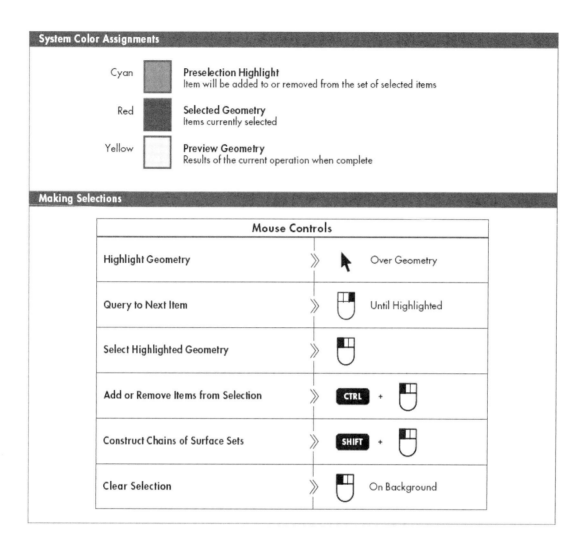

USING FILTERS

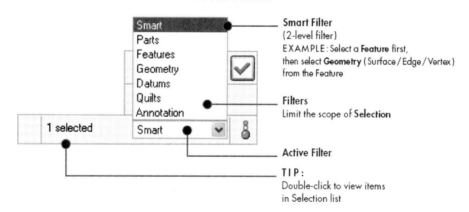

See www.ptc.com for PDF download of Quick Reference Cards.

ADVANCED SELECTION: Chain and Surface Set Construction

DEFINITIONS

General Definitions

Chain
A collection of adjacent edges and curves that share common endpoints. Chains can be open-ended or closed-loop, but they are always defined by two ends.

Surface Set
A collection of surface patches from solids or quilts. The patches do not need to be adjacent.

Methods of Construction

Individual
Constructed by selecting individual entities (edges, curves, or surface patches) one at a time. This is also called the One-by-One method.

Rule-Based
Constructed by first selecting an anchor entity (edge, curve, or surface patch), and then automatically selecting its neighbors (a range of additional edges, curves, or surface patches) based on a rule. This is also called the Anchor/Neighbor method.

CONSTRUCTING CHAINS

Individual Chains

One-by-One
To select adjacent edges one at a time along a continuous path:

1. Select an edge
2. Hold down SHIFT
3. Select the edge again
4. Select adjacent edges
5. Release SHIFT

Rule-Based Chains

Tangent
To select all the edges that are tangent to an anchor edge:

1. Select an edge
2. Hold down SHIFT
3. Highlight **Tangent** chain (Query may be required)
4. Select tangent chain
5. Release SHIFT

Boundary
To select the outermost boundary edges of a quilt:

1. Select a one-sided edge of a quilt
2. Hold down SHIFT
3. Highlight **Boundary** chain (Query may be required)
4. Select boundary chain
5. Release SHIFT

Surface Loop
To select a loop of edges on a surface patch:

1. Select an edge
2. Hold down SHIFT
3. Highlight **Surface** chain (Query may be required)
4. Select surface loop
5. Release SHIFT

From-To
To select a range of edges from a surface patch or a quilt:

1. Select the **From** edge
2. Hold down SHIFT
3. Query to highlight the desired **From-To** chain
4. Select From-To chain
5. Release SHIFT

Multiple Chains

1. Construct initial chain
2. Hold down CTRL
3. Select an edge for new chain
4. Release CTRL
5. Hold down SHIFT
6. Complete new chain from selected edge

See www.ptc.com for PDF download of Quick Reference Cards.

CONSTRUCTING SURFACE SETS

Individual Surface Sets

Single Surfaces
To select multiple surface patches from solids or quilts one at a time:

1 Select a surface patch 2 Hold down CTRL 3 Select additional patches (Query may be required) 4 Release CTRL

Rule-Based Surface Sets

Solid Surfaces
To select all the surface patches of solid geometry in a model:

1 Select a surface patch on solid geometry
2 Right-click and select **Solid Surfaces**

Quilt Surfaces
To select all the surface patches of a quilt:

1 Select a surface feature
2 Select the corresponding quilt

Loop Surfaces
To select all the surface patches that are adjacent to the edges of a surface patch:

1 Select a surface patch
2 Hold down SHIFT
3 Place the pointer over an edge of the patch to highlight the **Loop Surfaces**
4 Select the Loop Surfaces (The initial surface patch is de-selected)
5 Release SHIFT

Seed and Boundary Surfaces
To select all surface patches, from a **Seed** surface patch up to a set of **Boundary** surface patches:

1 Select the **Seed** surface patch 2 Hold down SHIFT 3 Select one or more surface patches to be used as boundaries 4 Release SHIFT (All surfaces from the Seed up to the Boundaries are selected)

Excluding Surface Patches from Surface Sets
To exclude surface patches during or after construction of a surface set:

1 Construct a surface set 2 Hold down CTRL 3 Highlight a patch from the surface set 4 Select the patch to de-select it

5 Release CTRL

CONSTRUCTING CHAINS AND SURFACE SETS USING DIALOG BOXES

To explicitly construct and edit Chains and Surface Sets, click **Details** next to a collector:

Chain Dialog Box  **Dashboard Collector**   **Surface Set Dialog Box**

See www.ptc.com for PDF download of Quick Reference Cards.

Index

A

Activate Window 341
Add Arrows (Drawing) 422
Add Component 43, 266
Add Inner Faces 550
Add New Parameter 463
Adds Counterbore 312
Adds Countersink 250, 609
Adds Tapping 609
Align 51, 311, 324
Align Offset 51
allow_3d_dimensions 414
Allows Assumptions 281
Analysis 248, 335, 546
Angle (Sectioning) 485
Angularity 442
Annotation(s) 205, 585
Appearance and Color 672
Appearance Editor 672
Arc 194, 202, 234, 236
Arithmetic Operators 227
Arrow Head (Drawing) 502
Asm 54, 458
Assemblies 13, 265
Assembly 41, 265, 271
Assembly Components 41, 265, 271
Assembly Constraints 265
Assembly Drawings 449-450, 475
Assembly Explode State 347, 349, 359, 372
Assembly Mode 265
Assembly Window 266, 275
Assignment Operators 227
Attributes (Table) 454
Automatic (Assembly) 43, 266
Auxiliary View (Drawing) 382, 421
Axial Pattern 512-513, 544, 567
Axis (Drawing) 63, 432, 491
Axis of Revolution 179, 189, 193, 213

B

Balloons (Drawing) 449
Base Feature 3
Bill of Materials (BOM) 302, 339
Blank 368
Blend(s) 120, 529, 531-532, 538
Blend Sections 529
Blended Background 84
Blending Palette 148
Blind 542
BOM (Assembly) 14, 302, 339-340, 449, 476
BOM (Drawing) 449, 476
BOM Balloon (Drawing) 476, 487
Bottom-up Design 267, 306
Broken View 383

Browser 65, 668
Browser Window 65-66
Button Editor 675

C

Capturing Design Intent 1
CARR Lane 265
Catalog 65
Catalog Parts 67-72
Centerline 128, 135, 189, 235
Chamfer(s) 179-180
Chamfer Tool 196, 208, 249
Child 2, 8, 13
Children 2, 8-9
Choose Button Image 674
Circle 24
Clean Dimensions 391, 426
Coaxial 169
Coaxial Hole 169
Coincident 334
Color and Appearance 39, 147, 670
Color Editor 147-149
Color Palette 147
Color Wheel 148
Colors 147
Combined State 498
Comparison Operators 227
Component Constraints 265-266, 272, 284, 305, 327
Component Create 307
Component Display 349, 495
Component Operations 337
Component Options 307
Component Placement 266
Components 43, 265-266, 272, 284, 305, 327
Configuration Files 183-187, 198, 209
Constant Section Sweep 549-550
Constraint(s) 127, 130
Context Sensitive Help 80, 601
Continue 26
Coord Sys 276
Coordinate systems off 146, 156, 323
Coordinate systems on 14
Copy 119, 225, 644
Copy Paste 225
Cosmetic Feature 216-218
Cosmetic Thread 216-218
Counterbore 312
Countersink 606
Create 29
Create a General View 416
Create Arc 236, 509, 539
Create Balloon 487, 492
Create Centerline 235
Create Circle 574
Create Cross Section 546
Create Ellipse 113
Create Fillet 103, 108, 145, 156

Create Flange Wall 521
Create Flat Pattern 527
Create Flat Wall 517, 520
Create Macros 666
Create Note 410
Create Points 163
Create Rectangle 308
Create Snap Lines 391
Create Standard Hole 250
Create Text 655
Create 2 Point Lines 541
crossec_arrow_length 435, 476
crossec_arrow_width 435, 476
Cross Sections 171-173, 402
 2D Cross Section 402, 422
C Size 60
Ctrl key 36, 83
Ctrl+Alt MMB 297, 331
Ctrl+Alt RMB 333
Ctrl+C (Copy) 225, 300, 514
Ctrl+D (Default Orientation) 237, 240, 299, 328, 334
Ctrl+G (Regenerate) 260
Ctrl+MMB 239
Ctrl+R (Repaint) 165, 299
Ctrl+S (Save) 233, 321
Ctrl+V 300
Current Sheet 446
Customize 664
Customize Screen 663, 674
Customize Toolbars 663, 665
Customizing 663
Cut 246

D

d symbol 174, 226-227, 257
Dashboard 34, 92, 119
datum_axes 186
Datum(s) 119, 141, 442
Datum Axes 157, 417
Datum Axis Tool 156
Datum Features 3, 7, 159
Datum Plane Tool 156, 261, 647
Datum Plane(s) 6, 157-158, 482
default_datum_planes 185
default_dec_places 535, 597
default_font 414
default_font_filled 446
def_layer 186
Default Constraint 275
Default Coordinate System 6
Default Datum Planes 6, 21
Default Exploded Views 360
Default Orientation (Drawing) 416
Define Internal Sketch 31, 154, 202, 238
Delete 393
Delete Line (Drawing) 439, 485

Delete Old Versions 281, 436, 445
Delete View 419
Dependent Copy 232
Design 271
Design Intent 10, 121
Design Part 230
Design Process 230
Designated 470
Detailed View 382, 430
Diameter 38, 167, 169, 179, 190, 546
Diameter Dimension 190, 195
Dimension(s) 145-146, 152, 154, 165, 220, 222, 239, 425
Dimension Properties 206, 257, 439
Dimension Tab 320
Direction of Rotation 622
Direction Reference 232
Display 126, 304
Display Settings 84
Display Style (Drawing) 424
Distance 248
Draft 102, 621-622
Draft Angle 102, 622
Draft Dialog Box 102
Draft Geometry 102, 622
Draft Hinges 102, 622, 630
Draft Surfaces 102, 622
Draft Tool 102, 622, 629, 636
Drag Handle 98, 106, 111
draw_arrow_length 428, 446, 452
draw_arrow_width 428, 446, 452
draw_arrow_style 414, 446, 452
drawing_text_height 414, 446, 452
Drawing Display 495
Drawing Formats 413
Drawing Mode 15, 54
Drawing Options 414, 490
Drawing Templates 380
Drawing View 389, 402, 421, 435, 444, 477
Drawings 15, 53, 376, 475
Drill 38, 607
.dtl files 414, 425, 446
DXF 377
Dynamically Trim 202

E
ECO 547
Edge Display 429
Edit 256, 259
Edit Attachment 488
Edit Button Image 674
Edit Definition 90-91, 289, 315, 341, 638
Edit Internal Sketch 639
Edit Style 369
Ellipse 113
Email 65, 78, 92
Environment 30, 119, 126, 250

Environment Settings 126
Equal Spacing 560
Erase 407
Erase All (Drawing Mode) 479
Excl Comp 494
Expert Machinist 264
Explicit Relations 226
Explode Position 362, 363
Exploded Assemblies 347
Exploded Assembly (Drawing) 498
Exploded States 347
Exploded View (Drawing) 498
Exploded Views 347-348
External Cosmetic Thread 179, 216-219
Extrude 120
Extrude Tool 22, 119, 133, 308
Extrusions 119

F
Failed Features 229
Failure Diagnostics 646
Failures 228
Family Tables 12, 229, 251-254
FEAT# 203
FEAT ID 203
Feature Info 86
Feature # 203
Feature-based Modeling 1
Feature Control Frame 441
Feature Depth 147
Feature Info 85, 177, 263, 339
Feature List 178
Feature Operations 644
Feature Preview 147
Features 2
File 78
File Functions 78
Fill (Section) 485
Fill Tool 651
Filled Dot (Drawing) 503
Fillet 145
First Wall 518
Fix Component (Assembly) 329
Fix Index 495
fixture_lib 268
Flat 492
Folder Browser 181-182
Folders 182
Font 82, 441
Format(s) 56, 376
Format Size 376, 378
Free Ends 550
Front Datum Plane 23
Full View 383
Fully Constrained 281, 288, 291, 325
Fundamentals 358

G
General Blends 532
General Tolerances 161
General View (Drawing) 382, 388
Geometric Dimensioning & Tolerancing 161, 197, 442
Geometry Check 662
Geometry Tolerances (Geom Tol) 129, 197, 442-443
Geom Tol 205, 262
Global Interface 335
Global Reference Viewer 314, 318
Graphics Window 66
Grid 150
Grid Intersection 484
Grid Params (Parameters) 565
Group 515, 644
Grouped Features 515
gtol_ansi 421
gtol_asme 421
gtol_datums 421

H
Hatch 485
Hatching 172, 304, 440
Helical Sweep 569-570, 574
Help 79, 385
Help Center 79, 358
Hidden Line 35, 81, 313, 368, 429
Hide 317
Hide Grid 453
Hole Pattern 589
Hole Preview 608
Hole Tool 35, 167, 204, 310, 510
Holes 35, 167, 169, 204, 310, 510
Hook Ends 579
Hyperlink 352, 586

I
IGES 377
Index 458
Info 85, 177, 263, 339
Information Areas 66
Information Tools 85, 219
Inner Faces 549
Insert (Assembly) 48, 277, 287
Insert Auxiliary View 420
Insert Balloon 487, 492
Insert Blend 538
Insert Dimension 438
Insert Drawing View 57, 416, 418, 420, 444
Insert Features 591
Insert General View 388, 444
Insert Helical Sweep 574
Insert Here 516
Insert Mode 516, 591-592
Insert Projection View 418, 480
Insert Sheet 443, 497

Insert Sweep 556
Insert Table 454
Instance 251
Isometric 134, 275, 389

L

Landscape 379
layer_axis 186
Layer(s) 141, 170
Layer Names 159-160
Layer Properties 159-160
Layer Tree 143, 273
Leader 502
Library Parts 267
Line(s) 152, 164
Local Parameters 464
Lock Scale 130
Lock View Movement 392, 408, 477

M

Main Window 65
Make Note 410, 446, 502
Manufacturing Model 230
Mapkey(s) 669-671, 673
Master Representation 350
Mate (Assembly) 342
Mate (Offset) 276, 280, 287
Material(s) 122, 210-211, 535
Material Condition 211, 442
Material Condition Symbol 442
Material Definition 123, 211
Material Setup 210-211
Mathematical Functions 228
max_balloon_rad 476
Mbr 459
Measure 546
Measuring Geometry 248
Menu Bar 66
Menu Mapper 79
Merge Ends 551
Message Area 66
mfglib 268-269
Millimeter Newton Second 121
min_balloon_rad 476
Mirror 644
Mirror Tool 140
MMB 25, 83, 165
Model Analysis 661
ModelCHECK 662
Model Information (Info) 212, 223
Model Notes 569, 571, 582
Model Parameters 468
Model Player 220-222
Model Sectioning 171-173
Model Size 526
Model Tree 88, 199, 273, 372, 616
Model Tree Columns 201, 272, 372
Model Tree Filters 88
Model Tree Items 42, 272

Modify 146, 191, 308
Modify Dimensions 131, 136, 146, 166, 191
Mouse Buttons 83
Move Component 293
Move Datum Tag 322
Move Dimension 207
Move Item to View 395, 427, 435, 487
Move Many 363
Move With Children 365
Multiple Sheets 449, 453, 497

N

Navigation tabs 667
Navigation Window 65-66, 181
Navigator 143
NC Check 230
New Constraint 47-50, 278, 324
New Dialog Box 21, 42, 54
New Drawing 378-379, 385
New Layer 158-159, 170
New Mapkey 669, 674, 676
New Presentation State 371
Next Xsec 485, 494
No Combined State 477
No Cross Section 304
No Duplicates 455
No Hidden 81, 368, 483
No Scale 383, 632
Non-default Thickness 590
Note(s) 20-205, 354, 571
Note Properties 456
Note Tab 204

O

Octagon 32
Offset 37
Old Versions 410
On Surface 504
Open URL 352
Operators and Functions 227
Options 187
Options (Drawings) 415
Orient 163
Orient Mode 80, 660
Orientation 81
Oriented 291

P

Page Setup 55-56, 60, 386-387, 398, 443
Palette 32
Pan 83
Parallel Blends 531, 538
Parameter Properties 471
Parameter Symbols 227
Parameters 177, 227, 461, 560
Parametric 1
Parametric Design 1

Parametric Dimensions 2
Parametric Modeling 2
Parent-Child Relationships 8, 13
Parent Feature 13
Parents 13
Part Design 5
Part Mode 5
Part Number 462
Partial View 383
Parts List 449, 460
Paste Special 225, 232
Pattern 320, 510, 610
Pattern (Axial) 512-513, 544, 567
Pattern (Directional) 518, 525
Pattern (Fill) 510-511
Pattern Table 610-611
Pattern Tool 320-321, 510, 612
Patterns 11, 321, 508, 510
Perspective View 383
Place View 406
Placement 45, 138, 332
Placement Tab 22, 36
Placing Components 266
Planar 564
Plane Tag Display 665
Plotting 224
Polar Grid 538
Preferences 356
Preview 46, 59, 276, 326, 542
Previewed Pattern 613
Primary Datum Planes 3
Printing 224
Pro/ASSEMBLY 265
Pro/DETAIL 275
Pro/ENGINEER Fundamentals 79
Pro/HELP 79, 358
Pro/LIBRARY 267-269
Pro/MANUFACTURING 264
Pro/NC 264
Pro/SHEETMETAL 518
Product View Lite 73-76
Profile Tab 522
Projection View 382
Properties 147-148, 206-207, 402, 421
PTC Catalog 267

Q

Qty 458
Quantity 339
Quick Fix 646
Quick Reference Cards 18, 677

R

Record (Table) 495
Record Mapkey 670-671
Rectangle 520
Redefine 646

Redraw 82, 336
Reference Dimension(s) 427
Referenced Features 4
References 238
Refit 27, 82, 396
Regenerate 130, 191, 247, 259, 281, 321
Regenerates Models 233, 281-283, 321, 337
Regeneration Manager 283
Relation(s) 12, 175, 177, 226, 258, 463, 473
Relations and Parameters 226, 474
Remove Material 138, 154, 243
Remove Surface 590
Rename 523
Reorder 589, 591
Reorient View 359
Repaint 390
Repeat Region (Table) 454
Report Symbols 456-458
Resolve Feature 228
Resolve Sketch 136
Resume 170, 516, 592, 621, 623, 658
Resuming Features 170, 623
Revolve 120
Revolve Tool 179, 202, 213
Revolved Cut 115
Revolved Extrusion 110, 184, 192, 214
Revolved Feature(s) 112, 115, 179
Revolved Protrusion 179
Rib Tool 237
Ribs 226
Right 82
RMB 55, 85
Rotate (Assembly) 336-337
Rotating Components 337
Rotational Blends 533
Round(s) 215
Round Edges 241, 615
Round Tool 215, 241-242
Rpt 458

S
Save 27
Saved View 359
Scale 431, 446
Scale (Drawing) 399
Scale Model 517, 520
Scale Rotate 33
Scale View 381, 433
Scheme 84
Screen Tip 6, 352
Secondary Reference Collector 36, 250, 313
Section(s) 171-173, 481
Section Lines 171-172, 494, 495
Section View (Assembly) 449
Section Views 383, 481

Sections (Drawing) 449, 481
Select Instance 268, 385
Select items 146
Selected Trajectory 549
Send Email 65, 78
Separate Window 266, 274
Set Datum 161, 197, 262, 597
Set Layers 170
Set Region 487
Set Working Directory 20, 78, 182, 268
Settings 88, 272
Shading 39, 81, 137, 368
Shape Tab 310, 520
Sheet (Drawing) 446
Sheetmetal 518
Shell 93-94, 99, 100, 516, 545, 590, 601
Shell Thickness 631
Shell Tool 99, 100, 516, 545, 589, 601, 631
Show All 63
Show Components 266
Show Dimensions 173, 175
Show/Erase dialog box 63, 390, 400, 409, 425
Sketch 188, 190, 238, 244, 538, 655-656
Sketch in 2D 245
Sketch in 3D 188-189
Sketch Options 538
Sketch Tool 24
Sketch Traj 557
Sketch Trajectory 549, 557
Sketch View Direction 238
Sketched Features 4
Sketched Formats 376
Sketched Hole 167-169
Sketcher 1, 167, 188-189
Sketcher Constraints 127, 130, 168
Sketcher Palette 32
Sketcher Point 163, 168
Sketcher Preferences 190, 484, 565
sketcher_dec_places 186, 535, 597
sketcher_starts_in_2d 186
Snap to Grid 538, 560
Spacing (Sectioning) 486
Spin 83
Spin Center 203
Spline 430
Split Areas 622
Standard Formats 376, 413
Standard Hole 180
Standard Orient 321
Standard Orientation 26, 82, 260, 308, 321, 325
Start Points 532
Start The View Manager 303, 401, 647
Status Bar 66

Steel 471
String 462
Strong 152
Student Edition 76
Style 368-369
Sub-Assembly 267, 271-272, 339
Suppress 170, 612, 623, 653
Suppressed Objects 170
Suppressing Features 170, 623
Surface of Revolution 572
Sweep 120, 549, 556
Sweep Section 549, 560
Sweep Tool 549, 557
Switch Dimensions 174-176, 259
Symmetric 133
System Colors 84
System Display Settings 84
System Formats 61, 446

T
Table(s) 455, 487
Tan Edge Display 406
Tangent Dimmed 491
Technical Support Info 79
Template 405
Template View 380, 405-406
Text 433, 624
Text Extrusions 624, 656
Text Style 207, 433-434, 440-441, 456-457, 571
Thicken Protrusion 107
Thin Protrusion 107
Thread(s) 179-180, 216-218
Through All 244
Thru Axis 574
Title Block 376
Toggle Section 541
Toggle the grid off 202
Toggle the grid on 202
Tolerance Value 442
tol_mode 535
Toolbar(s) 66, 666
Tool Tips 352, 354
Tools 30, 134, 202, 255, 258, 472
Top-down Design 267, 306
Trajectory 549
Transform 337
Transformations 232, 514
Translate 328
Tree Columns 272, 468
Tree Filters 42, 88
Trim 194, 202
Trimetric 305
Tryout Edition 76
Turn 83

U
Undo 194
Undo Changes 247, 259

Undo Sketcher Operations 194
Unexploded View 60
Units 122, 198, 209, 512
Units Manager 122, 518
Update 370, 400
Update Sheet 390, 445
URL 352, 354
Use 2D Sketcher 188
Use Default Template 397
User Defined 459
User Interface 663

V

Velocity 660
Verify Instances 254
View 39, 81
View Manager 304, 349, 360-361, 546
View Name 381
View Orientation 381, 478
View Scale 406
View Shade 263
View States (Drawing) 381
View Style 368
View Type (Drawing) 477
Views 358, 382, 476
Views (Drawings) 376, 476
Visibility 304

W

Weak Dimensions 129, 155
What's This? 79
Windchill 182
Window Activate 29
Wireframe 81, 368
Working Directory 20, 95, 267
Workpiece 230

X

X-Hatching 171-173, 546
X-Sec 303, 648
X&Y Spacing (Drawing) 446

Z

Z Axis 337
Zoom 83, 284, 289, 297
Zoom In 82, 490
Zoom Out 82, 132